# Tall Building Foundation Design

**Harry G. Poulos**

CRC Press is an imprint of the
Taylor & Francis Group, an **informa** business

CRC Press
Taylor & Francis Group
6000 Broken Sound Parkway NW, Suite 300
Boca Raton, FL 33487-2742



Printed and bound by CPI Group (UK) Ltd, Croydon, CR0 4YY

International Standard Book Number-13: 978-1-4987-9607-1 (Hardback)
978-1-1387-4803-3 (Paperback)



**Library of Congress Cataloging-in-Publication Data**

Names: Poulos, H. G., 1940- author.
Title: Tall building foundation design / by Harry Poulos.
Description: Boca Raton : CRC Press, [2017] | Includes bibliographical references and index.
Identifiers: LCCN 2017003312| ISBN 9781498796071 (hardback : alk. paper) |
ISBN 9781498796088 (ebook)
Subjects: LCSH: Tall buildings--Foundations.Classification: LCC
TH5201 .P735 2017 | DDC 721/.1--dc23
LC record available at https://lccn.loc.gov/2017003312

**Visit the Taylor & Francis Web site at**
**http://www.taylorandfrancis.com**

**and the CRC Press Web site at**
**http://www.crcpress.com**

# Contents

# Tall Building Foundation Design

# Preface

I have been involved in the design of tall building foundations for almost three decades, during which time building height records have been broken several times. The demands on foundation design for these buildings have increased accordingly, and have resulted in significant changes in design practice. State-of-the-art techniques are now becoming routine and the amount and quality of geotechnical data being acquired for the design process has also improved substantially. As I advance well into my eighth decade, I felt that it might be useful to set out an approach to tall building foundation design that I and my consulting and academic colleagues have developed and employed on a significant number of projects.

The approach that is described in this book employs a three-stage process, starting with a preliminary or concept design phase, followed by a detailed phase and then a final phase in which all aspects are checked, the construction drawings are finalised and the design is ready for implementation. The level of computational sophistication and the amount of geotechnical data available will generally increase as each phase proceeds, often starting with a relatively sparse amount of data, and then increasing in detail and quantity as the design process proceeds. This second phase would normally consist of detailed drilling, geophysics and in situ and laboratory testing. The final phase would incorporate filed element testing, usually on piles and perhaps shallow foundations, to enable optimisation and final "tuning" of the design.

The objectives of this book are as follows:

1. To clarify the issues that need to be considered in design, not only from the structural loadings, but also from loadings that arise from the ground in which the foundations are located.
2. To summarise some of the available information on geotechnical design techniques, for each of the three phases of design. These techniques range from empirical approaches that can be used as a first rough estimate of design requirement, through simplified but sound methods that may be employed for the detailed design, to detailed numerical analyses suitable for the final stages of the design process.
3. To set out methods by which the relevant foundation design parameters can be assessed.
4. To discuss procedures for pile testing and then for monitoring the performance of the foundation during and after construction.
5. To present some details of a limited number of case histories in which the various analysis and design techniques described in this book have been applied. These case histories are limited to ones in which I have been involved, but there are other publications which discuss many more case histories, for example, Hemsley (2000) and Katzenbach et al. (2016).

While this book focusses on the geotechnical design of foundations, there are brief sections on other aspects of tall building design which the geotechnical designer should be familiar with. These include the various structural forms of tall buildings, the options available for the foundation system, the various sources of load on the foundations and some very basic aspects of the structural design of the foundation system.

Because of the diverse sources of information contained in this book, no attempt has been made to unify the notation. Rather, the notation of the original source has generally been retained, with some minor departures. As a consequence, there are many over-worked symbols whose meaning is very much dictated by the context in which they appear. Accordingly, the definition of such symbols as $\alpha$, $\beta$, a, b, c, d, D, e, L, k, K, P, p, R, u and V, among many others, must be examined carefully before use.

The field of geotechnical engineering and its sub-field, foundation design, continues to develop and progress and so some of the material in this book may become dated in time. However, the general principles set out in this book are philosophically rather basic, and it is hoped that they will stand the test of time more durably than some of the empirical and simplified calculation methods.

**H.G. Poulos**
*Sydney, Australia*
*January 2017*

# Acknowledgements

The author and publisher gratefully acknowledge the organisations and individuals listed below, who have given permission for the reproduction of their material. The figures and tables listed are those from this book, while the reference from which the material is sourced is provided in the figure and table titles within the text of this book.

1. The American Concrete Institute for Figure 5.1.
2. Dr. A. Amini for Figures 13.18 and 13.19.
3. The Australian Geomechanics Society for Figure 9.23.
4. The American Society of Civil Engineers (ASCE) for Figures 8.2, 8.6, 8.18a and b, 8.30, 9.9, 9.10, 9.11, 9.12, 9.14, 9.15, 9.16, 9.17, 9.19, 9.20, 9.21, 9.22, 9.41, 9.46, 9.47, 10.8, 10.12, 11.5, 11.25, 11.33a and b, 11.16, 12.10, 15.27, 15.28, 15.29, 15.30, 15.31, 15.32, 15.33, 15.34, 15.35, Tables 10.8, 15.16, 15.17, 15.18 and 15.19.
5. The Canadian Geotechnical Society for Figures 15.4, 15.5, 15.6 and 15.7.
6. The Council for Tall Buildings and the Urban Habitat (CTBUH) for Figures 15.21, 15.22 and 15.23, and Tables 15.15 and 15.16.
7. The Deep Foundations Institute (DFI) for Figures 9.2, 9.3 and 9.4.
8. Elsevier for Figures 8.12, 8.16, 8.17 and 11.28.
9. International Association for Earthquake Engineering (IAEE) for Figures 11.38 and 11.39.
10. International Federation of Surveyors (FIG) for Figures 14.9 and 14.10.
11. International Society for Soil Mechanics and Geotechnical Engineering (ISSMGE) for Figure 12.15.
12. ICE Publishing for Figures 3.10, 7.4a and b, 7.13, 7.14, 7.15, 7.16, 7.17, 7.18, 7.19, 7.20, 7.21, 7.22, 7.23, 11.27, 11.29, 11.30, 14.2, 14.4, 14.5, 14.6, 14.7 and 14.8 and Table 8.3.
13. IoS Press for Tables 12.5 and 14.2.
14. Kinemetrics Open Systems and Services for Figure 14.1.
15. Kluwer Academic Publishers for Tables 10.4 and 10.5.
16. McGraw-Hill for Figures 5.2, 5.3, 5.4 and 12.1.
17. The Curators of the University of Missouri for Figures 15.9, 15.10, 15.11, 15.12, 15.13, 15.14, and Tables 15.3, 15.4, 15.5, 15.6, 15.7 and 15.8.
18. Nicholson Construction Company for Figure 3.9.
19. Queensland Government Department of Transport and Main Roads for Table 12.3.
20. Revista Geologica de America Central for Figure 11.3.
21. Dr. Peter Robertson for Figure 11.13.
22. Scientific and Academic Publishing for Figure 2.2.

23. The Southeast Asian Geotechnical Society for Figures 7.7, 8.31, 9.24, 9.25, 9.26, 9.27, 9.28, 9.29, 9.30, 9.31, 9.32, 9.33, 9.34, 9.35, 9.36, 15.18, 15.20, Tables 15.9, 15.12, 15.13 and 15.14.
24. Standards Australia for Figure 11.11 and Tables 5.1, 7.10, 7.11 and 11.8.
25. The Taylor & Francis Group for Figures 2.3, 2.4, 2.5, 2.8, 3.7, 6.2, 6.6, 11.36, 13.1, 13.6, 13.9, 13.10, 13.12, 13.13 and13.17 and Tables 6.3, 7.3, 13.1 and 13.2.
26. The U.S. Society on Dams for Figures 11.15 and 11.19.
27. U.S. Naval Facilities for Figures 11.25, 12.13 and 12.14.

The author gratefully acknowledges the contributions of many of his present and former colleagues and students to the material in this book, including the late Professors E.H. Davis, M. Novak and F.H. Kulhawy, Professors M.F. Randolph, P.W. Mayne, Charles Ng and J.M. Duncan, Drs. L.T. Chen, N. Loganathan, T.S. Hull, N.S. Mattes, A. Tabesh, H. Chow, S. Liyanapathirana and professional colleagues including Dr. A.J. Davids, Mr G. Bunce, Dr. A. Abdelrazaq, Mr S.H. Kim, Mr Patrick Wong, Mr Pawan Sethi and Ms F. Badelow. I am also indebted to Professors John Small and George Gazetas, and Lonnie Pack, for their helpful comments on some of the chapters.

I am indebted to Annette Wilson, who has assisted greatly with the preparation and modification of many of the figures. The management of the Coffey Group and its parent company, Tetra Tech, have generously provided the time and resources for me to compile this book. I am also appreciative of the assistance provided by the editorial staff at Taylor & Francis, especially Tony Moore, Ariel Crockett and Scott Oakley.

Finally, I am grateful to my wife Maria, and my children and grandchildren, who have provided moral support and encouragement, and who have had to suffer many episodes of my physical and mental absences during the compilation of this book.

If you have built castles in the air, your work need not be lost; that is where they should be. Now put the foundations under them.

Henry David Thoreau, *Walden*, 1854

# Author

**Professor Harry G. Poulos** AM FAA FTSE NAE DScEng is a Senior Principal with the geotechnical consulting company of Coffey Geotechnics in Sydney Australia, and an Emeritus Professor at the University of Sydney. He has been involved in a large number of major projects in Australia and overseas including the Egnatia Odos highway project in Greece, the Burj Khalifa tower in Dubai, and the Dubai tower in Doha, Qatar. He was elected a Fellow of the Australian Academy of Science in 1988 and a Fellow of The Australian Academy of Technological Sciences and Engineering in 1996. He has received a number of awards and prizes, including the Kevin Nash Gold Medal of the International Society of Soil Mechanics and Geotechnical Engineering in 2005. He was the Rankine Lecturer in 1989 and the Terzaghi Lecturer in 2004, and was selected as the Australian Civil Engineer of the Year for 2003 by the Institution of Engineers Australia. In 1993, he was made a Member of the Order of Australia for his services to engineering, and in 2010, he was elected a Distinguished Member of the American Society of Civil Engineers. In 2014, he was elected as a Foreign Member of the U.S. National Academy of Engineering.

# Chapter 1

# Introduction

## 1.1 DEFINITION OF A TALL BUILDING

Tall buildings, also referred to as high-rise buildings or skyscrapers, are an integral component of modern cities, and these terms will be used synonymously in this book. A definition of a 'tall building' as given by Craighead (2009: 1) is as follows: 'a multi-story structure in which most occupants depend on elevators (lifts) to reach their destinations'. Among the characteristics of such buildings are that the height can have a serious impact on evacuation, and that they extend to a height greater than the maximum reach of available fire-fighting equipment.

The height at which a building becomes a tall building is not universally defined, although it is generally accepted that buildings in excess of about 40 stories can be considered as high-rise. Relatively slender buildings with a smaller number of stories, or buildings within some European cities, may also be considered to be tall.

With the development of very tall buildings over the past two decades, the terms 'super-tall' or 'super high-rise' denote a building with a height of 300 m or greater. More recently, the term 'megatall' has been applied to buildings over 600 m high. The Council for Tall Buildings and the Urban Habitat (CTBUH) has accordingly indicated the following ranges of building height:

- 200–300 m: tall buildings
- 300–600 m: super-tall buildings
- >600 m: megatall buildings

In 2016, there were over 100 super-tall buildings and 2 megatall buildings fully completed and occupied globally.

## 1.2 EVOLUTION OF TALL BUILDINGS

Modern tall buildings evolved in the United States in the 1880s in Chicago, and over the ensuing decades, the majority of tall buildings were constructed there. Early tall buildings were constructed using load bearing walls, but with the advent of structural steel, tall buildings emerged in such cities as New York, Philadelphia and London. From the end of the nineteenth century until the latter part of the twentieth century, New York took the lead in the construction of tall buildings, and held the world record for building height, first with the Chrysler building, then the Empire State building and then the World Trade Center towers. In 1974, the Sears Tower in Chicago took over the mantle until 1998, when the title

*Table 1.1* Tallest buildings since 1885

| *Year* | *Location* | *Building* | *Height (m)* |
|---|---|---|---|
| 1885 | Chicago | Home Insurance | 55 |
| 1890 | New York | World | 94 |
| 1894 | New York | Manhattan Life | 106 |
| 1899 | New York | Park Row | 119 |
| 1908 | New York | Singer | 187 |
| 1909 | New York | Met Life | 213 |
| 1913 | New York | Woolworth | 241 |
| 1930 | New York | Bank of Manhattan | 283 |
| 1930 | New York | Chrysler | 319 |
| 1931 | New York | Empire State | 381 |
| 1972 | New York | 1 World Trade Center | 417 |
| 1974 | Chicago | Sears | 442 |
| 1998 | Kuala Lumpur | Petronas 1 & 2 | 452 |
| 2004 | Taipei | Taipei 101 | 508 |
| 2010 | Dubai | Burj Khalifa | 828 |

Source: Based on Parker, D. and Wood, A. 2013. *The Tall Buildings Reference Book*. Routledge, New York.

of tallest building moved to Kuala Lumpur Malaysia, with the completion of the Petronas Towers. Subsequently, this has been taken over by Taipei, and then in 2010 by Dubai.

Tall buildings have become a worldwide phenomenon over the past three decades, but with particular concentrations in Asia and the Middle East. Table 1.1 summarises the tallest modern buildings in the world, starting in 1885, based on the data of Parker and Wood (2013).

The 12 tallest buildings in the world, as at the end of 2015, are listed in Table 1.2.

## 1.3 BUILDING COMPONENTS

The main components of a tall building include the following:

1. The foundations: These generally consist of piles and a raft mat or slab.
2. The structure: This must be designed to resist very large wind and seismic loads, and usually requires a combination of basic structural systems such as a reinforced concrete core wall, a structural steel frame and prestressed concrete elements.
3. The façade: The façade and its fixing accessories have to resist the same wind and seismic loads as the structure.
4. Architectural works and finishes: These elements are related primarily to the building function rather than building height. In prestige buildings, the internal finishings may be at the upper end of the cost scale.
5. Elevators/lifts: The cost of elevators (or lifts) may be high due to the number required and their necessary high speed. There may be multiple lift zones, with a shuttle system to take passengers to mid or upper levels for transfer to the top zone.
6. Building services: In most cases, the services systems are divided into separate zones, each with its plant located on intermediate plant floors.

This book will focus on the design of the foundation system for tall buildings, bearing in mind that the structure and the foundation system are not independent components but an

*Table 1.2* The world's 12 tallest buildings, as of 2015

| No. | Building | City | Ht. (m) | Floors | Yr. compl. | Construction | Use |
|---|---|---|---|---|---|---|---|
| 1 | Burj Khalifa | Dubai | 828 | 163 | 2010 | Steel/ concrete | Office/resdl./ hotel |
| 2 | Shanghai Tower | Shanghai | 652 | 128 | 2015 | Composite | Hotel/office |
| 3 | Makkah Royal Clock | Makkah | 601 | 120 | 2012 | Steel/ concrete | Other/hotel |
| 4 | One World Trade Center | New York | 541 | 94 | 2014 | Composite | Office |
| 5 | Taipei 101 | Taipei | 508 | 101 | 2004 | Composite | Office |
| 6 | Shanghai World Fin. Center | Shanghai | 492 | 101 | 2008 | Composite | Hotel/office |
| 7 | Intl. Commerce Centre | Hong Kong | 484 | 108 | 2010 | Composite | Hotel/office |
| 8 | Petronas Twin Towers 1 | Kuala Lumpur | 452 | 88 | 1998 | Composite | Office |
| 9 | Petronas Twin Towers 2 | Kuala Lumpur | 452 | 88 | 1998 | Composite | Office |
| 10 | Zifeng Tower | Nanjing | 450 | 66 | 2010 | Composite | Office |
| 11 | Willis Tower | Chicago | 442 | 108 | 1974 | Steel | Office |
| 12 | KK100 | Shenzhen | 442 | 100 | 2011 | Composite | Hotel/office |

integral interactive system. A useful overview of the components of a tall building is given by Ascher (2011), while more detailed information on various aspects of the design and construction of tall buildings is provided by Chew (2012) and Tamboli (2014).

## 1.4 BUILDING COMPONENT COSTS

In most cities, the tall buildings are usually located in the Central Business District (CBD). The availability of development site areas is generally limited and the footprint of a building now commonly occupies 80%–100% of the site area. With the design of central cores for most of the tall buildings, the lettable area can be as high as 80% of the total floor area.

It is interesting to be aware of the main components of cost of a tall building. Elnimeiri and Almusharaf (2010) indicate the following breakdown of costs:

- Architecture: 40%
- Structure: 30%
- Mechanical, electrical, plumbing (MEP): 25%
- Elevator system: 5%

The cost of the foundation system is thus contained within the 30% of the structure cost. Ding and Wu (2014) indicate that civil engineering works for tall buildings in China (up to 300 m tall) make up 30%–35% of the total cost, although this percentage tends to increase for buildings over 600 m tall. They give the relative costs of the various components of the civil engineering works as follows:

- Above ground civil works: 64%
- Below ground works: 36%, which incorporates
  - Basement works: 16%
  - Basic maintenance (earth and rock): 13%
  - Pile foundations: 7%

## 1.5 SOCIAL ASPECTS OF TALL BUILDINGS

Ali and Al-Kodmany (2012) discuss the pros and cons of tall buildings from a global sociological perspective. The arguments *for* tall buildings are as follows:

1. They help to accommodate the rapidly increasing urban population within cities.
2. They reflect the increasing impact of global competition on the development of the world's major cities.
3. They can promote and stimulate urban regeneration of economically depressed areas of a city.
4. The agglomeration afforded by a series of tall buildings can lead to economy of scale and can increase productivity through access to denser markets.
5. They are an effective economic response to the remarkably high property values in major cities, values that continue to increase with time.
6. They promote sustainability via compact urban living and preservation of areas of open space.
7. They have the potential to consume less energy than low-rise buildings via energy-effective attributes such as agglomeration, auto fuel savings, savings in travel time and the potential for energy generation via the use of the foundation system.
8. Tall buildings can project a sense of socio-economic power and promote the city as a modern leading commercial centre.
9. The demand for high-quality tall buildings has led to advances in science, technology and engineering.

In contrast, arguments *against* tall buildings can be summarised as follows:

1. They involve a cost premium because of their complexity.
2. They can have adverse environmental effects on the microclimate, especially in terms of wind funneling and casting shadows.
3. They can lead to civic problems, such as overcrowding and traffic.
4. There may be socio-psychological impacts of living or working in high-rise buildings.
5. Aesthetically unpleasant buildings can harm the image of a city, although aesthetic perceptions can change with time. Inappropriately located tall buildings can also change the historic fabric and 'feel' of older areas within a city.
6. There are concerns about safety, including public safety, in relation to fires and terrorist actions.
7. They can reduce the potential for positive human interaction.

Overall, the positive aspects of tall buildings tend to outweigh the negative aspects, and the continued quest for more tall buildings, of ever-increasing height, attest to this direction of development of modern civilisation.

## 1.6 THE SPREAD OF TALL BUILDINGS

The last two decades have seen a remarkable increase in the rate of construction of 'super-tall' buildings in excess of 300 m in height. Figure 1.1 shows the significant growth in the number of such buildings either constructed or projected, to 2015 and beyond. A large number of these buildings are in the Middle East or in China. Dubai has now the tallest building in the world, the Burj Khalifa, which is 828 m in height, while in Jeddah Saudi Arabia, the Kingdom Tower is under construction as at the end of 2016, and will eventually exceed 1000 m in height.

'Super-tall' buildings in excess of 300 m in height, and 'megatall' buildings in excess of 600 m in height, are presenting new challenges to engineers, particularly in relation to structural and geotechnical design. Many of the traditional design methods cannot be applied with any confidence since they require extrapolation well beyond the realms of prior experience, and accordingly, structural and geotechnical designers are being forced to utilise more sophisticated methods of analysis and design. In particular, geotechnical engineers involved in the design of foundations for super-tall buildings are leaving behind empirical methods and are increasingly employing state-of-the art methods.

The trends in tall buildings have been discussed by Parker and Wood (2013), and may be summarised as follows:

1. Increase in number.
2. Increase in height.
3. Change in location from United States to Asia and the Middle East.
4. Change in function, from office to mixed use and residential.
5. Change in structural material, from all steel structures to composite steel and concrete structures.

They list, as the key drivers of the spread of tall buildings, the following:

1. Increasing land prices.
2. The desire for improved return on investment by developers.
3. The notion of a building as corporate branding.
4. The notion of a skyline dominated by tall buildings as global branding.
5. Rapid urbanisation in almost all parts of the world.
6. The potential effects of climate change.

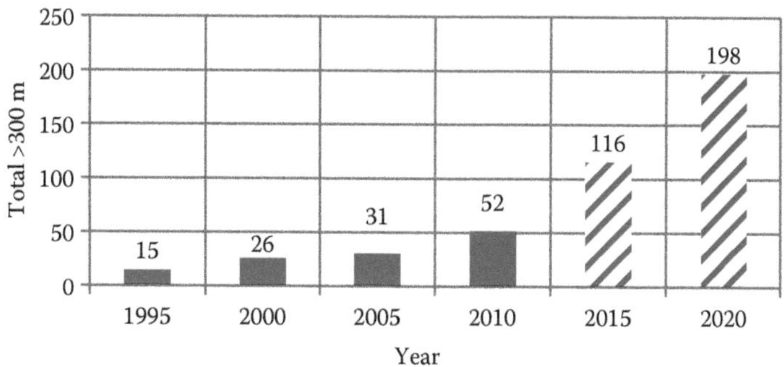


*Figure 1.1* Total number of buildings in excess of 300 m tall. (Based on data from CTBUH.)

## 1.7 LIMITS TO HEIGHT OF TALL BUILDINGS

There has been considerable speculation about the limits to the height of a tall building, and the following technical factors appear to be influential:

1. The limitations of current vertical transportation systems; there will be limits to the number of stages that can be employed in extremely tall buildings without hampering their functionality. Currently, elevators still rely on cables, and they have both a limited capacity and require a size of housing which can become excessive.
2. The limitation of structural systems and the construction materials. While stiffer and more robust structural systems can be developed, the strength of available structural materials does impose limits on building height.
3. The limitations of current construction processes. Improved means of lifting and placing the materials of which the structure is to be constructed would need to be developed.
4. Possible limitation of building height in relation to aircraft flight paths near city centres.
5. The requirements of large building footprint to facilitate stability of the tower. In major urban areas, with very expensive real estate, the land area required could become impractical.
6. The limitations of the loadings that the ground and foundation system could carry. The strength of both the ground and the foundation materials would limit the loadings, and hence the height, of the buildings that could be supported.

Considering only the latter foundation limitations, if, for example, a maximum feasible bearing pressure of 5 MPa was to be considered, with an average loading of 10 kPa per floor, the limit would be 500 floors, corresponding to a height in the order of 2000 m.

There are also a series of psychological factors for building occupants that may limit the utility of extremely tall buildings for many people, and there is a view that human response may be the most compelling factor limiting the height of buildings. These psychological factors include the following:

1. Coping with the building movements developed in the upper floors due to wind and seismic events. Research has indicated that humans may be very sensitive to low-frequency narrow band random motions, especially within the range of frequencies of 0.25–0.5 Hz.
2. There are also indications that people walking tend to experience less motion sensation than those standing, perhaps because the effects of self-motion tend to reduce the sense of vibration.
3. The atmospheric pressure decreases with increasing height, with a reduction of about 4% every 300 m above ground. This can lead to the 'popping' of ears as a person ascends in an elevator.
4. Beyond a certain height, there is a limit to the improvement of the view that humans enjoy, with a possibility that the view can become somewhat distorted.

## 1.8 FOUNDATION REQUIREMENTS

A foundation system is required to safely support the large lateral and vertical loads associated with high-rise buildings and to control total and differential movements of the foundation to within tolerable limits.

Buildings and their foundations are highly interactive systems; movements of the foundation result from the building loads, and these movements in turn influence the behaviour of the building and the consequent building loads. The behaviour of the foundation is mainly governed by the prevailing ground conditions, the foundation type and the magnitude and distribution of the building loads. The foundation design for high-rise buildings should therefore be considered as a performance-based soil–structure interaction (SSI) issue and not limited to traditional empirically based design methods, such as a traditional bearing capacity approach with an applied factor of safety. With such an interactive system, it is important to recognise that the foundation design requires cooperative and interactive input from both the geotechnical and structural engineers involved in the project.

Often the subsurface conditions at high-rise building sites are far from ideal, and geotechnical uncertainty is one of the greatest risks in the foundation design and construction process. Establishing an accurate knowledge of the ground conditions is essential in the development of economical foundation systems which perform to expectations.

The type of foundation system for a high-rise building is determined by the main design elements such as the building loads, the ground conditions and the required building performance as well as other important factors like local construction conditions, cost and project program requirements. When establishing the geotechnical model, local experience coupled with a detailed site investigation program is required. The site investigation is likely to include a comprehensive borehole drilling and in situ testing program together with a suite of laboratory tests to characterise strength and stiffness properties of the subsurface conditions. Based on the findings of the site investigation, the geotechnical model and associated design parameters are developed for the site, which are then used in the foundation design process.

Modern site investigation works are usually supplemented with a program of instrumented vertical and lateral load testing of prototype piles, for example, bi-directional load cell (Osterberg cell) tests, to allow calibration of the foundation design parameters and hence to allow better prediction of the foundation performance under loading. Completing the load tests on prototype piles prior to final design can provide confirmation of performance, including the integrity of pile construction, pile load–settlement behaviour and of the ground model and its engineering properties.

## 1.9 SCOPE OF THIS BOOK

This book aims to present a rational approach to the design of foundations for high-rise buildings. It discusses the important design aspects and presents a sequential foundation design process. It also highlights the need for close cooperation and interaction between the geotechnical designers and the structural designers throughout the design process in order to achieve good design outcomes.

Chapter 2 sets out some of the characteristics of high-rise buildings, while Chapter 3 discusses the available foundation types and the factors that affect their selection. The general process of foundation design is set out in Chapter 4, where a distinction is made between various categories of design and analysis methods, depending on the stage of design. Chapter 5 discusses the loads that are imposed on the structure and which have to be supported by the foundation system. Chapter 6 outlines procedures for the critical process of characterisation of the ground conditions and the assessment of the relevant geotechnical design parameters.

Chapters 7 to 11 discuss in detail the procedures for foundation design for various loadings and conditions. Chapter 12 reviews the procedures for the design of basement walls, and the control of groundwater. Load testing of piles is addressed in Chapter 13, while

Chapter 14 discusses performance monitoring of the foundation and the supported structure. Finally, Chapter 15 gives some details of case histories of foundation design for tall buildings.

The underlying philosophy in this book is to present the key aspects that require consideration in foundation design, and to outline some available methods for undertaking the calculations that form part of the design process, especially in the earlier stages of design. The requirements for detailed design are discussed, and the principles of applying the often complex numerical analyses involved are set out.

It will be assumed that the reader has a reasonable knowledge of the fundamentals of soil mechanics, and the principle of effective stress in particular. Reference can be made to a number of excellent texts, for example, Lambe and Whitman (1979); Atkinson (2007); Budhu (2011); Knappett and Craig (2012) and Holtz and Kovacs (2010) for further details. In addition, it is assumed that the reader has some familiarity with the principles of numerical analysis, and the finite element method in particular. Useful references for these aspects are Potts and Zdravkovic (2001); Lees (2016); Katzenbach et al. (2016) and Small (2016). The latter is particularly useful as it covers several aspects of geotechnical and foundation engineering which are addressed in this book.

Chapter 2

# Characteristics of tall buildings

## 2.1 SOME KEY FEATURES OF TALL BUILDINGS

There are a number of characteristics of tall buildings that can have a significant influence on foundation design, including the following:

1. The building weight, and thus the vertical load to be supported by the foundation, can be substantial. Moreover, the building weight increases non-linearly with height, as illustrated in Figure 2.1 (Moon, 2008). Thus, both ultimate bearing capacity and settlement need to be considered carefully.
2. High-rise buildings are often surrounded by low-rise podium structures which are subjected to much smaller loadings. Thus, differential settlements between the high-rise and low-rise portions need to be controlled.
3. The lateral forces imposed by wind loading, and the consequent moments applied to the foundation system, can be very high. These moments can impose increased vertical loads on parts of the foundation, especially on the outer piles within the foundation system. The structural design of the piles needs to take account of these increased vertical loads that act in conjunction with the lateral forces and moments.
4. The wind loads developed on the structure are dependent on the shape and form of the structure, and also on its dynamic response characteristics. Thus, changes in architectural form or structural design can lead to changes in the wind loads acting on the structure.
5. The wind-induced lateral loads and moments are cyclic in nature. Thus, consideration needs to be given to the influence of cyclic vertical and lateral loading on the foundation system, as cyclic loading has the potential to degrade foundation capacity and cause increased settlements.
6. Seismic action will induce additional lateral forces in the structure and also induce lateral motions in the ground supporting the structure. Thus, additional lateral forces and moments can be induced in the foundation system via two mechanisms:
   a. Inertial forces and moments developed by the lateral excitation of the structure.
   b. Kinematic forces and moments induced in the foundation piles by the action of ground movements acting against the piles.
7. The wind-induced and seismically induced loads are dynamic in nature, and as such, their potential to give rise to resonance within the structure needs to be assessed. The risk of dynamic resonance depends on a number of factors, including the predominant period of the dynamic loading, the natural periods of the structure (several modes often need to be considered), the structural damping, and the stiffness and damping of the foundation system.

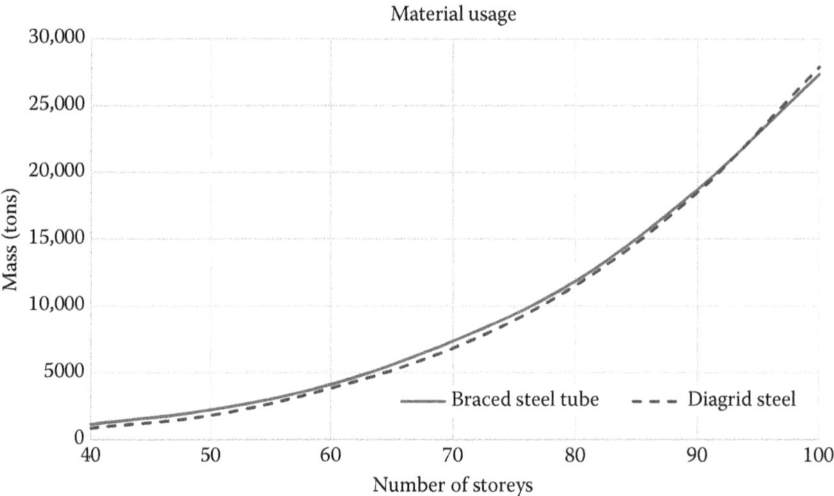


*Figure 2.1* Example of non-linear increase in building weight with increasing height. (Adapted from Moon, K.S. 2008. Material saving design strategies for tall building structures. *CTBUH 8th World Congress*, Dubai (available on CTBUH website).)

## 2.2 ARCHITECTURAL FORMS

The structural form of a building is closely related to its architectural form and is dictated by it. Figure 2.2 shows examples of aerodynamic form and base plan for tall buildings (Alaghmandan et al., 2016). The aerodynamic form clearly has a major influence on the wind loadings induced in the structure, and one of the design requirements is to consider

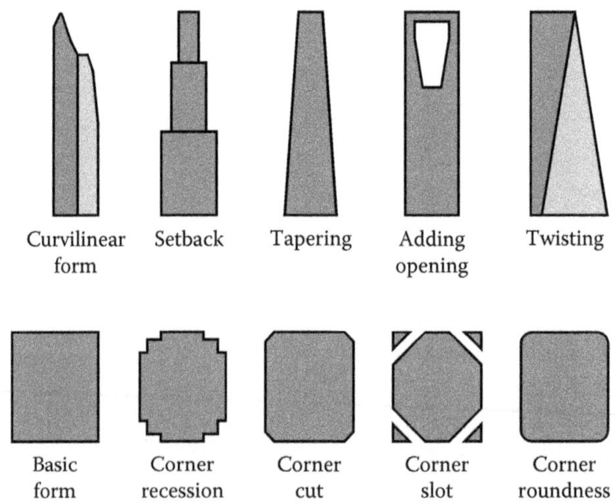


*Figure 2.2* Tall building forms. (Adapted from Alaghmandan, M., Bahrami, P. and Elnimeiri, M. 2014. *Architecture Research*, 4(3): 55–62; courtesy Scientific and Academic Publishing.)

strategies that can reduce the impact of wind. Aerodynamic modifications can be divided into two main categories:

1. Macro, which influences the main geometry, and which includes tapering, setback and twisting.
2. Micro, which includes relatively minor changes to the base plan.

Aerodynamically favourable forms can lead to significant reductions in wind forces. For example, the Taipei 101 building had corner modifications which provided a 25% reduction in the base moment as compared to the original square section.

An important architectural factor representing the geometry and form of a tall building is its base plan shape. Most early tall buildings were either a square or rectangular shape in plan, but in recent years, a variety of other shapes have been used. Alaghmandan et al. (2014) indicate the following percentages of base forms within the years 2000–2012:

- Rectangular: 58%
- Elliptical or circular: 16%
- Triangular: 9%
- Curvilinear: 7%
- Polygonal: 7%
- Parallelogram: 3%

The future trend appears to be toward aerodynamic and curvilinear shapes and forms. An interesting development is the 'twisting' building, which is one that progressively rotates its floor plates or its façade as it gains height. Usually, but not always, each plate is shaped similarly in plan, and is turned on a shared axis a consistent number of degrees from the floor below. In 2016, the tallest twisting tower is the Shanghai Tower, completed in 2015, which is 632 m tall, has 128 floors, and a total rotation of 120°. Another twisted tower, currently under construction (2017), is the Diamond Tower in Jeddah, Saudi Arabia, which is planned to be 432 m tall. This case is discussed in Chapter 15.

There is also a trend toward the use of 'super-slim' buildings, which have a very high slenderness ratio, in excess of 10 and in some cases, greater than 15. Whittle (2016) and Marfella et al. (2016) discuss factors of technical innovation in the process of designing and constructing super-slim towers. Controlling rotational movements and swaying under wind loading are particularly challenging for structural designers.

The architects' quest for individuality of a structure can place significant demands on the structural and foundation engineers who have to design the building. While the engineers need to preserve the vision of the architect, the architect in turn should appreciate the challenges that may be imposed on the engineers by their designs. Clearly, a collaborative approach between the architect and the structural engineer, and between the structural and the foundation engineer, is highly desirable.

## 2.3 STRUCTURAL FORMS

Khan (1969) classified structural systems for tall buildings, and Ali and Moon (2007) produced an extended classification for both steel and concrete structures which is shown in Figure 2.3. They modified Khan's classification system and distinguished between interior and exterior structures. Figures 2.4 and 2.5 illustrate these structural classifications.

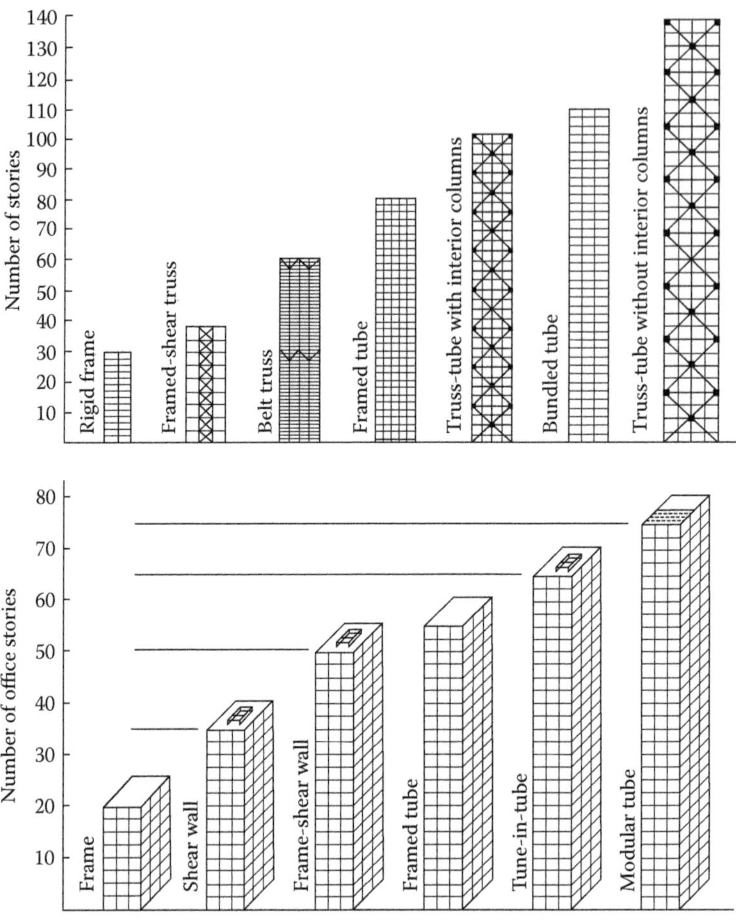


*Figure 2.3* Classification of tall building structural systems. (After Khan, F.R. 1969. Recent structural systems in steel for high-rise buildings. *Proceedings of the British Constructional Steelwork Association Conference on Steel in Architecture*. British Constructional Steelwork Association, London, pp. 24–26; Ali, M.M. and Moon, K.S. 2007. *Architectural Science Review*, 50.3: 205–223. Courtesy of Taylor & Francis.)

### 2.3.1 Interior structures

For interior structures (Figure 2.4), there are three broad types of load-resisting systems:

1. *Moment resisting frames*: These resist load primarily through the flexural stiffness of the members
2. *Shear trusses/shear walls*: These are generally located around service and elevator cores and stairwells. This type of system is very commonly used and the shear walls are generally treated as vertical cantilevers fixed at the base. They stabilise and stiffen the building against lateral loads.
3. *Core-supported outrigger systems*: In this system, the core is analogous to the mast of a ship, with outriggers acting as the spreaders and the exterior columns as the stays. Outriggers reduce the overturning moment in the core that would otherwise act as a pure cantilever, and transfer the reduced moment to the outer columns through the outriggers connecting the core to these columns.

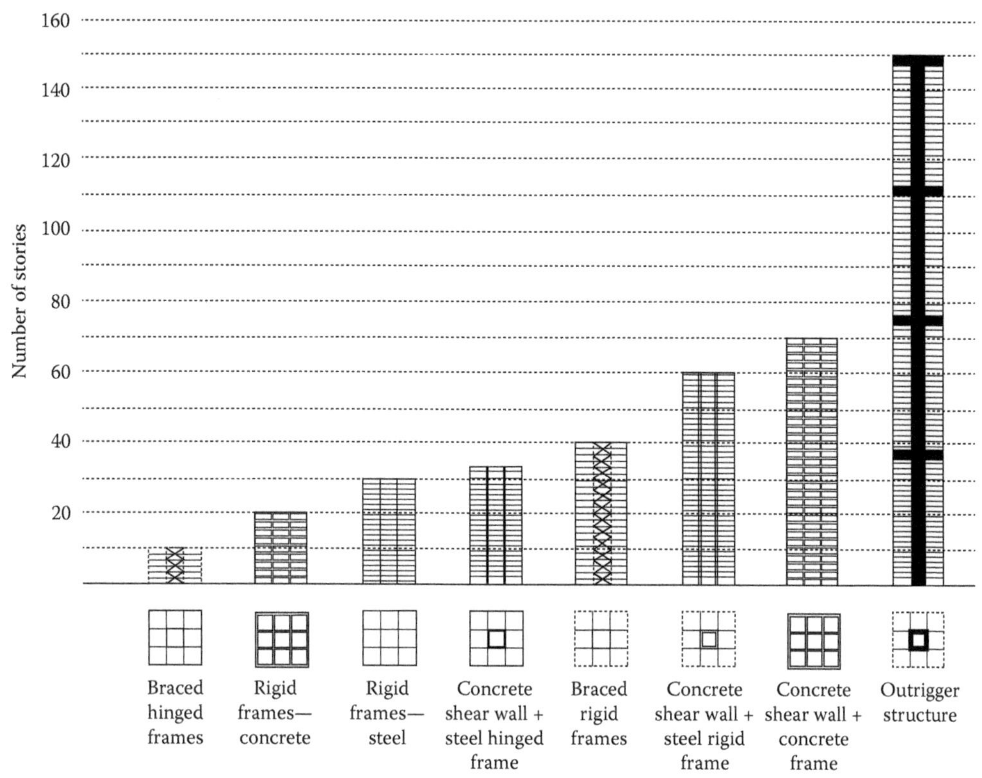


*Figure 2.4* Interior structure types. (Adapted from Ali, M.M. and Moon, K.S. 2007. *Architectural Science Review*, 50.3: 205–223. Courtesy of Taylor & Francis.)

### 2.3.2 Exterior structures

As pointed out by Ali and Moon (2007), the nature of the building perimeters has more structural significance in tall buildings than in any other building type, because of their vulnerability to lateral forces arising from wind and earthquake loads. It is desirable to concentrate as much lateral load resistance as possible on the perimeter of tall buildings to increase their structural depth and thus their resistance to lateral loads.

Alaghmandan et al. (2016) discuss the use of two main solutions, architectural and structural, to mitigate the effects of wind on tall buildings by designing the form aerodynamically, at least by using tapering and setbacks. Structural systems, such as tubes and diagrid systems, are efficient and help to reduce wind effects.

Figure 2.5 shows various types of exterior structures. The tube is one of the most typical exterior structures, and is, in effect, a three-dimensional structural system utilising the entire building perimeter to resist lateral loads. Examples of this system are the 110 storey Sears (now Willis) Tower, the 100 storey John Hancock Centre, and the 83 storey Amoco building, all in Chicago.

The framed tube system has closely spaced columns and deep spandrel beams rigidly connected together throughout the exterior frames. However, this system is said to become progressively inefficient over 60 storeys.

The braced tube is a variation of the framed tube but it uses widely spaced columns and diagonal braces instead of closely spaced columns.

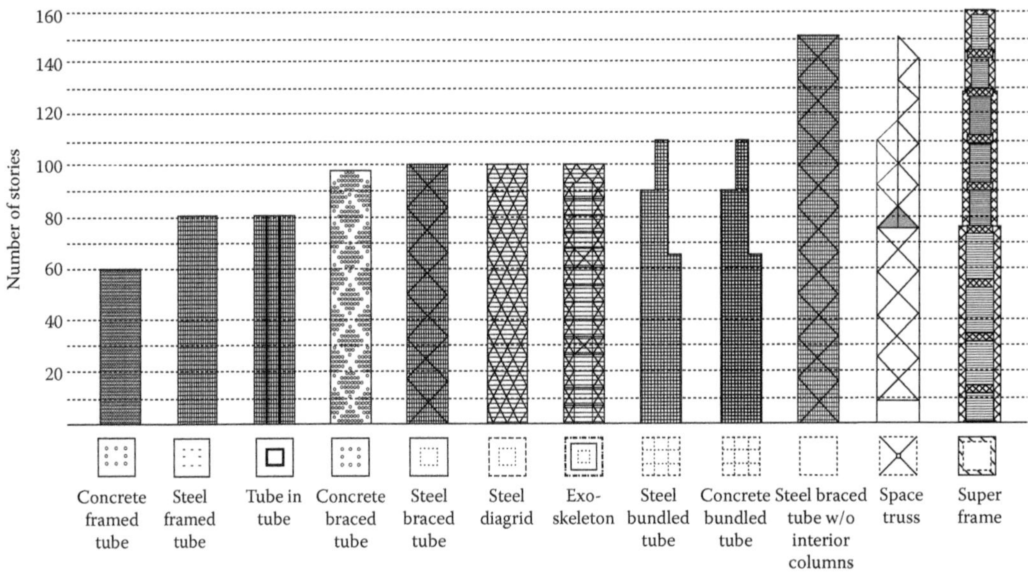


*Figure 2.5* Exterior structure types. (Adapted from Ali, M.M. and Moon, K.S. 2007. *Architectural Science Review*, 50.3: 205–223. Courtesy of Taylor & Francis.)

The bundled tube is a cluster of individual tubes connected together to form a single unit. The bundled tube system can have tubes of different shapes, terminating at different heights, and diagonals can also be added to increase the efficient height limit.

The tube in tube system can enhance the stiffness of a framed tube system. The floor diaphragm connecting the core and outer tube transfers the lateral loads to both systems.

The diagrid system utilises the effectiveness of diagonal bracing members in resisting lateral forces. Such structures provide both bending and shear rigidity, and so do not need high shear rigidity cores. Buildings with a diagrid system can be further strengthened and stiffened by engaging the core, in a similar manner to the tube in tube system.

Space truss structures are modified braced tubes with diagonals connecting the exterior to the interior. In some space structures, the diagonals penetrate the interior of the building, for example, the Bank of China building in Hong Kong.

A superframe is composed of megacolumns comprising braced frames of large dimensions at building corners, linked by multistorey trusses at about every 15–20 storeys.

In exoskeleton structures, lateral load-resisting systems are placed outside the building lines, away from their facades. An issue that has to be considered with such systems is the thermal expansion and contraction of the system.

Zhou et al. (2014) undertook a study of the structural efficiency of super-tall buildings, that is, the relative stiffness of the structure for a given height and plan area. They used an example of a 405 m structure, and examined the effects of plan shape, plan layout, elevation shape, elevation layout and outriggers. They reached the following conclusions:

1. To achieve the maximum lateral resistance, the plan shape should be triangular, and the use of mega columns around the periphery was desirable.
2. Tapering the elevation shape can improve structural efficiency.
3. The arrangement of outriggers between the core and flange frame can reduce the shear lag effect and improve lateral resistance efficiency.

## 2.4 STRUCTURAL MATERIALS

The structural materials used for tall buildings are primarily steel and reinforced concrete, with composite steel and concrete structures being very common in exploiting the favourable qualities of each material type. Typically, the trend has been to use high-strength concrete with 80 MPa or greater compressive strength. For example, the Kingdom Tower in Jeddah, Saudi Arabia, has been designed with 85 MPa for the lower parts of the tower, and 75 MPa and 65 MPa concrete for the upper floors and the spire respectively (Sim and Weismantle, 2014). The reinforcing bars are 420 and 520 MPa yield strength steel, with diameters up to 40 mm.

Another development is the re-emergence of timber as a structural building material. It is clearly not feasible to use timber for buildings within the 'tall' category (>200 m), and the tallest timber building in 2016 is a 14-storey apartment block in Bergen Norway, but a 34-storey tower in planned for completion in Stockholm Sweden in 2023. A design for a very tall and slender timber tower in London has recently been developed, and is purported to be cost-effective and faster to build. In addition, timber has an ecological advantage in that it can store $CO_2$, whereas more conventional materials such as steel and concrete emit $CO_2$.

## 2.5 SOME OTHER FACTORS CONTROLLING BUILDING PERFORMANCE

Apart from the nature of the structural system, other factors that may control the performance of a structure include the following (Choi, 2009):

1. The aspect ratio, that is, the ratio of the height to the footprint width. This ratio is preferably 6 or less, but may be in excess of 10 if special features are incorporated to improve human comfort during wind loading. As mentioned above, aspect ratios in excess of 15 have been used in recent years.
2. The building height. Relatively short buildings tend to be governed by strength considerations, while taller buildings are governed by lateral drift and wind-induced building motions (see below). Thus, the dynamic response of a tall building is of major importance in ensuring that it functions satisfactorily.

## 2.6 INTER-STOREY DRIFT

Inter-storey drift is the relative horizontal displacement of two adjacent floors in a building, and can also be expressed as a percentage of the storey height separating the two adjacent floors, this latter value often being termed the 'drift index' ($\delta$) or 'drift ratio'. It is a key indicator of the structural performance of tall buildings, and the larger the value, the greater the risk of structural damage or inadequate performance of the structure.

Scholl (1984) has proposed the following criteria for damage potential in relation to the drift index, $\delta$:

1. $\delta = 0.001$: non-structural damage is probable
2. $\delta = 0.002$: non-structural damage is likely
3. $\delta = 0.007$: non-structural damage is relatively certain, and structural damage is likely
4. $\delta = 0.015$: non-structural damage is certain, and structural damage is likely

Drift criteria are included in the design provisions of most building codes, with typical allowable values varying between 0.002 and 0.003.

## 2.7 DIFFERENTIAL SHORTENING

Kayvani (2014) has indicated that tall buildings may be affected by the differential shortening of the vertical structural elements under gravity loads, especially when these elements are of reinforced concrete construction. Such elements shorten differentially over time due to differences in elastic strains, shrinkage and creep effects. In particular, the core walls tend to shorten less than the tower columns, due to the fact that the permanent gravity stresses and strains in a core wall are often much lower than those for columns whose primary function is to resist gravity loads. Creep strains will also be less for the elements with lower stress. Moreover, the core walls are typically constructed ahead of the columns, and so the shrinkage and creep strains begin earlier. The net result may be a significant long-term differential shortening between the core walls and columns. This in turn can lead to a redistribution of loads, which may change over time. Kayvani suggests that structures should be modelled using non-linear finite element methods which can incorporate concrete shrinkage and creep as well as the construction sequence.

The redistribution of load will also be influenced by the stiffness of the foundation elements, as well as the time-dependent behaviour of the structural elements. Thus, proper consideration of the interaction between the superstructure and the foundation system is essential to develop an effective design outcome.

## 2.8 DYNAMIC CHARACTERISTICS

The dynamic response of tall buildings poses some interesting structural and foundation design challenges. In particular, the fundamental period of vibration of a very tall structure can be large, and conventional dynamic loading sources such as wind and earthquakes have a much lower predominant period and will generally not excite the structure via the fundamental mode of vibration. However, some of the higher modes of vibration will have significantly lower natural periods and may well be excited by wind or seismic action. These higher periods will depend primarily on the structural characteristics but may also be influenced by the foundation response characteristics.

As an example, the hypothetical case of a 1600 m tall concrete tower is considered. The tower is assumed to have a mass of 1.5 million tonnes, a base diameter of 120 m and a top diameter of 30 m. Figure 2.6 shows the natural frequencies computed from a finite element analysis (Irvine, M. 2008, private communication). The first mode has a natural period in excess of 20 s, but higher modes have an increasingly small natural period, and the higher axial, lateral and torsional modes have natural frequencies of 1 s or less. Such frequencies are not dissimilar to those induced by wind and seismic action.

It is interesting to note that a tall building such as the one considered here cannot accurately be considered as a simple flexural member or as a shear beam for the purposes of assessing natural frequencies. Figure 2.7 compares the ratio of the natural frequency to the fundamental frequency, and clearly demonstrates the substantial reduction in natural frequency for the higher modes. It also shows that the actual natural frequency lies between those for the flexural beam and the shear beam in this case.

### 2.8.1 Approximations for natural period of structure

It is often useful to have a quick means of estimating the natural period of a structure, prior to undertaking a more detailed structural analysis of its dynamic characteristics. A very

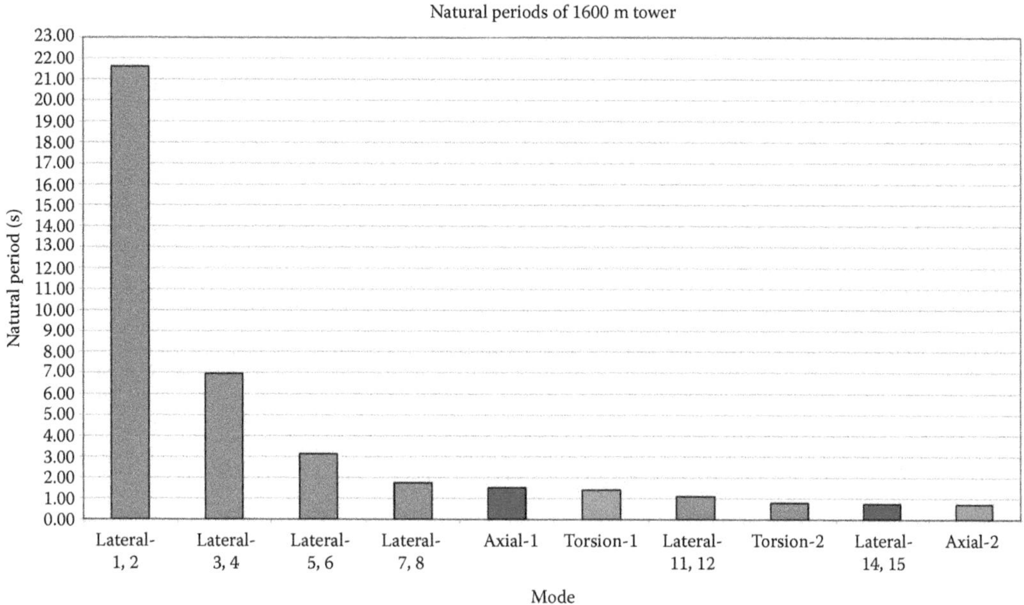


*Figure 2.6* Natural periods for various modes of vibration.

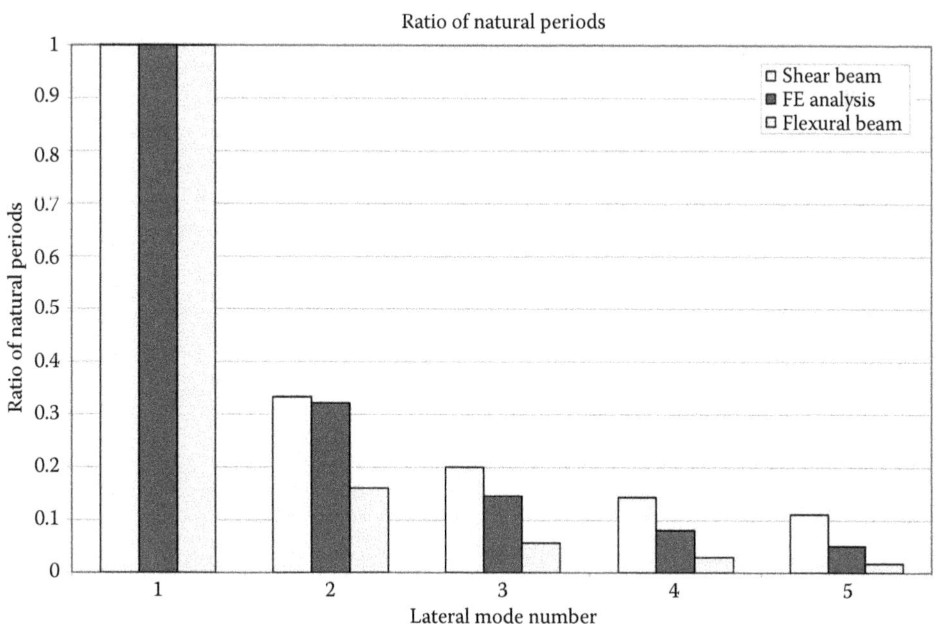


*Figure 2.7* Ratio of natural period to fundamental period, from various methods.

approximate relationship between the natural period, T, of a building and the number of stories, N, is

$$T \approx 0.1\ N\ s \tag{2.1}$$

However, a more accurate relationship for a variety of structures is given in the Australian Standard AS1170.4-2007, as follows:

$$T = 1.25\ k_t h_n^{0.75} s \tag{2.2}$$

where $k_t$ = 0.11 for moment-resisting steel frames, 0.075 for moment resisting concrete frames, 0.06 for eccentrically braced steel frames, and 0.05 for all other structures; $h_n$ is the height from base of structure to uppermost seismic weight or mass, in metres.

## 2.9 DAMPING SYSTEMS

With the use of higher strength materials and consequent lighter sections, the criterion of serviceability has become a governing factor in the design of many tall buildings. The control of structural motions should be considered for both static and dynamic loads. With wind-induced loads, the cross-wind response caused by vortex shedding can be as important as, or even more important than, the windward response. The responses can be reduced by either increasing the structural stiffness, increasing the damping associated with the structure or changing the shape of the structure. The stiffness is related to the structural system that is adopted, and more recent trends using tubes, diagrids and core-supported outrigger systems can achieve a much higher stiffness than traditional rigid frame structures.

The damping achieved by the primary structural system may be uncertain until the structure is completed, and a more reliable approach is to install auxiliary damping devices within the primary structural system. Such devices can be classified into two categories: 'passive' and 'active' systems. 'Passive' systems have fixed properties and do not require energy to perform as intended. 'Active' systems require an actuator or some form of active control mechanism relying on an energy source to modify the system response to continuously changing loads. Active systems are generally more effective than passive systems, but because of their economy and reliability, passive systems are more commonly used in building structures.

Figure 2.8 illustrates some of the passive and active damping systems available.

### 2.9.1 Passive systems

Passive systems can be considered in two categories:

- Energy-dissipating material-based systems such as viscous dampers and visco-elastic dampers
- Auxiliary mass systems that generate counteracting inertia forces, such as tuned mass damper (TMD) and tuned liquid damper (TLD).

A TMD is composed of a very large counteracting-inertia-force-generating mass accompanying relatively complicated mechanical devices that allow and support the intended

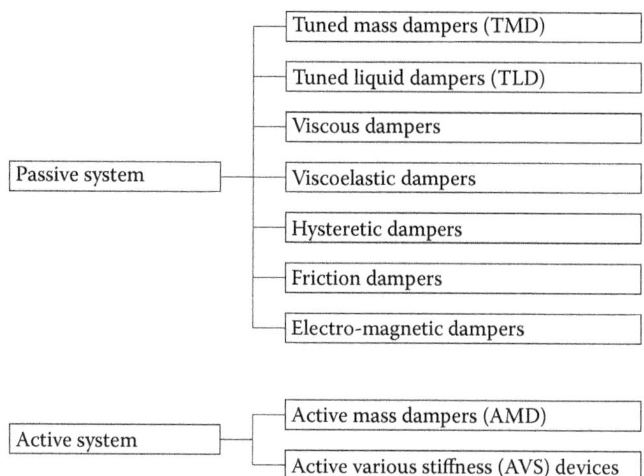


*Figure 2.8* Auxiliary damping systems for tall buildings. (Adapted from Ali, M.M. and Moon, K.S. 2007. *Architectural Science Review*, 50.3: 205–223. Courtesy of Taylor & Francis.)

performance of the mass. The frequency of the TMD mass is generally tuned to the fundamental frequency of the primary structure. The TMD is generally located near the top of the building and oscillates out of phase with the primary structure, generating a counteracting inertial force.

A TLD uses a waving water mass as a counteracting inertial force generator. The 'sloshing' frequency can be tuned by adjusting the dimensions of the water container and depth of water.

### 2.9.2 Active systems

Active systems have the ability to decide on, and carry out, a set of actions that will improve the present state of a structure, in a controlled manner and within a short period of time. Active systems can perform over a wide range of frequencies. Examples are active mass damper (AMD) and active variable stiffness device (AVSD). AMDs are similar to TMDs (tuned mass dampers) in appearance, but the vibration of the building is picked up by a sensor, the optimum power control power is calculated via a computer, and the movement of the building is reduced by shifting a moveable mass with an actuator.

AVSDs continuously alter the building's stiffness to keep the building frequency away from that of the external forces and so avoid a condition of resonance.

## 2.10 BUILDING HEIGHT RELATED TO NUMBER OF STOREYS

The CTBUH has provided approximate expressions relating the height, H, of various types of tall buildings to the number of storeys, s, as follows:

- For an office tall tower:

$$H = 3.9s + 11.7 + 3.9\left(\frac{s}{20}\right)m \qquad (2.3)$$

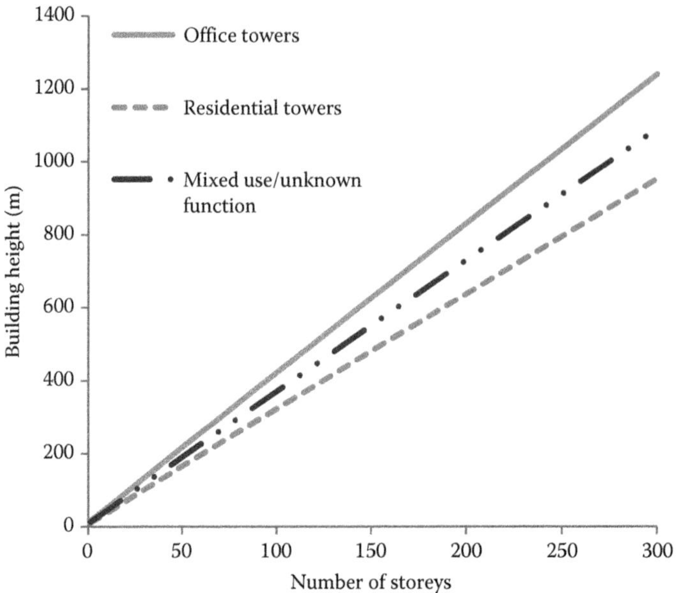


*Figure 2.9* Relationships between number of storeys and building height.

- For a residential/hotel tall building:

$$H = 3.1s + 7.75 + 1.55\left(\frac{s}{30}\right)m \tag{2.4}$$

- For a mixed use or unknown function tall building:

$$H = 3.5s + 9.625 + 2.625\left(\frac{s}{25}\right)m \tag{2.5}$$

Figure 2.9 plots the relationship between height H and number of storeys for these three cases.

Chapter 3

# Selection of foundation type

## 3.1 INTRODUCTION

The foundation is the medium through which the building loads are transferred from the superstructure to the ground, and the ground deformation is transferred to the superstructure.

The following foundation types are typically used for high-rise buildings:

- Shallow raft/mat foundation
- Compensated raft foundation
- Pile or barrette foundation
- Piled raft foundation
- Compensated piled raft foundation

Each of these foundation types is discussed below. Examples of the use of each of these foundations are described briefly. Attention is also given to some innovative foundation solutions which may be suitable for difficult ground conditions or for controlling the load distribution within the foundation system.

Good foundation design requires close collaboration between the structural and geotechnical engineers as the behaviour of both the superstructure and the foundation system needs to be adequately captured in the structural design, which in turn needs to be based on the foundation response provided by the geotechnical engineer. The design should ideally be an iterative process in order to establish compatible structural loadings and foundation deformations.

## 3.2 FACTORS AFFECTING FOUNDATION SELECTION

The factors that may influence the type of foundation selected to support a tall building include the following:

- Location and type of structure
- Magnitude and distribution of loadings
- Ground conditions
- Access for construction equipment
- Durability requirements
- Effects of installation on adjacent foundations, structures and people
- Relative costs
- Issues of sustainability
- Local construction practices and availability of construction materials

The various types of foundation are discussed below.

## 3.3 RAFT/MAT FOUNDATION

### 3.3.1 Concepts

High-rise developments often contain a multi-level basement which results in the base of the development being founded close to, or even embedded into, competent rock. This provides the opportunity for a shallow foundation to be constructed to support the development. In some cases, it is possible to have individual pad footings to support the columns and the core walls. However, as the gravity loads increase with building height, the use of individual footings becomes problematical, and it is then more common to adopt a raft or mat foundation to support the entire structure. The term 'raft' will be used in this book although 'mat' is commonly employed in some countries, including the United States.

Figure 3.1 illustrates several types of raft foundation, several of which involve thickened sections below heavily loaded columns.

A raft foundation is appropriate when suitable bearing material is very close to or at the lowest basement level and a wider spread of loads onto the underlying soils is required, or where better control over the differential settlements is required. It is also more robust and redundant than individual pads, with respect to the presence of local pockets of loose and soft soils underneath the foundation.

Raft foundations are relatively large in size, and hence the foundation bearing capacity is generally not the controlling factor in the design. The bearing capacity, derived from classical soil mechanics, should nevertheless be estimated as it will generally be used in considering the non-linear foundation settlement behaviour for high-rise building design. The effects of lateral and moment loading should be incorporated into the assessment of ultimate

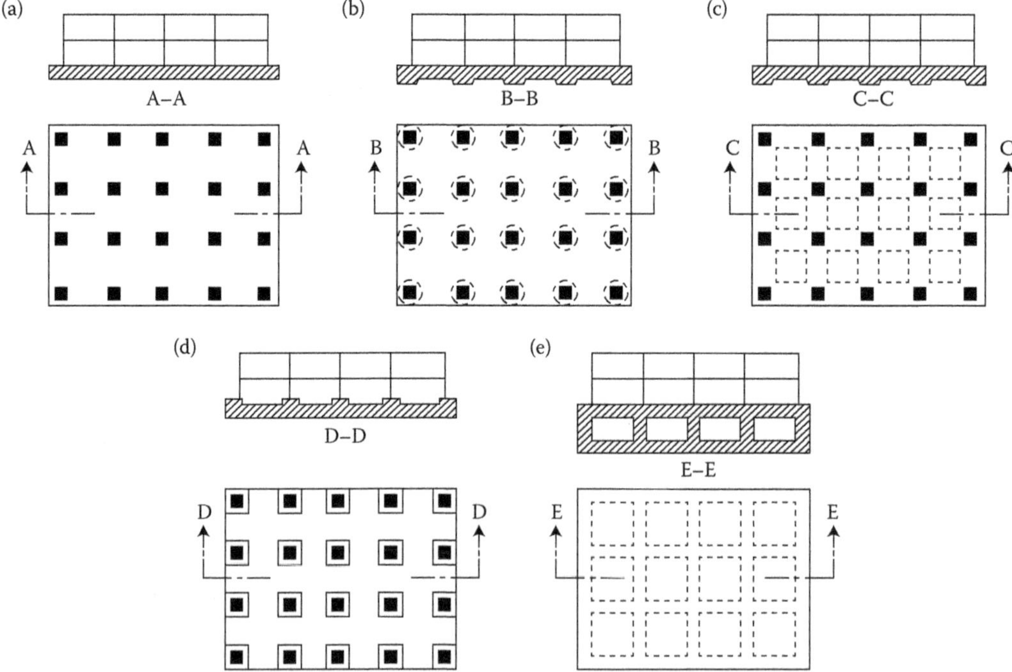


*Figure 3.1* Types of raft/mat foundation. (a) flat slab; (b) local thickening below columns; (c) thickened strips below column lines; (d) pedestals below columns and above main raft slab; (e) basement structure incorporating the raft.

bearing pressure. Soil stiffness is important in the design to understand the load distribution in the raft and for evaluating bending moment and shears. More detailed discussion of the geotechnical design of rafts is given in Chapters 7 and 8.

### 3.3.2 Examples of raft foundations

*Al Faisaliah building complex, Riyadh, Saudi Arabia*

Morrison (2000) has described the foundations for this 40-storey office tower block. The raft is 3–4 m thick and has a plan area of 1600 m$^2$. The thicker part of the raft is located centrally below the main tower footprint, with 3 m sections below large raking exterior columns. Outside of the main footprint, the slab thickness reduces to 0.6 m and supports a series of individual columns. The ground conditions comprise 1–6 m of silty sand and gravel overlying Jurassic limestone which was found to be fissured. The design was based on permissible pressures of 850 kPa for the core and 1000 kPa for the corner columns.

*Le Royal Hotel complex, Amman, Jordan*

This case has also been discussed by Morrison (2000) and involves a building with the superstructure in the form of a ziggurat, and elliptical in plan. The major and minor axis dimensions are 104 × 80 m. The raft thickness varies in steps from 1.5 m to 2.5 m to 3 m in accordance with the variation in column loads. The founding level for the raft is 23 m below street level, and the total height above street level is 108 m. The average bearing pressure is about 400 kPa.

*Buildings in Frankfurt, Germany*

Katzenbach, R. (personal communication, 2016) mentions a number of 'first generation' buildings in Frankfurt that were founded on shallow rafts in Frankfurt Clay. Typically, they were founded at depths of 12–20 m below ground surface, and included the following buildings:

- Dresdner Bank Tower (now the German Railway Authorities Building)
- The Deutsche Bank Twin Towers
- The Marriott Hotel

These towers were 140–160 m tall and had measured settlements ranging between 120 and 340 mm, together with differential settlements of 100–150 mm. These differential settlements gave rise to severe problems with tilting of the buildings. The experiences with these buildings led to the realisation that shallow raft foundations would not be adequate for buildings in excess of 160 m high in Frankfurt. This stimulated the development of piled raft foundations in that city.

## 3.4 COMPENSATED RAFT FOUNDATION

### 3.4.1 Concepts

Tall buildings very frequently have one or more basements to cater for car parking and/or commercial and retail space. In such cases, the construction of the raft involves excavation of the soil prior to construction of the foundation and the superstructure. Because of the stress reduction in the underlying ground caused by excavation, the net increase in ground stress due to the structure will be decreased, and hence it may be expected that the settlement and differential settlement of the foundation will also be decreased as compared with the corresponding foundation at ground level. The resulting foundation is often termed a compensated or buoyancy raft, and an example is illustrated in Figure 3.2. Such a foundation system can be very beneficial when constructing buildings on soft clay, as the settlements that occur can be significantly less than those if the foundation was located at or near the ground surface. However, there will be limitations to the height of buildings that can be constructed on such soils because of bearing capacity and settlement considerations.

### 3.4.2 Examples of compensated rafts

*Shell building, Houston, Texas*

The Shell building is a 50-storey tower which rises 218 m above street level. It is founded on a 2.5 m thick raft which is 52 m wide and 71 m long, with four levels of basement that extend beyond the raft periphery. The ground conditions include sand, silt, loam and clay, which are subject to shrink–swell movements. The gross average pressure was 405 kPa, but the net pressure reduced to 36 kPa as a consequence of the deep excavation. In the basement area outside the raft, there was a high upward pressure of 261 kPa due to buoyancy.

The maximum settlement observed was about 125 mm with a maximum angular distortion of 1/170 after 8 years. These are both significantly larger than normally accepted criteria, but the building appears to have functioned adequately. Further details are given by Hemsley (2000).

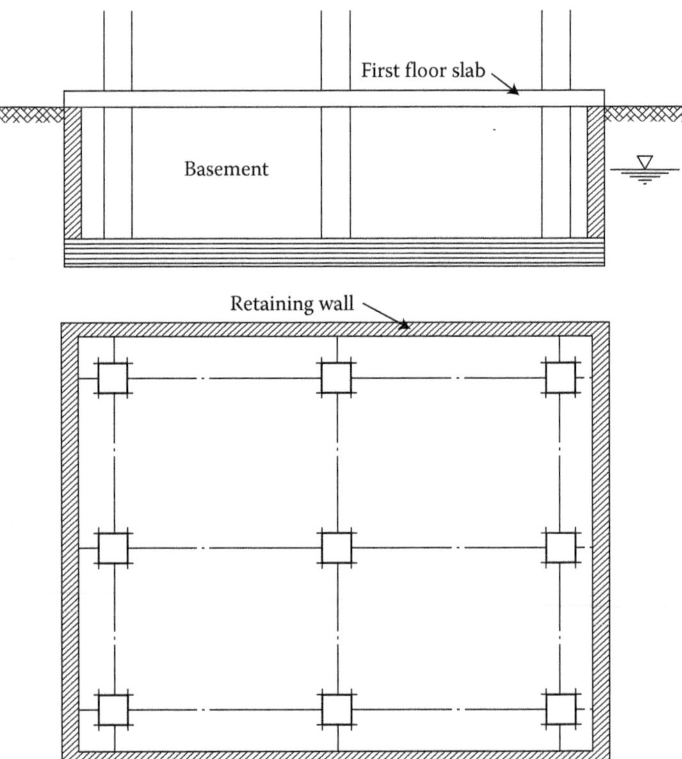


*Figure 3.2* Compensated raft foundation.

*Tower Palace III, Seoul, Korea*

Abdelrazaq et al. (2004) describe the case of the Tower Palace II, a 73-storey tower with an adjacent eight storey sports centre and six levels of basement. This building was, at the time of construction, the tallest building in Korea, and was located in an area of low seismicity and on rock. The tower was founded on a 3.5-m thick raft whose construction involved the pouring of 8000 $m^3$ of concrete. There were fault and shear zones identified in the rock about 7 m north of the foundation raft to approximately 150 m below the south edge of the raft. However, these defects did not appear to adversely influence the performance of the foundation as the measured foundation settlement was stated to be less than the anticipated value of 15 mm. In this case, the reduction of net stress due to the excavation did not appear to be a major factor in the small settlements experienced. However, the basement walls would have played a vital role in resisting lateral wind loads.

No quantitative details have been provided by the authors on the nature of the rock on which the raft was founded, but this case suggests that the rock was of very good quality, despite the apparent defects, and that compensated raft foundations founded on good quality rock can perform satisfactorily, provided that the lateral wind and seismic forces can be adequately catered for. However, for less competent founding conditions, rafts or compensated rafts are likely to experience excessive settlement and perhaps tilt, and may also not be able to provide adequate resistance to lateral and moment loadings. In such cases, consideration needs to be given to deep foundation options.

## 3.5 PILE FOUNDATIONS

### 3.5.1 Concepts

Often the ground conditions at a site are not suitable for a shallow raft foundation system, especially for high-rise buildings where the vertical and lateral loadings imposed on the foundation are significant. In these circumstances, it is necessary to support the building loads on deep foundations or piles, either single units or in groups, generally located beneath columns and load bearing walls.

Figure 3.3 illustrates a typical deep foundation arrangement in which several pile groups are employed to support the structural columns. The floor slab in this case is not designed to support any of the column loadings.

The key design issues in relation to pile foundations for high-rise buildings are set out in Chapter 4, together with a description of the foundation design process.

The selection of appropriate pile types will be influenced by several factors including ground conditions, available piling equipment, local construction experience and expertise, required structural and geotechnical capacity and site constraints. For high-rise buildings, bored or cast-in situ piles are very widely adopted. Driven piles are not commonly used in modern urban high-rise construction, due to the limitations in load capacity and the issues of noise and vibration that usually prevail in urban environments. In some cases involving only moderately tall buildings on sandy soils, continuous flight auger (or auger cast) piles may be an option, given that they are relatively quick to install and almost vibration-free. However, bored piles constructed under bentonite, or preferably with the aid of polymer drilling fluid rather than bentonite, are the most common form of piling in contemporary high-rise construction. Diameters in excess of 3 m can be achieved and can be designed for serviceability loads in excess of 50 MN in appropriate ground conditions.

Once the geotechnical model is developed and the pile type is selected, the designer can assess the axial capacity (both in compression and tension) of a range of pile diameters and

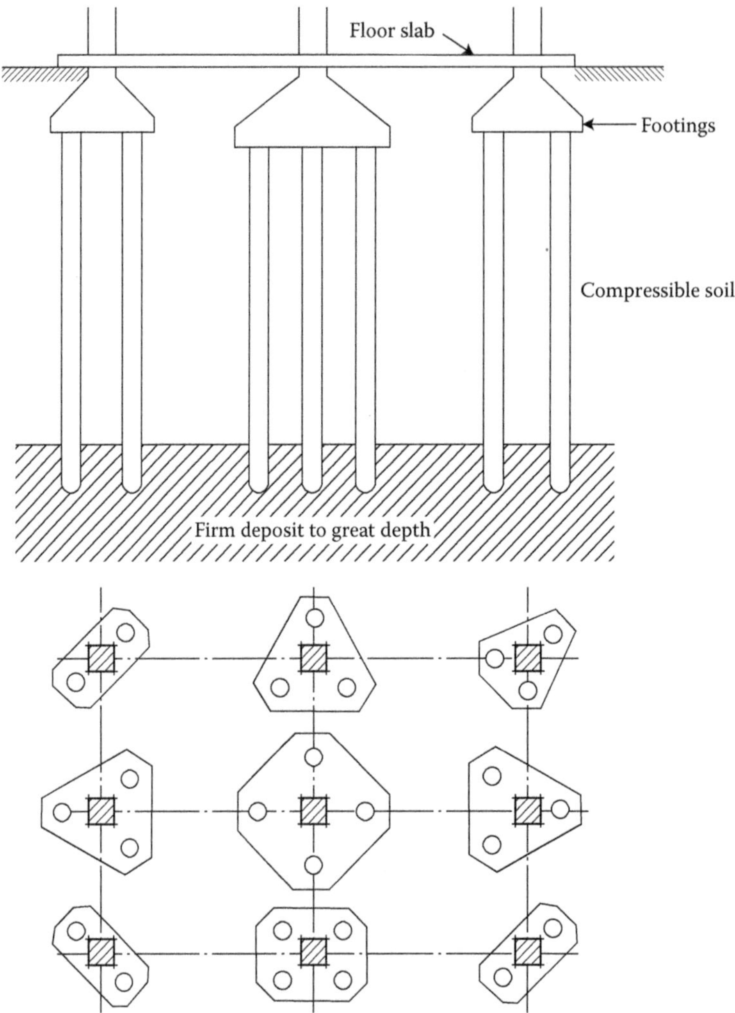


*Figure 3.3* Typical pile group arrangement.

types using static methods of analysis (see Section 4.8). Using this information, the designer can develop suitable pile size and arrangement options for the tower foundation.

A piled foundation for towers often comprises a large numbers of piles and therefore the challenge in the design is to capture the effects of the group interaction. It is well recognised that the settlement of a pile group can differ significantly from that of a single pile at the same average load level due to group effects, as shown in Figure 3.4. Also, the ultimate load that can be supported by a group of piles may not be equal to the sum of the ultimate load which can be carried by each pile within the group, and therefore consideration must be given to the pile group efficiency.

More detailed discussions of the design of pile foundations are given in Chapters 7 and 8.

### 3.5.2 Barrettes

Barrettes are, in effect, large rectangular piles. Similar to rotary bored piles, barrettes are installed using drilling techniques under bentonite fluid or polymers that assist in maintaining

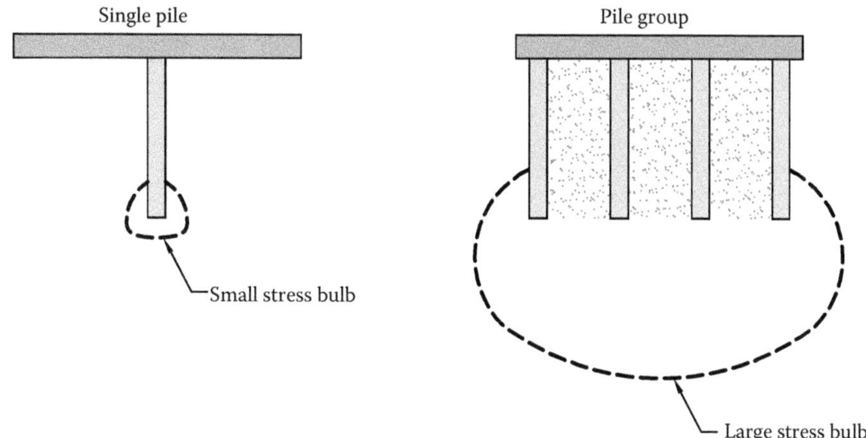


*Figure 3.4* Pile group effect.

the stability of the bore during drilling. Unlike bored piles, barrettes are not installed using rotary drilling rigs but rather using a diaphragm wall cutter such as a hydrofraise, or via the use of a clamshell. The construction methods used for barrettes have been used widely for constructing diaphragm walls, but have also been used very effectively as the foundations of high-rise buildings to carry heavy loads and high bending moments. Barrettes have the advantage over traditional bored piles in that they can be installed to larger depths (over 100 m), with a greater degree of verticality accuracy, and are able to sustain larger loads. Barrettes can also be used to create composite sections such as T-sections or cruciform sections. Submaneewong and Teparaksa (2009) have described the results of lateral load tests on T-shaped barrettes in Bangkok, which showed a significantly larger lateral resistance than a conventional circular bored pile.

Due to their large size, barrettes lend themselves to either conventional basement construction or top down methods. The latter can be achieved by extending the barrette to ground surface level or, if the barrette sections are considered too large, plunge columns of smaller sections can be placed into the barrettes.

Design methods for barrettes are essentially the same as for bored piles, except that, under lateral loading, the relevant stiffness for the load direction in question must be used. It is often convenient, albeit approximate, to represent a barrette as an equivalent diameter bored pile with equal circumference and axial stiffness, and with a lateral stiffness relevant to the direction of the lateral shear and moment loading. An advantage of barrettes is that their strong axis can be aligned in the wind direction which produces the largest lateral loads.

Thasnanipan et al. (2000) provide useful information on construction times for the various processes involved in barrette construction in Bangkok in the latter part of the twentieth century, and these are given in Table 3.1.

### 3.5.3 Examples of piled and barrette foundations

*Sears Tower, Chicago*

Reinforced drilled concrete piles (or caissons in the local Chicago terminology) bearing on rock have been used to support the 442 m tall Sears Tower in Chicago. Diameters ranged between 1.83 and 2.13 m under the tubular line columns, with two 3.05 m diameter

Table 3.1 Average construction time for typical barrettes in Bangkok (1.2 × 3.0 × 44.5 m)

| *Activity* | *Construction time (h)* |
|---|---|
| Excavation | 16.5 |
| Checking verticality with Koden | 2 |
| De-sanding | 5 |
| Cage installation | 4.5 |
| Stanchion installation | 3 |
| Tremie preparation | 3.5 |
| Concrete pouring | 3.5 |

caissons supporting two gravity only columns located near the mid-point of the centre tube. The average length of the caissons was 18.3 m below a 1.53 m thick concrete mat, and socketed into the underlying bedrock. Permanent steel liners were used for all caissons, with a thickness of 32 mm. An allowable bearing pressure of 10 MPa was used in design.

### *Commerzbank Tower, Frankfurt, Germany*

The 302 m tall Commerzbank building is located on a layer of stiff Frankfurt Clay about 33 m thick, underlain by Frankfurt limestone. Because of the significant weight of the building, a combined piled raft foundation was not found to be economically feasible, and so a fully piled foundation system was adopted, with piles founded in the underlying limestone. The piles were 1.8 m in diameter for the upper 20 m of the pile, then reducing to 1.5 m diameter over the remaining length, which was up to 48.5 m. The raft (not taken into account in the geotechnical design) varied in thickness from 2.2 to 4.5 m.

### *Taipei 101, Taipei, Taiwan*

The foundations for the 508 m tall Taipei 101 tower consisted of cast-in-place bored piles, selected on the basis of an extensive series of pile load tests. Under the main tower, 1.5 m diameter bored piles were used under a 1.5-m thick concrete raft. The piles extended to bedrock 40–65 m below grade, through soft clay, colluvium and weathered rock, and were socketed an additional 15–30 m into solid sandstone to cater for allowable loads of 10.7–14.2 MN in compression, and about half these values for uplift. A total of 380 piles were used under the tower raft footprint.

### *Petronas Towers, Kuala Lumpur, Malaysia*

The twin 452 m tall towers are located over deep Kenny Hill formation residual soil, with limestone 60 m to more than 180 m below grade, or 35–155 m below foundation raft level. Each tower sits on a heavily reinforced concrete raft, 4.5 m thick, below which are a series of barrettes, rectangular in plan. The barrettes vary in length from 20 to 105 m, such that the depth of soil between the barrette bases and the steeply dipping limestone bedrock is reasonably constant. The limestone was found to be cavernous in places, with some clay-filled pockets close to the limestone surface. An extensive pressure grouting program was included in the design to densify local soft spots in or near the rock surface.

## 3.6 PILED RAFT FOUNDATIONS

### 3.6.1 Concepts

Many high-rise buildings are constructed with thick basement slabs, but when piles are used in the foundation, it is often assumed that the basement slab does not carry any of the foundation loads. In some cases, it is possible to utilise the basement slab in conjunction with the piles to obtain a foundation that satisfies both bearing capacity and settlement criteria.

A piled raft foundation is a composite system in which both the piles and the raft share the applied structural loadings, as shown in Figure 3.5. Within a conventional piled foundation, it may be possible for the number of piles to be reduced significantly by considering the contribution of the raft to the overall foundation capacity. In such cases, the piles provide the majority of the foundation stiffness while the raft provides a reserve of load capacity. In situations where a raft foundation alone might be used, but does not satisfy the design requirements (in particular the total and differential settlement requirements), it may be possible to enhance the performance of the raft by the addition of piles. In such cases, the use of a limited number of piles, strategically located, may improve both the ultimate load capacity and the settlement and differential settlement performance of the raft, such that the design requirements can be met.

The main advantages of adopting a piled raft foundation are

- As piles need not be designed to carry all the load, there is the potential for substantial savings in the cost of the foundations.
- Piles may be located strategically beneath the raft so that differential settlements can be controlled.
- Piles of different length and/or diameter can be used at different locations to optimise the foundation design.
- Varying raft thicknesses can be used at different locations to optimise the foundation design.
- Piles can be designed to carry a load approaching (or equal to) their ultimate geotechnical load, provided that the raft can develop an adequate proportion of the required ultimate load capacity.

The most effective application of piled rafts occurs when the raft can provide adequate load capacity, but the settlement and/or differential settlements of the raft alone exceed

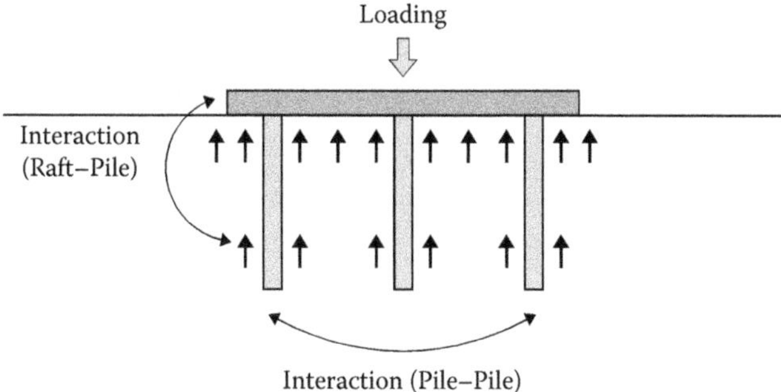


*Figure 3.5* Load distribution in piled raft system.

the allowable values. Poulos (2001b) has examined a number of idealised soil profiles, and found that the following situations may be favourable:

- Soil profiles consisting of relatively stiff clays
- Soil profiles consisting of relatively dense sands or weak rocks

In both circumstances, the raft can provide a significant proportion of the required load capacity and can also contribute to the foundation stiffness, especially after the pile capacity has been fully mobilised (Viggiani, 1998; Mandolini et al., 2005; Viggiani et al., 2011).

It has also been found that the performance of a piled raft foundation can be optimised by selecting suitable locations for the piles below the raft. In general, the piles should be concentrated in the most heavily loaded areas, while the number of piles can be reduced, or even eliminated, in less heavily loaded areas (Horikoshi and Randolph, 1998).

There are soil profiles in which piled rafts may not provide much, if any, advantage over a conventional piled foundation. These include

- Profiles with very soft clays at or near the surface of the raft, where the raft can contribute only a relatively small proportion of the required ultimate load capacity.
- Profiles which may be subjected to long-term consolidation settlement; in this case, the soil may tend to lose contact with the raft and transfer all the load to the piles.
- Profiles which may be subjected to expansive (upward) movements; in this case, the soil movements will result in increased contact pressures on the raft and the consequent development of tensile forces in the piles.

Broad design guidelines for piled rafts have been developed by ISSMGE (2013). More detailed discussion of the geotechnical design of piled rafts is given in Chapters 7 and 8.

### 3.6.2 Examples of piled raft foundations

*Emirates Twin Towers, Dubai, UAE*

The twin towers, 355 and 305 m in height, were founded on a raft 1.5 m thick, without any basements. The foundations consisted of bored piles 1.2 and 1.5 m diameter and about 40–45 m long. The measured settlement performance of the towers during construction was far more favourable than the design expectations. A more detailed description of these structures and their foundation design is given in Chapter 15.

*Burj Khalifa, Dubai, UAE*

The foundations of the 828 m tall tower consisted of a pile-supported raft 3.7 m thick. 196 bored cast in place piles were used, 1.5 m in diameter, 43–47 m long, with a working load of about 30 MN each. The ground conditions below the base of the raft consisted of interbedded layers of cemented calcareous sediments. A more detailed description of this structure and its foundations is given in Chapter 15.

*Shanghai Tower, Shanghai, China*

This tower is 632 m in height and has 128 storeys. The ground conditions consisted primarily of alluvial clays, and 955 reinforced concrete bored piles, 52–56 m long, were used to support the tower, connected by a 6 m thick concrete mat foundation. The tower was completed early in 2016.

*Kingdom Tower, Jeddah, Saudi Arabia*

Currently under construction, this tower is planned to be in excess of 1000 m tall and will thus be the tallest building in the world when completed in 2017–2018. The ground conditions are difficult and consist of a series of layers of weak rock together with highly permeable coral rock. Sound sandstone rock is located about 122 m below the ground surface. The foundation system is similar in concept to that of the Burj Khalifa, and consists of a combined piled raft system. A matrix of 270 bored piles, 1.5 and 1.8 m in diameter, are arranged over the tower footprint. The piles extend to a depth of 108 m near the centre of the tower, reducing to 45 m deep near the end of the three wings. The raft slab thickness varies from 5 m at the ends of the wings to 4.5 m near the centre of the tower. The piles have been constructed under polymer slurry and are designed to provide the majority of axial load resistance via skin friction, with some of the load also being carried by the raft slab. A comprehensive series of pile load tests verified the load carrying and settlement characteristics of the piles. The estimated long-term settlement of the tower is in the order of 100 mm, with differential settlements between the centre and the ends of the wings being about 20 mm.

## 3.7 COMPENSATED PILED RAFT FOUNDATION

### 3.7.1 Concepts

Figure 3.6 illustrates a compensated piled raft foundation, in which piles are placed below a compensated raft. This foundation option is the one most likely to experience the least settlement and highest capacity of any of the foundation options, since the compensation effect of the excavated ground and the additional vertical and lateral resistance of the basement walls both contribute to both the stiffness and capacity of the foundation system.

Poulos (2005c) and Sales et al. (2010) discuss this foundation type and demonstrate that it has the potential to significantly reduce the settlement of a raft alone, or of a piled or piled raft system with no basement. Section 8.10.4 of Chapter 8 gives further details of the design process.

### 3.7.2 Examples of compensated raft foundations

Many of the buildings designed as piled rafts have some measure of compensation because of the presence of basements which extend some depth below the ground. In many cases,

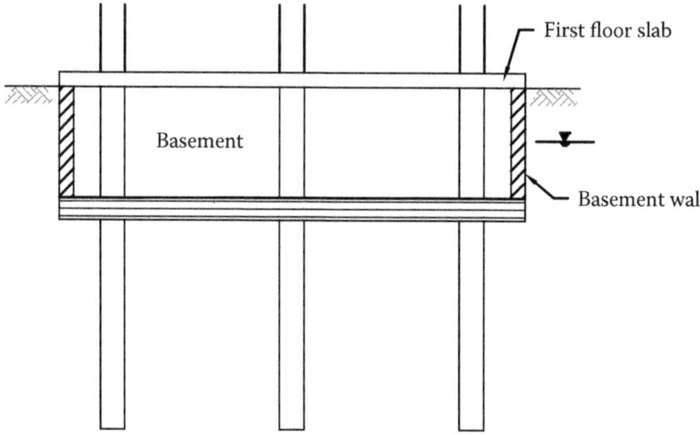


*Figure 3.6* Compensated piled raft foundation system.

given the significant net pressures exerted by very tall buildings, the effects of compensation tend to be ignored or else are not particularly significant.

*La Azteca building, Mexico City, Mexico*

An outstanding early application of the compensated piled raft principle is the La Azteca building in Mexico City (Zeevaert, 1957). The use of compensation reduced the settlements substantially from those that would have been experienced without compensation. The ground conditions consisted of a very deep deposit of sift Mexico clay. This case is considered in more detail in Chapter 15.

*Messe Torhaus, Frankfurt, Germany*

The Messe Torhaus in Frankfurt has been discussed by Sommer et al. (1985) and Franke et al. (2000), while Katzenbach et al. (2000) have presented the project and performance measurements in detail. The foundation consists of two similar piled rafts, 10 m apart, each having an area of 17.5 × 24.5 m in plan. Each raft is 2.5 m thick, and is supported by a group of 6 × 7 bored piles that are 20 m long and 0.9 m in diameter. Only 3 m of soil was excavated to install the raft and the groundwater level is just below the raft. The weight of excavated soil was about 23.1 MN, the raft weight was about 26.8 MN, and the total applied load was about 200 MN for each raft.

*Messe Turm, Frankfurt, Germany*

The 256 m tall Messe Turm is probably the most analysed piled raft case, since it was well instrumented and was the highest European building for many years. Sommer et al. (1985), Sommer (1993), Franke et al. (2000) and Katzenbach et al. (2000) have described the details of the piled raft foundation. The square raft, with a side length of 58.8 m, and a variable thickness from 3 m at the edge to 6 m at the centre, was founded on 64 piles having a diameter of 1.3 m. The piles were distributed in three rings (28 piles with 26.9 m in length in the outer ring, 20 piles with 30.9 m in the middle ring and another 16 piles, 34.9 m long, in the inner ring). Sommer (1993) reported that the soil was initially excavated about 7 m, the piles were installed and then the excavation was completed. The pile load measurements only started at the end of excavation with installation of the top load cells, so the stresses generated in the piles due to the excavation were not registered. Another important fact was that there were changes in groundwater level during the construction. The first lowering was done to allow the raft concreting to be carried out, and was interrupted just after raft completion. The groundwater drawdown continued for 2 years, was suspended for another 2 years, and then activated again for about another 2.5 years. This resulted in a change in the total load of almost 300 MN during each change in groundwater level. Sales et al. (2010) estimated that the reduction in loads on the foundation system due to the excavation would have been as follows:

Excavation = 616 MN
Maximum buoyant force = 311 MN

The excavation process induced tensile forces in the piles during the early stages of construction, but these forces changed to compression as the building load came on.

## 3.8 FOUNDATION SELECTION GUIDE

### 3.8.1 Influence of foundation width and safety factor

Mandolini (2003) provides a useful chart to assist in the selection of an appropriate foundation type, considering the conventional range of foundation options. This chart is shown in Figure 3.7, and contains four regions:

1. Region I, for which a raft foundation may be relevant. In this region, the computed factor of safety against failure may be large, and the estimated maximum settlement may be less than the specified value (8 cm in this case).
2. Region II, in which a 'small' piled raft may be appropriate. In such cases, the settlement criteria may be satisfied but the factor of safety is too low. Thus, the design of a foundation in this category is governed by load capacity considerations.
3. Region III, in which both the settlement and the capacity criteria require attention.
4. Region IV, where the foundation design is controlled by settlement, rather than load capacity, considerations.

It is Region IV in which many high-rise building foundations lie, and for which a piled raft foundation may be a suitable option.

### 3.8.2 Suitability of foundation systems for various ground conditions

An approximate guide to which foundation systems may be suitable to adopt for tall buildings on various ground profiles is given in Table 3.2. It should be emphasised that this is a rough guide only, and that a proper decision on the type of foundation requires careful consideration of the factors set out in this chapter.

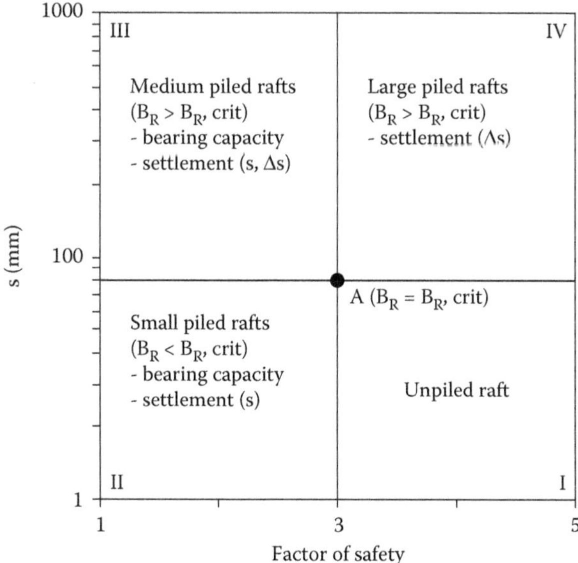


*Figure 3.7* Example of chart for selection of foundation type (Mandolini, 2003). Targets: minimum factor of safety (FS) = 3, maximum settlement = 8 cm.

*Table 3.2* Suitability of foundation types for various ground conditions

| *Ground conditions* | *Suitable foundation types* | *Notes* |
|---|---|---|
| Deep soft clay | Compensated piled raft; piled raft; piled foundation | Raft may provide only limited resistance and stiffness |
| Shallow clay layers overlying bedrock | Piled foundation to rock | Little is generally to be gained by using alternative system such as piled raft |
| Deep stiff clay | Piled raft | Many examples in Frankfurt |
| Deep medium-dense sand | Piled raft | Examples in Japan, Germany |
| Strong rock | Raft; piled raft with limited number of short piles | Piles may only be necessary under very heavily loaded areas, if at all |
| Weak rock to depth | Piled raft; compensated piled raft | Bored piles can develop considerable skin friction in such strata |
| Karstic limestone | Piled raft | A piled raft provides redundancy in case some piles encounter karstic conditions |
| Rock strata becoming weaker with depth, or overlying weak strata | Raft; piled raft | Keep piles well above weak strata, and design for considerable settlement |

## 3.9 SOME INNOVATIVE FOUNDATION OPTIONS

While most tall buildings employ the foundation systems discussed above, there are some innovative options that may be worthy of consideration in some special cases. Two of these, shell foundations and micropiles, are outlined briefly below. Also discussed are controlled stiffness inserts (CSIs) to provide a means of evening out loads and settlements, and energy piles, which can serve a dual purpose in providing both load-carrying capacity and energy generation.

### 3.9.1 Shell foundations

This foundation type has been discussed in detail by Kurian (2006). An example of such a foundation is shown in Figure 3.8, which appears to be the simplest form of shell that can be used. The advantages of such a foundation type are that the loads can be transferred through the often weaker near-surface soils to more competent bearing strata, and that significant vertical and lateral foundation capacity can thus be generated. Shell foundations are potentially economical because of the savings in materials they offer, and so may be attractive in developing countries where material costs are high and labour costs are relatively low. Kurian (2006) provides a very detailed exposition of the structural and geotechnical design aspects of shell foundations.

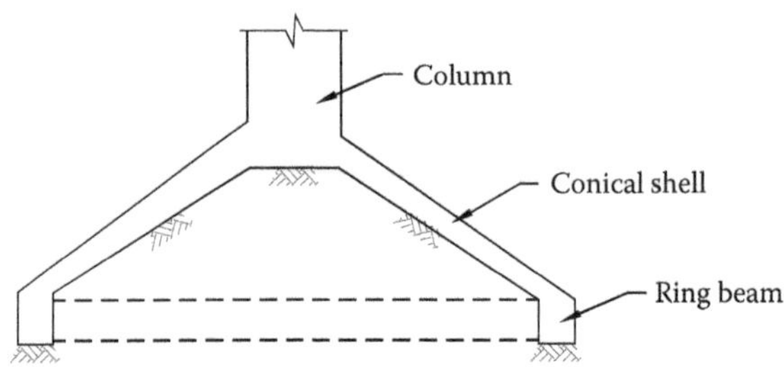


*Figure 3.8* Typical shell foundation for column. (Adapted from Kurian, N.P. 2006. *Shell Foundations*. Universities Press, Hyderabad.)

### 3.9.2 Foundations for difficult ground conditions

Cases exist where it is not practical or economical to use conventional deep foundation solutions. These include ground where obstructions, boulders or solution cavities exist. Of particular importance are the latter, which occur in regions containing karstic limestone. Knott et al. (1993) have discussed some foundation solutions for such ground conditions, which include the following:

1. Shallow foundations, generally mats or rafts, which are placed on the rock where the rock is relatively shallow and cavities in the rock surface are not too deep to clean and backfill with concrete. Permeation grouting, pressure grouting, compaction grouting or jet grouting may be used to fill voids, displace soft soil in the voids or solidify soil within the voids. The purpose of such treatment is to provide uniform stiffness below the foundation and to help prevent loss of support due to void collapse or soil migration.
2. Driven steel H-piles. These can be driven to refusal, but may not be a reliable option due to the possibility of the tip bearing on steeply dipping rock, and of possible further voids below tip level.
3. Bored piles (drilled piers) or barrettes, which offer the ability to drill probe holes in the base of the shaft to assess whether major defects exist within the zone of support. They are also easily adaptable to varying length requirements, but sound construction procedures are essential for their success.
4. Minipiles or micropiles, which are drilled and grouted and can be adjusted to variable lengths. They also offer the opportunity to fill voids that are encountered during construction. Their design often involves the assumption that only shaft friction can contribute to the pile capacity, as end bearing may be unreliable if an underlying void collapses.

In relation to the latter solution, Dotson and Tarquino (2003) propose the use of micropiles which are installed via rotary eccentric percussive duplex drilling. This method uses an inner rod and an outer casing, with the spoil flushed inside the casing. The bit on the inner drill rod uses a down-hole hammer, with the bit specially designed to open up during drilling to a diameter slightly larger than the outside diameter of the drill casing. A slightly oversized hole is thus created through obstructions or rock and this allows the casing to follow the bit down. Compressed air is used to drive the hammer and also acts as the drilling fluid to lift the cuttings. This drilling method is effective in soils with large amounts of obstructions (e.g. cobbles, boulders or demolition waste) and is also very effective in advancing a drill casing through highly fractured zones in karst. Intimate contact between the casing and the surrounding ground is maintained. Tremie grouting is used to place grout in a wet hole, being pumped through the tube as it is slowly lifted from the hole. When working in highly broken, fractured or karstic rock, grout loss is possible and it may be desirable to perform water testing and seal grouting.

The response of the down-hole hammer indicates whether rock of sufficient quality has been penetrated. Once a competent zone is established for the bond zone, the casing is withdrawn to the top of the bond zone and the pile is filled internally with grout. Once the grout level has stabilised in the bond zone, the reinforcing steel (typically a central rod in small diameter micropiles) is inserted. Figure 3.9 shows typical cross-sections of such a pile.

The design of micropiles follows similar principles to the design of bored piles. The main difference is that the axial capacity of micropiles can often be governed by the structural strength, rather than the geotechnical capacity, and that being relatively small in diameter,

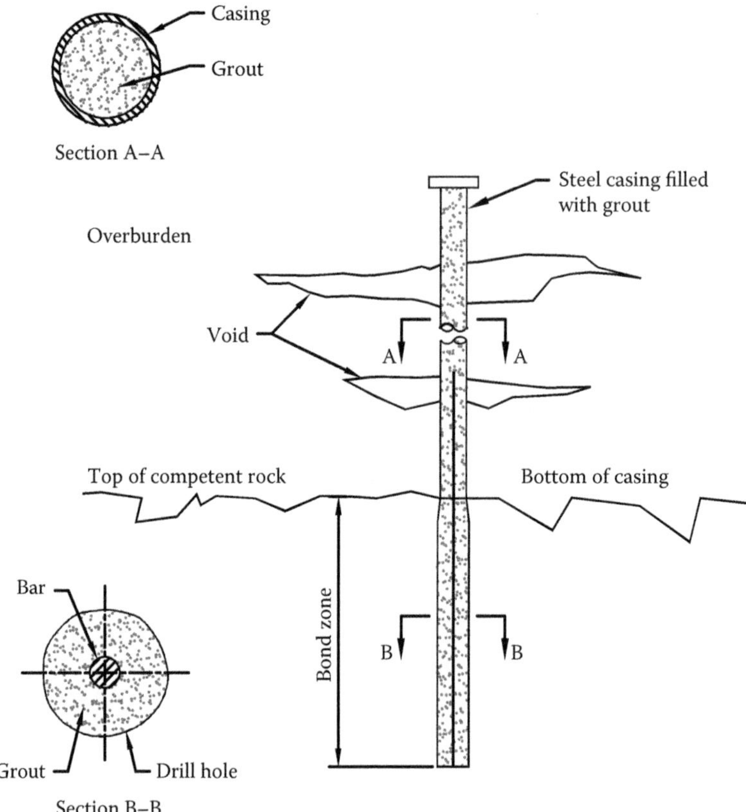


*Figure 3.9* Typical cross section of micropile in difficult ground. (Adapted from Dotson, D. and Tarquino, F. 2003. A creative solution to problems with foundation construction in karst. *Proceedings of the 9th Multidisciplinary Conference on Sinkholes and the Engineering and Environment Impacts of Karst*, Huntsville, AL, September. Courtesy of Nicholson Construction Company.)

the lateral capacity and stiffness of individual micropiles may be relatively small. The micropiles discussed by Dotson and Tarquino (2003) typically range between 50 and 300 m in diameter, but in principle, larger diameters could be used.

### 3.9.3 Controlled stiffness inserts

In all but small symmetrical pile group configurations, the axial load distribution will be non-uniform, with a tendency for larger loads to be carried by piles near the perimeter of the pile group, while the inner piles may carry a significantly smaller load. Consequently, the load that can be applied to a pile group may be limited by the most-heavily loaded pile in that group. This may lead to an over-conservative design in many cases. Clearly, it would be useful to have a means of reducing or eliminating the non-uniformity of load distribution within a group while also controlling the settlement.

Poulos (2006c) has described the use of stiffness inserts at the head of the more heavily loaded piles in a group to achieve the above objectives. Figure 3.10 shows a schematic diagram of a controlled stiffness insert (CSI) together with a photograph of one such insert applied to the head of an H-pile in a building project in Hong Kong.

(a)

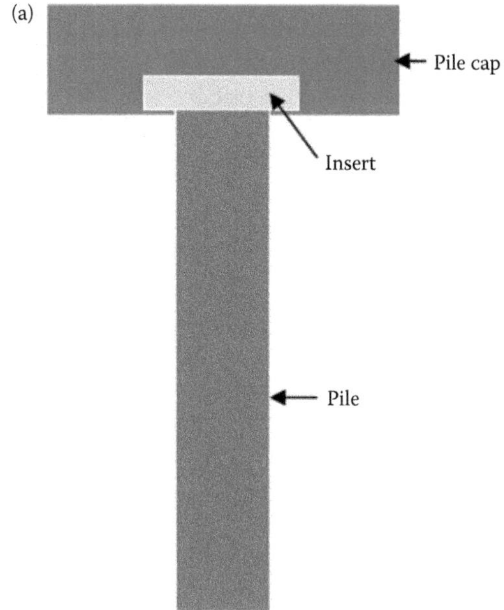


(b)

*Figure 3.10* Controlled stiffness inserts: (a) schematic diagram and (b) example of use of CSI on precast piles. (Adapted from Poulos, H.G. 2006c. Use of stiffness inserts in pile groups and piled rafts. *Geotechnical Engineering, ICE*, 159(GE3): 153–160. Courtesy of ICE Publishing.)

The concept of the CSI is to reduce the stiffness of the pile on which it is attached and so to increase the settlement and/or reduce the axial load in that pile, in a controlled fashion. The stiffness KI of the CSI is given simply as

$$KI = EA/t \tag{3.1}$$

where E is the Young's modulus of insert, A the cross-sectional area of insert and T is the thickness of insert.

By selecting appropriate stiffnesses for these inserts, it is possible to obtain an almost uniform distribution of axial load within the group, together with a specified settlement.

A relatively simple method for assessing the required stiffness of the inserts is provided by Poulos (2006c), who gives examples of the difference in the load–settlement characteristics of groups with and without stiffness inserts for two different cases in which non-uniform load and/or settlement behaviour would occur without the inserts.

### 3.9.4 Energy piles

An energy pile is a pile that uses the thermal capacity of the ground around the pile to store heat. Heat is pumped from the ground during winter, and is replenished during the summer months. Pipes within the piles circulate a heat transfer fluid which transports the ground temperature to the building's central control system. There, a heat pump is used to increase the temperature if heating is required, or to decrease the temperature if cooling is required, similar to a refrigerator or a standard central heating system.

Figure 3.11 shows the principles of an energy pile. Considerable research has been undertaken to develop both the concept and suitable design approaches to the design of energy piles, for example, Brandl (2006), Bourne-Webb et al. (2009).

Some important factors that influence the efficiency of energy piles include

- Unit weight of soil
- Water content
- Void ratio
- Mineralogy
- Thermal conductivity
- Specific heat

It can be noted that thermal properties of soils are less variable than many other engineering parameters.

The most favourable ground conditions for energy piles are saturated sands and clays, especially if there is flow of groundwater. Under favourable ground conditions, a typical

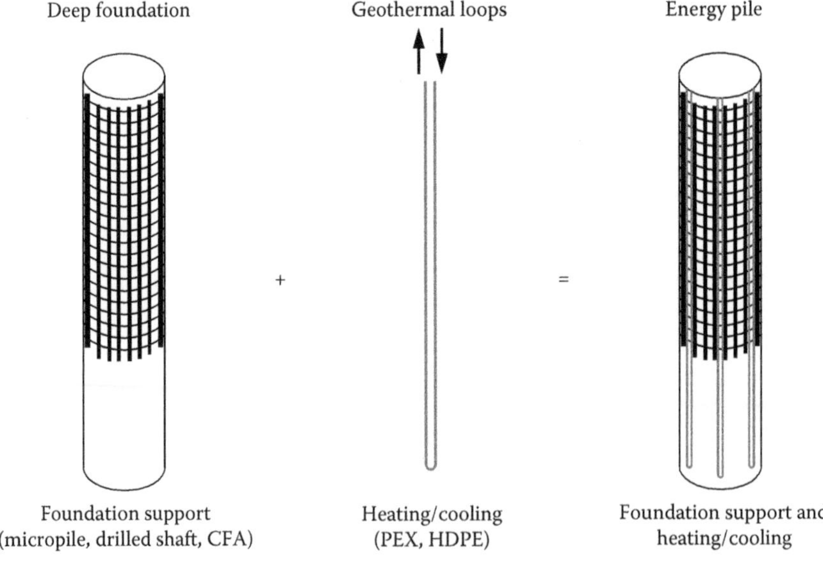


*Figure 3.11* Energy piles.

energy pile can provide about 80 W/m run of pile. Olgun (2013) quotes the case of the Main Tower in Frankfurt, where a foundation with 223 energy piles provided a power output of about 500 kW.

An aspect that must be considered in design is that the thermal effects associated with the energy piles cause internal movements within the pile itself, and these movements give rise to induced internal stresses. When a pile is heating, the expansion of the concrete gives rise to additional compressive stresses, while conversely, when the pile is cooling, the concrete contracts and is subjected to tensile stresses.

Energy piles have a number of positive aspects, including the fact that they use the building substructure, they lead to significantly decreased costs and they do not require any extra land. On the other hand, there are also some drawbacks, namely, the ground may not offer optimum conduction, their use may affect the pile cap design and installation methods, additional heat exchanger bores may be required and last but not least, there is limited (if any) experience of their use in many countries. Nevertheless, with further research and demonstrations of their effectiveness, energy piles are likely to become a very viable option for future tall building projects.

Chapter 4

# The foundation design process

## 4.1 INTRODUCTION

The design of tall building foundations involves the consideration of several aspects which require input from both geotechnical and structural experts. The structural experts are usually responsible for the assessment of the loads applied to the foundation, while the geotechnical expert focusses on the foundation resistance and the movements arising from the applied loads. The foundation process is generally carried out in a number of stages, and this chapter will summarise the key design issues that must be addressed, and then the stages of design which culminate in the final design. The design criteria associated with the key design issues are dealt with in Chapters 7 through 11. These chapters also discuss in detail the analyses that are involved in each of the design stages.

## 4.2 GENERAL DESIGN REQUIREMENTS

The foundation for any structure, but particularly a high-rise structure, must be designed to satisfy the following broad criteria:

- Design so that the structure–foundation system is stable, and safety is secured under all forms of loading.
- Design for serviceability, so that settlements, differential settlements and lateral movements and strains do not impair the function of the structure.
- Design for human comfort, so that the vibrations of the building are sufficiently small that the building occupants are not inhibited from carrying out their intended activities.
- Design for durability, so that the foundations remain durable and functional throughout the design life of the building.
- Design for sustainability.

Design for safety and durability is covered in detail in Chapter 7, and also in Chapter 11 for seismic conditions. Design for serviceability is addressed in Chapters 8 and 9, while design for human comfort is dealt with in Chapters 10 and 11. Design for sustainability is addressed at the end of this chapter.

## 4.3 KEY DESIGN ISSUES

The following issues will generally need to be addressed in the design of foundations for high-rise buildings:

1. Ultimate capacity of the foundation under vertical, lateral and moment loading combinations.
2. The influence of the cyclic nature of wind, earthquakes and wave loadings (if present) on foundation capacity and movements.
3. Overall settlements.
4. Differential settlements, both within the high-rise footprint, and between high-rise and low-rise areas.
5. Possible effects of externally imposed ground movements on the foundation system, for example, movements arising from construction activities such as excavations for pile caps or adjacent facilities.
6. Earthquake effects, including the response of the structure–foundation system to earthquake excitation, and the possibility of liquefaction in the soil surrounding and/or supporting the foundation.
7. Dynamic response of the structure–foundation system to wind-induced (and, if appropriate, wave-induced) forces.
8. Structural design of the foundation system, including the load sharing among the various components of the system, for example, the piles and the supporting raft, and the distribution of loads within the piles.
9. Durability of the foundation system over the design life of the structure.

## 4.4 CATEGORIES OF DESIGN/ANALYSIS

Poulos (1989) has suggested that methods of analysis and design can be classified into three broad categories:

- *Category 1:* Empirical or semi-empirical methods.
- *Category 2:* Soundly based methods, employing simplified theory and/or charts.
- *Category 3:* Relatively advanced methods incorporating more realistic site-specific soil profiles and more realistic soil models.

Category 1 and Category 2 methods are useful for the earlier stages of design, and for checking more complex analysis methods for the later design stages. Category 3 methods are increasingly being used for the final detailed design stage, and it is now not uncommon for three-dimensional (3D) finite element (or finite difference) methods to be employed with relatively advanced soil constitutive models that can reflect such characteristics as non-linearity, dilatancy, changing shear strength and stiffness as a function of strain levels and/or stress path, and strength and stiffness dependency on cyclic loading.

## 4.5 THE OVERALL FOUNDATION DESIGN PROCESS

The process of foundation design is well established, and generally involves the following aspects:

1. A desk study and a study of the geology and hydrogeology of the area in which the site is located.

2. Planning and execution of the site investigation to assess site stratigraphy and variability.
3. In situ testing to assess appropriate engineering properties of the key strata.
4. Laboratory testing to supplement the in situ testing and to obtain more detailed information on the behaviour of the key strata than may be possible with in situ testing.
5. The formulation of a geotechnical model for the site, incorporating the key strata and their engineering properties. In some cases where ground conditions are variable, a number of models for different parts of the site may be necessary to allow proper consideration of the variability over the site.
6. Preliminary assessment of foundation requirements, based upon a combination of experience and relatively simple methods of analysis and design. In this assessment, considerable simplification of both the geotechnical profile(s) and the structural loadings is necessary.
7. Refinement of the design, based on more accurate representations of the structural layout, the applied loadings, and the ground conditions. From this stage and beyond, close interaction with the structural designer is an important component of successful foundation design.
8. Detailed design, in conjunction with the structural designer. As the foundation system is modified, so too are the loads that are computed by the structural designer, and it is generally necessary to iterate toward a compatible set of loads and foundation deformations.
9. In situ foundation testing at or before this stage is highly desirable, if not essential, in order to demonstrate that the actual foundation behaviour is consistent with the design assumptions. This usually takes the form of testing of prototype or near-prototype piles. If the behaviour deviates from that expected, then the foundation design may need to be revised. Such a revision may be either positive (a reduction in foundation requirements) or negative (an increase in foundation requirements). In making this decision, the foundation engineer must be aware that foundation testing involves only individual elements of the foundation system, and that the piles and the raft within the system will interact. The overall foundation behaviour may thus not be able to be assessed directly from the foundation test results without consideration of the foundation–soil interaction effects. More details are provided in Chapter 13.
10. Monitoring of the performance of the building during and after construction. At the very least, settlements at a number of locations around the foundation should be monitored, and ideally, some of the piles and sections of the raft should also be monitored to measure the sharing of load among the foundation elements. Such monitoring is becoming more accepted as standard practice for high-rise buildings, but not always for more conventional structures. As with any application of the observational method, if the measured behaviour departs significantly from the design expectations, then a contingency plan should be implemented to address such departures. It should be pointed out that departures may involve not only settlements and differential settlements that are greater than expected, but also those that are smaller than expected. Chapter 14 provides more details on the monitoring process.

The components of the overall foundation design process are illustrated in Figure 4.1. It can be noted that the assessment of sustainability is included in modern design processes, both for environmental and economic reasons.

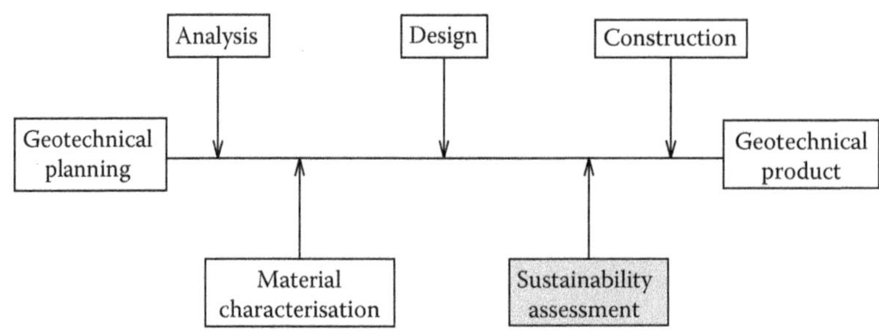


*Figure 4.1* The overall foundation design process.

## 4.6 STAGES IN FOUNDATION DESIGN AND DESIGN

The foundation design process set out above is usually undertaken in three broad stages:

1. A preliminary design, which provides an initial basis for the development of foundation concepts and costing. This may involve some or all of the first six aspects in the overall process set out above.
2. A detailed design stage, in which the selected foundation concept is analysed and progressive refinements are made to the layout and details of the foundation system. This stage is desirably undertaken collaboratively with the structural designer, given that the structure and the foundation act as an interactive system. This stage will generally focus on the detailed design and design refinement processes, together with additional field investigations, in situ and laboratory testing data, and perhaps some preliminary load testing.
3. A final design phase, in which both the analyses and the parameters employed in them are finalised. This stage focusses on items 8–10 of the list above.

It should be noted that the geotechnical parameters used for each stage may change as more knowledge of the ground conditions, and the results of in situ and laboratory testing, become available. The parameters for the final design stage should also incorporate the results of foundation load tests.

It is recommended that each stage is reviewed by an experienced peer reviewer in the foundation engineering field, working in conjunction and cooperation with the foundation designer.

### 4.6.1 Preliminary design

The aim of the preliminary concept/schematic design stage is to assess the feasibility of alternative foundation options and develop an initial design. This involves the evaluation of the approximate behaviour of various foundation options, based on simplified ground models developed from the available geotechnical data. The methods of analysis should also be relatively simplified and consistent with the data that is available at this stage. Such methods would most likely be categorised as Category 1 or 2 methods as set out in Section 4.4. Clearly, it would be inappropriate to launch into detailed Category 3 finite element analyses if only preliminary SPT or CPT data were available.

From this design step, the following foundation design details could be provided to the structural designers with a recommended foundation option for preliminary design purposes:

### Option 1: Raft option

- Bearing capacity
- Vertical ground stiffness values

### Option 2: Deep foundation option

- Pile capacities (geotechnical and structural) for a range of pile diameters
- Geotechnical capacity: compression/tension/lateral pile resistance
- Structural capacity: compression/tension/bending moment/shear
- Horizontal and vertical pile stiffness values (single pile and group) for a range of pile diameters

### Option 3: Piled raft foundation

- Information as for the raft and pile options

Using this information the structural designer can commence the preliminary structural design of the superstructure, considering the feasible foundation options, and including the effects of soil–structure interaction. The foundation system development includes the following:

- Preliminary selection of raft size/thickness
- Development of pile layout options for various pile diameters
- Preliminary evaluation of building performance, under gravity and lateral load effects
- Assessment of the pile group efficiency for vertical and lateral loads
- Assessment of the foundation stiffness and its impact on the behaviour of the superstructure
- Assessment of the stiffening effects of the superstructure, such as core walls, on the load distribution at the top of the foundation

Table 4.1 summarises the process and outputs of this phase of the foundation design and provides a checklist of issues to be considered.

*Table 4.1* Checklist for preliminary foundation design

| *Activity* | *Requirements* |
|---|---|
| Develop initial geotechnical model | Use available ground investigation data to obtain a stratigraphic model and assign preliminary design parameters |
| Consider alternative foundation options | Raft, piles, piled raft, with or without compensation, depending on basement requirements |
| Compute raft bearing capacity | If raft or piled raft option is to be considered. A hand calculation is generally adequate |
| Compute pile capacity versus pile length | Consider practical range of shaft (and base) diameters. Both compression and tension capacities to be computed. Hand or spreadsheet calculations are generally adequate |
| Estimate foundation requirements for ultimate vertical loadings | Compute required raft area (if relevant) or number of piles for various pile diameter and length combinations for the vertical loads supplied by the structural engineer |
| Estimate final settlement for serviceability vertical loadings | Use a simplified approach, depending on the foundation type and soil type. Category 1 or 2 hand or spreadsheet calculations are generally adequate |

### 4.6.2 Detailed design

Once the three components of loading, foundation type and ground conditions are reasonably well defined, the detailed design of the foundation can be undertaken. The key issues in Section 4.3 then need to be addressed, using methods of analysis which are consistent with the nature of the available geotechnical and load data. Such methods would almost certainly fall into Category 2 and/or Category 3 methods set out in Section 4.4.

The main objective of this phase is to develop and optimise the foundation scheme, and provide individual pile stiffness values to the structural engineer. The following information should be developed during this phase of design:

1. The number, size and length of the piles. It may be appropriate to have piles of different length within the system to improve the settlement and differential settlement performance.
2. The location of the piles.
3. The thickness of the raft. This may also vary, depending on the loading and differential settlement criteria and the moments generated within the raft.
4. The axial and lateral loads, and the maximum moment, in each of the piles.
5. The bending moments and shears within the raft.
6. The distribution of settlement and differential settlement across the foundation system.
7. Values of vertical and lateral stiffness for each pile (or barrette) within the system. These values must include the effects of pile–soil–pile interaction (see Chapter 8), and can be incorporated by the structural designer into the overall structural model, thus enabling a consistent set of loads and foundation responses to be developed.

Table 4.2 provides a checklist for the detailed phase of the design process.

### 4.6.3 Final design and post design study

The main objective of this phase is to finalise the foundation scheme and check for the effects of refinements such as basement wall resistance which may provide some additional scope for foundation design optimisation.

During both the detailed and final design stages, the foundation design is generally based on numerical analyses and previous experience in similar conditions. Results from large-scale testing, for example, pile load testing, are invaluable in confirming design assumptions. Load testing is discussed in detail in Chapter 13. It is recommended that large-scale testing be carried out during the detailed design stage as the results may allow confirmation or modification of the design, which may in turn lead to a more cost-effective design. Site presence by the geotechnical designer during construction of both prototype and production piles, to observe the construction methodology and as-revealed ground conditions, is an important element in the confirmation/adjustment of the original design assumptions.

Where rock is exposed in basement excavations, face mapping should be carried out as this will provide valuable information on the rock mass characteristics, and will allow for checking of the influence of method of excavation and groundwater conditions. These aspects are often difficult to assess from discrete boreholes. Based on the mapping results, the geotechnical model(s) adopted for the final foundation design should be reviewed and updated and, if necessary, the design modified. Ideally this should be carried out before the start of foundation construction.

The output from the final design stage will be appropriate drawings and specifications which will allow the foundation system to be tendered and then constructed. In addition,

*Table 4.2* Checklist for detailed design phase

| *Activity* | *Requirements* |
|---|---|
| Refine geotechnical model | Obtain additional investigation data to refine stratigraphic model and assign final design parameters. |
| Develop initial foundation layout | Adopt a factor of safety for the piles (e.g. 2 if piles only, 1.5 if piled raft) and calculate the axial pile capacity and thus how many piles are required for vertical loading at and around each column location. This may require considering more than one pile diameter. See Chapter 7. |
| Obtain individual pile stiffness values | For the chosen pile layout, obtain the axial and lateral stiffness for each pile within the group. This is to be supplied to the structural designer, and is best done by applying the nominal working load to each pile within the group. Details are given in Chapter 8. |
| Check ultimate geotechnical capacity | Factor down geotechnical vertical and lateral resistances; use ultimate limit state load combinations. Apply all required load combinations (vertical, lateral, moment, torsional loads) to check if foundation system is stable. If not, increase number or size of piles, and re-compute. Details are given in Chapter 7. |
| Check ultimate limit state pile loads (and raft pressures, if appropriate) | Do not factor down geotechnical vertical and lateral resistances, and use ultimate limit state load combinations. Apply all required load combinations to check individual pile loads and moments and raft pressures and moments. For large pile groups, the assumption of a rigid raft may give very large loads on outer piles. If this occurs, re-analyse for the vertical and moment loads only, using a program that can incorporate raft flexibility. |
| Check cyclic axial load in each pile | Use analysis for ultimate limit state pile loads to obtain the cyclic component of axial load in each pile. To avoid unduly large vertical loads in outer piles, it is preferable to use a program that can account for raft flexibility. If cyclic load is less than 50% of axial shaft capacity, then cyclic degradation should not occur (see Chapter 7). If this condition is not met, then it may be necessary to either add more piles, or to reduce the axial capacity of the affected pile(s) in the analysis and then re-analyse. |
| Estimate settlement and lateral movements for serviceability loadings | Use unfactored geotechnical vertical and lateral resistances, and serviceability load combinations. Long-term analyses for vertical loadings should be carried out using a program that can incorporate raft flexibility and employing long-term soil modulus values. Check that maximum settlement and angular rotation (or differential settlement) are within acceptable limits. If not, then either increase the number or size of piles, re-arrange the piles, or increase the raft thickness. Short-term analyses, e.g. for wind or seismic loads, should use short-term soil modulus values. All components of loading should be applied in this case. See Chapter 8 for a detailed treatment. |
| Seismic design issues | Carry out a seismic site assessment to estimate<br>• The peak bedrock acceleration for the required earthquake return period. This may require input from an expert in seismology.<br>• Site amplification effects—this may involve running a site response program such as SHAKE.<br>• Carry out a liquefaction potential assessment. SPT, CPT or shear wave velocity may be used as a basis.<br>• Estimate the shear force and bending moments in the pile due to inertial and kinematic effects. Simplified methods are usually adequate.<br>• Details are given in Chapter 11. |
| Estimate dynamic response characteristics of the foundation | This requires assessment of the dynamic stiffness and damping of the foundation system. These values are used by the structural engineer to incorporate into the dynamic structural analysis to estimate amplitudes of dynamic motion.<br>In the absence of other information, the foundation stiffness can often be taken as the static stiffness of the foundation system, using short-term deformation parameters.<br>For lateral motions, the radiation damping is generally small for large pile groups, and the damping ratio can be estimated conservatively as the internal damping ratio of the soil or rock. Typically, this may range between about 0.01 and 0.05.<br>As a check, it is often adequate to idealise the foundation as an equivalent pier and use the solutions for stiffness and damping of an embedded shallow foundation. See Chapter 10. |

*(Continued)*

*Table 4.2 (Continued)* Checklist for detailed design phase

| *Activity* | *Requirements* |
|---|---|
| Check for possible effects of external ground movements | Such ground movements may arise from construction operations, dewatering, excavation and nearby tunnel construction. Approximate methods can be employed to assess<br>• Additional axial forces and additional settlements arising from ground settlements. Conventional calculation methods for negative skin friction can usually be employed.<br>• Additional bending moments and shears in the piles due to the lateral ground movements. Design charts are available for excavation and tunnel-induced movements, as well as for simplified distributions of lateral ground movement.<br>Group effects can be ignored for such calculations as the single pile case is generally the critical case. See Chapter 9. |

the anticipated foundation settlements and movements, during and after construction, should be documented.

Evaluation of the measured foundation performance by the monitoring of the piles and/ or raft during construction of the superstructure is strongly recommended. While pile load tests indicate the individual pile performance, only monitoring of the piles and raft will provide the behaviour of the foundation as a whole. Monitoring will allow assessment of the overall behaviour of the foundation and comparison with predicted performance, as well as providing valuable information for the structural designer regarding the behaviour of the superstructure itself. Chapter 14 provides further details on foundation monitoring.

It is also important to precisely monitor and record the construction activity, in particular increments in structural loads during construction and post-construction phases. These data can be used to further understand and interpret the instrumentation records of the foundation.

A checklist for this third and final phase of the design process is given in Table 4.3. The aspects considered are similar to those in Table 4.2, but will generally involve the most refined level of analysis, usually a Category 3 approach.

### 4.6.4 Some practical design issues

#### *4.6.4.1 Pile spacing*

Fellenius (2016) has pointed out that the size of the pile cap or raft is a part of the design, and that this size is governed by the pile diameter and the spacing of the piles. In general, the greater the spacing, the greater the required thickness of the pile cap or raft, but the use of close spacings can give rise to the risk of interference between the piles during installation, whether the piles are driven or bored. Accordingly, for a building foundation, the pile spacing criterion should involve both the pile diameter (d) and length (L), and the suggested minimum spacing, $s_{min}$, is as follows:

$$s_{min} = 2.5\,d + 0.02\,L \qquad (4.1)$$

Fellenius does however point out that spacing is not an absolute and that there may be occasions in which the piles can be touching to form components of a larger foundation element or of a retaining wall.

#### *4.6.4.2 Pile arrangement*

In general, piles should be placed where they will be of most benefit, usually under heavily loaded areas (e.g. lift core walls) and columns. Having a regular pattern of piles may not be

*Table 4.3* Checklist for final phase of foundation design

| *Activity* | *Requirements* |
|---|---|
| Refine geotechnical model | Check on stratigraphic model and final design parameters, using pile load test data, if available. |
| Set up computational model | This is generally best done using a 3D finite element analysis such as PLAXIS 3D or FLAC3D. |
| Check geotechnical ultimate limit state | Apply the most critical load combination(s), as assessed from the detailed design stage, and apply reduction factors to the pile vertical and lateral resistances. Check that the foundation system does not collapse. |
| Check structural limit state | Apply the most critical load combination(s), as assessed from the detailed design stage, but do not apply reduction factors to the pile vertical and lateral resistances. Check that the pile loads and moments are within acceptable limits for structural design. |
| Check cyclic axial load in each pile | Use analysis for ultimate limit state pile loads to obtain the cyclic component of axial load in each pile. Check that the cyclic load is less than 50% of axial shaft capacity, so that cyclic degradation is avoided. If this condition is not met, then it may be necessary to either add more piles, or to reduce the axial capacity of the affected pile(s) and re-analyse. |
| Estimate settlement and lateral movements for serviceability loadings | Use unfactored geotechnical vertical and lateral resistances, and serviceability load combinations. Long-term analyses for vertical loadings should be carried out using long-term soil modulus values. Check that maximum settlement and angular rotation (or differential settlement) are within acceptable limits. If not, then either increase the number or size of piles, rearrange the piles, or increase the raft thickness. Short-term analyses should be carried out using short-term soil modulus values. |

effective or economical, as columns and piles that are offset can result in very high moments in the basement slab or raft.

As a simple procedure for preliminary design, it can be useful to consider only the vertical loads, and assign a factor of safety of between 1.5 and 2 to the piles being considered. The number of piles required to sustain the applied load within that area can then be obtained by dividing the load by the reduced pile capacity.

#### 4.6.4.3 *Pile size*

Modern equipment has enabled piles of very large diameter, or barrettes of significant size, to be constructed. Careful consideration needs to be given to the chosen pile size to try and optimise the foundation costs, taking into account the cost of installation of a smaller number of very large foundation units versus a larger number of smaller units. As pointed out by Fellenius (2016), smaller diameter piles are usually cheaper and require thinner pile caps or rafts, while the construction is faster, and verifying the pile capacity and integrity is less costly. In addition, distributing the loads on a larger number of piles increases the system redundancy and reduces the risks of problems with foundation imperfections.

#### 4.6.4.4 *Pile verticality*

Pile construction specifications generally include a tolerance for pile verticality. 1 in 75 is often used as a criterion for bored piles, although 1 in 100 may be specified for more stringent control. For barrettes, more control of verticality is usually possible, and criteria as stringent as 1 in 400 have been specified (Thasnanipan et al., 2000). However, such criteria may not be achievable for very long barrettes, for example, in excess of 50 m long.

In addition, an allowable eccentricity from the plan position of the pile head of about 75 mm is often quoted. Allowance must then be made in the structural design of the piles for the bending moments arising from such an eccentricity.

## 4.7 DESIGN FOR SUSTAINABILITY

There are many definitions of the term 'sustainability', but in relation to foundation design, sustainability can be considered to be the integrated process that balances the environmental and financial aspects of planning, design and construction of foundations, while reducing risk and also improving safety, quality and durability without limiting opportunities for future generations (after DFI, 2015).

Sustainable design involves the balancing of environmental, economic and social factors, so that the foundation system developed satisfies the safety and code requirements and can be constructed within the allotted time frame for a competitive price. Such a design involves efficient scheduling, the use of available local resources, and the avoidance of excess material and labour costs that may be inherent in over-conservative design. Thus, a key aspect of sustainability with respect to foundations for tall buildings is minimisation of resource use, costs and environmental impacts.

Sustainability is incorporated as an additional aspect to consider in the foundation design process, as illustrated in Figure 4.1.

Saravanan (2011) has given an example of typical carbon dioxide ($CO_2$) emissions in the construction of a foundation with 24 bored piles, 13 m long and 0.3 m in diameter, as follows (in tonnes):

- Concrete and steel reinforcement: 14.7
- Transportation: 2.5
- Installation: 5.0

This then gives a total of 21.2 tonnes of $CO_2$ for this modest foundation system.

For a 10 m deep basement wall, Table 4.4 reproduces the $CO_2$ emissions for three wall thicknesses.

The potential savings in $CO_2$ emissions are as follows:

- For each cubic metre of concrete: 0.46t
- For each tonne of reinforcing steel: 1.77t

Sustainability can be enhanced by the following procedures:

- Minimising the use of concrete
- Minimising construction time
- Exploring the potential for energy generation
- Re-using existing foundations (where appropriate)

*Table 4.4* $CO_2$ emissions for a 10 m deep wall

| *Wall thickness (m)* | *$CO_2$ emissions (tonnes/m run)* |
|---|---|
| 0.8 | 13 |
| 1.0 | 16 |
| 1.5 | 25 |

Source: Saravanan, V.K. 2011. Cost effective and sustainable practices for piling construction in the UAE. MSc thesis, Heriot Watt University.

Saravanan (2011) suggests the following broad strategies to assist in sustainable practices in deep foundation design and construction:

- Adequate geotechnical investigation
- Efficient pile design for materials construction
- Selection of the appropriate pile type
- Improved energy and material management
- Use of sustainable materials
- Pile testing
- The use of combined piled-raft foundations
- Employing multi-use piles (for both load bearing and energy generation)

## 4.8 AN EXAMPLE OF PRELIMINARY FOUNDATION DESIGN

The design of the foundation system will be examined in detail in Chapters 6 through 11. However, as an early example of the process and outcomes of a preliminary foundation design, a hypothetical case of a site at which the subsoil conditions consist of the strata outlined in Table 4.5 will be considered. The key pile design parameters, namely the ultimate shaft friction and the ultimate end bearing capacity, are shown in this table, while the ultimate raft bearing capacity is assumed to be 300 kPa. Chapters 6 and 7 will deal in more detail with the estimation of these parameters. A limit state design approach will be adopted in this example.

A preliminary design will be undertaken for a tall building which has a 50 m × 50 m footprint, and which will be subjected to an average design pressure (dead plus live load) of 800 kPa. Adopting an average load factor of 1.4, the design loading is then 1120 kPa. The total load over the entire building footprint is then 2500 × 1120 = 2,800,000 kN = 2800 MN.

For the piles and the raft, the geotechnical reduction factor (see Section 7.3) will be taken as 0.6, and the design concrete strength will be taken as 20 MPa. A preliminary estimate is required of options for the number of piles, and the necessary length and diameter of the piles. In this preliminary phase of the design process, consideration would be focussed on the requirement to resist axial compression loads. Using a limit state design approach, the design criteria would be as set out in Chapter 7, Equations 7.2 and 7.3.

### 4.8.1 Estimation of required size and number of piles

Figure 4.2 plots the computed design axial capacity versus length of piles with diameters of 1.0, 1.5 and 2 m. A 2.8 m × 1.0 m rectangular barrette is also considered, using the same design parameters as for the bored piles. The following points can be noted:

- As would be expected, the design capacity increases with increasing pile diameter
- The design capacity increases with increasing length, until the design geotechnical capacity reaches the design structural strength; thereafter, it becomes constant,

*Table 4.5* Summary of geotechnical profile and model, for example, case

| *Depth range (below basement) (m)* | *Stratum type* | *Ultimate shaft friction, fs (MPa)* | *Ultimate end bearing capacity, fb (MPa)* |
|---|---|---|---|
| 0–30 | Stiff clay | 0.10 | 1.2 |
| 30–50 | Dense sand | 0.15 | 5.0 |
| 50+ | Weak rock | 0.50 | 10.0 |

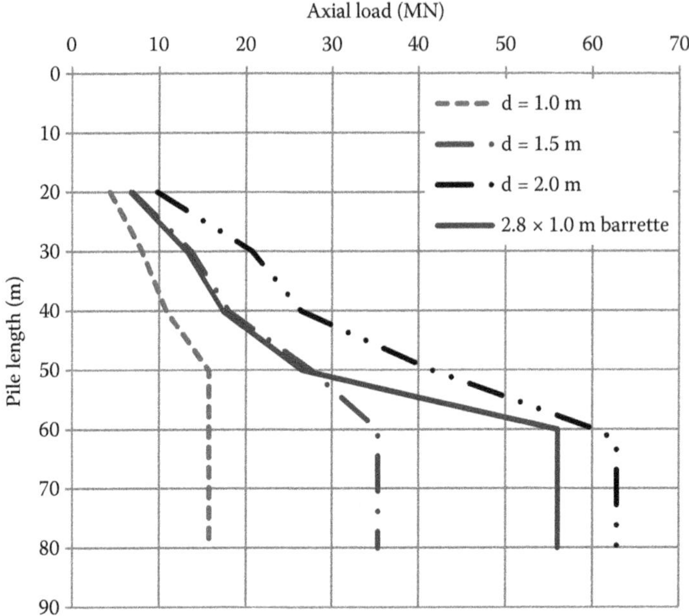


*Figure 4.2* Axial capacity versus pile length for example case.

indicating that the pile design is governed by the structural, rather than the geotechnical, strength

- The design axial capacity of the barrette is similar to that of a 1.5 m diameter pile until the length exceeds 50 m, after which the barrette bears on the lowermost and stronger stratum and picks up additional end bearing

It would be desirable for the deep foundation to be founded into the lowermost stratum, so as to reduce the settlement, and so a reasonable pile length appears to be 55 m. At this length, the calculated design axial capacity of the piles is shown in Table 4.6.

Considering now the effect of having a piled raft, rather than a fully piled foundation, the design bearing capacity of the raft is 0.6 × 300 = 180 kPa, and adopting 80% of the gross raft area to allow (approximately) for the area occupied by the piles, the design axial capacity of the raft is then 0.8 × 2500 × 180 = 360,000 kN = 360 MN. In this case, the piles will then be required to contribute a design axial capacity of 2800 – 360 = 2440 MN.

*Table 4.6* Summary of design capacity of piles and required number of piles

| *Pile type* | *Design capacity per pile MN (55 m long)* | *Design capacity per unit volume (MN/m³)* | *Required number of piles—fully piled option* | *Required number of piles—piled raft option* |
|---|---|---|---|---|
| 1.0 m diameter | 15.7 | 0.36 | 178 | 155 |
| 1.5 m diameter | 31.5 | 0.32 | 89 | 77 |
| 2.0 m diameter | 50.9 | 0.29 | 55 | 48 |
| 2.8 × 1.0 m barrette | 41.2 | 0.27 | 68 | 59 |

Table 4.6 shows the resulting number of piles required for both the fully piled option and the piled raft option. Also shown is the design capacity of each foundation option per unit volume of concrete. It can be seen that

1. As would be expected, the larger the diameter or size of pile, the less piles are required
2. Taking account of the raft in the piled raft option decreases the required number of piles. In this initial assessment, the saving is in the order of 16%
3. The 1.0 m diameter piles provide the best solution from the point of view of sustainability, giving an axial load capacity of 0.36 $MN/m^3$
4. The barrette in this case gives the least sustainability performance, at 0.27 $MN/m^3$

From an overall design perspective, 1.0 m diameter piles may not be the best solution, as their structural capacity is reached at a length of 50 m, and if subjected to additional loading due to wind or seismic loadings, they would tend to be loaded beyond their design load. Thus, the preferable choice would be among the remaining three options, with the decision depending on the construction costs and the location and magnitude of the column and core loadings. From the point of view of sustainability, 1.5 m diameter piles would have an advantage.

The above approach provides a preliminary estimate of the pile foundation requirements from the very limited point of view of the static compressive axial capacity of a single isolated pile. However, once these requirements are established, more detailed assessments must be made in which the following issues are examined:

1. The effects of having a group of piles need to be considered. If the piles are closely spaced, the capacity of the group may be less than the sum of the individual capacities of the piles within the group.
2. A check on axial capacity in tension may also be necessary in some cases, although with many tall buildings, the piles below the tower footprint would be unlikely to be subjected to a net tension loading. However, such loadings could well occur in low-rise podium areas adjacent to the tower.
3. Possible effects of cyclic loading on the axial capacity will need to be assessed.
4. The effects of lateral loading, and the consequent induced bending moments and shears in the piles.
5. The settlement performance of the pile group or piled raft. Settlement and differential settlement, rather than axial capacity, is frequently the most critical aspect of the foundation design for tall buildings.
6. The effects of dynamic loading.
7. The effects of seismic events on the foundation system and also on the site in general, particularly the potential for liquefaction.

Chapter 5

# Building loads

## 5.1 SOURCES OF LOADING

Building loads are normally provided to the foundation designer by the structural designer and include the major load combinations relevant to foundation design and building performance. These can be classified according to their source or loading characteristics with direction, and the usual sources of loading that may need to be considered for foundation design include the following:

- Dead loads
- Live loads
- Wind loads
- Earthquake loads
- Loads arising from earth pressures
- Loads arising from ground movements
- Loads from other sources, such as snow and ice

Dead and live loads are always significant, but also of particular importance for high-rise buildings is wind loading, together with seismic loading in active earthquake zones. These latter actions generate a large eccentric loading on the foundation plan and are very important two load cases for both the building foundation and the superstructure.

Figure 5.1 (Boggs and Dragovich, 2006) provides an indication of the frequency of the dynamic loadings that arise from wind and earthquake loadings, together with typical ranges of the natural period of low-rise and high-rise buildings. It can be seen that the typical frequency of wind loadings is significantly less than that of earthquake loadings, and that, as a consequence, wind loading is likely to be critical for high-rise buildings, whereas earthquake loading may be more critical for lower-rise buildings.

A range of loading conditions, including static, transient and cyclic loading, should be considered in the geotechnical design of the foundation system.

## 5.2 IMPORTANCE LEVELS AND ANNUAL PROBABILITY OF EXCEEDANCE OF DESIGN EVENTS

In many recent standards, loadings on structures have been related to the consequences of failure for the structure. These consequences have in turn been expressed in terms of 'importance levels', which range from 1 when the consequences of failure are low, to 5 for very high consequences of failure of exceptional structures. For tall towers, where there would be a high consequence for loss of life or very great economic, social or environmental

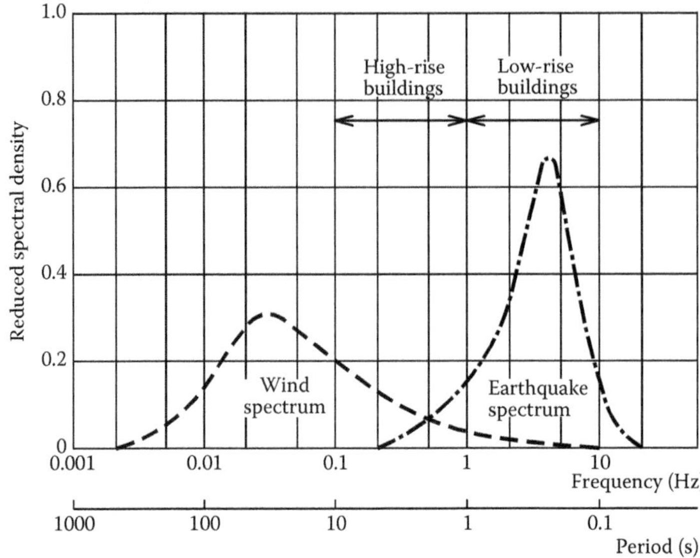


*Figure 5.1* Frequency ranges of wind and earthquake loadings. (Boggs, D. and Dragovich, J. 2006. *Performance-Based Design of Concrete Buildings for Wind Loads*. American Concrete Institute, Montreal, Quebec, pp. 15–44; courtesy of American Concrete Institute.)

consequences, the relevant importance level would be 4 or perhaps 5 if the structure is deemed to be 'exceptional'. In the latter case, the reliability of the structure, and hence the return period for various loadings, would need to be set on a case by case basis.

For structures assessed to have an importance level of 4, the annual probability of exceedance of the design events for the ultimate limit states would be as shown in Table 5.1.

## 5.3 LOAD FACTORS AND LOAD COMBINATIONS

Adoption and application of load factors will be dependent on the design method adopted (e.g. working stress with factors of safety, or limit state), and needs to be consistent with the basis of both the structural and geotechnical design. The current design practice in many countries employs a limit state approach.

*Table 5.1* Annual probability of exceedance of design events for ultimate limit states for importance level = 4

| *Design working life (years)* | *Annual probability of exceedance* | | |
|---|---|---|---|
| | *Wind* | *Snow* | *Earthquake* |
| 25 | 1/1000 | 1/250 | 1/1000 |
| 50 | 1/2500 | 1/500 | 1/2500 |
| ≥100 | <1/2500[a] | <1/500[a] | <1/2500[a] |

*Source:* AS1170.0. 2002. *Structural Design Actions—Part 0: General Principles*. Standards Australia, Amendment No. 5 (2011).

[a] To be determined by a risk analysis.

*Table 5.2* Load combinations for ultimate limit state: Stability

| Load combination | Load factor for load type | | | | |
|---|---|---|---|---|---|
| | $G$ | $Q$ | $W_u$ | $E_u$ | $S_u$ |
| 1 | 1.35 | 0 | 0 | 0 | 0 |
| 2 | 1.2 | 1.5 | 0 | 0 | 0 |
| 3 | 1.2 | $\psi_c = 0.4$ | 1.0 | 0 | 0 |
| 4 | 1.0 | $\psi_E = 0.3$ | 0 | 1.0 | 0 |
| 5 | 1.2 | $\psi_s = 0.7$ | 0 | 0 | 1.0 |

Note: For cases where the dead load produces a stabilising effect, a load factor of 0.9 is applied to the dead load G.

*Table 5.3* Load combinations for ultimate limit state: Strength

| Load combination | Load factor for load type | | | | |
|---|---|---|---|---|---|
| | $G$ | $Q$ | $W_u$ | $E_u$ | $S_u$ |
| 1 | 1.35 | 0 | 0 | 0 | 0 |
| 2 | 1.2 | 1.5 | 0 | 0 | 0 |
| 3 | 1.2 | 0.6 | 0 | 0 | 0 |
| 4 | 1.2 | $\psi_E = 0.4$ | 1.0 | 1.0 | 0 |
| 5 | 0.9 | 0 | 1.0 | 0 | 1.0 |
| 6 | 1.0 | $\psi_E = 0.3$ | 0 | 1.0 | 0 |
| 7 | 1.2 | $\psi_c = 0.4$ | 0 | 0 | 1.0 |

A series of load combinations is generally considered for the ultimate limit state design, as set out in the relevant standard being employed. Serviceability limit state loads should be adopted for settlement analysis of the foundation, completed as part of the performance-based design of the structure.

Tables 5.2 and 5.3 show typical load factors and load combinations for the ultimate limit state, based on values from Australian Standard AS1170.0:2002 for residential and office buildings. In this context, 'stability' refers to a limit state of equilibrium or of gross displacements or deformations, while 'strength' refers to collapse, rupture or excessive deformations of the structure, section, member or connection within the structure. In relation to the foundation system, it is likely that the stability will be the most significant issue to consider, as this incorporates both the bearing capacity of the foundation under vertical load, and the overturning and lateral resistances of the foundation system when subjected to lateral wind or earthquake loads.

For the serviceability limit state, combinations of loads appropriate to the serviceability condition need to be considered. The usual combinations include those shown in Table 5.4. If there are other sources of loading, such as earth pressures or water pressures, then these should also be considered with a load factor of 1.0 applied.

In Tables 5.2 to 5.4, the following symbols apply:

$G$ = permanent dead load
$Q$ = imposed live load
$W_u$ = ultimate wind load
$W_s$ = serviceability wind load
$E_u$ = ultimate earthquake load
$E_s$ = serviceability earthquake load

*Table 5.4* Common load combinations for serviceability

| Load combination | Load factor for load type | | | | |
|---|---|---|---|---|---|
| | $G$ | $Q$ | $W_u$ | $E_u$ | $S_u$ |
| 1 | 1.0 | $\psi_s = 0.7$ | 0 | 0 | 0 |
| 2 | 1.0 | $\psi_l = 0.4$ | 0 | 0 | 0 |
| 3 | 1.0 | $\psi_s = 0.7$ | 1.0 | 0 | 0 |
| 4 | 1.0 | $\psi_s = 0.7$ | 0.0 | 1.0 | 0 |

$S_u$ = ultimate values of other various load types
$\psi_c$ = combination factor for live loads
$\psi_E$ = combination factor for earthquake loads
$\psi_s$ = combination factor for other short-term loads
$\psi_l$ = combination factor for other long-term (quasi-permanent) loads

## 5.4 APPROXIMATE INITIAL ASSESSMENT OF VERTICAL DEAD AND LIVE LOADINGS FOR SERVICEABILITY

It is often useful to have a preliminary estimate of building loads, prior to detailed information being provided by the structural designer. For preliminary estimates of serviceability loadings, and in the absence of other information, typical vertical average service loadings for high-rise buildings tend to be between 10 and 15 kPa per storey.

Ohsaki (1976) provides a more detailed summary of typical dead loadings due to self-weight with suggested typical values being as follows:

- Weight of superstructure: 6.5 kPa/floor
- Weight of basement: 17.5 kPa/basement level

The self-weight of the foundation will depend on the raft thickness, the number and dimensions of the piles used and the unit weight of the concrete.

Table 5.5 lists typical live floor loadings for various functional spaces within a building.

*Table 5.5* Common floor loads (based on BS 6399: Part1:1984)

| *Space* | *Unit load (kPa)* |
|---|---|
| Art gallery | 4.0 |
| Bars | 5.0 |
| Parking structures | 2.5 |
| Classrooms | 3.0 |
| Dance halls | 5.0 |
| Offices | 5.0 |
| Private home | 1.5 |
| Theatres (fixed seats) | 4.0 |

## 5.5 WIND LOADING

### 5.5.1 The nature of wind loading

Wind loading is usually the critical source of lateral and moment loading on tall buildings. Wind interacts with the terrain and the ground roughness profile to create turbulent air flow whose character changes with increasing height above the ground. Comprehensive information on wind loadings is provided by Holmes (2015), Davies et al. (2014) and Burton et al. (2014), and only a brief summary of key points is given here.

A schematic diagram of the wind-induced pressures on a building is shown in Figure 5.2 (Davies et al., 2014). The pressure on the windward face is positive while on the leeward face, it is negative. The two components combine to impose a drag force on the building face. On the sides of the building, the wind flow separates, creating suctions with the largest magnitude closest to the separation point.

A wake is formed when the wind flows past the building, and the wind organises into vortices, as illustrated in Figure 5.3 (Davies et al., 2014). This phenomenon is termed 'vortex shedding' and occurs from alternating sides of a building at a frequency, $f_s$, that follows the Strouhal relationship:

$$f_s = SU/D \tag{5.1}$$

where

U is the mean wind speed,

D the building width

and S is the Strouhal number = 0.12 for a square section and 0.20 for a circular section.

At some particular wind speed, the frequency of vortex shedding $f_s$ may align with the fundamental vibration frequency of the building, and may consequently induce a large resonant response. The building acceleration due to wind loading increases as the wind speed increases, but the onset of vortex shedding can cause a relatively sudden increase in acceleration when the vortex shedding frequency approaches the natural frequency of the building,

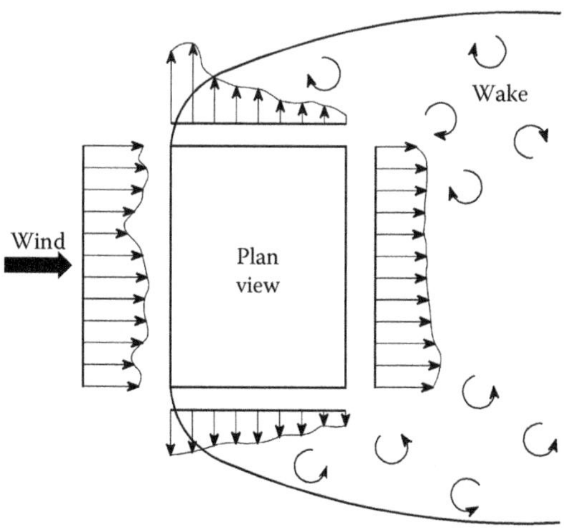


*Figure 5.2* Wind-induced pressures on a building. (Adapted from Davies, A. et al. 2014., *Tall and Supertall Buildings, Planning and Design*. McGraw-Hill Education, New York. Courtesy of McGraw-Hill.)

*Figure 5.3* Vortex shedding from a building. (Adapted from Davies, A. et al. 2014. *Tall and Supertall Buildings, Planning and Design*. McGraw-Hill Education, New York. Courtesy of McGraw-Hill.)

as shown in Figure 5.4 (Davies et al., 2014). The acceleration decreases again when the vortex shedding frequency is greater than the natural frequency of the building. If the building stiffness is increased such that the natural frequency increases from 0.12 to 0.14 Hz, the maximum acceleration due to the vortex shedding increases.

The vortex shedding phenomenon is dependent on the shape of the structure, so that by adjusting the shape, the effects of vortex shedding can be reduced or mitigated, as shown in Figure 5.4.

### 5.5.2 Procedure for estimating wind actions

As an example of the procedure for estimating wind actions on a structure, the Australian–New Zealand Standard AS1170.2–2011 for wind loading defines four steps in estimating the wind actions on structures:

1. Determine the site wind speeds
2. Determine the design wind speed from the site wind speeds

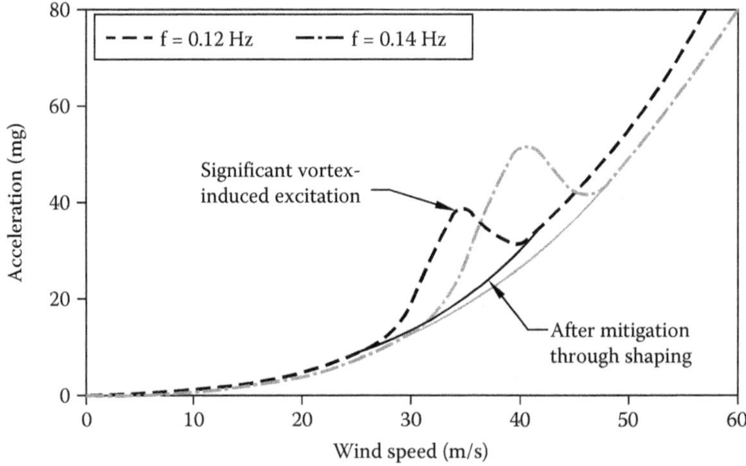


*Figure 5.4* Acceleration response due to vortex excitation. (Adapted from Davies, A. et al. 2014. *Tall and Supertall Buildings, Planning and Design*. McGraw-Hill Education, New York. Courtesy of McGraw-Hill.)

3. Determine the design wind pressures and distributed forces
4. Calculate the wind actions on the structure

Procedures for each of these steps are set out in the Standard, but the provisions are limited to buildings no taller than 200 m. For taller structures, it is necessary to carry out wind tunnel testing, as discussed in Section 5.5.5. The site wind speeds depend on the geographic location of the site, the wind direction, the nature of the terrain and the topography. The design wind speed is then obtained from the maximum cardinal direction site wind speed.

The design wind pressures are obtained from the design wind speeds, the density of air, and factors for the aerodynamic shape and dynamic response. Finally, the design wind actions are obtained from the design wind pressures and a reference area on which the wind pressure at that height acts.

In the conventional method of estimating the wind pressure, the pressure increases with the square of the design wind speed, and the wind speed increases with height above ground. However, as pointed out by Davies et al. (2014), the variation of wind speed with height can depend on the type of event that gives rise to the wind; for example, the distribution for a tropical hurricane can be very different to that for a thunderstorm.

### 5.5.3 Rules of thumb

For initial hand calculations, and to obtain a first idea of the order of the loading, Bull (2012) suggests the following rules of thumb:

1. The lateral load can be approximated as 1.5% of the dead load
2. 1 kPa pressure for roof structures
3. 2 kPa pressure for tall buildings

### 5.5.4 Preliminary approximations

In the absence of other information or code requirements, the distribution of wind loading, $q_w$, at a height z above the surface, can be approximated as follows:

$$q_w = C\, z^{0.25} \text{ kPa} \tag{5.2}$$

where C = wind pressure coefficient.

Ohsaki (1976) adopts a value of C = 1.2, and on this basis, by integrating the pressure over the number (N) of storeys, and assuming a height of 3.5 m per storey, the total wind force, $H_w$, can be estimated as

$$H_w = 1.44\, B\, N^{1.25} \text{ kN} \tag{5.3}$$

The overturning moment, $M_w$, is approximated as

$$M_w = 1.07\, B\, N^{2.25} \text{ kNm} \tag{5.4}$$

where B = width of superstructure on which wind acts.

### 5.5.5 Wind tunnel testing

For most contemporary major tall buildings, the final design values of wind pressures and actions are obtained via the use of wind tunnel testing, rather than by conventional methods of calculation. Boundary layer wind tunnels are designed to simulate the natural wind approaching the building site. A scale model of the structure is placed inside the tunnel which has a floor roughness and surrounding topography which closely replicates that at the actual site. In the area surrounding the model structure, the buildings and local topographic features are also modelled in detail.

The choice of a model scale is very important, and requires a compromise between a model that is small enough to allow simulation of the environment around the tested structure, but large enough to allow accurate representation of the features of the structure. Typical model scales for tall buildings lie between 1:300 and 1:500.

The structure itself is modelled aeroelastically, such that the mass per unit length and the mode shapes for the first three modes of vibration are matched to the target structure. The ratio of the natural frequencies for each pair of modes will also be matched. As pointed out by Burton et al. (2014), the inherent level of structural damping in the aeroelastic models needs to be kept low, and should be less than the expected level of structural damping; typically, it is possible to achieve 0.5% damping in such models.

More detailed information on wind tunnel testing is provided by Holmes (2015), Burton et al. (2014) and Davies et al. (2014). A standard on wind load testing of structures is provided in ASCE/SEI 49-12 (2012).

### 5.5.6 Methods of reducing wind forces

There are at least three broad strategies by which wind forces on a structure can be reduced:

- Increasing the mass and stiffness of the structure so that the natural frequency is increased and the dynamic response under the wind loading is reduced
- Modifying the shape of the building to control the wind pressures and wind actions. This can be achieved by
  - Rounding, chamfering or stepping in the corners
  - Tapering the width of the building with increasing height
  - Using a different cross section at various heights
  - Creating openings for the wind to pass through
  - Using aerodynamic spoilers that protrude from the exterior of the building and break up the vortices
  - Employing a 'twisting' building shape; the use of such a shape can reduce wind loadings by up to 25%
  - Providing additional damping to increase the dissipation of energy imparted by the wind forces. Available damping measures have been discussed briefly in Chapter 2

Further details are provided by Davies et al. (2014).

## 5.6 EARTHQUAKE LOADINGS

For tall buildings, it is generally the case that loads induced on the structure by earthquakes are less critical than those imposed by wind. Nevertheless, it is necessary to include the earthquake-induced loads as part of the sequence of load combinations that have to be considered in structural and foundation design.

Loadings arising from earthquakes are usually assumed to be predominantly horizontal, although recent seismic events in New Zealand have demonstrated that substantial vertical loadings can also be generated. Such loadings are generally estimated via the use of response spectra for at least the early stages of design, and the use of such spectra is discussed in Chapter 11. More detailed analyses of earthquake loadings can be made via dynamic structural analyses in which the response of the structure–foundation–soil system to representative earthquake excitation time histories can be evaluated.

## 5.7 LOADINGS FROM EARTH PRESSURE

Loadings due to earth pressures are generally relevant to the design of the basement walls and sub-structure system. Such loads are generally estimated from some form of earth pressure theory in the early stages of design, and then from more detailed soil–structure interaction analyses when considering the detailed and final design stages. These methods are discussed more fully in Chapter 12.

For very preliminary assessments, Bull (2012) suggests the following preliminary values of earth pressure for walls that are laterally supported at the top and the bottom so that the lateral strain in the soil is approximately zero, and assuming a uniform surcharge loading of 10 kPa:

- A pressure of $0.5(20z + 10)$ kPa at a depth z, when z is above the water table
- A pressure of $0.5(20z - 10(z - 2) + 10)$ kPa when z is below the water table

The above values imply a coefficient of earth pressure at rest of 0.5, which may be rather low for stiff or overconsolidated soils.

For cantilever retaining walls, the pressure can be assumed to be active earth pressure, and approximated as follows:

- $0.33(20z + 10)$ kPa above the water table
- $0.33(20z - 10(z - 2) + 10)$ kPa below the water table

In most cases involving tall building foundations, a diaphragm wall will be constructed to retain the basement excavation, and in some cases, will form part of the permanent foundation system. Such a wall normally encircles the deep foundations and the lateral pressures on the wall are then balanced. However, it is possible in some cases that the wall may not completely encircle the foundation system, and there will then be an out of balance lateral earth force that has to be resisted by the deep foundations. The consequence of additional lateral loads due to out-of-balance earth forces should then be considered.

## 5.8 OTHER LOADS

The influence of ground movements is treated in detail in Chapter 9. It is preferable to consider the interaction between the foundation system and the source of ground movement via the magnitude of the ground movements, rather than trying to directly convert the ground movement to an equivalent force. Such attempts can often lead to very misleading outcomes, due to the differences in the nature of loads arising from ground movements and those arising from direct structural loadings.

Other sources of loading that may need to be considered include snow, ice, thermal effects, major impacts and explosions. Requirements to consider such loads are set out in the relevant standards that govern the structural design of buildings.

Chapter 6

# Ground characterisation

## 6.1 INTRODUCTION

The characterisation and quantification of the ground conditions is a critical part of the foundation design process, and is based upon the execution of an appropriate geotechnical investigation programme. The main aim of such a programme is clearly set out in Eurocode EN1997 s3.2.1(1)P, as follows:

> Geotechnical investigations shall provide sufficient data concerning the ground and groundwater conditions at and around the construction site for a proper description of the essential ground properties and a reliable assessment of the characteristic values of the ground parameters to be used in design calculations. (p. 22)

Regrettably, there remains a consistent tendency among project principals and design and construct teams to try and economise on the geotechnical investigation in order to save money and time for the project. The often articulated plea by the geotechnical engineer: 'You pay for a site investigation, whether you have one or not' (ICE, 1991), is frequently not heeded. As a consequence, problems related to the ground conditions can result in a much greater final project cost than if a proper investigation had been done in the first place. Fortunately, for most modern high-rise developments, there is now an increased appreciation of the necessity for proper ground investigation and characterisation.

Proper planning is essential for effective geotechnical investigations. Ground characterisation is usually based on the results of in situ geotechnical investigations which are generally undertaken in three stages:

1. A preliminary desk study to assess the geological conditions and to identify any previous geotechnical information at or near the site.
2. A preliminary programme of site drilling, for preliminary assessment purposes.
3. A detailed programme of site drilling and in situ testing to obtain information for design and construction.

This chapter will give a brief summary of some of the main aspects relating to the ground characterisation and the development of a geotechnical model for design. Detailed descriptions of the ground investigation process and the various techniques involved are available in several texts, for example, Simons et al. (2002); Clayton et al. (2005); Hunt (2005); Look (2007) and Schnaid (2009).

## 6.2 KEY ASPECTS OF GROUND CHARACTERISATION

Characterisation of the ground conditions at a site involves a number of stages:

1. An understanding of the geology of the site
2. A study of the history of the site, that is, any previous construction or other engineering activities, such as filling, de-watering, building construction, etc.
3. A site investigation, to assess the nature and characteristics of the subsurface strata which exist, and also of the groundwater conditions at the site;
4. Development of a geotechnical model (or models) for the site, involving the representation of the stratigraphy, the selection of a model of soil behaviour and the quantification of the associated engineering parameters.
5. Appropriate in situ and laboratory testing to refine estimates of the appropriate engineering properties required for foundation design.

Ideally, input should be obtained from the engineering geologist (for item 1 above), and from local information and records (if available) for item 2. Input is also highly desirable from a geophysicist (for item 3) and a hydrogeologist (for an assessment of the groundwater conditions in item 3), as well as from the geotechnical engineer. The latter will generally be responsible for item 4, which is an essential and critical precursor to the foundation design, and for specifying item 5 and interpreting the results.

The importance of a proper understanding of the site geology (item 1) has been emphasised time and time again, from the earlier days of soil mechanics (Terzaghi and Peck, 1948) to more recent times (Fookes, 1997), and cannot be overstated. In the following sections, attention will focus primarily on aspects related to the latter four stages of characterisation.

## 6.3 GROUND INVESTIGATION METHODS AND GUIDELINES

### 6.3.1 Desk study

The desk study consists of the collection of available documentation relevant to the site and the structure to be constructed. The documentation may consist of one or more of the following:

1. Previous ground investigations
2. Topographical maps
3. Aerial photos
4. Geological maps
5. Historical maps
6. Rainfall and climate records

An important aspect that requires early definition is the site history. It is not uncommon for the geotechnical engineer to commence work on a site, and to implicitly assume that no activity had occurred prior to his/her arrival. However, 'time zero' rarely starts at this point, and the history of the site may involve the previous existence of buildings (now demolished) on the site, or processes such as filling and/or dewatering. In all cases, this previous site history will have at least two important influences:

- The effective stress state may not be laterally uniform, and the soils may have been preloaded to different intensities across the site
- The prior processes may still be causing settlements at the site

Some useful indications of the current conditions at the site may be inferred via measurement of piezometric levels at various depths within the soil profile. The presence of excess pore water pressures may reveal ongoing effects from prior site loading or dewatering. In turn this may indicate the potential for continuing settlement, which may in turn lead to the development of negative friction in pile foundations, or excessive long-term settlements.

In conjunction with piezometric levels, it is essential that an indication be obtained of preconsolidation pressures with the various soil strata. Laboratory oedometer tests are usually used for this purpose, but alternative methods can now be used, including interpretations from piezocone tests (Mayne and Holtz, 1988). Inappropriate assumptions regarding preconsolidation pressures (e.g. assuming that the entire soil profile is normally consolidated) may lead to gross over-estimates of foundation settlement.

### 6.3.2 Investigation techniques

#### 6.3.2.1 Rotary drilling

The customary approach to site investigation is to carry out a programme of drilling of boreholes and sampling and testing of the main strata. The number and location of the boreholes is dependent on the size and nature of the structure, and the geological conditions at the site. A great deal has been written about this aspect of ground characterisation and detailed descriptions appear in almost all texts dealing with soil mechanics and foundation engineering. It is obvious that boreholes need to be located not only within the footprint of a tall building but also outside the footprint, where low-rise structures may be constructed. Because drilling of boreholes is generally an expensive aspect of site investigation, it is often very cost-effective to consider the possibility of using engineering geophysical techniques as a supplement to conventional drilling.

#### 6.3.2.2 Geophysical techniques

In modern well-planned investigation programmes, rotatry drilling is supplemented by geophysical testing, which can be divided into two broad categories: invasive tests, and non-invasive tests. The former require a borehole, and cross-hole, down-hole, up-hole, and seismic penetration tests are examples of such invasive techniques. Non-invasive techniques are undertaken at the ground surface, and include seismic reflection, seismic refraction, spectral analysis of surface waves (SASW) and multistation analysis of surface waves (MASW). Details of these techniques are given by Foti et al. (2014).

Data from geophysical techniques, when properly interpreted, provide a number of major benefits, including

- They provide a means of identifying the stratigraphy between boreholes.
- They can identify localised anomalies in the ground profile, for example, cavities, sinkholes or localised pockets of softer or harder material.
- They can identify bedrock levels.
- They provide quantitative measurements for the shear wave and compression wave velocities within the ground profile. This information can be used to estimate the in situ values of soil stiffness at small strains, and hence to provide a basis for quantifying the deformation properties of the soil strata.

A number of geophysical techniques have been developed to specifically address one or more of the above issues. Whiteley (1983) describes some of the techniques available, including seismic refraction which is used to estimate the depth and nature of the bedrock, gravity methods to detect buried channels and rock depths, and down-hole and cross-hole logging to assess a profile of seismic velocity with depth, thus enabling an interpretation of the stratigraphic conditions. Matthews et al. (1996) have described a technique for using surface waves to estimate stiffness versus depth profiles in near-surface soil and rock, without the need for boreholes. The equipment required includes an energy source (hammer or vibrator), two or more receivers (usually geophones), a recording device (typically a spectrum analyser or a seismograph) and a portable computer for data processing. There is clearly a limit to the depth of investigation with this technique, ranging from about 8 m for clays to about 20 m for rock. A powerful and versatile approach involving seismic tomographic imaging to map the subsurface conditions has been developed in recent years. This approach can address all four of the above issues, and is described in more detail below.

#### *6.3.2.3 Site uniformity borehole seismic and seismic tomographic imaging*

This method has been described by Whiteley and Pedler (1994), and involves lowering a 12-channel seismic detector array with hydrophones at 2 m intervals into a PVC cased borehole. Seismic energy is generated at numerous locations using an impact mass operated by a pneumatic ram. The seismic data is collected on a seismograph and stored on computer disk for later processing. The seismic first arrival data from the sources close to the borehole give a vertical seismic profile (VSP) which can be correlated with the borehole logs to allow the computation of subsurface seismic velocities. First arrivals from the various offset seismic sources are interactively picked from the seismic records or computer display. These data are replotted at each detector depth and processed using a tomographic procedure which produces a contour plot of seismic velocity along each traverse. The travel times of the first arrival P waves from the surface source to a down-hole detector are controlled by the elastic properties of the earth materials and the distribution of subsurface interfaces. Any condition which 'weakens' the soil, such as a cavity or an isolated deposit of soft or loose material, will normally lead to a scattering of the seismic wave and a delay in its travel time, producing a region of low seismic velocity on the seismic tomographic image.

The non-destructive nature of seismic tomographic imaging (STI) allows testing of a location to be undertaken a number of times, thus allowing assessment of changes in ground conditions due to construction processes such as grouting, and changes in groundwater conditions.

The initial (small-strain) stiffness of the strata can be obtained directly from the measured seismic velocities. For application to practical foundation design, the small-strain shear modulus derived from STI needs to be corrected for the appropriate strain level. Atkinson (2000) gives a detailed discussion of this issue, and it is addressed later in this chapter.

Whiteley (2000) and Powell and Whiteley (2008) describe the use of STI for imaging of cavities in karst limestone. Figure 6.1 shows an example of the tomographic image obtained, and clearly reveals the low velocity zone which is the cavity.

A further application of STI to detect bedrock levels occurred in a project in the Genting Highlands in Malaysia. From initial borehole drilling and seismic tomographic imaging, founding levels for the support of a multistorey residential complex were assessed. Figure 6.2 shows a plan and section of the foundations of one of the high-rise blocks. The predicted founding levels were subsequently compared with the actual pile founding levels, and were found to be in very good agreement, generally within 1–2 m.

In summary, modern seismic imaging techniques provide a valuable and convenient means of filling-in the gaps between conventional boreholes, and of defining irregularities within the geotechnical profile. Despite their availability, there is still a reluctance to employ

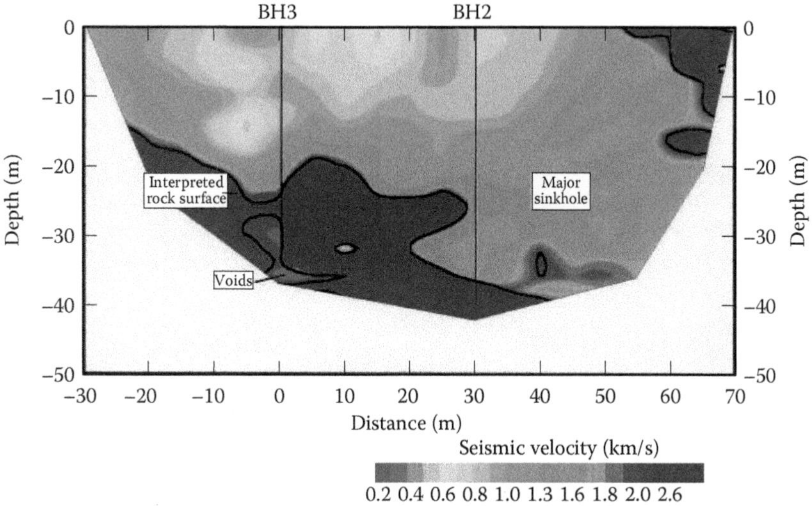


*Figure 6.1* Tomographic image for centre in limestone. (Adapted from Whiteley, R.J. 2000. Seismic imaging of ground conditions from buried conduits & boreholes. In: *Proceedings of Paris 2000: Petrophysics Meets Geophysics*, Paper B-15, Paris, 6–8 November, 2000, EAGE.)

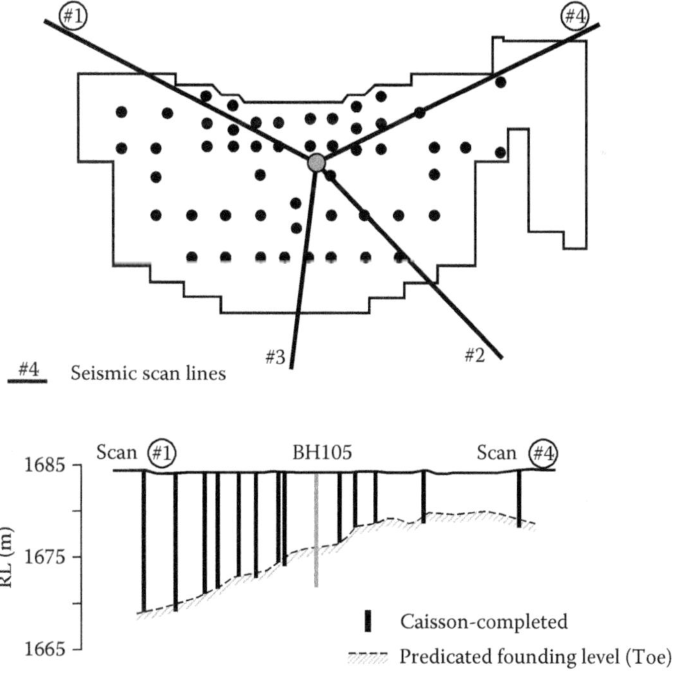


*Figure 6.2* Plan and section for high-rise blocks on bored piles. (Adapted from Powell, G.E. and Whiteley, R.J. 2008. Prediction of piling conditions in Genting Highlands granites Malaysia with borehole seismic imaging. Geotechnical and Geophysical Site Characterization, Huang and Mayne (eds), Taipei, Taiwan, pp. 925–928. Courtesy of Taylor & Francis.)

seismic techniques, due to a combination of at least three factors: unfamiliarity on the part of the geotechnical engineer, the perceived cost (which is, in fact, not high) and the limited availability of competent practitioners who can undertake and properly interpret the STI process.

### 6.3.3 Extent of investigations

For high-rise structures, typical guidelines for the extent of the ground investigation for detailed design are summarised in Table 6.1. The investigation should not be confined to the footprint of the structure, but should extend outside the footprint over an area that is likely to be influenced by the structure. It may be important to assess the ground conditions below adjacent structures and facilities that may be affected by the new construction, including the effects of construction operations.

As pointed out by Haberfield (2013), the behaviour of a high-rise foundation system requires consideration of the ground behaviour in the following critical locations:

1. Immediately below the raft foundation or basement slab, where the important factors are the strength for bearing capacity and the stiffness for settlement and interaction effects
2. Along the pile shaft, where the strength and stiffness are important for the pile shaft resistance, settlement and interaction effects, and also for excavatability and stability
3. At and just below the pile tip level, where the strength and stiffness play an important role in the base bearing capacity and stiffness of individual piles and the pile group
4. Beneath the pile tips, where the ground stiffness can influence the foundation settlement for depths of up to twice the building width

### 6.3.4 Portrayal of ground investigation data

It is a good practice to assimilate all the available ground investigation data in as compact a form as is feasible. For example, borehole data can be imported into programmes such as gINT, which may then be used to develop cross sections for various parts of the site. Such cross sections are a valuable aid to developing appropriate stratigraphic models. There are also some 3D portrayal techniques becoming available, so that a ready appreciation of site uniformity and possible anomalies can be made.

A useful procedure for the geotechnical designer is to prepare, preferably by hand, a single page summary of the stratigraphy and the key design parameters. Such a process can assist greatly in developing a 'feel' for the ground conditions and how they can be modelled.

*Table 6.1* Guideline to extent of investigations

| *Foundation type* | *Spacing of investigation points* | *Approximate depth of investigation* |
|---|---|---|
| Shallow | 15–40 m | $2B_f$ (pad), $4B_f$ (strip) where $B_f$ = footing width |
| Raft, closely spaced footings | " | 1.5B, where B = building width |
| Piles to rock | " | 3 m or 3d below pile founding level, where d = pile diameter |
| Floating piles or piled rafts | " | 1.5B below 2L/3, where B = building width, L = pile length |

Source: Look, B. 2007. *Handbook of Geotechnical Investigations and Design Tables*. Taylor & Francis, London.

### 6.3.5 Possible reasons for failures in the ground characterisation process

Duncan (1979) suggests that the site characterisation process serves two main purposes:

- The anticipation of problems and effects
- Quantification of the site geometric characteristics and material properties

Osterberg (1979) contends that the site exploration process can be considered a failure if it does not reveal subsurface conditions needed for safe economical design of foundations or earth structures, or fails to allow proper assessment of excavation, bracing and other construction operations. Osterberg attributes such failures to the following five general reasons:

- Failure to use general knowledge of geologic processes in planning the exploration programme and in evaluation of the findings of the investigation
- A preconceived notion of what the site evaluation should be and a reluctance (or even refusal) to consider evidence which contradicts the preconceived ideas
- Failure to use all available tools for site evaluation, even though they may be simple and obvious
- Failure to properly discuss the goals of the exploration programme with all the persons involved
- A failure to set up open and free lines of communication

Of course, these reasons presuppose that there is no undue limitation on the expenditure for the investigation, and regrettably, such limitations remain, and can add a further risk, that of failure to do adequate investigation to properly define the site characteristics. Goldsworthy (2006) discusses a quantitative approach for estimating the optimum expenditure for ground investigation, based on an appreciation of the financial consequences of unforeseen events or requirements and the sensitivity of the construction to ground conditions. He indicates that optimum costs of an investigation programme may range from about 1% of the project cost, for a low sensitivity project, to between 6% and 8% for a high-sensitivity project. Typical current levels of expenditure are generally well below 1%, indicating that an inadequate amount is spent on the ground investigation.

This view is supported by Clayton (2001), who has summarised data which indicates that, with the traditional levels of expenditure on ground investigation (typically less than 1%) cost overruns on highway projects were as much as 100%, while expenditure of about 6% of construction cost appeared to be necessary to reduce the cost overrun to 10% or less. It was also found that a significant proportion of the cost overruns could be attributed to just two factors: inadequate planning of site investigations, and the inadequate interpretation of these investigations.

Clayton (2001) has suggested that there will never be sufficient time and money available to investigate with sufficient thoroughness the properties of all the ground to be affected by construction. To combat this reality, he suggests a risk management strategy which involves a proper identification of geotechnical hazards and their associated risks. The optimum level and sophistication of the investigation and design process can then be selected to match the assessed level of risk.

## 6.4 KEY FOUNDATION DESIGN PARAMETERS

The fundamental soil classification, strength and deformation parameters are always important for the design of foundations and the associated geotechnical works. However, from the viewpoint of foundation design, most contemporary foundation systems incorporate both piles and a raft, and in such cases, the following foundation design parameters require assessment:

- The ultimate skin friction for piles in the various strata along the pile
- The ultimate end bearing resistance for the founding stratum
- The ultimate lateral pile–soil pressure for the various strata along the piles
- The ultimate bearing capacity of the raft
- The stiffness of the soil strata supporting the piles, in the vertical direction
- The stiffness of the soil strata supporting the piles, in the horizontal direction
- The stiffness of the soil strata supporting the raft

It should be noted that the soil stiffness values are not unique but will vary, depending on whether the designer requires long-term values (for long-term settlement estimates) or short-term values (for dynamic response to wind and seismic forces). The soil stiffness values will generally vary with applied stress or strain level, and will tend to decrease as these levels increase. For dynamic response of the structure–foundation system, an estimate of the internal damping of the ground is also required, as it may provide the main source of foundation damping.

### 6.4.1 Principles of selection of design parameters

The selection of design parameters involves the exercising of a considerable amount of experience and judgement, as the natural variability of the soil and rock strata generally precludes the application of hard and fast methods. Various approaches have been adopted, many of which involve statistical techniques which are not always applicable, and which when applied, can lead in some cases to over-conservatism. A useful approach has been suggested by Schneider (1997) in which the design value $x_d$ of a parameter, x, is estimated as follows:

$$x_d = x_m \cdot \left( \frac{1 - V_x}{2} \right) \tag{6.1}$$

where $x_m$ is the mean value of x and $V_x$ = the coefficient of variation of x.

As a guide, in the absence of information on $V_x$, Schneider suggests the values shown in Table 6.2.

Caution must be exercised in using this approach when the soil is imposing load on the foundation, rather than supporting it. In such a case, a value of soil strength higher than the mean should be adopted, so that the negative sign in Equation 6.1 should then become positive.

*Table 6.2* Typical values of coefficient of variation $V_x$ for geotechnical parameters

| *Parameter* | *Range of values of coefficient of variation $V_x$* | *Recommended average $V_x$* |
|---|---|---|
| Density | 0.01–0.10 | 0 |
| Angle of internal friction | 0.05–0.15 | 0.1 |
| Cohesion | 0.3–0.5 | 0.4 |
| Compressibility | 0.2–0.7 | 0.4 |

Source: Schneider, H.R. 1997. Definition and determination of characteristic soil properties. In: *Proceedings of the 14th International Conference on Soil Mechanics and Foundation Engineering*, Hamburg. Balkema, Rotterdam, Vol. 4, pp. 2271–2274.

## 6.5 IN SITU TESTING TECHNIQUES

There is a wide variety of types of in situ testing that can be usefully undertaken to assist in ground characterisation for tall building foundation design. Detailed descriptions of such techniques can be found in many references, including Mayne et al. (2009) and Schnaid (2009).

Conventional in situ testing techniques include the following:

1. *Standard penetration test (SPT).* This is a crude test but one that is almost always carried out and which provides a useful preliminary basis for assessing the general ground characteristics. In some countries, it forms one of the main sources of geotechnical information, with several key geotechnical parameters being correlated with the SPT value.
2. *Cone penetration test (CPT).* This test is very useful in clay deposits or in relatively loose sand strata. It has the great advantage of being a continuous test and so can identify thin layers of weaker material within thicker strata of better quality. Empirical correlations exist between most of the key geotechnical parameters and the measured cone resistance values.
3. *Piezocone test (CPTu).* This is a development of the CPT test which also allows the rate of dissipation of excess pore pressures developed during penetration to be measured. It thus provides a means of assessing the in situ coefficient of consolidation of the soil.
4. *Seismic cone test (SCPT).* This is also a development of the CPT test, but has the added advantage that it may be used to measure the shear wave velocity in the soil being tested. This in turn can be used to estimate the small-strain shear modulus ($G_0$) which, as will be demonstrated later in this chapter, is a very valuable parameter for assessing foundation movements. The following basic expression is employed:

   $$G_0 = \rho v_s^2 \tag{6.2}$$

   where $\rho$ is the soil mass density and $v_s$ is the shear wave velocity.
5. *Field shear vane (FVT).* This measures the undrained shear strength of a clay soil, on the assumption that the clay is saturated and exhibits undrained behaviour during the test. It is not suitable for sandy soils or relatively stiff clays.
6. *Dilatometer test (DMT).* This test enables the soil stiffness and strength to be estimated.
7. *Pressuremeter test (PMT).* This test involves the insertion and expansion of a flexible membrane into the ground at various depths. Via appropriate interpretation (usually based on cavity expansion theory), estimates may be made of the in situ lateral stress, and the strength and stiffness of the soil in which the test was carried out. It is customary to carry out load–unload–reload cycles, with the latter cycles being used to obtain more relevant values of soil stiffness for foundation design.
8. *Cross-hole and/or down-hole geophysical tests.* These enable tomographic images to be developed of the ground adjacent to and between the boreholes in which the tests are undertaken. Shear wave velocity values can be interpreted from these data, enabling values of small-strain shear modulus $G_0$ to be obtained from Equation 6.2.
9. *Plate load tests (PLT).* These are generally used to provide load-settlement data from which the stiffness of the soil below the plate may be assessed, via the use of elastic theory. Such a test is generally only carried out under vertical loading, and is only relevant to shallow or raft foundations. Horizontal PLT can also be carried out, and these are potentially useful for assessing the lateral stiffness of near-surface soils, which may in turn be very useful in estimating the lateral movements of pile and piled raft foundations.
10. *Pile load tests.* These are discussed in detail in Chapter 13.

From the point of view of tall building foundations, some of the relevant techniques and how their results can be utilised are summarised in Table 6.3 (after Schnaid, 2009).

## 6.6 LABORATORY TESTING

### 6.6.1 Routine tests

A large number of laboratory tests for soil and rock are available, many of which are routine or basic tests that provide data on basic properties such as particle size, plasticity characteristics, moisture content, density and specific gravity. Chemical tests may also be useful, for example, carbonate content and sulphate content tests may give indications of potential problems for load capacity and concrete durability, respectively. When such tests are specified, there should also be a clear statement of the purpose of the tests for the project in hand. Too often, tests are undertaken as a matter of course, with little or no thought given to the application of the resulting data.

For foundation design, more detailed testing for geotechnical strength and deformation parameters will generally be required. Conventional triaxial testing is of limited value for assessing design parameters for pile foundations, as the method of stress application does not reflect the way in which load transfer occurs from the piles to the surrounding soil. Conventional tests such as unconsolidated undrained tests are of very limited value, and may even be misleading because of the sample disturbance that may be involved in obtaining and setting up the test specimens. However, it should be borne in mind that some of the available Category 1 correlations of design parameters with undrained shear strength of clays are based on this simple type of triaxial test.

### 6.6.2 Triaxial and stress path tests

More sophisticated triaxial and stress path testing can provide stiffness parameters over a range of stress appropriate to the foundation system, and can be used to compare with values from other means of assessment. These tests include the following:

1. *Consolidated undrained triaxial test.* This test is usually carried out to obtain undrained shear strength values ($s_u$) for a clay soil, once the soil sample has been re-consolidated to some relevant hydrostatic pressure. If pore pressure measurements are taken, then effective stress strength parameters can also be derived, as long as two or more tests are carried out on either the same sample, or on very similar samples taken from the same depth in the same borehole. Values of undrained Young's modulus can be derived from these tests, but should not be used directly for foundation design as they are unlikely to represent the in situ values due to differences in the stress history and the stress path to which the sample has been subjected.
2. *Drained triaxial test.* This test is usually carried out on samples of coarse-grained soils although fine-grained soils can also be tested, but may take a long time to complete. The samples are consolidated to a relevant lateral effective stress and then sheared sufficiently slowly that no excess pore pressures are developed during the test. Drained strength parameters ($c'$ and $\phi'$) can be derived from these results, together with the drained Young's modulus $E'$. However, the latter value should not be used directly for deep foundation design, because of the differences in the stress history and the stress path to which the laboratory sample has been subjected.

*Table 6.3* Application of in situ testing techniques to tall building foundation design

| *Category* | *Test* | *Measurements* | *General applications* | *Application to foundation design* |
|---|---|---|---|---|
| Non-destructive geophysical tests | Seismic refraction | P waves from surface | Ground characterisation | Ground characterisation and correlations with ground stiffness |
| | Surface waves (SASW) | R waves from surface | Small-strain stiffness | |
| | Cross-hole | P and S waves in borehole | Small-strain stiffness | |
| | Down hole | P and S waves with depth | Small-strain stiffness | |
| Penetration tests | SPT | SPT-N value | Soil profiling | Correlations with design parameters |
| | CPT | Cone resistance and sleeve friction | Soil profiling, correlations with strength and compressibility | Correlations with design parameters |
| | Piezocone (CPTu) | Cone resistance, sleeve friction, excess pore pressure | As for CPT, and permeability characteristics | Correlations with design parameters, rate of consolidation |
| | SCPT | As for CPT, plus shear wave velocity | As for CPT, plus deformation properties | Correlations with design parameters, small-strain shear modulus |
| Dilatometer | Flat DMT | In situ lateral stress, pressure to expand membrane | Strength and stiffness | Correlations with design parameters |
| | Seismic dilatometer | As for DMT, plus shear wave velocity | Profiling, strength, stiffness, deformation properties | Correlations with design parameters, small-strain shear modulus |
| Pressuremeter | Pressuremeter (pre-bored) | Pressure versus volume | Strength, stiffness, initial stress state, time and rate effects | Correlations with design parameters |
| | Pressuremeter, self-boring | Pressure versus volume | Strength, stiffness, initial stress state, time and rate effects | Correlations with design parameters |
| Foundation tests | Plate load test | Load versus settlement | Deformation and bearing capacity | Near-surface strength and stiffness for shallow foundation |
| | Pile load test, axial | Load versus settlement, distribution of axial load along shaft | Pile capacity and stiffness, and distribution of shaft friction and end capacity | Applicable to single pile behaviour |
| | Pile load test, lateral | Load versus pile head deflection and rotation | Lateral capacity and stiffness | Applicable to single pile behaviour |

Source: Schnaid, F. 2009. *In situ Testing in Geomechanics*. Taylor & Francis, Abingdon, UK.

3. *Cyclic triaxial test.* This test generally involves the application of a mean deviator stress to the consolidated soil sample, the subsequent application of a number of cycles of deviator stress (usually under undrained conditions), and then a final monotonic undrained loading to failure. The main objectives of such tests are to estimate the accumulation of strain during the cyclic loading and the post-cyclic undrained shear strength as a percentage of the initial static undrained shear strength (as measured from a separate test on a nominally identical sample). Cyclic triaxial testing can also be useful in providing an indication of the degradation effects on the stiffness/strength properties of the foundation ground material due to cyclic loading. For example, for the Burj Khalifa project, cyclic triaxial test results indicated that a degree of degradation was possible in the mass ground strength/stiffness properties, but that under the anticipated applied loading, the foundations would be loaded to sufficiently small-strain levels such that potential degradation of strength and stiffness would be limited.
4. *Stress path test.* Stress path testing was promoted by Lambe (1964) and Davis and Poulos (1963), and is used to obtain measurements of the undrained and drained deformation parameters using a stress history and stress path relevant to that to be experienced by the soil below the foundation system. This test involves triaxial testing with the following steps:
   a. Initial reconsolidation of the soil sample to the estimated in situ effective vertical and horizontal stress state. This is most effectively done via $K_0$ testing, following the estimated stress history of the sample.
   b. The application of increments of vertical and lateral stress to the sample under undrained conditions. These increments are representative of the stresses anticipated to be applied to the soil by the foundation at the depth from which the sample was taken. The undrained Young's modulus may be obtained from this stage.
   c. The sample is allowed to drain by opening the drainage valves on the triaxial apparatus. From the measured volume change, and the overall measured axial strain, both the drained Young's modulus $E'$ and the drained Poisson's ratio $\nu'$, can be computed.

This test may provide relevant values of the modulus values for shallow foundations and rafts, but may not be directly relevant to pile foundations.

### 6.6.3 Simple shear test

This test has been described in detail by Bjerrum and Landva (1966) and Dyvik et al. (1987). It is used primarily to obtain values of undrained shear strength. Cyclic tests may also be carried out to assess the potential for cyclic degradation or loss of undrained shear strength following cyclic loading. Static and cyclic loading can be performed under either stress-controlled or strain-controlled conditions. The specimen can be subjected to varying cyclic stress/strain levels and frequencies. It is also possible to perform undrained or drained creep tests by having a sustained horizontal shear stress on the specimen and measuring shear strain versus time.

### 6.6.4 Resonant column testing

The resonant column test is commonly used for laboratory measurement of the low-strain properties of soils. It subjects solid or hollow cylindrical specimens to torsional or axial loading by an electromagnetic loading system, usually harmonic loads for which frequency and amplitude can be controlled. It can be used to measure the small-strain shear modulus and damping ratio of a soil or rock sample, and the variation of modulus and damping ratio with increasing shear strain level. Such data are valuable for carrying out dynamic response analyses of the foundation system.

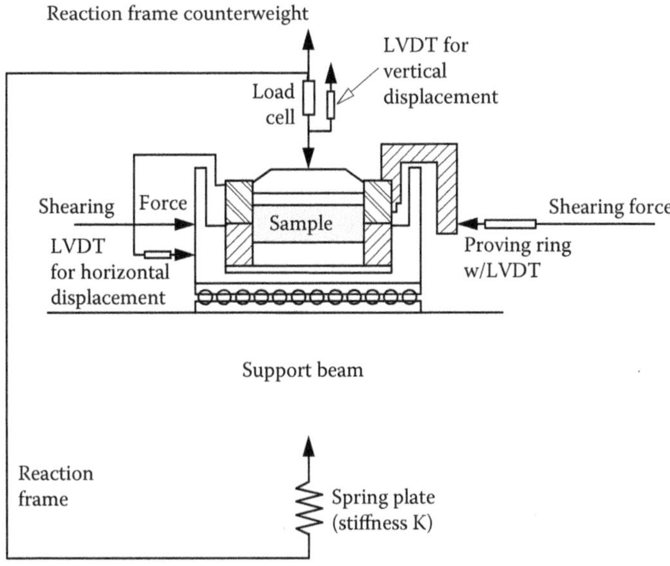


*Figure 6.3* Constant normal stiffness test setup.

### 6.6.5 Constant normal stiffness testing

It has generally been accepted by practitioners that there is no suitable laboratory test which can be used reliably to measure the ultimate shaft friction $f_s$. However, there has been a significant advance over the past three decades in direct shear testing of interfaces, with the development of the 'constant normal stiffness' (CNS) test (Lam and Johnston, 1982; Ooi and Carter, 1987). The basic concept of this test is illustrated in Figure 6.3, and involves the presence of a spring of appropriate stiffness against which the normal stress on the interface acts. This test provides a closer simulation of the conditions at a pile–soil interface than the conventional constant normal stress direct shear test. The normal stiffness $K_n$ represents the restraint of the soil surrounding the pile, and is given by

$$K_n = \frac{4G_s}{d} \tag{6.3}$$

where $G_s$ is the shear modulus of surrounding soil and *d* is the pile diameter.

The units of $K_n$ are stress per unit length. Equation 6.3 is based on the assumption that the soil or rock mass behaves as a linear elastic material, with a constant shear modulus $G_s$. In calculating an appropriate value of $K_n$, a suitable choice must be made for the elastic shear modulus of the formation material. If only relatively small dilations are expected, then a tangent value of the mass modulus $G_s$ may be used. If relatively large dilations take place, so that the response of the mass to an increase in radial pressure falls into the non-linear range, then it will be more prudent to adopt a secant value for $G_s$.

The effects of interface volume changes and dilatancy can be tracked in a CNS test, and the results are particularly enlightening when cyclic loading is applied, as they demonstrate that the cyclic degradation of interface frictional resistance is due to the reduction in normal stress arising from the volumetric contractions caused by the cyclic displacements applied to the interface.

## 6.7 PRELIMINARY ASSESSMENT OF BASIC SOIL PARAMETERS

### 6.7.1 Introduction

For preliminary design, it is useful to have a means of assessing the relevant geotechnical parameters before detailed in situ and/or laboratory test data are available. Such assessments are generally made via empirical correlations with relatively crude available data such as SPT values, particle size data or simply visual or descriptive information. Comprehensive collections of geotechnical correlations have been compiled by Kulhawy and Mayne (1990), Look (2007) and Ameratunga et al. (2016). This section summarises a very limited number of these correlations. It should be emphasised that such correlations should not be used for detailed design, but may be useful for preliminary design, and as checks for data obtained from more detailed and sophisticated testing procedures.

### 6.7.2 Drained shear strength parameters

The drained cohesion c′ of both fine-grained and coarse-grained soils is generally taken as zero for saturated soils that are not heavily over-consolidated or cemented. Representative values of the drained angle of internal friction $\phi'$ for relatively coarse-grained soils are shown in Table 6.4.

For fine-grained soils, $\phi'$ depends on the plasticity characteristics. Some information is provided by Wesley (2010) on typical values of $\phi'$ which tend to decrease with increasing plasticity index (PI) and can be lower than 10 degrees for extremely high PI values. Typical values of c′ range between 0 for soft normally consolidated clays, to 10–25 kPa for firm clays, and up to 100 kPa for hard clays. For compacted clays, c′ may range between about 10–25 kPa.

### 6.7.3 Undrained shear strength parameters

Undrained shear strength $s_u$ may be correlated with several parameters, including the following:

SPT-N:

$$\frac{s_u}{p_a} \approx 0.06\ N \tag{6.4}$$

where $p_a$ is the atmospheric pressure. Note that this is a very rough correlation.

*Table 6.4* Representative values of drained angle of internal friction, $\phi'$

| | *Drained angle of internal friction $\phi'$ (degrees)* | |
|---|---|---|
| *Soil type* | *Loose* | *Dense* |
| Sand, round grains, uniform | 27.5 | 34 |
| Angular sand, well graded | 33 | 45 |
| Sandy gravels | 35 | 50 |
| Silty sand | 27–33 | 30–34 |
| Inorganic silt | 27–30 | 30–35 |

Drained friction angle $\phi'$:
For direct shear conditions,

$$\frac{s_u}{\sigma'_{vo}} \approx 0.5 \sin\phi' \ \mathrm{OCR}^{0.8} \tag{6.5}$$

where OCR is the over-consolidation ratio, and $\sigma'_{v0}$ is the vertical effective stress.

### 6.7.4 Small-strain shear modulus, $G_0$

A number of correlations of the small-strain shear modulus $G_0$ are set out below. When attempting to employ small-strain modulus values for routine foundation design, allowance must be made for the effects of strain level in decreasing the operative modulus to be used. While some software packages may allow for such modulus 'degradation' with increasing strain level, most current packages do not. Methods of making allowance are discussed later in Section 6.9.

#### *6.7.4.1 Correlations with SPT*

Many correlations between shear wave velocity and SPT-N values have been summarised by Ohta and Goto (1978), while more direct correlations between $G_0$ and SPT-N values have been suggested by Hirayama (1994), as follows:

$$G_0 = 11.9 N^{0.78} \ \mathrm{MPa} \tag{6.6}$$

$$G_0 = 14.1 N^{0.68} \ \mathrm{MPa} \tag{6.7}$$

$$G_0 = 5 N \ \mathrm{MPa} \tag{6.8}$$

These three correlations give similar results for N values up to about 40, but for larger N values, Equation 6.7 gives smaller values than the other two equations.

Alternative correlations have been proposed to relate the small-strain shear modulus $G_0$ to the SPT-N value, which generally take the following form:

$$G_0 \approx X \, [N_{1(60)}]^y \ \mathrm{MPa} \tag{6.9}$$

where $[N_{1(60)}]$ is the SPT value, corrected for overburden pressure and hammer energy and X and y are parameters that may depend on soil type.

Typical values of X and y are shown in Table 6.5.

#### *6.7.4.2 Correlations with CPT*

Mayne and Rix (1993) have developed the following empirical relationships between $G_0$ and the static cone resistance for clay soils:

$$G_0 = 406 \, (q_c)^{0.695} / e_0^{1.130} \ \mathrm{kPa} \tag{6.10}$$

Table 6.5 Typical parameters for small-strain shear modulus correlations

| Soil type | X | y |
|---|---|---|
| Sandy soils | 90.8 | 0.32 |
| Clayey soils | 97.9 | 0.27 |
| All soils | 90.0 | 0.31 |

Source: Hasancebi, N. and Ulusay, R. 2007. *Bull Eng Geol Environ*, 66: 203–213.

$$G_0 = 2.87\,q_c^{\,1.335}\ \text{kPa} \tag{6.11}$$

where $q_c$ is the measured cone tip resistance, in kPa and $e_0$ is the initial void ratio.

Equation 6.10 is a better statistical fit to the data than Equation 6.11.

Lehane et al. (2005) have proposed the following alternative dimensionless relationship:

$$\frac{G_0}{q_c} = \frac{185\,(q_c/p_a)^{-0.7}}{(\sigma'_{vo}/p_a)^{-0.35}} \tag{6.12}$$

where $p_a$ is the atmospheric pressure and $\sigma'_{vo}$ is the initial vertical effective stress.

## 6.7.5 Consolidation parameters: Compression ratio

### 6.7.5.1 *Correlations with index properties*

There are a vast number of correlations between compression ratio (CR) (or compression index $C_c$) and various index properties, including liquid limit, natural water content, PI and initial void ratio. Some of these are summarised by Balasubramaniam and Brenner (1981). Among these correlations are the following, for soils in a normally consolidated state:

For marine clays of Southeast Asia (Cox, 1966)

$$CR = 0.0043\,w_n \tag{6.13}$$

For all clays, (Elnaggar and Krizek, 1970)

$$CR = 0.156e_0 + 0.0107 \tag{6.14}$$

For French clays with $w_n < 100\%$, (Vidalie, 1977)

$$CR = 0.0039\,w_n + 0.013 \tag{6.15}$$

For French clays, (Vidalie, 1977)

$$CR = 0.403\log(w_n) - 0.478 \tag{6.16}$$

*Table 6.6* Coefficient β for various soil types

| *Soil type* | *Coefficient* β |
|---|---|
| Coarse sand | 0.05–0.1 |
| Fine sand | 0.15–0.3 |
| Sandy clay | 0.2–0.4 |
| Pure clay | 0.4–0.8 |
| Peat | 0.8–1.6 |

Source: CUR. 1996. *Building on Soft Soils*. CUR Centre for Civil Engineering, CRC Press, The Netherlands.

Schofield and Wroth (1968)

$$CR = \frac{1.325\ I_p}{1 + e_0} \tag{6.17}$$

where $w_n$ is the natural water content (%), $e_0$ the initial void ratio and $I_p$ is the plasticity index.

#### 6.7.5.2 *Correlations with cone penetration resistance*

For a variety of soil types, CUR (1996) provides the following correlation between CR and $q_c$:

$$CR = \frac{2.3\beta\sigma'_{vo}}{q_c} \tag{6.18}$$

where $\sigma'_{vo}$ is the initial vertical effective stress and β is the coefficient depending on soil type, as shown in Table 6.6.

#### 6.7.5.3 *Effect of over-consolidation*

It is well-recognised that the CR for an over-consolidated soil is significantly less than the value for the soil in a normally consolidated state. Kulhawy and Mayne (1990) indicate that, on average, CR for a soil in an over-consolidated state is about 20% of the value for the normally consolidated state. This ratio can be as low as 10 for some soils.

### 6.7.6 Constrained modulus, D

#### 6.7.6.1 *Correlation with porosity*

The constrained modulus D can be related to the effective vertical stress via the following equation developed by Janbu (1963):

$$D = m \cdot p_a \left( \frac{\sigma'_v}{p_a} \right)^{\beta} \tag{6.19}$$

where $p_a$ is the atmospheric pressure, $\sigma'_v$ the vertical effective stress, m the modulus number, which is a function of porosity and β the exponent depending on soil type, as follows: $\beta = 1$ for normally consolidated clays, $\beta = 0.5$ for coarse-grained soils and $\beta = 0$ for heavily overconsolidated clays.

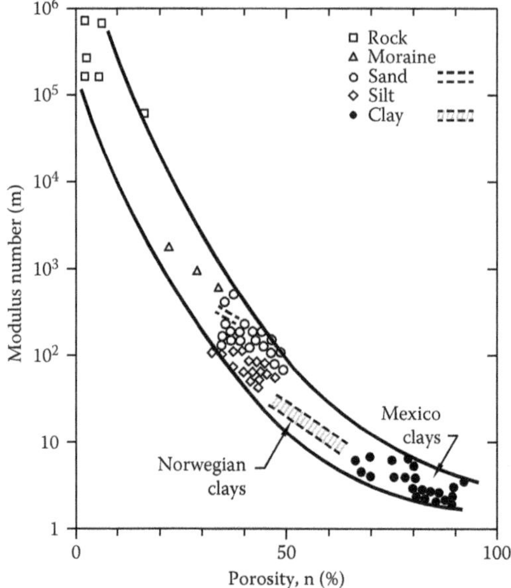


*Figure 6.4* Relationship between Janbu modulus number and porosity for normally consolidated geomaterials. (Adapted from Janbu, N. 1963. Soil compressibility as determined by oedometer and triaxial tests. *Proceedings of the European Conference on Soil Mechanics and Foundation Engineering*, Wiesbaden, Vol. 1, pp. 19–25.)

The relationship between modulus number m and porosity n is shown in Figure 6.4, which indicates that the value of m for rocks (where n is very low) is in the order of $10^5$–$10^6$, while for sands and silts, m is typically between about 50 and 200. Figure 6.5 shows a more detailed plot of m as a function of porosity for normally consolidated silts and sands. For over-consolidated silts and sands, the value of m should be multiplied by a factor of 5–10 to reflect the increase due to over-consolidation.

For soft clays, much lower values of m are relevant, generally less than 20. In such cases, it may be preferable to relate D to the compression ratio, CR, as in Equation 6.21.

#### 6.7.6.2 Correlation with CPT

Kulhawy and Mayne (1990) have presented the following empirical relationship between D and cone resistance $q_c$ for clays:

$$D = 8.25 \cdot (q_c - \sigma_{v0}) \tag{6.20}$$

where $\sigma_{v0}$ is the initial vertical total stress.

#### 6.7.6.3 Correlation with CR

For normally consolidated clay soils, the constrained modulus D can also be related to the CR via the following approximate expression:

$$D \approx \frac{2.3\sigma'_v}{CR} \tag{6.21}$$

where $\sigma'_v$ is the initial vertical effective stress.

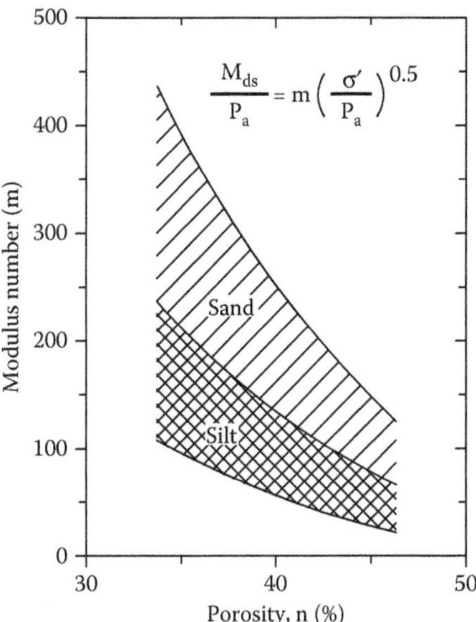


*Figure 6.5* Modulus number for normally consolidated silts and sands. (Adapted from Janbu, N. 1963. Soil compressibility as determined by oedometer and triaxial tests. *Proceedings of the European Conference on Soil Mechanics and Foundation Engineering*, Wiesbaden, Vol. 1, pp. 19–25.)

Thus, in this case, $\beta = 1$ and the modulus number m in Equation 6.19 is simply $m = 2.3/CR$.

### 6.7.7 Poisson's ratio $\nu_s$

Poisson's ratio for isotropic materials lies between 0.5 and 0. It can vary with such factors as stress level and over-consolidation ratio, but as a first approximation, the values in Table 6.7 can serve as a guide.

### 6.7.8 Coefficient of consolidation, $c_v$

Typical ranges of values of the coefficient of consolidation $c_v$, are shown in Table 6.8 (Duncan et al., 1990).

*Table 6.7* Poisson's ratio for geomaterials

| *Material* | *Poisson's ratio* $\nu$ |
|---|---|
| Saturated soils under undrained conditions | 0.5 |
| Soft clays under drained conditions | 0.3–0.4 |
| Stiff clays under drained conditions | 0.2–0.35 |
| Loose sands and silts under drained conditions | 0.3–0.4 |
| Dense sands and silts under drained conditions | 0.2–0.4 |
| Rocks | 0.10–0.35 (depends on rock type) |

*Table 6.8* Typical ranges of coefficient of consolidation $c_v$

| *Soil type* | *Coefficient of consolidation $c_v$ (m²/day)* |
|---|---|
| Coarse sand | >1000 |
| Fine sand | 10–1000 |
| Silty sand | 1–100 |
| Silt | 0.05–10 |
| Compacted clay | 0.005–0.5 |
| Soft clay | <0.02 |

Source: Duncan, J.M., Wright, S.G. and Wong, K.S. 1990. Slope stability during rapid drawdown. *Seed Memorial Symposium Proceedings*, BiTech Publishers, Vancouver, BC, Canada, Vol. 2, pp. 235–272.

## 6.8 PRELIMINARY ASSESSMENT OF FOUNDATION DEFORMATION PARAMETERS

Because of the differences in stress conditions around various types of foundations, care must be taken to use appropriate correlations for deformation parameters that reflect these stress conditions. For example, deformation parameters for vertically loaded shallow foundations will generally be different from those for vertically loaded pile foundations, while the values for laterally loaded foundations will differ from those for vertical loading.

Typical of the correlations that can be used for preliminary estimates of foundation deformation parameters are the following correlations derived from the work of Decourt (1982, 1995):

Soil Young's modulus below a raft or shallow foundation:

$$E_{sr} = 2N \text{ MPa} \tag{6.22}$$

Young's modulus along and below a pile (vertical loading):

$$E_s = 3N \text{ MPa} \tag{6.23}$$

where N is the SPT value.

Correlations of Young's modulus with CPT data have also been suggested. For example, for shallow foundations on sand, Schmertmann (1970) suggests the following relationship betwen Young's modulus $E_s$ and static cone resistance $q_c$:

$$E_s = 2q_c \tag{6.24}$$

For axially loaded piles, Poulos (1989) has suggested the following very rough approximation:

$$E_s = \eta \cdot q_c \tag{6.25}$$

where $\eta = 5$ for normally consolidated sands, 7.5 for over-consolidated sands and 15 for clays.

For lateral response analyses of piles, the above correlations need to be modified, and as a first approximation, the Young's modulus values for vertical loading should be reduced by multiplying by a factor of 0.7, to allow for the greater soil strain levels arising from lateral loading.

For piles in rock, it is common to correlate design parameters with the unconfined compressive strength, $q_u$, at least for preliminary purposes. Young's modulus for vertical loading, $E_{sv}$ can be roughly estimated as follows:

$$E_{sv} = a(q_u)^b \tag{6.26}$$

where a varies between about 100 and 500 for a wide range of rocks, and b is generally taken as 1.0.

In employing such correlations, it should be recognised that, in the field, they may be influenced by geological features and structure that cannot be captured by a small and generally intact rock sample. Nevertheless, in the absence of other information, such correlations provide at least an indication of the order of magnitude.

More detailed correlations for rock mass modulus are provided by Hoek and Diederichs (2006), who relate the rock mass modulus to the geological strength index, GSI, and a disturbance factor that reflects the geological structure.

Zhang (2010) has suggested that the effects of geological structure can be incorporated via a factor $\alpha_E$ related to the rock quality designation, RQD (%). The following relationship is proposed for the rock mass modulus, $E_m$:

$$E_m = \alpha_E \cdot E_r \tag{6.27}$$

where $E_r$ is the modulus of the intact rock, and $\alpha_E = 0.0231 \cdot \text{RQD} - 1.32 \geq 0.15$.

## 6.9 DEFORMATION PARAMETERS DERIVED FROM SMALL-STRAIN SHEAR MODULUS

### 6.9.1 Introduction

The small-strain value of shear modulus, $G_0$, is being recognised increasingly as a valuable indicator of the deformability of geomaterials (e.g., Mayne, 2001; Mayne et al., 2009). It is particularly attractive because it can be related to the in situ shear wave velocity, $V_s$, (Equation 6.2), which in turn can be measured directly by a variety of methods, including the seismic CPT test and cross-hole or down-hole geophysics.

For preliminary assessments of small-strain shear modulus, where measured values of shear wave velocity $V_s$ may not be available, Table 6.9 gives some typical ranges of values of $V_s$ for various geomaterials. Typical values of mass density are shown in Table 6.10. It should

*Table 6.9* Typical values of shear wave velocity $V_s$

| *Material type* | *Typical range of $V_s$ (m/s)* |
|---|---|
| Very soft soils | 50–100 |
| Soft soils | 100–200 |
| Stiff clays | 200–375 |
| Gravels, soft rocks | 375–700 |
| Weathered rocks | 700–1400 |
| Hard rocks | 1400+ |

*Table 6.10* Typical values of mass density

| | *Mass density (t/m³)* | |
|---|---|---|
| *Soil type* | *Poorly graded* | *Well graded* |
| Loose sand | 1.75 | 1.85 |
| Dense sand | 2.00 | 2.10 |
| Soft clay | 1.75 | 1.75 |
| Stiff clay | 2.05 | 2.05 |
| Silty soils | 1.75 | 1.75 |
| Gravelly soils | 2.05 | 2.15 |

be noted that, to derive the small-strain shear modulus in MPa from $V_s$, the mass density needs to be expressed in kt/m³, that is, the values in Table 6.10 must be divided by 1000.

This section indicates firstly how $G_0$ values can be modified to obtain a first estimate of secant modulus values of Young's modulus, which may be useful for Category 2 analysis and design procedures. It then describes how $G_0$ can be used to estimate the short-term and long-term values of Young's modulus of geomaterials, taking into account their non-linear behaviour. This latter approach may be relevant to both Category 2 and Category 3 analysis and design procedures.

### 6.9.2 Estimation of secant values of soil modulus for foundation analysis

For application to routine design, allowance must be made for the reduction in the shear modulus because of the relatively large strain levels that are relevant to foundations under normal serviceability conditions. As an example, Poulos et al. (2001) have suggested the reduction factors shown in Figure 6.6 for the case of clay soils where $G_0/s_u = 500$ ($s_u$ = undrained shear strength). This figure indicates that

- The secant modulus for axial loading may be about 20%–40% of the small-strain value for a practical range of factors of safety
- The secant modulus for lateral loading is smaller than that for axial loading, typically by about 30% for comparable factors of safety

Haberfield (2013) has demonstrated that, when allowance is made for strain level effects, modulus values derived from geophysical tests can correlate well with those from pressuremeter tests (PMTs). He shows an example from a project in Dubai in which a reduction factor of 0.2 was applied to the small-strain modulus values derived from cross-hole seismic test results. The modulus values so derived were found to be consistent with values obtained from subsequent pile load tests.

### 6.9.3 Development of parameter relationships

#### 6.9.3.1 Elastic materials

From the basic theory of elasticity, Young's modulus, E, can be related to the shear modulus G, as follows:

$$E = 2(1 + \nu)G \tag{6.28}$$

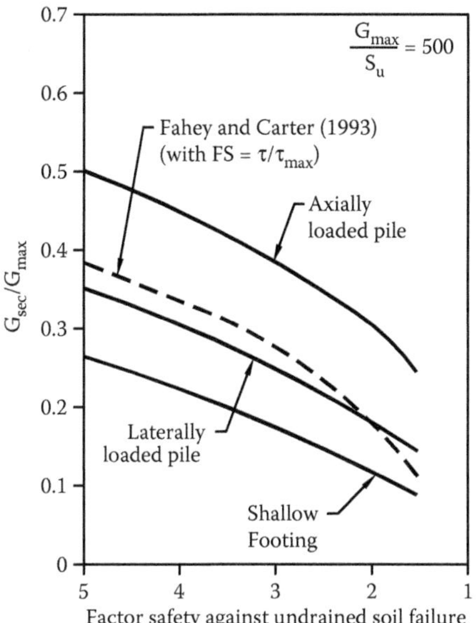


*Figure 6.6* Example of ratio of secant shear modulus to small-strain value. (Adapted from Poulos, H.G., Carter, J.C. and Small, J.C. 2001. Foundations and retaining structures – Research and practice. *Theme Lecture, Proceedings of the 15th International Conference on Soil Mechanics and Geotechnical Engineering*, Istanbul, Balkema, Vol. 4, pp. 2527–2606.)

Alternatively, E can be related to G and the bulk modulus K as follows:

$$E = \frac{9G}{[3 + G/K]} \tag{6.29}$$

The bulk modulus K can in turn be related to the constrained modulus D as

$$K = \frac{(1+\nu)D}{[3(1+\nu)]} \tag{6.30}$$

Thus, from the above equations, E can be related to G and D as

$$E = \frac{3G}{[1 + b \cdot G/D]} \tag{6.31}$$

where

$$b = \frac{(1-\nu)}{(1+\nu)} \tag{6.32}$$

For an ideal elastic soil, D can be related to Young's modulus and Poisson's ratio as

$$D = \frac{(1-\nu)E}{[(1+\nu)\cdot(1-2\nu)]} \tag{6.33}$$

For real soils, the effects of non-linear behaviour on G and D can be incorporated, as set out below.

## 6.9.4 Representation of non-linear behaviour

### 6.9.4.1 Shear modulus G

Non-linear soil behaviour can be introduced conveniently via the following approximate relationship between shear modulus G and shear stress level (Fahey and Carter, 1993; Mayne, 2001):

$$G = G_0 R_f \tag{6.34}$$

where

$$R_f \text{ is the stress level factor} = \left(1 - f \cdot \tau_{rat}^{g}\right), \tag{6.35}$$

$\tau_{rat}$ the shear stress ratio $= \tau/\tau_u$, $\tau$ the shear stress, $\tau_u$ the ultimate shear stress and f, g are empirical parameters, with typical values for many soils being $f = 1.0$ and $g = 0.3$.

### 6.9.4.2 Constrained modulus D

For soils and geomaterials in which the volume compressibility may be significant, the constrained modulus, D, can be expressed via Equation 6.19, developed by Janbu (1963).

Correlations between the modulus number, m, and porosity, n, are shown in Figures 6.4 and 6.5.

## 6.9.5 Undrained and drained Young's moduli

### 6.9.5.1 Undrained Young's modulus $E_u$

For a saturated soil under undrained conditions, the bulk modulus K and the constrained modulus D, are infinite, and Poisson's ratio, $\nu_u$, is 0.5. Thus, from Equations 6.34 and 6.35:

$$E_u = 3G_0 \cdot R_f \tag{6.36}$$

Figure 6.7 plots the ratio $E_u/G_0$ as a function of shear stress ratio, for the parameters $f = 1.0$ and $g = 0.3$. As would be expected, this ratio decreases as the shear stress level increase toward failure. At a shear stress ratio of 0.4, representing a factor of safety of 2.5, $E_u/G_0$ is about 0.7.

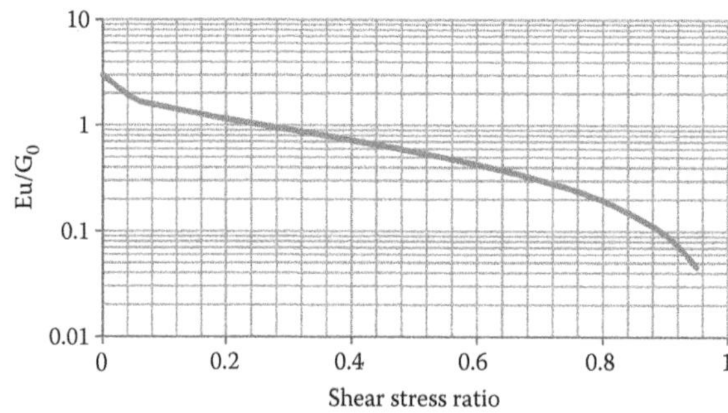


*Figure 6.7* Undrained Young's modulus ratio, $E_u/G_0$, related to shear stress level.

#### 6.9.5.2 *Drained Young's modulus* $E'$

Under drained conditions, Poisson's ratio $\nu = \nu'$ (the effective stress value of Poisson's ratio), and the drained Young's modulus can be related to $G_0$ as follows:

$$E' = \frac{3G_0 \cdot R_f}{[1 + [b \cdot Rf \cdot G_0/D]]} \quad (6.37)$$

Thus, via this formulation, non-linearity is expressed via the factor $R_f$, which reflects the non-linearity in shear behaviour, and via the constrained modulus, D, which reflects the dependence of the volumetric behaviour on the effective stress state.

### 6.9.6 Application to geomaterials

#### 6.9.6.1 *Coarse-grained soils*

For most real soils, it is generally inappropriate to approximate the stress–strain behaviour via a simple non-linear relationship, as the volumetric and shear deformations are not related directly via the elastic relationships. It is preferable to consider each component of deformation separately.

Thus, for coarse-grained soils, from Equations 6.31, 6.19 and 6.34, the drained Young's modulus $E'$ is given by

$$E' = \frac{3G_0 \cdot R_f}{[1 + \{bR_f G_0/m \cdot p_a(\sigma'_v/p_a)^{0.5}\}]} \quad (6.38)$$

Figure 6.8 shows the relationship between $E'/G_0$ and m, for a soil with $\nu = 0.3$, $G_0/p_a = 1000$, and $\sigma'_v/p_a = 1.0$. The following points may be noted:

1. The ratio $E'/G_0$ decreases as the modulus number, m, decreases, that is, as the volume compressibility of the soil increases. Clearly, volume compressibility plays a very important role.
2. $E'/G_0$ decreases as the shear stress ratio increases.
3. A soil does not become incompressible until the modulus number, m, is about 3000 or more, depending on the shear stress ratio.

#### 6.9.6.2 *Clays and fine-grained soils*

For fine-grained soils, from Equations 6.31, 6.36 and 6.21, the drained Young's modulus $E'$ can be related to $G_0$ as follows:

$$E' = \frac{3G_0 \cdot Rf}{[1 + \{b \cdot CR \cdot (G_0/\sigma'_{vo}) \cdot R_f/2.3\}]} \quad (6.39)$$

Figure 6.9a and b plots the ratio $E'/G_0$ for values of $G_0/\sigma'_{vo} = 500$ and 1000, respectively, and for various values of CR. The following characteristics can be noted:

1. As CR increases, the ratio $E'/G_0$ decreases.
2. $E'/G_0$ decreases as $G_0/\sigma'_{vo}$ increases.
3. As the shear stress ratio increases, $E'/G_0$ decreases, but the variation is relatively small, especially for larger values of CR.

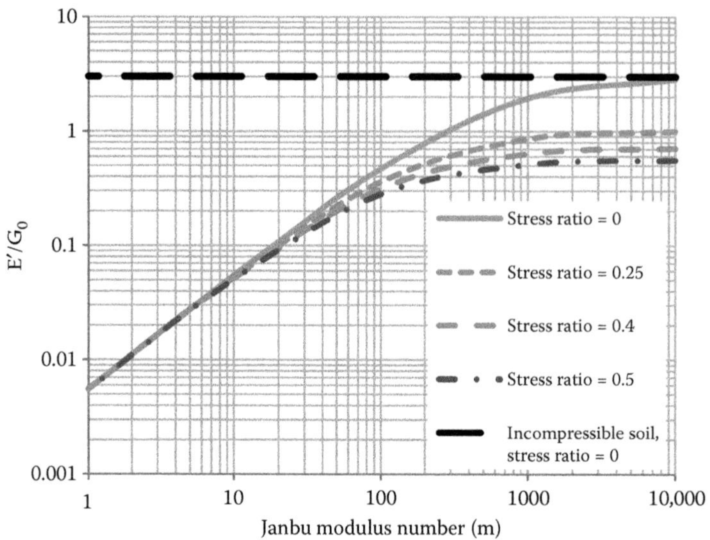


*Figure 6.8* Variation of Young's modulus with Janbu modulus number, m.

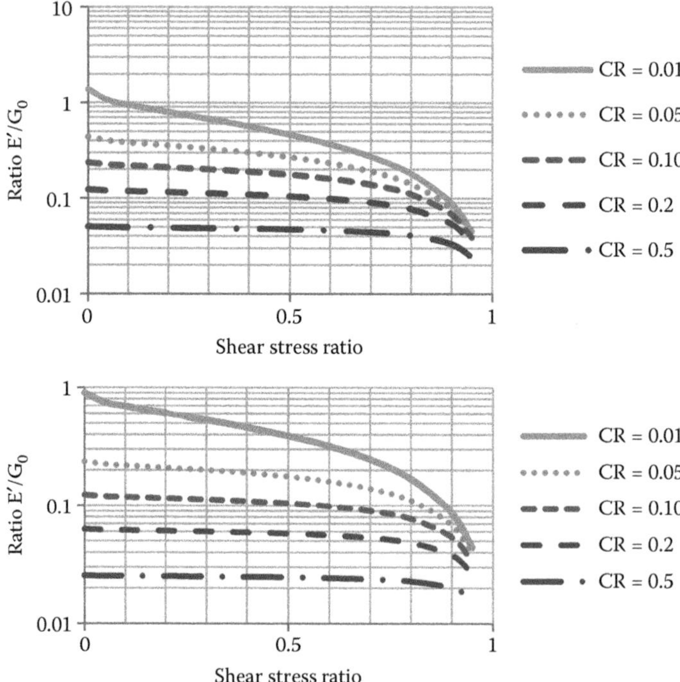


*Figure 6.9* Ratio of drained Young's modulus to small-strain shear modulus versus shear stress ratio. (a) $G_0/\sigma'_{vo} = 500$ and (b) $G_0/\sigma'_{vo} = 1000$.

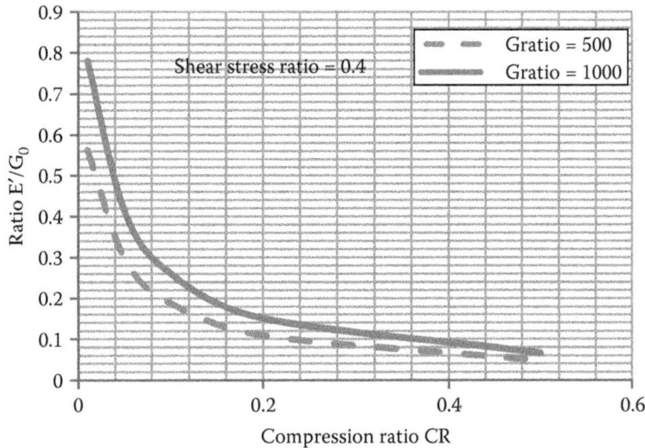


*Figure 6.10* Dependence of $G_0/\sigma'_{vo}$ on CR, for shear stress ratio = 0.4.

For a shear stress ratio of 0.4, corresponding to an overall factor of safety of 2.5, Figure 6.10 plots the variation of $E'/G_0$ with CR. The critical importance of CR is amply demonstrated in this figure.

#### *6.9.6.3 Ratio of drained to undrained moduli*

From Equations 6.36 and 6.39, the ratio of the drained to undrained moduli, $E'/E_u$, is given by

$$\frac{E'}{E_u} = \frac{1}{[1 + \{b \cdot CR \cdot (G_0/\sigma'_{vo}) \cdot R_f/2.3\}]} \tag{6.40}$$

For an ideal elastic soil, the ratio $E'/E_u$ can be shown to be as follows:

$$\frac{E'}{E_u} = \frac{(1+\nu')}{1.5} \tag{6.41}$$

Figure 6.11 plots the ratio $E'/E_u$ as a function of shear stress ratio, $\tau_{rat,}$ for various values of the CR, for $\nu' = 0.3$ and a $G_0/\sigma'_{vo} = 500$. Also shown is the relationship for an ideal elastic soil.

The following points can be noted:

1. $E'/E_u$ increases as the shear stress ratio increases, reflecting the marked decrease in $E_u$ as failure is approached.
2. $E'/E_u$ decreases substantially as the CR increases.
3. For an ideal elastic soil, by definition, the ratio $E'/E_u$ does not depend on the shear stress ratio, and at low shear stress levels, is larger than the other cases plotted in Figure 6.11.

#### *6.9.6.4 Rocks*

Rocks may be considered in a similar manner to coarse-grained soils. However, the volumetric component of deformation in rocks may often be small, and accordingly, it appears

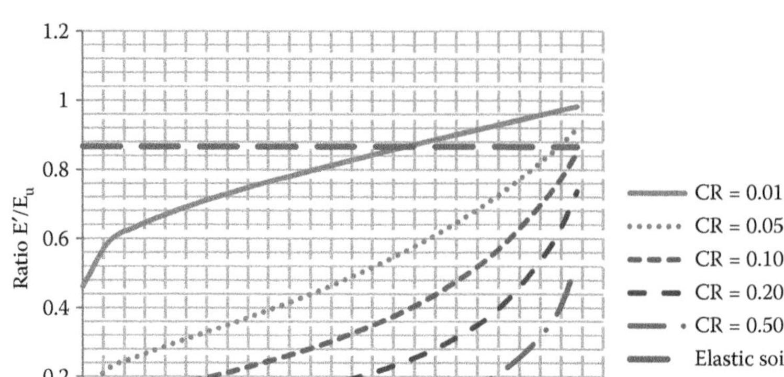


*Figure 6.11 Ratio of drained to undrained Young's moduli, for $G_0/\sigma'_{vo} = 500$ and $\nu' = 0.3$.*

reasonable to assume that the behaviour of rocks can be represented by a non-linear elastic model based on the shear deformations. The drained Young's modulus, E′, for such materials can then be approximated as follows:

$$E' = 2(1+\nu')G_0 \cdot R_f \tag{6.42}$$

The great advantage of using $G_0$ in this case is that defects in the rock mass will be reflected in the measured shear wave velocity, so that it is not then necessary to try and estimate the mass modulus of the rock by making corrections to laboratory-measured values of the material modulus of intact samples for the influence of joints, discontinuities and defects.

### 6.9.7 Application to foundation movement estimation

The relationships set out above enable an assessment to be made of values of both undrained and drained Young's modulus on the basis of assessed values of $G_0$ and D. Such relationships can be useful in practice as both $G_0$ and D are relatively 'stable' parameters that can be measured readily via field or laboratory tests. In addition, to incorporate the effects of non-linearity, an assessment of the shear strength is required so that the shear stress ratio can be estimated for the particular problem in hand.

Ideally, $G_0$ and D (together with CR and Poisson's ratio $\nu$) should be obtained from appropriate laboratory and/or field tests. However, there are many cases in which the results of such testing are not available, and in such cases, use can be made of empirical correlations with in situ penetration test data or laboratory index data. A limited number of these correlations have been summarised in Section 6.7.

One important implication of the results herein is that, for an over-consolidated soil, the ratio $E'/G_0$ will be greater than for the same soil in a normally consolidated state, because the constrained modulus for the over-consolidated state is typically 5–10 times that in the normally consolidated state.

In estimating the movement of various foundation types using the theory of elasticity as a basis, it is critical to estimate appropriate values of the Young's modulus of the soils supporting the foundation, taking into account the level of stress or strain within the supporting soil. There are at least two approaches that can be adopted in pursuing this objective:

1. Adopting overall average modulus values that reflect the average level of stress or strain within the soil. As an example, Mayne (2001) has reported consistent success in predicting the load–settlement behaviour of axially loaded piles by adopting the shear stress ratio as the inverse of the factor of safety against failure and applying the Fahey and Carter (1993) degradation expression (Equation 6.34). As an alternative, Poulos et al. (2001) developed an approach that used estimates of the strain levels imposed on the soil by the foundation to obtain values of shear modulus and Young's modulus for the level of applied load, using empirical representations of the degradation function for shear modulus, $G/G_0$.
2. Using an iterative procedure for estimating the detailed distribution of modulus within the soil mass as a function of stress or strain at various points within the soil. In this case, initial estimates of the modulus would be required (e.g. 70% of the small–strain values) to start the analysis, and these values would be progressively altered to correspond with the computed stress or strain levels within the soil. The process would be repeated until convergence is obtained between assumed modulus values and the stress or strain levels.

Clearly, the first alternative is the more attractive from a practical viewpoint because it can be undertaken without the detailed numerical analysis that the second approach would require.

#### 6.9.7.1 Estimation of applied stress ratio

As a very simplified approximation, when dealing with the settlement of near-surface foundations on or in a layered soil profile, the applied stress ratio may be estimated as the ratio of the additional vertical stress in the layer due to the foundation, $\delta\sigma_v$, divided by the ultimate vertical pressure that can be sustained in that layer, $p_{ult}$.

$\delta\sigma_v$ can be estimated conveniently from elastic theory (e.g., via solutions in Poulos and Davis, 1974). $p_{ult}$ can be estimated from empirical expressions for shallow foundation bearing capacity, or via the theoretical procedures, as set out in Chapter 7.

When applying the approach to axially loaded piles, it is convenient to assume that the applied stress ratio is the inverse of the factor of safety for all layers along the pile shaft, and immediately below the pile tip. For layers below the pile tip, the reduction of stress with depth can again be approximated via elastic theory. In this way, some account, albeit approximate, can be taken of the lessening of non-linear effects with increasing depth.

### 6.9.8 Examples of application of the approach

Two examples of the application of the above approach to deformation parameters assessment are described below. Although neither case is directly related to tall tower foundations, the principles involved are equally applicable to such cases.

#### 6.9.8.1 Footing tests on clay at Bothkennar, UK

Footing tests were carried out at the SERC Bothkennar site on the Forth Estuary, approximately midway between Edinburgh and Glasgow, Scotland. Full details of the geotechnical conditions at the site have been given in the *Eighth Geotechnique Symposium* in Print in 1992, and details of the tests on rigid footings have been given by Jardine et al. (1995). Two

*Table 6.11* Summary of typical ground conditions at the Bothkennar site

| *Stratum* | *Depth range (m)* | *Soil type* | *Water content (%)* | *Liquid limit (%)* | *Plastic limit (%)* |
|---|---|---|---|---|---|
| 1 | 0–1 | Silty clay crust | 40 | 45 | 25 |
| 2 | 1.0–1.3 | Shelly silty clay | – | – | – |
| 3 | 1.3–2.2 | Soft clayey silt, some shell fragments | 50 | 48–70 | 20–30 |
| 4 | 2.2–7.0 | Soft silty clay | 60–75 | 58–78 | 25 |
| 5 | 7.0–11.2 | Soft silty clay | 50–75 | 62–86 | 22–30 |
| 6 | 11.2–19.4 | Firm silty clay | 40–70 | 52–78 | 22–30 |
| 7 | 19.4– | Coarse silty sand with shells | 20–30 | – | – |

footings (A and B) were tested, with footing A (2.2 m square) being loaded to failure relatively rapidly, while footing B (2.4 m square) was loaded rapidly to about 67% of the failure pressure of footing A, and then allowed to consolidate for a period of more than 2 years.

Table 6.11 summarises the ground conditions at the site. The water table was about 0.9 m below ground surface

For the calculations of footing behaviour, the upper 14 m of the profile was divided into nine layers, and average values of shear wave velocity, cone resistance and CR were obtained from the available field and laboratory data. Details of the geotechnical model adopted are shown in Table 6.12. The calculation process was as follows:

1. The ultimate bearing capacity was estimated on the basis of static cone resistance values.
2. The constrained modulus D for each layer was obtained from values of CR measured in laboratory tests.
3. The undrained Young's modulus was obtained from Equation 6.36.
4. The drained Young's modulus was obtained from Equation 6.39.
5. The drained Poisson's ratio was taken to be 0.35.
6. Calculations were carried out for both the undrained and drained load–settlement relationships, for both footings, using the relevant modulus values at each load level.

Figure 6.12 shows the computed undrained and drained load–settlement relationships for footing A, which was loaded rapidly. Also shown is the measured load–settlement curve, and it will be seen that this curve agrees reasonably well with the computed undrained load–settlement relationship.

*Table 6.12* Geotechnical model for footing analysis

| *Layer* | *Thickness (m)* | *Average shear wave velocity $V_s$ (m/s)* | *Average cone resistance $q_c$ (MPa)* | *Average CR* |
|---|---|---|---|---|
| 1 | 1 | 90 | 0.50 | 0.05 |
| 2 | 1 | 90 | 0.30 | 0.08 |
| 3 | 1 | 76 | 0.35 | 0.35 |
| 4 | 1 | 88 | 0.30 | 0.32 |
| 5 | 2 | 100 | 0.35 | 0.38 |
| 6 | 2 | 110 | 0.45 | 0.40 |
| 7 | 2 | 120 | 0.52 | 0.42 |
| 8 | 2 | 130 | 0.65 | 0.43 |
| 9 | 2 | 140 | 0.75 | 0.45 |

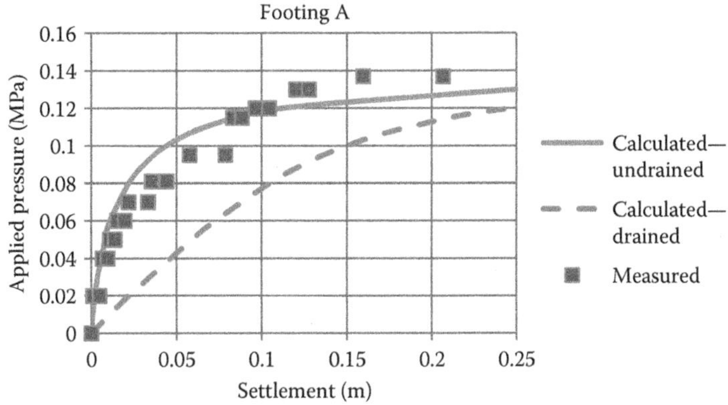


*Figure 6.12* Calculated and measured load–settlement behaviour for footing A.

Figure 6.13 shows the computed and measured load–settlement behaviour for footing B, which was loaded rapidly to an applied pressure of 90 kPa and then allowed to settle for an extended period. In this case, the measured settlement during the initial rapid loading is in quite good agreement with, but slightly larger than, the calculated undrained load–settlement behaviour, probably reflecting some small amount of consolidation settlement during the loading process. The measured settlement after 2 years is approaching the calculated total final settlement from the drained analysis. Jardine et al. (1995) judged that the end of primary consolidation would have been completed after about 10,000 h (about 14 months), but they did state that the primary consolidation and creep phases are hard to distinguish. Overall, the agreement between the calculated and measured behaviour of both footings at the Bothkennar site appears to be quite satisfactory.

#### 6.9.8.2 Pile load test, Opelika, USA

Mayne (2001) presented the results of a pile load test carried out at the Opelika test site in Alabama, USA. The test pile was a drilled shaft 11.0 m long and 0.914 m in diameter. The

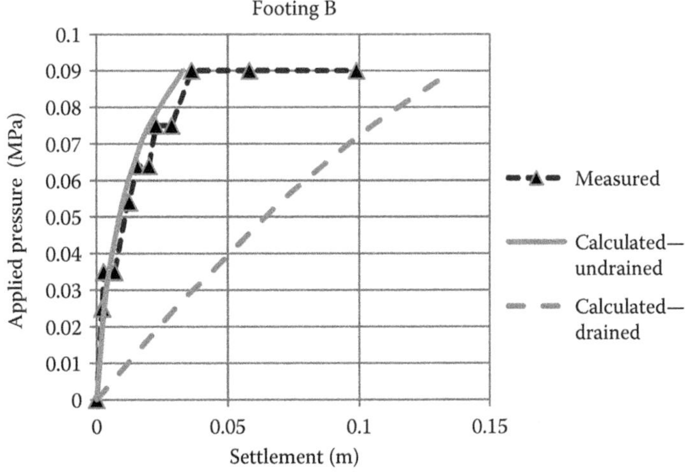


*Figure 6.13* Calculated and measured load–settlement behaviour for footing B.

*Table 6.13* Geotechnical model used for Opelika test pile

| *Layer* | *Depth range (m)* | *Soil description* | *Average Shear wave velocity $V_s$ (m/s)* | *Average cone resistance $q_c$ (MPa)* |
|---|---|---|---|---|
| 1 | 0–3 | Residual soil | 180 | 2.0 |
| 2 | 3–6 | " | 200 | 3.0 |
| 3 | 6–11 | " | 210 | 4.0 |
| 4 | 11–13 | " | 270 | 5.0 |

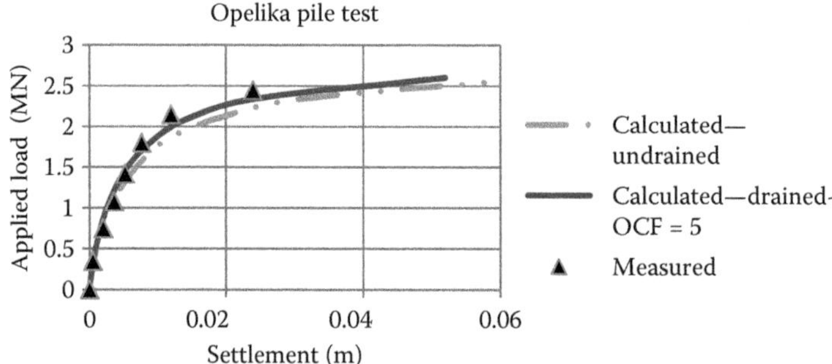


*Figure 6.14* Computed and measured load–settlement behaviour for Opelika pile load test.

ultimate axial capacity was estimated to be 2.9 MN, and this was assumed to remain the same for both undrained and drained loading.

The site consisted of silty to sandy residual soils that graded into partially weathered schist and gneiss. The water table was 2–3 m below ground surface, and seismic piezocone data was obtained at the site. From the cone resistance and shear wave velocity measurements, the simplified geotechnical model shown in Table 6.13 was derived.

From the material descriptions and the cone resistance data, the profile was clearly over-consolidated, and so it was assumed that the confined modulus of the over-consolidated soil strata was 5 times that of the soils in a normally consolidated state.

Calculations were carried out for the load–settlement relationship for the test pile, using the simplified expressions derived by Randolph and Wroth (1978) (see Chapter 8). Both undrained and drained conditions were considered, and in each case, the undrained Young's modulus was derived from the small-strain shear modulus, via the shear wave velocity, while the drained modulus also made use of the constrained modulus derived from the cone test data. These modulus values were used in the Randolph and Wroth equations for each of the applied loads considered.

Figure 6.14 shows the computed load–settlement curves and also the measured curve reported by Mayne (2001). It can be seen that there is relatively little difference between the undrained and drained load–settlement curves, and that both agree reasonably well with the measured relationship.

The closeness of the curves for undrained and drained loading is due largely to the over-consolidated state of the soil. Larger differences would be evident in the case of an initially normally consolidated soil, although possible installation effects on the state of the soil adjacent to the pile would then need to be considered.

Chapter 7

# Design for ultimate limit state

## 7.1 INTRODUCTION

A key aspect of the design of foundations for tall buildings is the assessment of requirements to ensure that the foundation system has adequate capacity and stability to withstand all the loads and load combinations that it may be subjected to. There must be adequate capacity to safely resist the imposed vertical and lateral loads, and adequate rotational resistance to withstand the applied moment and torsional loads. For low-rise buildings, emphasis has been based traditionally on vertical bearing capacity, and this too is a most important consideration for high-rise buildings. However, overall stability is also a critical issue in the latter case, and so must be given appropriately detailed consideration in design.

In this chapter, the framework for strength and stability design will be described, both in terms of the traditional approach that uses an overall factor of safety, and then via the limit state design approach which is now becoming more dominant. Consideration will then be given, in turn, to shallow foundations, pile foundations and piled raft foundations. In each case, Category 1, 2 and 3 methods will be described, with more detailed consideration being given to Category 2 methods. Finally, issues related to the durability design of foundations will be addressed.

## 7.2 TRADITIONAL FACTOR OF SAFETY APPROACH

In this approach, the geotechnical design criterion can be expressed as follows:

$$P_{all} = R_u/FS \tag{7.1}$$

where $P_{all}$ is the allowable load (for the applied loading being considered), $R_u$ the ultimate load capacity (for the applied loading being considered) and FS the overall factor of safety.

In estimating $R_u$, consideration must be given to the likely modes of failure of the foundation.

For a piled or piled raft foundation system, both the 'single pile' mode and the 'block' mode should be considered, and the lesser value taken.

In the traditional method, the elements of uncertainty are lumped into a single factor of safety FS, which is typically between 2 and 3. Table 7.1 summarises values that are often employed in practice in the United States and elsewhere.

Despite the many limitations of such an approach, it is still widely employed in engineering practice in many countries, and specific design values often appear in national codes or

*Table 7.1* Typical values of conventional factor of safety for pile design

| *Method of estimating capacity* | *Loading condition* | *Minimum factor of safety* | |
|---|---|---|---|
| | | *Compression* | *Tension* |
| Theoretical or empirical, to be verified by pile load test | Usual | 2.0 | 2.0 |
| | Unusual | 1.5 | 1.5 |
| | Extreme | 1.15 | 1.15 |
| Theoretical or empirical. To be verified by pile driving analyser | Usual | 2.5 | 3.0 |
| | Unusual | 1.9 | 2.25 |
| | Extreme | 1.4 | 1.7 |
| Theoretical or empirical, not verified by load test | Usual | 3.0 | 3.0 |
| | Unusual | 2.25 | 2.25 |
| | Extreme | 1.7 | 1.7 |

standards. However, it is generally unsuitable for tall building design as it cannot be readily applied to assess the overall stability of a foundation system subjected to a combination of vertical, lateral and moment loadings. For this reason, a limit state design approach is more appropriate, and is discussed below.

## 7.3 LIMIT STATE DESIGN APPROACH

There is an increasing trend for limit state design principles to be adopted in foundation design, for example, in the Eurocode 7 requirements and those of the Australian Piling Code (AS2159-2009). In terms of limit state design using a load and resistance factor design (LRFD) approach, the design criteria for the ultimate limit state, for both structural and geotechnical design, are as follows:

$$R_{ds} \geq E_d \tag{7.2}$$

$$R_{dg} \geq E_d \tag{7.3}$$

where $R_{ds}$ is the design structural strength $= \phi_s \cdot R_{us}$, $R_{dg}$ the design geotechnical strength $= \phi_g \cdot R_{ug}$, $R_{us}$ the ultimate structural strength, $R_{ug}$ the ultimate geotechnical strength (capacity), $\phi_s$ the structural reduction factor, $\phi_g$ the geotechnical reduction factor and $E_d$ the design action effect (factored load combinations).

The above criteria in Equations 7.2 and 7.3 are applied to the entire foundation system, while the structural strength criterion (Equation 7.2) is also applied to each individual pile. However, it is not considered to be a good practice to apply the geotechnical criterion (Equation 7.3) to each individual pile within the group, as this can lead to considerable overdesign (Poulos, 1999).

$R_{dg}$ and $R_{ds}$ can be obtained from the estimated ultimate structural and geotechnical capacities, multiplied by appropriate reduction factors. The estimation of the ultimate geotechnical capacity $R_{ug}$, for various foundation types, is set out later in this chapter.

The required load combinations for which the structure and foundation system have to be designed will usually be dictated by an appropriate structural loading code, as discussed in Chapter 5. In some cases, a large number of combinations may need to be considered.

### 7.3.1 Estimation of geotechnical reduction factor $\phi_g$

Values of the structural and geotechnical reduction factors are often specified in national codes or standards. The selection of suitable values of $\phi_g$ requires considerable judgement and should take into account a number of factors that may influence the foundation performance. As an example, the now-superseded Australian Piling Code AS2159-1995 specifies values of $\phi_g$ between 0.4 and 0.9, the lower values being associated with greater levels of uncertainty and the higher values being relevant when a significant amount of load testing is carried out.

A later version of this standard, AS2159-2009, employs a risk assessment approach to arrive at an appropriate geotechnical reduction factor, depending on a number of issues, as follows:

- The geological complexity of the site
- The extent of ground investigation
- The amount and quality of geotechnical data
- Experience with similar foundations in similar geological conditions
- The method of assessment of geotechnical parameters for design
- The design method adopted
- The method of utilising the results of in situ test data and pile installation data
- The level of construction control
- The level of performance monitoring of the supported structure during and after construction

Each of these factors is given a subjective risk rating, ranging between 1 for very low risk, to 5 for very high risk. The individual risk ratings are weighted via an importance factor for that factor, and then an average risk rating (again between 1 and 5) is computed from the sum of the individual weighted risk factors. The higher the average risk rating, the lower the geotechnical reduction factor. Some benefit is derived by having a high redundancy foundation system, for example, a large group of piles, or a piled raft foundation. Load testing provides further benefits and leads to a higher $\phi_g$ value, that is, a less conservative design.

$\phi_g$ can typically range between 0.4 for conservative designs involving little or no pile testing and where uncertain ground conditions prevail, to 0.8 for cases in which a significant amount of testing is carried out and the ground conditions and design parameters have been carefully assessed.

## 7.4 DESIGN FOR CYCLIC LOADING

In addition to the normal design criteria, as expressed by Equations 7.2 and 7.3, it is suggested that an additional criterion should be imposed for the piled foundation of a tall building to cope with the effects of repetitive loading from wind, seismic or wave action, as follows:

$$\eta R_{gs} \geq E_c \tag{7.4}$$

where $R_{gs}$ is the ultimate geotechnical shaft capacity, $E_c$ the maximum half-amplitude of cyclic wind loading and $\eta$ the cyclic load ratio.

This criterion attempts to avoid the full mobilisation of shaft friction along the piles, thus reducing the risk that cyclic loading will lead to a degradation of shaft capacity. $E_c$ can be obtained from computer analyses which gave the cyclic component of load on each pile, for various wind loading cases.

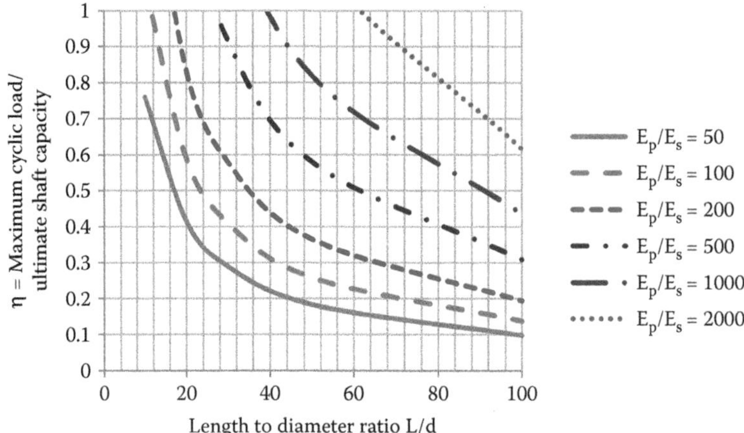


*Figure 7.1* Maximum cyclic load ratio η for piles. (Adapted from Randolph, M.F. 1983. Design considerations for offshore piles. *ASCE Special Conference on Geotechnical Practice in Offshore Engineering*, Austin, pp. 422–439.)

For the Emirates project in Dubai (Poulos and Davids, 2005), η was selected as 0.5, based on laboratory data from laboratory CNS tests. The simplified expression developed by Randolph (1983) may also be used to estimate the required value of η. Using this expression, the results shown in Figure 7.1 have been obtained. These results show that

1. As the length to diameter, L/d, of the piles increases, η tends to reduce
2. As the relative stiffness of the piles (expressed as the ratio of the Young's modulus of the pile material to the average Young's modulus of the soil layer) increases, η tends to increase

For typical cases of piles supporting tall buildings, L/d lies within the range 20–50, and the ratio of pile to soil modulus ($E_p/E_s$) lies within the range 100–500. Thus, from Figure 7.1, within these ranges of L/d and $E_p/E_s$, η lies between about 0.2 and 0.7. Thus, in the absence of a more specific evaluation, the value of 0.5 suggested above appears to be reasonable as an average value to employ.

## 7.5 SOIL–STRUCTURE INTERACTION ISSUES

When considering soil–structure interaction for the ultimate limit state (e.g. the bending moments in the raft of a piled raft foundation system), the worst response may not occur when the pile and raft capacities are factored downward. As a consequence, additional calculations may need to be carried out for geotechnical reduction factors both less than 1 and greater than 1. As an alternative to this duplication of analyses, it would seem reasonable to adopt a reduction factor of unity for the pile and raft resistances, and then factor up the computed moments and shears (e.g. by a factor of 1.5) to allow for the geotechnical uncertainties. The structural design of the raft and the piles will also incorporate appropriate reduction factors.

## 7.6 SUMMARY OF DESIGN ANALYSIS PROCESS

A summary of the analyses that are recommended to be carried out for building foundation design are shown in Table 7.2. These analyses involve various combinations of factored/

*Table 7.2* Summary of design analyses

| *Case* | *Purpose* | *Factor applied to geotechnical strength parameters* | *Load case* | *Comment* |
|---|---|---|---|---|
| i | Geotechnical design capacity | $\phi_g$ | ULS | Geotechnical reduction factor, $\phi_g$, applied to geotechnical capacity parameters to assess overall stability of the pile group |
| ii | Structural design capacity | 1.0 | ULS | Unfactored geotechnical parameters are adopted to assess maximum pile axial load and pile bending moment using short-term pile modulus |
| iii | Serviceability | 1.0 | SLS | Unfactored geotechnical parameters are adopted to assess pile head deflections and rotations |

unfactored geotechnical strengths and ultimate limit state (ULS) or serviceability limit state (SLS) loadings. SLS design is considered in detail in Chapter 8.

## 7.7 ESTIMATION OF ULTIMATE CAPACITY OF SHALLOW AND RAFT FOUNDATIONS

### 7.7.1 Category 1 (empirical) methods

Methods in this category include correlations of ultimate bearing capacity $p_{ur}$, under purely vertical loading, with SPT, including the following:

$$p_{ur} = K_1 \cdot N_r \text{ kPa} \tag{7.5}$$

where $N_r$ is the average SPT ($N_{60}$) value within depth of one-half of the footing or raft width and $K_1$ the factor shown in Table 7.3.

A further correlation, this time between the ultimate bearing capacity of a square or circular footing or raft and CPT data is as follows (MELT, 1993):

$$p_{ur} - a_1[1 + a_2 \cdot D/B]q_c + q_0 \tag{7.6}$$

where $a_1$, $a_2$ are parameters depending on soil type and condition (Table 7.4), $q_0$ the overburden pressure at level of base, $q_c$ the measured cone tip resistance, D the depth of embedment below surface and B the width of footing or raft.

*Table 7.3* Correlation factor $K_1$

| *Soil type* | *$K_1$ (Raft)* |
|---|---|
| Sand | 90 |
| Sandy silt | 80 |
| Clayey silt | 80 |
| Clay | 65 |

Source: After Decourt, L. 1995. Prediction of load-settlement relationships for foundations on the basis of the SPT-T. Ciclo de *Conferencias International "Leonardo Zeevaert"*, UNAM, Mexico, pp. 85–104.

*Table 7.4* Parameters $a_1$ and $a_2$ for ultimate bearing capacity of square shallow footings and rafts

| *Soil type* | *Condition* | $a_1$ | $a_2$ |
|---|---|---|---|
| Clay, silt | All | 0.32 | 0.35 |
| Sand, gravel | Loose | 0.14 | 0.35 |
| | Medium | 0.11 | 0.50 |
| | Dense | 0.08 | 0.85 |
| Chalk | – | 0.17 | 0.27 |

Source: MELT. 1993. Regles techniques de conception et de calcul des fondations des ouvrages de genie civil. Cahier des Clauses Techniques Generales applicables aux Marches Publics de Travaux, Fascicule 62 – Titre V, Ministere de l'Equipement du Logement et des Transports, Paris.

### 7.7.2 Category 2 (simplified) methods

The classical Terzaghi bearing capacity theory forms the basis of most Category 2 methods of estimating ultimate bearing capacity. An extension to the original theory was suggested by Vesic (1975) and has now gained widespread acceptance in foundation engineering practice. This method takes some account of the stress-deformation characteristics of the soil and is applicable over a wide range of soil behaviour. It is based on the solutions obtained from the theory of plasticity, but empiricism has been included in significant measure, to deal with the many complicating factors that make a rigorous solution for the capacity intractable.

For a shallow rectangular foundation the general bearing capacity equation, which is an extension of the expression first proposed by Terzaghi (1943) for the case of a central vertical load applied to a long strip footing, is usually written in the form:

$$q_u = \frac{Q_u}{BL} = cN_c\zeta_{cr}\zeta_{cs}\zeta_{ci}\zeta_{ct}\zeta_{cg}\zeta_{cd} + \frac{1}{2}B\gamma N_\gamma\zeta_{\gamma r}\zeta_{\gamma s}\zeta_{\gamma i}\zeta_{\gamma t}\zeta_{\gamma g}\zeta_{\gamma d} + qN_q\zeta_{qr}\zeta_{qs}\zeta_{qi}\zeta_{qt}\zeta_{qg}\zeta_{qd} \qquad (7.7)$$

where $q_u$ is the ultimate bearing pressure that the soil can sustain, $Q_u$ the corresponding ultimate load that the foundation can support, B the least plan dimension of the footing, L the length of the footing, c the cohesion of the soil, q the overburden pressure and $\gamma$ is the unit weight of the soil. It is assumed that the strength of the soil can be characterised by a cohesion *c* and an angle of friction $\phi$.

The parameters $N_c$, $N_\gamma$ and $N_q$ are the general bearing capacity factors which determine the capacity of a long strip footing acting on the surface of soil represented as a homogeneous half-space. The factors $\zeta$ allow for the influence of other complicating features. Each of these factors has double subscripts to indicate the term to which it applies (c, $\gamma$ or q) and which phenomenon it describes (r for rigidity of the soil, s for the shape of the foundation, i for inclination of the load, t for tilt of the foundation base, g for the ground surface inclination and d for the depth of the foundation). Most of these factors depend on the friction angle of the soil, $\phi$.

In Table 7.5, closed-form expressions are presented for the bearing capacity factors. As noted above, some are only approximations. In particular, there have been several different solutions proposed in the literature for the bearing capacity factors $N_\gamma$ and $N_q$. Solutions by Prandtl (1921) and Reissner (1924) are generally adopted for $N_c$, and $N_q$, although Davis and Booker (1971) produced rigorous plasticity solutions which indicate that the commonly adopted expression for $N_q$ (Table 7.5) is slightly non-conservative, but it is generally accurate

*Table 7.5* Bearing capacity factors for shallow foundations

| *Parameter* | *Cohesion* | *Self-weight* | *Surcharge* |
|---|---|---|---|
| Bearing capacity | $N_c = (N_q - 1)\cot\phi$<br>$N_c = 2 + \pi \quad \text{if } \phi = 0$ | $N_\gamma \approx 0.0663e^{9.3\phi}$ Smooth<br>$N_\gamma \approx 0.1054e^{9.6\phi}$ Rough $\phi > 0$ in radians<br>$N_\gamma = 0$ if $\phi = 0$ | $N_q = e^{\pi\tan\phi}\tan^2\left(45^\circ + \frac{\phi}{2}\right)$ |
| Rigidity[a,b] | $\zeta_{cr} = \zeta_{qr} - \left(\frac{1-\zeta_{qr}}{N_c\tan\phi}\right)$<br>or for $\phi = 0$<br>$\zeta_{cr} = 0.32 + 0.12\left(\frac{B}{L}\right) + 0.60\log_{10}I_r$ | $\zeta_{\gamma r} = \zeta_{qr}$ | $\zeta_{qr} = \exp\left(\left(-4.4 + 0.6\frac{B}{L}\right)\tan\phi + \left[\frac{3.07\sin\phi\log_{10}2I_r}{(1+\sin\phi)}\right]\right)$ |
| Shape | $\zeta_{cs} = 1 + \left(\frac{B}{L}\right)\left(\frac{N_q}{N_c}\right)$ | $\zeta_{\gamma s} = 1 - 0.4\left(\frac{B}{L}\right)$ | $\zeta_{qs} = 1 + \left(\frac{B}{L}\right)\tan\phi$ |
| Inclination[c] | $\zeta_{ci} = \zeta_{qi} - \left(\frac{1-\zeta_{qi}}{N_c\tan\phi}\right)$<br>or for $\phi = 0$<br>$\zeta_{ci} = 1 - \left(\frac{nT}{cN_cB'L'}\right)$ | $\zeta_{\gamma i} = \left[1 - \frac{T}{N + B'L'c\cot\phi}\right]^{n+1}$ | $\zeta_{qi} = \left[1 - \frac{T}{N + B'L'c\cot\phi}\right]^{n}$ |
| Foundn. tilt[d] | $\zeta_{ct} = \zeta_{qt} - \left(\frac{1-\zeta_{qt}}{N_c\tan\phi}\right)$<br>or for $\phi = 0$<br>$\zeta_{ct} = 1 - \left(\frac{2\alpha}{\pi + 2}\right)$ | $\zeta_{\gamma t} = (1 - \alpha\tan\phi)^2$ | $\zeta_{qt} \approx \zeta_{\gamma t}$ |

*(Continued)*

*Table 7.5 (Continued)* Bearing capacity factors for shallow foundations

| *Parameter* | *Cohesion* | *Self-weight* | *Surcharge* |
|---|---|---|---|
| Surface inclination[e] | $\zeta_{cg} = \zeta_{qt} - \left(\frac{1-\zeta_{qt}}{N_c \tan\phi}\right)$ <br> or for $\phi = 0$ <br> $\zeta_{cg} = 1 - \left(\frac{2\omega}{\pi + 2}\right)$ | $\zeta_{\gamma g} \approx \zeta_{qg}$ <br> or for $\phi = 0$ <br> $\zeta_{\gamma g} = 1$ | $\zeta_{qg} = (1 - \tan\omega)^2$ <br> or for $\phi = 0$ <br> $\zeta_{qg} = 1$ |
| Depth[f] | $\zeta_{cd} = \zeta_{qd} - \left(\frac{1-\zeta_{qd}}{N_c \tan\phi}\right)$ <br> or for $\phi = 0$ <br> $\zeta_{cd} = 1 + 0.33\tan^{-1}\left(\frac{D}{B}\right)$ | $\zeta_{\gamma d} = 1$ | $\zeta_{qd} = 1 + 2\tan\phi\,(1-\sin\phi)^2 \tan^{-1}\left(\frac{D}{B}\right)$ |

[a] The rigidity index is defined as $I_r = G/(c + q\tan\phi)$ in which G is the elastic shear modulus of the soil and the vertical overburden pressure, q, is evaluated at a depth of B/2 below the foundation level. The critical rigidity index is defined as

$$I_{rc} = \frac{1}{2}\exp\left[(3.30 - 0.45B/L)\cot\left(45^\circ - \frac{\phi}{2}\right)\right]$$

[b] When $I_r > I_{rc}$, the soil behaves, for all practical purposes, as a rigid plastic material and the modifying factors $\zeta_r$ all take the value 1. When $I_r < I_{rc}$, punching shear is likely to occur and the factors $\zeta_r$ may be computed from the expressions in the table.

[c] For inclined loading in the B direction ($\theta = 90°$), n is given by $n = n_B = (2 + B/L)/(1 + B/L)$. For inclined loading in the *L* direction ($\theta = 0°$), *n* is given by

$$n = n_L = (2 + L/B)/(1 + L/B)$$

For other loading directions, n is given by $n = n_\theta = n_L\cos^2\theta + n_B\sin^2\theta$. $\theta$ is the plan angle between the longer axis of the footing and the ray from its centre to the point of application of the loading. B′ and L′ are the effective dimensions of the rectangular foundation, allowing for eccentricity of the loading, and T and N are the horizontal and vertical components of the foundation load.

[d] $\alpha$ is the inclination from the horizontal of the underside of the footing.

[e] For the sloping ground case where $\phi = 0$, a non-zero value of the term $N_\gamma$ must be used. For this case is $N_\gamma$ negative and is given by

$$N_\gamma = -2\sin\omega$$

where $\omega$ is the inclination below horizontal of the ground surface away from the edge of the footing.

[f] D is the depth from the soil surface to the underside of the footing.

enough for most practical applications. However, significant discrepancies have been noted in the values proposed for $N_\gamma$. It has not been possible to obtain a rigorous closed-form expression for $N_\gamma$, but several authors have proposed approximations. For example, Terzaghi (1943) proposed a set of approximate values and Vesic (1975) suggested the approximation, $N_\gamma \approx 2(N_q + 1)\tan\phi$, which has been widely used in geotechnical practice, but is now known to be non-conservative with respect to more rigorous solutions obtained using the theory of plasticity for a rigid plastic body (Davis and Booker, 1971). For values of friction angle in the typical range from 30° to 40°, Terzaghi's solutions can overestimate this component of the bearing capacity by factors as large as 3.

Analytical approximations to the Davis and Booker solutions for $N_\gamma$ for both smooth and rough footings are presented in Table 7.5. These expressions are accurate for values of $\phi$ greater than about 10°, the usual range of practical interest. It is recommended that the expressions derived by Davis and Booker, or their analytical approximations presented in Table 7.5, be used in practice and the continued use of other inaccurate and non-conservative solutions should be discontinued.

Although for engineering purposes satisfactory estimates of load capacity can usually be achieved using Equation 7.7 and the factors provided in Table 7.5, this expression can be considered, at best, only an approximation. For example, it assumes that the effects of soil cohesion, surcharge pressure and self-weight are directly superposable, whereas soil behaviour is highly non-linear and thus superposition does not necessarily hold as the limiting condition of foundation failure is approached.

#### *7.7.2.1 Ultimate capacity under combined loadings*

Under horizontal lateral loading, the ultimate lateral resistance can be estimated, via a Category 2 approach, as the sum of the frictional resistance of the base area plus the passive resistance of the embedded depth of the footing or raft. Each of these components can be obtained from basic soil mechanics principles, employing undrained (total stress) strength parameters for saturated clays subjected to rapid loading (e.g., wind or earthquake forces) and effective stress strength parameters for other cases.

The ultimate capacity of a shallow foundation under moment loading is generally estimated by the effective width method, an example of which is illustrated in Figure 7.2. In this method, the bearing capacity of a foundation subjected to an eccentrically applied vertical loading is assumed to be equivalent to the bearing capacity of a foundation with a fictitious

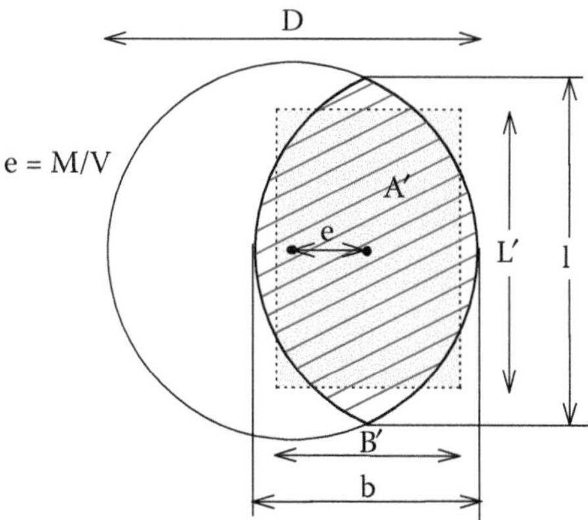


*Figure 7.2* Example of effective area method for a circular footing.

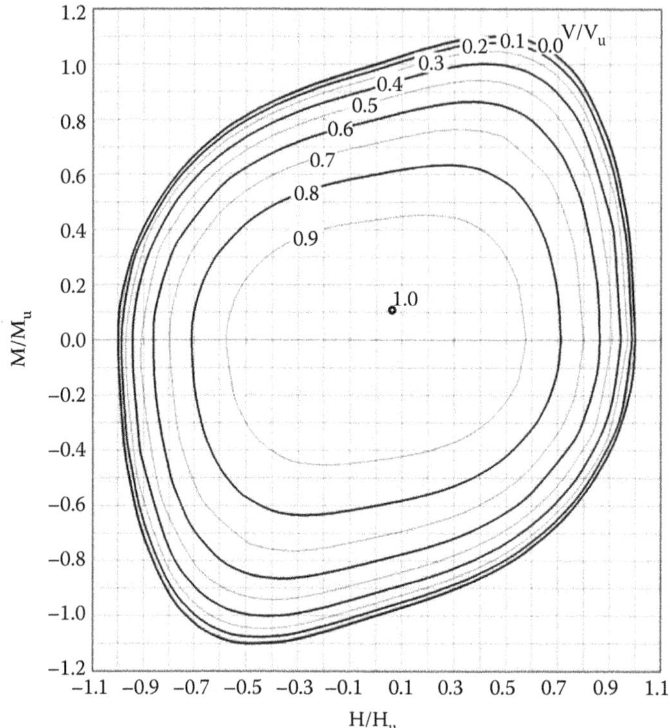


*Figure 7.3* Failure surfaces for footing subjected to combined loading.

effective area on which the vertical load is centrally applied. This approach, although approximate, has been found to be reasonable (Poulos et al., 2001).

Poulos et al. (2001) discuss various approaches for estimating the ultimate capacity of a shallow foundation subjected to combined vertical, lateral, and moment loading. An approximate failure equation was proposed by Taiebat and Carter (2000), as follows:

$$f = \left(\frac{V}{V_u}\right)^2 + \left(\frac{M}{M_u}\left(1 - \alpha_1 \frac{H \cdot M}{H_u |M|}\right)\right)^2 + \left|\left(\frac{H}{H_u}\right)^3\right| - 1 = 0 \quad (7.8)$$

where $V_u$, $H_u$ and $M_u$ are the ultimate resistances for vertical, horizontal and moment loading respectively, and $\alpha_1$ is a factor that depends on the soil profile. For a homogeneous soil a value of $\alpha_1 = 0.3$ provides a good fit to the bearing capacity predictions from the numerical analysis. Equation 7.8 is shown as a contour plot in Figure 7.3.

The less usual case of foundations subjected to a combination of a concentric vertical load and a torsional moment has also been studied by Perau (1997).

#### *7.7.2.2 Bearing capacity of non-homogeneous soils*

The ultimate bearing capacity of foundations on non-homogeneous soils has been examined for the important case that arises often in practice where the undrained shear strength of the soil varies approximately linearly with depth below the soil surface, that is,

$$s_u = c_0 + \rho z \quad (7.9)$$

or below a uniform crust, that is,

$$\begin{aligned} s_u &= c_0 \quad \text{for} \quad z < \frac{c_0}{\rho} \\ s_u &= \rho z \quad \text{for} \quad z > \frac{c_0}{\rho} \end{aligned} \tag{7.10}$$

in which $c_0$ is the undrained shear strength of the soil at the surface, $\rho$ is the strength gradient and z is the depth below the soil surface. Several theoretical approaches attempted to take account of this effect, most notably the work by Davis and Booker (1973) and Houlsby and Wroth (1983). Both used the method of stress characteristics from the theory of plasticity. Assuming the soil to obey the Tresca yield criterion, Davis and Booker's plane strain solution has been expressed as

$$q_u = \frac{Q_u}{B} = F\left[(2+\pi)c_o + \frac{\rho B}{4}\right] \tag{7.11}$$

where F is a function of the soil strength non-homogeneity ($\rho B/c_0$) and the roughness of the foundation–soil interface. Values of the bearing capacity factor F are reproduced in Figure 7.4 for the above two different undrained strength profiles. The solutions of Davis and Booker (1973) can also be adapted to circular footings, at least approximately, by the replacement of Equation 7.11 by

$$q_u = \frac{Q_u}{B} = F\left[6c_o + \frac{\rho B}{6}\right] \tag{7.12}$$

#### *7.7.2.3 Footings on layered soil profiles*

Methods for calculating the bearing capacity of multi-layer soils range from averaging the strength parameters (Bowles, 1988), using limit equilibrium considerations (Button, 1953; Reddy and Srinivasan, 1967; Meyerhof, 1974), to a more rigorous limit analysis approach based on the theory of plasticity (Chen and Davidson, 1973; Florkiewicz, 1989; Michalowski and Shi, 1995; Merifield et al., 1999). Semi-empirical approaches have also been proposed based on experimental studies (e.g. Brown and Meyerhof, 1969; Meyerhof and Hanna, 1978). Almadi and Kouhali (2016) have suggested simplified expressions for the bearing capacity of a strip footing on a two-layer clay profile, for both the case of a strong layer over a weak layer, and a weak layer over a strong layer.

Almost all of these studies are limited to footings resting on the surface of the soil and are based on the assumption that the displacement of the footing prior to attaining the ultimate load is relatively small. In some cases, such as those where the underlying soil is very soft, the footings will experience significant settlement, and sometimes even penetrate through the top layer into the deeper layer. In such cases, it may be necessary to adopt a Category 3 analysis.

### 7.7.3 Category 3 methods

More sophisticated Category 3 methods have been employed to examine cases that may not be amenable to accurate solution via Category 2 methods. As set out by Poulos et al. (2001), problems that have been addressed via Category 3 methods include the following:

1. Problems involving large strains, for example, footings on very soft clay.
2. Multi-layer problems.

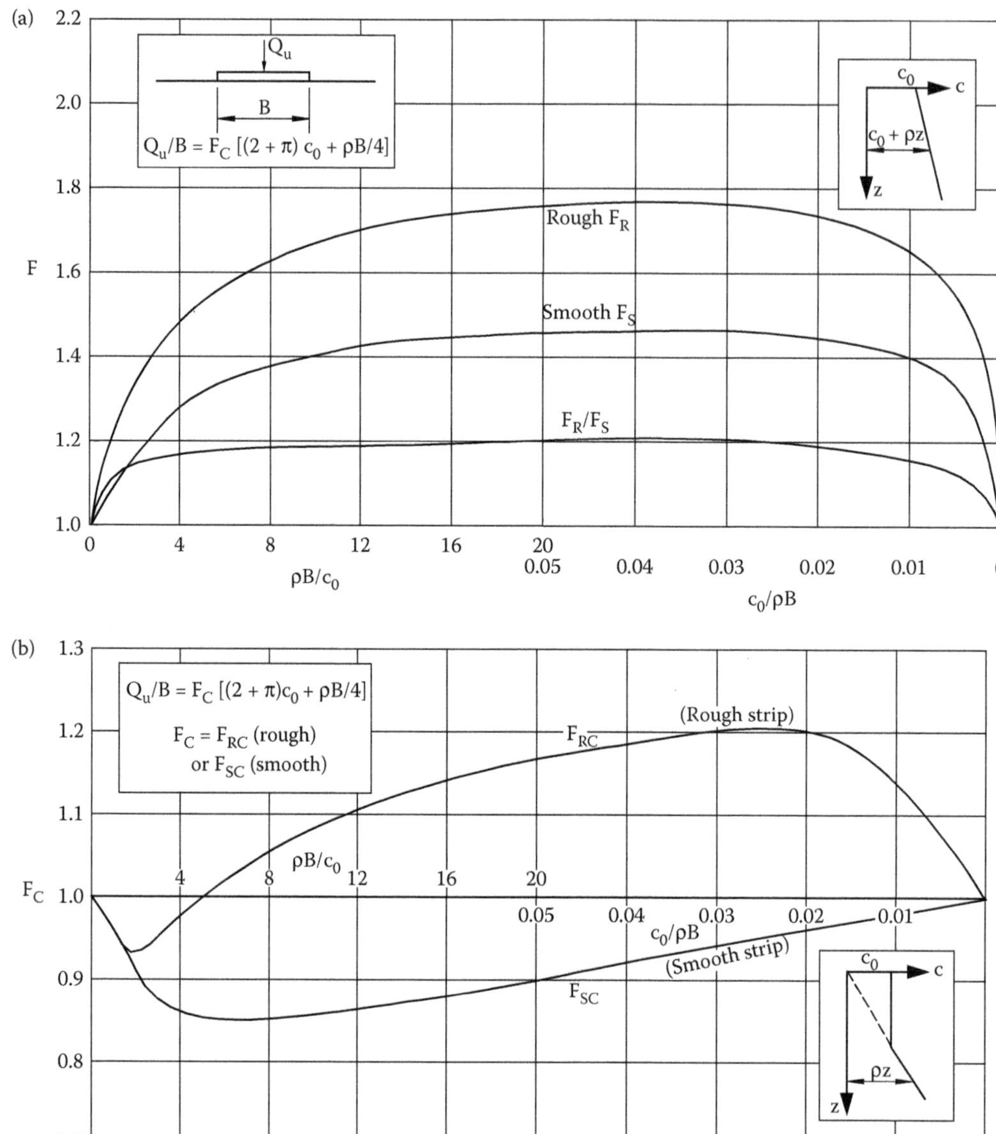


*Figure 7.4* Bearing capacity of shallow foundation on non-homogeneous clay. (a) for linearly increasing shear strength with depth from the surface; (b) for linearly increasing shear strength with a surface crust. (From Davis, E.H. and Booker, J.R. 1973. *Géotechnique*, 23: 551–563. Courtesy of ICE Publishing.)

3. Soil profiles with strength weakening with depth.
4. Profiles with a clay 'sandwich' between sandy layers.
5. Problems involving brittle soil behaviour.

The finite element method can handle very complex layered patterns, and has also been applied to such problems (e.g. Griffiths, 1982; Love et al., 1987; Burd and Frydman, 1997; Merifield et al., 1999). In employing advanced numerical methods, it is important to employ a realistic soil model and to be able to assign the necessary parameters using available in situ and laboratory data.

Sloan (2012) has described the use of the finite element method in conjunction with limit analysis to obtain solutions for a range of stability problems, including bearing capacity problems.

### 7.7.4 Summary

From the preceding discussion of the bearing capacity of shallow foundations, the following conclusions can be drawn:

1. The use of conventional theory, based on the original approach suggested by Terzaghi and extended by others such as Vesic, to calculate the bearing capacity of a foundation on homogeneous soil has stood the test of time, and is generally regarded as being adequately reliable for use in engineering practice. Although this approach is approximate and makes a number of simplifying assumptions, as identified above, it is considered acceptable for most practical problems of shallow footings on relatively homogeneous soils. However, the use of the outdated values of some of the bearing capacity factors, particularly the factor $N_\gamma$, should be discontinued. The use of the factors set out in Table 7.5 is recommended.
2. Significant developments have been made in recent times concerning methods for estimating the ultimate load capacity of footings subjected to combinations of vertical, horizontal and moment loading, and failure loci such as those expressed by Equation 7.8 can be employed.
3. The effective width method, commonly used in the analysis of foundations subjected to eccentric loading, provides good approximations to the collapse loads. Its continued use in practice therefore appears justified.
4. Improvements have also been made in the area of non-homogeneous and layered soils. Theoretically sound and relatively simple to use design methods are now available for cases involving the following:
   a. Clays where the undrained shear strength increases linearly with depth
   b. Two layers of clay
   c. A layer of sand overlying relatively soft clay
5. Sophisticated experimental and theoretical studies have highlighted the brittle nature of footing system behaviour that can occur when relatively thin stronger soils, loaded by surface footings, overlie much weaker materials.

It would appear that to date the problem of reliably predicting the bearing capacity of multiple layers of soil lying beneath a footing and within its zone of interest remains beyond the means of relatively simple Category 2 hand calculation methods. In such cases, the use of Category 3 methods would appear to be appropriate.

## 7.8 ESTIMATION OF ULTIMATE CAPACITY OF PILES AND PILE GROUPS

### 7.8.1 Axial capacity of single piles

#### *7.8.1.1 Basic approach*

A very widely used approach for estimating the ultimate load capacity is to use a static analysis that utilises conventional soil mechanics techniques in conjunction with measured soil strength properties. In this approach, the ultimate compressive load capacity $P_u$ is calculated as the sum of the ultimate shaft capacity $P_{su}$ and the ultimate base capacity $P_{bu}$, less the weight of the pile, $W_p$. In turn, $P_{su}$ and $P_{bu}$ are related to the unit ultimate shaft and base resistances, so that $P_u$ is calculated as follows:

$$P_u = \int f_s C\, dz + f_b A_b - W_p \quad (7.13)$$

where $f_s$ is the ultimate shaft friction; C the pile perimeter; dz is the increment of pile length along the shaft: $f_b$ the ultimate base resistance; $A_b$ the area of the pile base; $W_p$ the pile weight. The value of $f_s$ is usually obtained by use of Coulomb's equation while the ultimate base resistance $f_b$ is usually derived from bearing capacity theory. The usual principles of soil mechanics can be applied, that is, undrained analyses use total stresses and undrained values of $f_s$ and $f_b$, while long-term or drained analyses use effective stresses and drained values of $f_s$ and $f_b$.

Various approaches for estimating $f_s$ and $f_b$, mostly falling into Category 1 or Category 2, are discussed in the following sections.

#### *7.8.1.2 Category 1 methods: Correlations with SPT data*

Empirical correlations with the results of SPT data usually take the following form:

$$f_s = A_N + B_N N \text{ (kPa)} \quad (7.14)$$

where $A_N$ and $B_N$ are empirical numbers, and depend on the units of $f_s$, and N = SPT value at the point under consideration.

$$f_b = C_N N_b \text{ (MPa)} \quad (7.15)$$

where $C_N$ is an empirical factor; $N_b$ the average SPT within the effective depth of influence below the pile base (typically 1–3 pile base diameters).

The most widely used correlations are those developed originally by Meyerhof (1956) for driven piles in sand, in which $A_N = 0$, $B_N = 2$ for displacement piles or 1 for small-displacement piles, and $C_N = 0.3$. Limiting values of $f_s$ of about 100 kPa were recommended for displacement piles and 50 kPa for small-displacement piles. Some other correlations have included other soil types and both bored piles and driven piles. A number of these are summarised by Poulos (1989).

Decourt (1982, 1995) has developed correlations between $f_s$ and SPT, which take into account both the soil type and the methods of installation. For displacement piles, $A_N = 10$ and $B_N = 2.8$, while for non-displacement piles, $A_N = 5–6$ and $B_N = 1.4–1.7$. For the base, values of $C_N$ are shown in Table 7.6. Correlations for piles in gravel are discussed by Rollins et al. (1997).

*Table 7.6* Factor $C_N$ for base resistance

| *Soil type* | *Displacement piles* | *Non-displacement piles* |
|---|---|---|
| Sand | 0.325 | 0.165 |
| Sandy silt | 0.205 | 0.115 |
| Clayey silt | 0.165 | 0.100 |
| Clay | 0.100 | 0.080 |

Source: Decourt, L. 1995. Prediction of load-settlement relationships for foundations on the basis of the SPT-T. Ciclo de *Conferencias International "Leonardo Zeevaert"*, UNAM, Mexico, pp. 85–104.

It is emphasised that the correlations with SPT must be treated with caution as they are inevitably approximate, and not of universal applicability. For example, different correlations are used in Hong Kong to those indicated above, but have been developed for use in the prevailing local geological conditions (GEO, 1996).

#### *7.8.1.3 Category I methods: Correlations with CPT data*

Typical correlations with CPT data are as follows:

Pile ultimate shaft resistance (Bustamante and Gianeselli, 1982):

$$f_s = q_c / k_s \leq f_{sl} \tag{7.16}$$

Pile ultimate base capacity (Frank and Magnan, 1995):

$$f_b = k_b \cdot q_c \tag{7.17}$$

where $k_s$ is the shaft factor, $f_{sl}$ the limiting ultimate shaft friction and $k_b$ the base factor.

Table 7.7 gives recommended values of $k_s$ and $f_{sl}$, which depend on soil type and pile type. Values of $k_b$ are given in Table 7.8. Here, the value of $q_c$ used in Equation 7.16 should be the average value within a distance of 1.5 base diameters above and below the base. Excessively large and low values are excluded from the average (Bustamante and Gianeselli, 1982).

#### *7.8.1.4 Category I methods: Correlations with rock strength*

For piles in rock, it is common to correlate design parameters with the unconfined compressive strength, $q_u$, at least for preliminary purposes. Some of the available correlations are summarised in Table 7.9.

In employing such correlations, it should be recognised that, in the field, they may be influenced by geological features and structure that cannot be captured by a small and generally intact rock sample. Nevertheless, in the absence of other information, such correlations provide at least an indication of the order of magnitude.

*Table 7.7* Ultimate shaft friction correlation factors for CPT tests

| | | *Clay and silt* | | | | | *Sand and gravel* | | | *Chalk* | |
|---|---|---|---|---|---|---|---|---|---|---|---|
| *Pile type* | | *Soft* | *Stiff* | | *Hard* | | *Loose* | *Med.* | *Dense* | *Soft* | *Weathered* |
| Drilled | $k_s$ | – | – | 75[a] | – | – | 200 | 200 | 200 | 125 | 80 |
| | $f_{sl}$ (kPa) | 15 | 40 | 80 | 40 | 80 | – | – | 120 | 40 | 120 |
| Drilled, removed casing | $k_s$ | – | 100 | 100[b] | – | 100[b] | 250 | 250 | 300 | 125 | 100 |
| | $f_{sl}$ (kPa) | 15 | 40 | 60 | 40 | 80 | – | 40 | 120 | 40 | 80 |
| Steel-driven close ended | $k_s$ | – | 120 | | 150 | | 300 | 300 | 300 | | |
| | $f_{sl}$ (kPa) | 15 | 40 | | 80 | | – | – | 120 | | [c] |
| Driven concrete | $k_s$ | – | 75 | | – | | 150 | 150 | 150 | | |
| | $f_{sl}$ (kPa) | 15 | 80 | | 80 | | – | – | 120 | | [c] |

Source: MELT. 1993. Regles techniques de conception et de calcul des fondations des ouvrages de genie civil. Cahier des Clauses Techniques Generales applicables aux Marches Publics de Travaux, Fascicule 62 – Titre V, Ministere de l'Equipement du Logement et des Transports, Paris.

[a] Trimmed and grooved at the end of drilling.

[b] Dry excavation, no rotation of casing.

[c] In chalk, $f_s$ can be very low for some types of piles; a specific study is needed.

*Table 7.8* Base capacity factors for CPT

| *Soil type* | | | $q_c$ *(MPa)* | $k_b$ *(ND)* | $k_b$ *(D)* |
|---|---|---|---|---|---|
| Clay silt | A | Soft | <3 | 0.40 | 0.55 |
| | B | Stiff | 3–6 | | |
| | C | Hard | >6 | | |
| Sand gravel | A | Loose | <5 | 0.15 | 0.50 |
| | B | Medium | 8–15 | | |
| | C | Dense | >20 | | |
| Chalk | A | Soft | <5 | 0.20 | 0.30 |
| | B | Weathered | >5 | 0.30 | 0.45 |

Source: MELT. 1993. *Regles techniques de conception et de calcul des fondations des ouvrages de genie civil.* Cahier des Clauses Techniques Generales applicables aux Marches Publics de Travaux, Fascicule 62 – Titre V, Ministere de l'Equipement du Logement et des Transports, Paris.

Note: ND = non-displacement pile; D = displacement pile.

*Table 7.9* Correlations of design parameters for piles in rock

| *Parameter* | *Correlation* | *Remarks* |
|---|---|---|
| Ultimate bearing capacity (raft) | $p_{ur} = a_0 q_u$ | $a_0$ can vary from about 0.1 for extremely poor quality rock to 24 for intact high-strength rock (Merifield et al., 2006). A value of 2 is likely to be reasonable and conservative in many cases |
| Ultimate shaft friction, $f_s$ | $f_s = a(q_u)^b$ | a generally varies between 0.20 and 0.45<br>b in most correlations is 0.5 |
| Ultimate end bearing, $f_b$ | $f_b = a_1(q_u)b_1$ | $a_1$ generally varies between 3 and 5<br>$b_1$ in most correlations is 1.0, although Zhang and Einstein (1998) adopt $b_1 = 0.5$ and $a_1 = 4.8$ (average) |
| | $f_b = 6.39(q_{um})^{0.45}$ | $q_{um} = (\alpha_E)^{0.7}\, q_u$ (Zhang, 2010). $\alpha_E$ related to RQD as given in Equation 6.27 of Chapter 6 |

#### 7.8.1.5 Category 2 methods: Total stress method

The total stress ('alpha') method is possibly more appropriately categorised as a Category 1 method, as it involves the use of empirical correlations, but it nevertheless has a sound basis via the Coulomb equation for frictional resistance of a cohesive soil–pile surface.

For the estimation of undrained pile capacity in saturated clay soils, $f_s$ is usually related to the undrained shear strength via an adhesion factor $\alpha$, hence the terminology 'alpha method'. The ultimate shaft friction can then be approximated as follows:

$$f_s = \alpha s_u F_1 F_2 \tag{7.18}$$

where $\alpha$ is the adhesion factor, $s_u$ the undrained shear strength, $F_1$ the reduction factor for pile slenderness and $F_2$ the correction factor for method of installation.

The value of $\alpha$ is a function of $s_u$, and has been empirically derived by a number of investigators. Figure 7.5 shows a correlation based on Kulhawy and Phoon (1993). The considerable scatter in this correlation should be noted. For driven piles, Tomlinson (2004) relates the value of $\alpha$ to the nature of the soils above the clay layer and also to the length to diameter of the pile. Also for driven piles, an alternative approach was

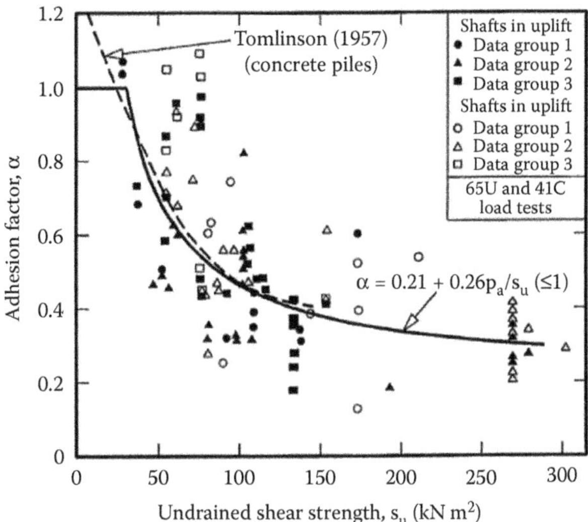


*Figure 7.5* Pile adhesion factor $\alpha$. (Based on Kulhawy, F.K. and Phoon, K.K. 1993. *Design and Performances of Deep Foundations, ASCE, Special Publication*, 38: 172–183.)

developed by Fleming et al. (1985) who related $\alpha$ to the normalised undrained shear strength as follows:

$$\alpha = 0.5/(s_u/\sigma'_v)^{1/2} \quad (\text{for } s_u/\sigma'_v < 1) \tag{7.19}$$

$$\alpha = 0.5/(s_u/\sigma'_v)^{1/4} \quad (\text{for } s_u/\sigma'_v > 1) \tag{7.20}$$

where $\sigma'_v$ is the vertical effective stress at the point in question along the pile shaft.

The reduction factor for pile slenderness $F_1$ has been suggested by Semple and Rigden (1984) to be related to the pile length-to-diameter ratio L/d as follows:

$$F_1 = 1 \quad \text{for} \quad L/d < 50 \tag{7.21}$$

$$F_1 = 0.7 \quad \text{for} \quad L/d > 120 \tag{7.22}$$

$F_1$ is interpolated linearly between 1.0 and 0.7 for $120 > L/d > 50$.

The factor $F_2$ can be taken as 1.0 for driven piles, but is generally accepted to be less than 1 for bored piles. From Fleming et al. (1992) and Viggiani et al. (2011) the value of $F_2$ can be taken to be between about 0.7–0.8 for bored piles.

$f_b$ is obtained from undrained bearing theory as follows:

$$f_b = N_c s_{ub} + \sigma_{vb} \tag{7.23}$$

where $N_c$ is the bearing capacity factor, $s_{ub}$ the average undrained shear strength within a depth of two base diameters below the pile base and $\sigma_{vb}$ is the total overburden stress at the level of the pile base.

$N_c$ is related to the relative depth of the pile base, and can be approximated as follows:

$$N_c = 6 + L/d_b \leq 9 \tag{7.24}$$

where L is the pile length and $d_b$ the pile base diameter.

#### 7.8.1.6 Category 2 methods: Effective stress method

There has been a tendency in recent years toward the use of a more broadly-applicable effective stress approach, rather than a total stress approach, to the calculation of pile shaft capacity. It is argued that the high pore pressure gradient set up near the pile shaft by loading the pile causes rapid dissipation of excess pore pressures, and thus drained conditions prevail at the shaft–soil interface. The ultimate shaft friction $f_s$ is then given as

$$f_s = \sigma'_{hf} \tan\delta \tag{7.25}$$

where $\sigma'_{hf}$ is the normal effective stress at failure on the pile–soil interface and $\delta$ the effective interface friction angle between soil and shaft.

Following Mandolini (2012), $\sigma'_{hf}$ can be related to the vertical effective stress, $\sigma'_v$ and the radial displacement at the pile–soil interface, $u_r$, as follows:

$$\sigma'_h = (K_s\sigma'_v + 4G \cdot u_r/d) \tag{7.26}$$

where $K_s$ is the lateral stress coefficient, G the shear modulus of the surrounding soil and d the pile diameter. Thus, $f_s$ can be expressed as

$$f_s = (K_s\sigma'_v + 4G \cdot u_r/d)\tan\delta \tag{7.27}$$

In turn, $u_r$ can be approximated as follows:

$$u_r = 0.6 \cdot n \cdot d_{50} \cdot \tan\psi_p \tag{7.28}$$

where n is a factor between 5 and 20, $d_{50}$ the mean particle size (50% passing) and $\psi_p$ is the dilatancy angle at pile–soil interface. For clay soils, $u_r$ is very small, and the second term in Equation 7.26 can usually be neglected. For sandy soils, the term involving $u_r$ may be more significant, especially for smaller diameter piles, but it tends to become small for larger diameter piles and is therefore often ignored. Thus, in common practice, $f_s$ is estimated as

$$f_s = K_s\sigma'_v \tan\delta \tag{7.29}$$

$$= \beta \cdot \sigma'_v \tag{7.30}$$

where $\beta$ is the shaft friction coefficient $= K_s \tan\delta$.

Burland (1973) demonstrated that a lower limit of $K_s$ is $K_o$, the coefficient of earth pressure at rest, and so $\beta$ for driven piles in normally consolidated clay is $\beta_{nc}$, where

$$\beta_{nc} = (1 - \sin\phi')\tan\phi' \tag{7.31}$$

and $\phi'$ is the effective angle of internal friction of the clay.

For typical values of $\phi'$ in the range of 20–30°, $\beta_{nc}$ varies only between 0.24 and 0.29. Such values are in accordance with measurements of both positive and negative friction on driven piles in soft clay. For overconsolidated clays, $\beta$ is greater than $\beta_{nc}$, and if the over-consolidation ratio (OCR) is known, Meyerhof (1976) suggests that $\beta$ can be estimated as follows:

$$\beta = \beta_{nc}(OCR)^{0.5} \tag{7.32}$$

Patrizi and Burland (2001) suggested the following useful approximation for driven piles in clay:

$$\beta = 0.1 + 0.4s_u/\sigma'_v \tag{7.33}$$

This expression implicitly takes the effects of over-consolidation into account via the undrained shear strength ratio, which typically is about 0.22 for normally consolidated soils and increases with increasing OCR.

For bored piles, provided that the pile is formed promptly after excavation of the shaft, little change in the in situ effective stress state in the soil should occur, and Equation 7.29 may be used, with $K_s = K_o$, the coefficient of earth pressure at rest. However, in heavily overconsolidated clay, some allowance for stress relaxation is recommended by Fleming et al. (1985), who suggest two alternatives: a reduction of the value of $K_s$ by 20% (i.e. to $0.8K_o$), or using the mean stress between the in situ horizontal stress and that due to the wet concrete in the pile shaft, that is, $K_s = (1 + K_o)/2$.

Despite the use of effective stresses to compute the shaft capacity of piles in clay, the ultimate base capacity $f_b$ of such piles is usually still calculated from the total stress approach in Equation 7.23.

For piles in sand or gravel, or piles in saturated clays under long-term drained conditions, effective stress analysis of ultimate load capacity is appropriate. If the cohesive component of drained strength is ignored, the ultimate shaft friction $f_s$ and ultimate base resistance $f_b$ are usually expressed as follows:

$$f_s = K_s\,\sigma'_v \tan\delta \tag{7.34}$$

$$f_b = N_q\sigma'_{vb} \tag{7.35}$$

where $N_q$ is the bearing capacity factor and $\sigma'_{vb}$ the effective vertical overburden stress at the level of the pile base.

Equation 7.34 is the same equation as that in the $\beta$ method for clays (Equation 7.29). Conventional methods of calculation for piles in sand and gravel (e.g., Broms 1966; Nordlund 1963) assume that both $f_s$ and $f_b$ increase, more or less linearly, with depth. However, initial interpretations of the research by Vesic (1969) and others suggested that the average shaft friction and base resistance did not increase linearly with depth, but that they reached limiting values at depths of between 5 and 20 diameters, depending on the relative density of the soils. This phenomenon was attributed by some to a form of arching, together with the effects of soil compressibility and the reduction of friction angle with increasing stress level. While there is still some controversy over this hypothesis (Kulhawy, 1984), there is now a prevailing view that $f_s$ and $f_b$ may continue to increase with increasing depth, but at a decreasing rate. For practical design purposes, it is not uncommon to adopt limiting values of both $f_s$ and $f_b$ for piles in sands and gravels.

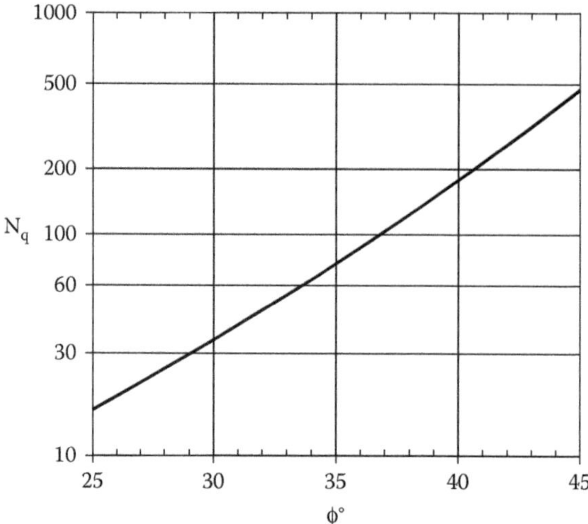


*Figure 7.6* Base bearing capacity factor $N_q$. (Adapted from Berezantzev, V.G., Khristoforov, V. and Golubkov, V. 1961. Load bearing capacity and deformation of piled foundations. *Proceedings of the, 5th International Conference on Soil Mechanics and Foundation Engineering*, Paris, Vol. 2, pp. 11–15.)

Figure 7.6 shows a commonly used solution for the bearing capacity factor $N_q$ as a function of the angle of internal friction $\phi$. Poulos and Davis (1980) suggest that, in using this chart, to allow for the effects of installation of the pile, the value of $\phi$ for driven piles should be taken as $(\phi_0 + 40)/2°$, while for bored piles, a value of $\phi_0 - 3°$ should be adopted, where $\phi_0$ is the initial angle of friction of the soil in the vicinity of the pile base prior to pile installation. More comprehensive solutions for $N_q$, incorporating the effects of the relative compressibility of the soil and relative embedment depth, are presented by Kulhawy (1984) and Fleming et al. (1992).

Tables 7.10 through 7.12 give some correlations for the parameters $\delta$, $K_s$ and the limiting values of shaft friction ($f_{sl}$) and end bearing ($f_{bl}$) for piles in sands.

#### 7.8.1.7 Uplift resistance of piles

For piles without an enlarged base, the ultimate uplift capacity $Q_{ut}$ can be estimated as follows:

$$Q_{ut} = f_{stav} \cdot A_s + W \tag{7.36}$$

*Table 7.10* Interface friction angle $\delta$ for piles in sand

| *Interface materials* | *Ratio of interface friction angle to soil friction angle* $\delta/\phi$ | *Typical field analogy* |
|---|---|---|
| Sand/rough concrete | 1.0 | Cast in place |
| Sand/smooth concrete | 0.8–1.0 | Precast |
| Sand/rough steel | 0.7–0.9 | Corrugated |
| Sand/smooth steel | 0.5–0.7 | Coated |
| Sand/timber | 0.8–0.9 | Pressure treated |

Source: Stas, C.V. and Kulhawy, F.H. 1984. Critical evaluation of design methods for foundations under axial uplift and compression loading. Report for EPRI, No. EL-3771, Cornell University.

*Table 7.11* Horizontal stress coefficients

| *Foundation type and method of installation* | *Ratio of horizontal stress coefficient to in situ value, $K_s/K_o$* |
|---|---|
| Jetted pile | 0.5–0.67 |
| Drilled shaft, case-in-place | 0.67–1.0 |
| Driven pile, small displacement | 0.75–1.25 |
| Driven pile, large displacement | 1–2 |

Source: Stas, C.V. and Kulhawy, F.H. 1984. Critical evaluation of design methods for foundations under axial uplift and compression loading. Report for EPRI, No. EL-3771, Cornell University.

where $f_{stav}$ is the average uplift shaft resistance, $A_s$ the shaft surface area and W the weight of pile, (allowing for buoyancy where applicable).

In relatively soft or loose soils, $f_{stav}$ can be taken to be the same as for downward loading. However, in sands, De Nicola and Randolph (1993) have demonstrated that the ratio of the shaft resistance in tension ($f_{st}$) and compression ($f_s$) are related to the relative compressibility of the pile, as follows:

$$\frac{f_{st}}{f_s} = \{1 - 0.2\log_{10}[100/(L/d)]\}(1 - 8\eta + 25\eta^2) \tag{7.37}$$

and

$$\eta = v_p\left(\frac{L}{d}\right)\left(\frac{G_{av}}{E_p}\right)\tan\delta \tag{7.38}$$

where L is the pile length, d the pile diameter, $\eta$ the dimensionless pile compressibility factor, $\nu_p$ the Poisson's ratio of pile material, $\delta$ the pile–soil interface friction angle, $G_{av}$ the

*Table 7.12* Limiting values of ultimate shaft and base resistance – Piles in sand

| *Soil type and condition* | *Limiting shaft friction $f_{sl}$ (kPa)* | *Limiting base resistance $f_{bl}$ (MPa)* | *Source* |
|---|---|---|---|
| V. loose sand<br>Loose sand/silt<br>M. dense silt | 48 | 1.9 | API (1984) |
| Loose sand<br>M. dense sand/silt<br>Dense silt | 67 | 2.9 | API (1984) |
| M. dense sand<br>Dense sand/silt | 80 | 4.8 | API (1984) |
| Dense sand<br>V. dense sand/silt | 96 | 9.6 | API (1984) |
| Dense gravel<br>V. dense sand | 115 | 12.0 | API (1984) |
| Calcareous sand (uncemented) | | | |
| Driven piles | 10–20 | 1–4[a] | Nauroy et al. (1986) |
| Bored piles | 60–100[a] | 0.5–3[a] | Poulos (1989) |

[a] Depends on soil compressibility – smaller values for more compressible soils.

average value of soil shear modulus along shaft and $E_p$ the Young's modulus of pile material. $f_{st}$ is generally less than $f_s$ unless the soil is relatively weak or the pile is relatively short.

For piles with an enlarged base, $Q_{ut}$ can be taken as the lesser of

1. The sum of the weight of the pile and the ultimate uplift resistance of the entire base area (similar to an anchor pull-out)
2. The sum of the weight of the pile, the frictional force along the shaft, and the ultimate uplift resistance of the net base area.

In saturated clays and silts, the long-term uplift capacity may be considerably less than the short-term capacity because of the dissipation of negative pore water pressures (Meyerhof and Adams, 1968). For calculation of the ultimate uplift resistance of the base, the solutions of Rowe and Davis (1982) for anchor plates may be used.

#### *7.8.1.8 Other category 2 methods*

A number of approaches have been developed for offshore piling applications, in an attempt to improve on the simple effective stress approach for driven piles. Many of these methods have incorporated correlations with CPT data (essentially a Category 1 approach) within a Category 2 effective stress analysis framework. Typical of these methods are

1. The Imperial College (ICP-05) Method (Jardine et al., 2005)
2. The University of Western Australia (UWA-05) Method (Lehane et al., 2005)
3. The Norwegian Geotechnical Institute (NGI-05) Method (Clausen et al., 2005)
4. The Fugro Method (Kolk et al., 2005)

Schneider et al. (2008) have provided a useful summary of the above methods. They have also examined the predictive performance of these methods and concluded that the best performing of the above approaches was substantially better than the API (2000) method or the total stress alpha method. Xu et al. (2008) found that the UWA-05 method was a more reliable method for predicting end bearing capacity, as it incorporated the effects of partial plugging during the installation of steel pipe piles and allowed for variations in the CPT resistance in the vicinity of the pile tip.

Because these methods are generally confined to driven piles, they may not be directly applicable to the majority of piles for tall buildings, which tend to be large-diameter bored piles.

#### *7.8.1.9 Other factors influencing pile axial capacity*

The axial capacity of piles, and in particular, the shaft friction capacity, are influenced by a number of factors in addition to the properties of the supporting ground and the interface conditions. Among these factors are the following:

1. *Time effects.* it is well-recognised that the shaft capacity of driven piles in clay tends to increase with time, due to dissipation of excess pores that are developed during installation. There are also significant time effects for driven piles in sand, despite the fact that excess pore pressure dissipation is not generally a factor. Much field data indicates that the load-carrying capacity of piles driven into sand may increase significantly over months, long after pore pressures have dissipated, for example, Chow et al. (1997, 1998); Bowman and Soga (2003); Jardine and Chow (1996); Jardine et al. (2006). The magnitude of the increase is variable, but most of it is due to increased

shaft resistance rather than tip resistance. Piles driven into silts and fine sands set up proportionately more than those in coarse sands and gravels. Both driven and jacked piles exhibit setup, whereas bored piles do not. Mitchell (2008) considers the mechanisms responsible for these time-dependent effects and finds that physical–mechanical processes involving particle rearrangement and internal stress redistribution under the action of the new in situ stress conditions play the dominant, if not the only, role in producing sand aging effects. The associated void ratio decreases play only a minor role in accounting for the property changes. In addition, chemical and microbiological processes that lead to interparticle bonding and cementation are not considered to be the major contributor to property changes during sand aging.

2. *Effects of installation process and drilling fluid.* Fleming and Sliwinski (1977) suggested that the shaft resistance for bored piles constructed using bentonite slurry should be reduced by 10%–30% for prudence, although comparative studies of the ultimate shaft resistance of bored piles installed with or without bentonite slurry in granular and cohesive soils have shown no significant difference in performance with the two methods of installation. Experience with large-diameter bored piles and barrettes in saprolites in Hong Kong indicates that the use of bentonite slurry may not produce detrimental effects on pile performance, provided that its properties are strictly controlled. Van Impe (1991) has indicated that the shaft resistance may also be affected by the concrete fluidity and pressure. The method and speed of casting, together with the quality of the concrete, can have a significant effect on the lateral stress on the pile–soil interface and can thus influence shaft capacity. Bernal and Reese (1984) reported that unless the slump of concrete is at least 175 mm and the rate of placement is at least 12 m/hour and a concrete mix with small-size aggregates is used, the pressures exerted by the fluid concrete will be less than the hydrostatic pressure, which can result in lower shaft resistance particularly in soils with high $K_0$ values. Recent investigations (Lam and Jefferis, 2014; Lam et al., 2014) have revealed that the use of polymer drilling fluid, rather than bentonite, generally leads to an increased shaft friction, and the use of polymer fluids is now becoming much more prevalent in high capacity pile foundation construction.
3. *Diameter effects.* inspection of Equation 7.27 reveals that the shaft diameter plays a role in the development of shaft friction. For small diameter piles, the second term related to the radial displacement at the interface, may play a dominant role, whereas it becomes far less significant for large-diameter piles and is generally ignored. For this reason, caution must be exercised in extrapolating the results of tests on small diameter piles to large-diameter piles, without proper consideration of the second term in Equation 7.27. Ignoring this diameter effect can lead to an over-estimation of shaft friction.

#### 7.8.1.10 Category 3 methods

Some efforts have been made to carry out numerical simulations of the entire construction and loading history of a single pile, but such methods remain largely within the realm of research, and the benefits of such analyses in practice have yet to be demonstrated.

### 7.8.2 Axial capacity of pile groups and piled rafts

#### 7.8.2.1 Category 2 methods

For Category 2 analysis of pile groups and piled rafts, it is convenient to simplify the proposed foundation system into an equivalent pier and then examine the overall stability and settlement of this pier. For the ultimate limit state, the bearing capacity under compressive vertical loading can be estimated from the classical approach in which the lesser of the following two values is adopted:

1. The sum of the ultimate capacities of the piles plus the net area of the raft (if in contact with the soil)
2. The capacity of the equivalent pier containing the piles and the soil between them, plus the capacity of the portions of the raft outside the equivalent pier which is in contact with the ground

It is often convenient to consider the group efficiency factor, Y, defined as

$$Y = \frac{\text{ultimate capacity of pile group}}{\text{sum of ultimate capacities of individual piles}} \tag{7.39}$$

Early methods of estimating Y were often based purely on geometry, for example, the Converse Labarre formulae and Feld's rule (Poulos and Davis, 1980), but it is now recognised that such approaches are deficient in not considering the soil and pile characteristics. Table 7.13 summarises recommendations for the estimation of Y for various situations. It should be emphasised that these approaches are approximate only.

For piled rafts, De Sanctis and Mandolini (2006) have proposed the following convenient approximation for the ultimate axial capacity, $P_{upr}$:

$$P_{upr} = P_{up} + \alpha_{uR} \cdot P_{ur} \tag{7.40}$$

where $P_{up}$ is the ultimate axial capacity of the piles (as a group), $\alpha_{uR}$ the proportion of raft capacity mobilised at failure ($0 \le \alpha_{uR} \le 1.0$) and $P_{ur}$ the ultimate axial capacity of the raft alone.

$\alpha_{uR}$ can be approximated as follows:

$$\alpha_{uR} = 1 - 3(A_G/A) \cdot (d/s) \tag{7.41}$$

where $A_G$ area occupied by piles A total raft area d pile diameter s average centre-to-centre spacing of piles in the foundation system.

*Table 7.13* Typical pile group efficiency factors

| *Case* | *Group efficiency Y* | *Remarks* |
|---|---|---|
| Driven piles in loose-to-medium dense sand | 1.0 | Y may be considerably greater than 1: adopt 1 for design |
| End bearing piles on rock, dense sand or gravel | 1.0 | Base resistance is not much effected by group action, even at small spacings (Meyerhof, 1976) |
| Bored friction piles in sand | 0.67 | For 'customary spacings': that is, 3 ± 1 diameters (Meyerhof, 1976) |
| Friction piles in clay-cap above surface | Lesser of $P_B/\Sigma P_u$, or 1.0 | Terzaghi and Peck (1967). Make allowance for any soft layers below base |
| | $\left\{\frac{1}{1+\left(\frac{\Sigma P_u}{P_B}\right)^2}\right\}^{0.5}$ | Poulos and Davis (1980). Make allowance for any soft layers below base |

Note: $P_B$ = ultimate load capacity of block containing piles and soil, $\Sigma P_u$ sum of ultimate capacities of individual piles.

For relatively large spacings, $\alpha_{uR}$ approaches unity, but as the relative spacing s/d decreases, $\alpha_{uR}$ decreases.

#### 7.8.2.2 Category 3 methods

Category 3 methods can be used to facilitate pile group and piled raft design. If a finite element analysis is undertaken with a suitable non-linear soil model, assumptions regarding pile group efficiency are not required as the analysis should be able to identify the critical mechanism and provide reasonable quantitative estimates of group capacity, as long as the details of the group and the soil profile, and the applied loading system, are modelled reasonably accurately. An early example of such an analysis is provided by Pressley and Poulos (1986) who present solutions for the non-linear load–settlement behaviour of square-configuration pile groups. The groups are represented by an equivalent axially symmetric model, and a non-linear finite element analysis is used to examine the mechanisms of group behaviour and their variation with pile spacing. It is shown that, at close spacings, the block failure mechanism occurs, with significant plastic zones being developed below the group and full pile–soil slip only being developed along the outer piles. As the pile spacing increases, the failure mechanism gradually changes to the 'single-pile' mode, whereby full pile–soil slip occurs along all piles. Within the limitations of accuracy of the finite element solution, the values of group settlement ratio and efficiency are in reasonable agreement with values derived from existing theories.

De Sanctis and Mandolini (2006) have demonstrated the effective use of finite element analyses to carry out parametric studies and develop the approximate approach for ultimate capacity of piled rafts in Equation 7.40.

### 7.8.3 Ultimate capacity under lateral loading

#### 7.8.3.1 Modes of failure and pile head conditions

The calculation of the ultimate lateral capacity of piles usually involves consideration of the statics of a pile under lateral loading. This requires specification of the distribution of ultimate lateral pile–soil pressure with depth, the structural strength of the pile in bending, and the postulated failure mode of the pile–soil system. The conditions usually examined are

- Failure of the soil supporting the pile (termed 'short-pile failure' by Broms (1964a,b)
- Structural failure of the pile itself (termed 'long-pile failure' by Broms)

In addition, the two limiting pile head conditions that are usually considered are a free (unrestrained) head and a fixed head (restrained against rotation).

In the context of tall building foundations, where piles are connected by a raft or basement slab, it is likely that the appropriate pile head condition would be the fixed head case.

#### 7.8.3.2 Estimation of ultimate lateral pile–soil pressures

Estimation of the ultimate lateral capacity of piles requires an estimate to be made of the ultimate lateral pressure that can be developed between a pile and the surrounding soil. Following the work of Broms (1964a,b), it has been common for the ultimate lateral pressure, $p_y$, to be estimated as follows:

- For clays under undrained conditions

$$p_y = N_c \cdot s_u \tag{7.42}$$

where $N_c$ is a lateral capacity factor and $s_u$ is the undrained shear strength.

Various solutions have been developed for $N_c$, but in general, it is found to increase from 2 at the ground surface to about 9–12 at a depth of 3–4 diameters, then remaining constant for greater depths. In many practical applications, a value of 9 is adopted (as for the undrained end bearing capacity of piles in clay).

- For sands

$$P_y = N_s p_p \tag{7.43}$$

where $N_s$ is a multiplying factor and $p_p$ the Rankine passive pressure. $N_s$ is usually within the range 3–5, with 3 being a common design value. Fleming et al. (1992) suggest that an appropriate value of $N_s$ is $K_p^2$, where $K_p$ is the Rankine passive pressure coefficient $= \tan^2(45 + \phi/2)$ and $\phi$ the angle of internal friction of the soil.

A number of alternative approaches are available, and Kulhawy and Chen (1993) have compared three of the available distributions. They have concluded that Broms' approach appears to be very conservative at depth, but that conversely, the approach of Reese et al. (1974) appears to be quite bold.

#### *7.8.3.3 Single pile theories*

The classical work in this area has been published by Broms (1964a,b), and this work continues to be the cornerstone of many practical assessments today. Kulhawy and Chen (1993) have carried out an assessment of the applicability of Broms' method, based on comparisons with the results of a number of laboratory and field tests on bored piles. For both undrained and drained lateral load capacities, Broms' method tended to underestimate the ultimate lateral load by about 15%–20%. They concluded that, while Broms' method was conservative overall, it provided as good an approach as any other method, and could give good results if empirical adjustments were made to the values computed from the theory. It should be stated that Broms himself acknowledged that his assumed ultimate lateral pressure distributions were conservative.

Despite the widespread use of Broms' method, it must be recognised that it has a number of practical limitations, among which are the following:

1. It assumes that the soil layer is homogeneous with depth
2. It considers only a pure sand (frictional soil), or a clay under undrained conditions having a constant strength with depth
3. It considers only a single pile, and not a group of piles directly

Owing to these limitations, the practitioner must exercise considerable judgement in applying Broms' theory.

Fleming et al. (1992) have provided modified versions of Broms' solutions which avoid some of the approximations of the original theory, and have presented their results in the form of charts for free head and fixed head piles in both cohesive and cohesionless soils. On the basis of curve fitting of the Fleming et al. solutions for the case of a fixed head pile, approximate expressions for the ultimate lateral capacity of a single pile are given in Table 7.14. The ultimate lateral load capacity will be the lesser of the values for the short-pile and long-pile mode.

Meyerhof (1995) has provided a summary of an alternative approach to the estimation of ultimate lateral capacity that incorporates the effects of load eccentricity and inclination.

*Table 7.14* Approximate expressions for ultimate lateral capacity of a single fixed head pile

| *Soil type* | *Failure mode* | *Expression* | *Definitions* | |
|---|---|---|---|---|
| Cohesive | Short pile | $H_1 = 0.5x_1^2 + 4.25x_1$ | $H_1 = H/c_ud^2$ | $x_1 = L/d$ |
| Cohesive | Long Pile | $H_2 = 4.08x_2^{0.544}$ | $H_2 = H/c_ud^2$ | $x_2 = M_y/c_ud^3$ |
| Cohesionless | Short pile | $H_3 = 0.495x_3^2 + 0.010x_3$ | $H_3 = H/K_p^2\gamma d^3$ | $x_3 = L/d$ |
| Cohesionless | Long Pile | $H_4 = 1.652x_4^{0.668}$ | $H_4 = H/K_p^2\gamma d^3$ | $X_4 = M_y/K_p^2\gamma d^4$ |

Note: H = ultimate lateral capacity; $c_u$ = undrained shear strength; $M_y$ = yield moment of pile section; $K_p$ = Rankine passive pressure coefficient; $\gamma$ = unit weight of soil; L = pile length; d = pile diameter or width.

The solutions provided by Broms, Fleming et al. and Meyerhof assume a single layer of soil, and so their solutions cannot be applied directly to the following problems:

1. Layered soil profiles
2. Profiles containing both sand and clay layers
3. Uniform sand profiles in which the water table is not at the surface or below the pile tip
4. Groups of piles

#### *7.8.3.4 Layered and non-homogeneous soils*

Some extensions to Broms' theory have been made in an attempt to overcome some of the above limitations. For example, Poulos (1985) has developed a general solution for piles in a two-layer cohesive soil. This solution involves the solution of a quadratic equation of the form:

$$aH^{*2} + bH^* + c = 0 \tag{7.44}$$

where $H^* = H/p_2dL$, $p_2$ is the ultimate lateral pile–soil pressure of lower layer, L the pile length and d the pile diameter or width. The coefficients a, b and *c* depend on the relative thickness of the two layers, the relative strength of the layers, the eccentricity of loading, and the characteristics of the pile. Figure 7.7 shows some typical results derived from that analysis. The ratio of the lateral capacity for the two-layer soil to that for a homogeneous layer is plotted against the relative thickness of the upper layer, for both the short- and long-pile failure modes. This figure highlights that the near-surface layer has a very important effect on the ultimate lateral capacity.

For multi-layered soil profiles, closed-form solutions and design charts are not feasible, and a simple computer-based analysis is required. Such an analysis is based on the simple principles of statics used by Broms and others, and requires the estimation of the ultimate lateral pile–soil pressures.

#### *7.8.3.5 Effects of inclined loading*

Meyerhof (1995) has given detailed consideration to the effects of inclined loading on a pile, and has developed practical procedures for combining axial and lateral capacities in such cases. A simple alternative approach has been suggested by Cho and Kulhawy (1995), who have obtained correction factors based on the results of undrained load tests on bored piles (drilled shafts) in clay, to modify the axial and lateral pile capacities. The notation adopted

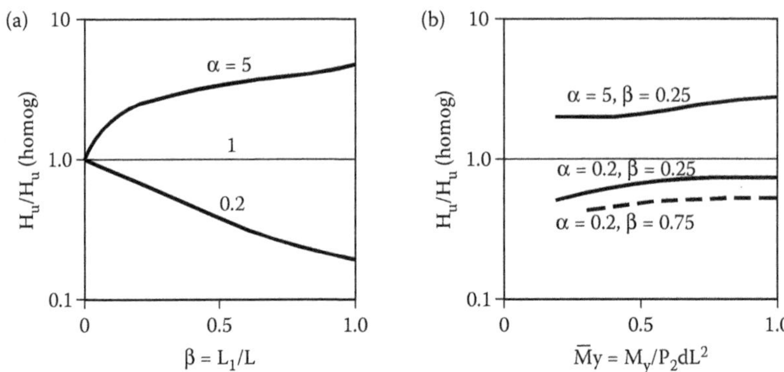


*Figure 7.7* Effect of layering on ultimate lateral resistance. (a) Short pile failure and (b) long pile failure. (From Poulos, H.G. 1985. Ultimate lateral pile capacity in two-layer soil. *Geotechnical Engineering, SEAGS*, 16(1): 25–38. Courtesy of SEAGS.)

by these authors is illustrated in Figure 7.8, and the following expressions were developed for the vertical and horizontal components of the inclined failure load:

1. For inclined uplift

$$\text{Vertical component: } P_O = Q_{su}(1 - \Psi/90) + W_s \tag{7.45}$$

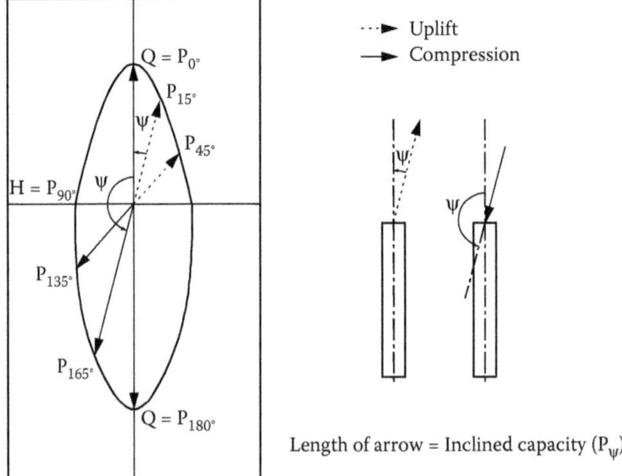


*Figure 7.8* Polar capacity diagram for inclined loading on pile. (Adapted from Cho, N-J. and Kulhawy, F.H. 1995. *Journal of the Korean Geotechnical Society*, 11(3): 91–111.)

Horizontal component: $P_{90} = H_u(\sin\Psi)^{0.5}$ (7.46)

2. For inclined compression

Vertical component: $P_{180} = Q_{sc}(\Psi/90 - 1) + Q_{tc}[(\Psi/90) - 1]^{7.3}$ (7.47)

Horizontal component: $P_{90} = H_u(\sin\Psi)^{0.5}$ (7.48)

where $Q_{su}$ is the uplift capacity of shaft, $W_s$ the weight of pile, $H_u$ the ultimate lateral load capacity, $Q_{sc}$ the ultimate shaft capacity in compression, $Q_{tc}$ the ultimate base capacity in compression and $\psi$ the angle of inclination of load to vertical, in degrees.

#### *7.8.3.6 Pile groups*

For practical purposes, the ultimate lateral capacity of a pile group can be estimated as the lesser of

- The sum of the ultimate lateral capacities of the individual piles in the group
- The ultimate lateral capacity of an equivalent block containing the piles and the soil

In the latter case, only the 'short-pile' case need be considered. Care should be exercised when applying Broms' theory directly to groups in clay, as it implicitly assumes that there is a 'dead' zone from the surface to a depth of 1.5 diameters in which the soil contributes no lateral resistance. This assumption is inappropriate for large-diameter blocks with a low length-to-diameter ratio. A more rational approach is to consider the block via statics, taking a smaller 'dead zone' equal to 1.5 times the individual pile diameter. An alternative approach for block failure of groups, including combined loadings, is set out below.

### 7.8.4 Capacity of pile groups and piled rafts under combined loadings

#### *7.8.4.1 Category 2 approach*

The preceding discussion has focussed on the capacity of a foundation system to resist either vertical loading or lateral loading, but for tall buildings, substantial horizontal, moment and torsional loads may be applied to the foundation simultaneously with vertical loads. A Category 2 analysis may be used, very approximately, to examine this case. To do so, the foundation system must first be idealised as a single equivalent pier, as shown in Figure 7.9. Clearly, if there are only a few piles within the foundation system, this approach is unlikely to provide useful results unless modified to incorporate the raft resistance separately.

The resistance of the equivalent pier to the separate load components now needs to be computed, and then combined, as follows:

1. For vertical loading, the ultimate load capacity can readily be computed by summing the shaft and base resistances of the equivalent pier, as for a single pile (see Section 7.8.1). In this calculation, the shaft friction may be increased as compared with that of a single pile, because the failure surface will generally involve soil-to-soil shearing, rather than pile-to-soil shearing. For bored concrete piles, there may however be little difference between the frictional resistances of the two cases.

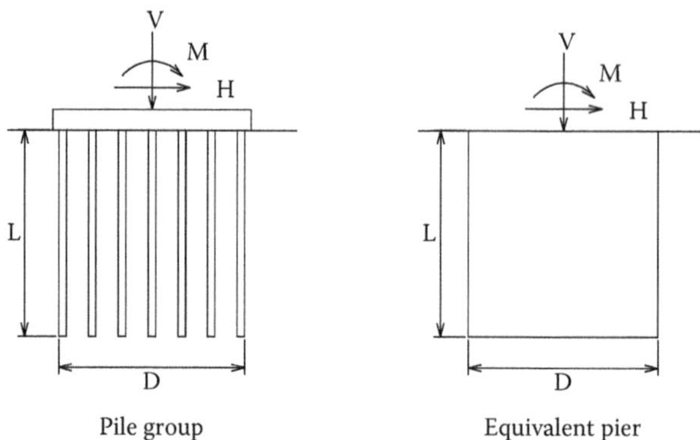


*Figure 7.9* Equivalent pier approximation for Category 2 analysis of pile group stability.

2. For horizontal loading, in conjunction with moment loading in the same direction, the ultimate lateral capacity of the side of the equivalent pier can be estimated from consideration of statics. For simple uniform and linearly increasing profiles of ultimate lateral pile–soil pressure with depth, the solution presented by Poulos and Davis (1980) may be used. Figure 7.10 shows the solutions for a pier of diameter d and length L, subjected to a horizontal load H which is applied at an eccentricity of e above the ground surface (e = M/H, where M is the applied moment and H is the applied load). The diameter 'd' is the width of the equivalent pier in the direction of loading.

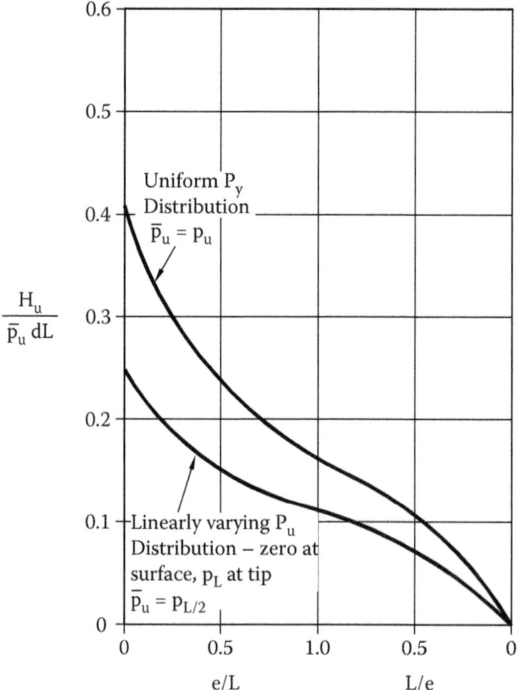


*Figure 7.10* Ultimate lateral resistance of a free-headed rigid pier. (Adapted from Poulos, H.G. and Davis, E.H. 1980. *Pile Foundation Analysis and Design*. John Wiley, New York.)

3. For the ultimate limit state case, it appears reasonable to adopt a length of the equivalent pier equal to the pile length, given that the very large lateral movements can occur in the ultimate case. However, a shorter length may be relevant to the serviceability limit case, and as set out in Chapter 8, the length of the equivalent pier should then be taken as the shorter of the actual length or the critical length of a single pile under lateral loading.
4. For the case of a soil with a uniform ultimate lateral pressure $p_u$ with depth, the ultimate lateral load capacity can be expressed conveniently as

$$H_u = p_u DL[(1+2e/L)^2+1)^{0.5}-(1+2e/L)] \quad (7.49)$$

To estimate the ultimate lateral pier–soil pressure $p_u$ in these solutions, the following approximations may be made:

For purely cohesive soils: $p_u = N_c \cdot s_u$ (7.50)

For cohesionless soils, $p_u = 3K_p\sigma'_v$ (7.51)

where $N_c = 2 + 3.5z/D \leq 9$, $s_u$ is the undrained shear strength, $K_p$ the Rankine passive pressure coefficient $= \tan^2(45 + \phi/2)$, $\phi$ the angle of internal friction, $\sigma'_v$ the vertical effective stress, z the depth below surface and D the pier diameter $\equiv$ d in Figure 7.10.

It may also be reasonable to add the shear resistance of the base of the pier to the horizontal resistance of the side of the pier. The base shear resistance can be estimated as

$$H_{ub} = V \cdot \tan\delta \quad (7.52)$$

where V is the applied vertical load applied plus pier weight and $\delta$ the friction angle between pier base and underlying soil. This can be taken equal to the angle of internal friction $\phi$ of the underlying soil.
5. For torsional loading, the ultimate torsional load may be taken as the shaft friction around the perimeter of the equivalent pier. It would seem reasonable to assume that the torsional shaft friction is the same as the shaft friction for vertical loading.
6. Following on the concepts of combined loading for shallow foundations (Section 7.7.2.1), but modifying it to simplify the analysis, the approximate failure criterion for the equivalent pier may then be expressed as follows:

$$f = (V/V_u)^2 + (H_x/H_{ux})^2 + (H_y/H_{uy})^2 + (T/T_u)^2 \leq 1.0 \quad (7.53)$$

where V is the applied vertical load, $V_u$ the design ultimate vertical load capacity, $H_x$ the applied horizontal load in x-direction, $H_{ux}$ the design ultimate lateral load capacity in x-direction, $H_y$ the applied horizontal load in y-direction, $H_{uy}$ the design ultimate lateral load capacity in x-direction, T the applied torsion and $T_u$ the design ultimate torsional capacity.

The effects of applied moment can be accommodated by computing the equivalent eccentricity of horizontal load and incorporating it into the calculation of the ultimate lateral load capacities. The design values of capacity in the above equation will be the computed ultimate values multiplied by the geotechnical reduction factor $\phi_g$.

The above approach is tentative and its reliability has yet to be proven. It may therefore be desirable to use a Category 3 method to obtain more reliable results.

#### 7.8.4.2 Category 3 approach

The most effective means of carrying out a Category 3 analysis is to employ a pile group analysis program that considers non-linear pile–soil response and pile–soil–pile interaction, and is capable of handling all six components of loading simultaneously (e.g. the commercially available programs REPUTE or GROUP8). In employing such programs, the following procedure can be used:

1. First, to check whether the design criterion for geotechnical strength (Equation 7.3) for the whole foundation system is satisfied:
   a. The axial and lateral capacities of the piles, and the raft for a piled raft, are reduced by the geotechnical reduction factor $\phi_g$ (see Section 7.3.1).
   b. The specified ultimate limit state load combinations are applied in turn, and the analysis is run to see if the foundation system with the reduced axial and lateral capacities can sustain the loads without failure. If so, then the geotechnical strength criterion in Equation 7.3 is satisfied. If not, then the foundation system has to be fortified in some way until it is found to be adequate. It should be emphasised that the geotechnical capacity of the individual components of the foundation system, particularly the piles, do not need to be checked, as such a process is unnecessarily onerous and could result in a very uneconomical design.
2. Second, to check whether the design criterion for structural strength (Equation 7.2) is satisfied:
   a. The axial and lateral capacities of the piles, and the raft if present, are left unfactored. By so doing, the loads and moments developed within the system are not artificially limited by factoring down.
   b. The specified ultimate limit state load combinations are applied in turn, and the analysis is run. Now, the computed loads and moments in each of the piles, and the moments and shear forces in the raft, can be multiplied by an appropriate load factor (e.g. 1.5), and then these values are checked to see if they are below the design structural strength of each component. If so, then the structural strength criterion in Equation 7.2 can be considered to be satisfied. If not, then the elements which do not satisfy the structural strength criterion need to be fortified until they are found to be adequate.

The suggested procedure therefore reflects the approach set out in Table 7.2.

In carrying out the analyses described above, there are a number of computational issues that need to be considered when calculating the loads and moments in the piles and raft, as set out below.

#### 7.8.4.3 Effect of soil modulus values used

In principle, the following values of soil Young's modulus should be employed in the Category 3 pile group analyses:

1. For sustained dead plus live load combinations, long-term (drained) modulus should be used.
2. For load combinations involving wind and earthquake loadings, short-term modulus values should be used.

It has been found that the computed pile head loads and moments are not particularly sensitive to the soil modulus values used. As an example, the case of a pile group with a square array of 16 piles in a 4 × 4 configuration was analysed. Each pile was 45 m long and 2.5 m in diameter, and the centre-to-centre spacing between the piles was 8 m. For the ground profile considered, similar to that encountered at the site of the Incheon 151 Tower (see Chapter 15), each pile had an ultimate axial capacity of 49 MN. The following (factored) load combination was applied:

Vertical load = 320 MN
Lateral wind load = 16 MN
Lateral wind moment = 800 MN m

Three sets of soil properties were used in the pile group analyses, which were undertaken with the proprietary program CLAP (Coffey, 2007):

1. The short-term Young's modulus values, together with the unfactored ultimate pile shaft friction and end bearing values
2. The long-term soil modulus values, together with the unfactored ultimate pile shaft friction and end bearing values
3. The long-term soil modulus values, together with the ultimate pile shaft friction and end bearing values factored by a geotechnical reduction factor $\phi_g$ of 0.6

The following observations were made from these analyses:

1. The vertical and horizontal pile head loads were affected by only a few % by the choice of soil modulus value, when unfactored shaft friction and end bearing values were used. In the more heavily loaded piles, the largest values tended to occur when the long-term modulus values were used.
2. The vertical and horizontal pile head loads were more affected when the shaft friction and end bearing loads were factored down. The effect was typically in the order of 10%–15% in this case.
3. The computed pile head moments were greatly affected by the choice of soil parameters. With unfactored shaft and end bearing resistances, the computed moments in all piles were significantly greater when the long-term modulus values were used than when short-term values were used. With factored shaft and end bearing resistances, some of the moments even changed sign as compared with the unfactored resistances.

On the basis of these analyses, it would appear prudent to employ the following guidelines when computing structural actions within the piles and raft:

1. Employ long-term soil modulus values in the analyses
2. Do not factor down the pile ultimate shaft and end bearing resistances

#### *7.8.4.4 Effect of raft flexibility*

Most of the commercially available pile analysis programs assume that the pile cap or raft is rigid. While this may be a reasonable assumption for a small group of piles, it becomes increasingly inaccurate as the size of the raft increases. The flexibility of the raft can have a significant influence on the computed load distribution, as evidenced by the following example (Chow and Poulos, 2015).

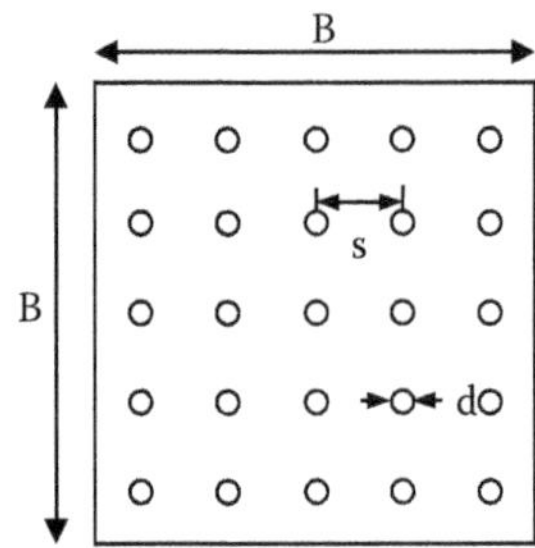


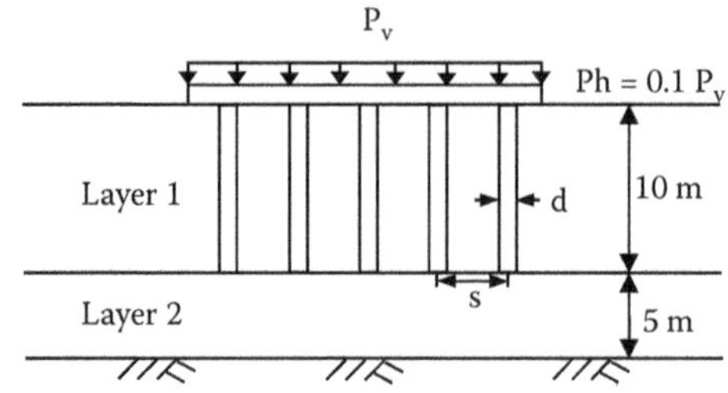


*Figure 7.11* Foundation embedded in a two-layered soil. (Adapted from Chow, H.S.W. and Poulos, H.G. 2015. The significance of raft flexibility in pile group and piled raft design. *Proceedings of the 15th ANZ Conference Geomechanics*, Wellington.)

In this simple example, pile groups of different sizes are embedded in a two-layered soil model as shown in Figure 7.11, and the properties of soil materials are summarised in Table 7.15. The piles have a diameter of 0.6 m, with a length of 15 m and a Young's modulus of 30,000 MPa. The spacing (s) between the piles is taken as three times the pile diameter, d (i.e. s = 3 × d).

Three cases are analysed, 3 × 3, 5 × 5 and 7 × 7 square groups. For each case, both a relatively rigid cap (5 m thick) and a relatively flexible cap (0.5 m thick) are considered. Analyses have been carried out for the pile groups subjected to (a) a uniform axial load and (b) a uniform horizontal load, equal to 10% of the uniform vertical load. The magnitude of the axial load ($P_v$) is taken as the total ultimate capacity ($P_u$) of the pile group divided by an overall factor of safety (FOS) of 2.5 (i.e. $P_v = P_u/2.5$).

Figure 7.12a and b shows, for the central central and corner piles respectively, normalised axial load (P/Pav) versus number of piles in the foundation with rigid and flexible rafts where P is the load on pile and Pav the average load on pile (total applied load/number of piles) for the centre and corner piles.

For a 3 × 3 pile group with a rigid raft, the axial loads in the centre and corner piles are similar to those with a flexible raft. However, as the size of the pile group increases, the centre piles with the flexible raft tend to carry higher loads than the case with the rigid raft, while the corner piles with a flexible raft are carrying lower loads than in the case of a rigid raft.

As shown in Figure 7.12a and b, the axial loads in the pile group with a flexible raft are relatively uniformly distributed, but with a rigid raft, the corner piles are carrying higher loads than the centre piles. As the size of pile group with a rigid raft increases, the difference between the loads carried by the centre and corner piles increases.

Based on the three cases considered, for a small pile group, the assumption of a rigid raft will generally be adequate for assessing the axial loads in the pile. However, as the size of pile group increases, a flexible raft assumption should be used for the pile axial load assessment.

In most cases involving tall buildings, the loading will not be uniformly distributed, but will involve concentrated column loads. In such cases, proper modelling of the raft flexibility may be even more important than with a uniformly distributed loading.

*Table 7.15* Soil properties for example

| *Material* | *Thickness (m)* | *Young's Modulus, $E_s$ (MPa)* | *Skin friction, $f_s$ (kPa)* | *End bearing, $f_b$ (MPa)* |
|---|---|---|---|---|
| Layer 1 | 10 | 20 | 25 | – |
| Layer 2 | 5 | 100 | 120 | 6 |

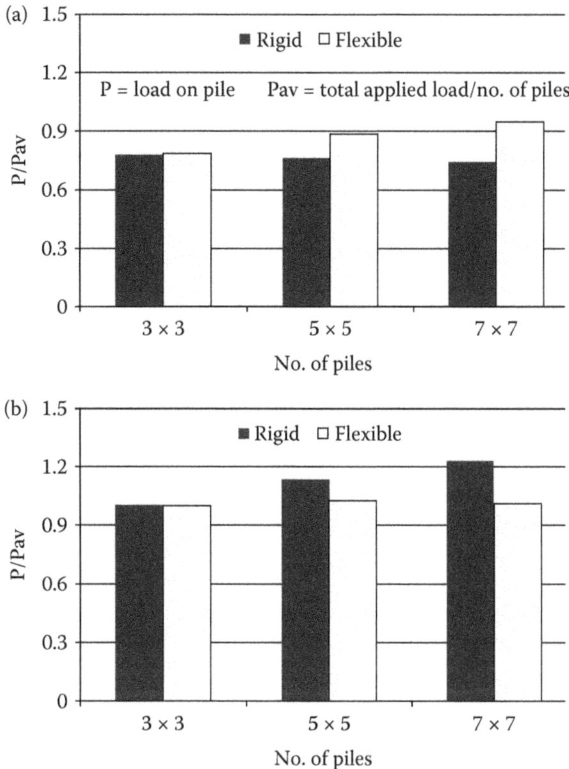


*Figure 7.12* (a) Effect of raft flexibility on axial load on central piles. (b) Effect of raft flexibility on axial load on corner piles. (Adapted from Chow, H.S.W. and Poulos, H.G. 2015. The significance of raft flexibility in pile group and piled raft design. *Proceedings of the 15th ANZ Conference Geomechanics*, Wellington.)

## 7.9 NEGATIVE SKIN FRICTION: A PRACTICAL DESIGN APPROACH

### 7.9.1 Introduction

It has long been recognised that piles located within a settling soil profile will be subjected to negative skin friction. Despite the widespread recognition of negative skin friction, there remains a misconception that this phenomenon will reduce the ultimate geotechnical axial load capacity of a pile (termed here the geotechnical capacity). As pointed out by Fellenius (1991) and Poulos (1997a), among many others, this concept is not valid. Because geotechnical failure of a pile requires that the pile moves (or 'plunges') past the soil, negative skin friction cannot be present when this happens, and so the geotechnical capacity will not be reduced by negative skin friction unless there is strain softening at the pile–soil interface. This is unlikely to occur in soft clays, for which the problem of negative skin friction is most prevalent.

The key issues related to negative skin friction are as follows:

- It will induce additional axial forces in the pile. Fellenius (1991, 2004) has suggested the terminology 'drag force' for this induced force.
- It will cause additional settlement of the pile which Fellenius (1991, 2004) has termed 'downdrag'. However, to avoid confusion with other connotations of the term 'downdrag', the term 'drag settlement' will be used herein to refer to this additional settlement induced by negative skin friction.

The issue of settlement when negative friction acts will be considered in Chapter 9, as it is essentially a serviceability problem, and so this present section will focus on the additional axial forces which are induced in the pile and which have to be taken into account in the structural design of the pile for the ultimate limit state condition. However, it is still necessary to address the issue of ground movements when dealing with this problem. The design approach described below falls into Category 2. Category 3 methods could also be employed using, for example, finite element simulation, but such analyses would generally require a full 3D treatment, and this level of complexity may not always be warranted because of the often limited available information related to ground movements and how they are developed.

### 7.9.2 The negative friction problem

The general problem of negative skin friction acting on a single pile is illustrated in Figure 7.13, where a pile is situated within a soil layer or layers which are settling, and below which there are one or more layers which are not settling. The upper layer will be termed the 'settling layer' and the underlying layer(s) will be termed the 'stable layer'. For simplicity, only a single settling layer and a single stable layer are shown in Figure 7.13. The pile is loaded by an axial force $P_A$ and, again for simplicity, the settlement profile is assumed to decrease linearly with depth from a maximum value So at the ground surface to zero at the base of the settling layer.

### 7.9.3 Design for geotechnical capacity

The presence of negative skin friction does not generally reduce the geotechnical capacity of a pile, and so, for the conventional design approach involving an overall factor of safety, the design requirement for geotechnical capacity may be expressed as follows:

$$R_{ug} = FS \cdot P_w \tag{7.54}$$

where $R_{ug}$ is the ultimate geotechnical capacity of pile (making no allowance for negative friction), FS the overall factor of safety and $P_w$ the working load applied to pile.

In terms of limit state design, the criterion set out in Equation 7.3 remains.

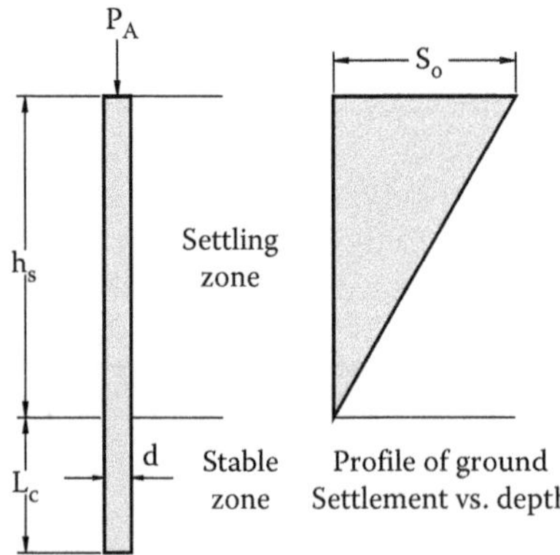


*Figure 7.13* Pile subjected to negative friction. (Adapted from Poulos, H.G. 2008a. A practical design approach for piles with negative friction. *Proceedings of the Institution of Civil Engineers, Geotechnical Engineering*, 161(GE1): 19–27.)

### 7.9.4 Design for structural capacity

In terms of overall factor of safety:

$$R_{us} = FS_s \cdot P_{max} \tag{7.55}$$

where $R_{us}$ is the ultimate structural strength, $FS_s$ the factor of safety for structural strength and $P_{max}$ the maximum axial force in pile, including the applied load and the drag force.

In terms of limit state design, the criterion in Equation 7.2 remains.

In computing $P_{max}$, it is usual to consider various combinations of the applied dead, live, wind and earthquake loads, in addition to the maximum drag force $P_{Nmax}$. Typical load factors would be 1.25–1.3 for dead load, 1.5 for live load, 1.0 for wind loading and earthquake loading and 1.2 for the drag force. Most codes will have specific combinations of these loads and forces that have to be considered.

The value of $P_{Nmax}$ can be computed as the drag force at the neutral plane, which is the depth ($z_N$) at which the friction changes from negative to positive, and which is also the depth at which the soil settlement and the pile settlement are equal. Conservatively, this depth can often be taken as the depth of soil movement, that is, at the base of the settling soil layer(s). Alternatively, a more detailed estimation of $z_N$ can be made, using, for example, the approach described by Poulos (1997a).

$P_{Nmax}$ can be estimated on the assumption that full mobilisation of negative skin friction above the neutral plane has occurred, so that

$$P_{Nmax} = \sum f_N \cdot C\,\delta l \tag{7.56}$$

where $f_N$ is the negative skin friction (usually taken to be equal to the positive skin friction), C the pile circumference and $\delta l$ the length increment along pile, and the summation is carried out from the top of the pile to the neutral plane.

### 7.9.5 Design approach considering the pile head settlement

The estimation of settlement of a pile subjected to negative friction has been considered by Fellenius (1991), Poulos (2008a) and several other authors, and will be discussed in more detail in Chapter 9. However, in many cases, designing the pile so that it does not continue to settle with increasing ground settlement may be a desirable condition, and leads to an alternative design criterion which is described in more detail below.

To avoid having the pile settle continually as the ground settles, Poulos (2008a) has shown that the portion of the pile located below the depth of ground movement should be designed to have an adequate margin of safety against the combined effects of the applied loads and the maximum drag force. It can be shown that under these circumstances, the depth of the neutral plane then lies below the depth of soil movement. This criterion may be expressed as follows, in terms of the conventional factor of safety:

$$R_{ug2} \geq FS_2(P_A + P_{Nmax}) \tag{7.57}$$

where $R_{ug2}$ is the ultimate geotechnical capacity of the pile in the stable zone below the depth of soil settlement and $FS_2$ the factor of safety for that portion of pile in the stable zone.

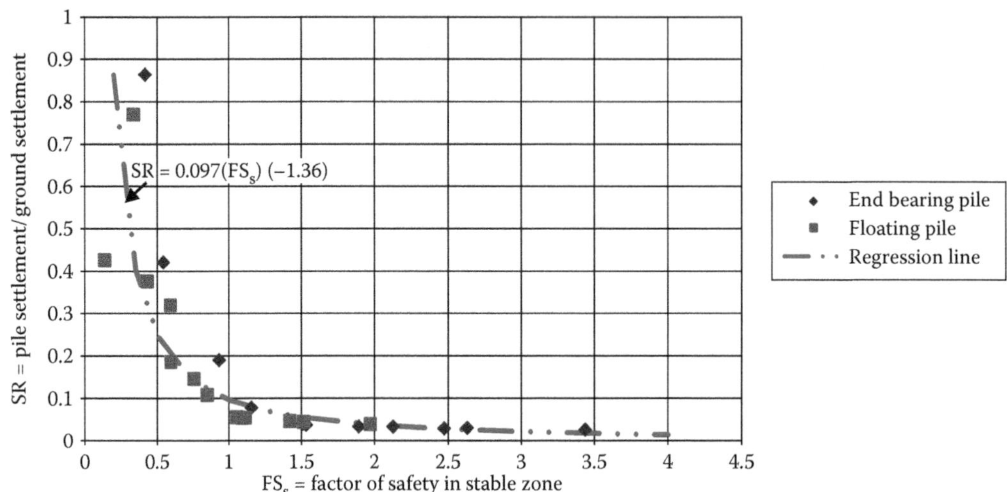


*Figure 7.14* Computed dimensionless drag settlement versus factor of safety in stable zone. (Adapted from Poulos, H.G. 2008a. A practical design approach for piles with negative friction. *Proceedings of the Institution of Civil Engineers, Geotechnical Engineering,* 161(GE1): 19–27.)

To evaluate the proposed alternative design criterion, two hypothetical but typical problems have been analysed by Poulos (2008a). The first involves a single 'end bearing' pile located in a 20 m thick soft clay layer which will experience a ground surface settlement of 100 mm, underlain by a stiff clay layer. The second case involves an identical settling layer as for the end bearing pile, but the underlying layer is a medium clay layer with considerably smaller strength and stiffness than in the first case. This will be denoted as the 'floating pile' case. For simplicity, the results will be described in terms of the conventional factor of safety concept.

For cases involving both friction piles and end bearing piles, Figure 7.14 plots the ratio of the drag settlement $S_D$ to the ground surface settlement So versus $FS_2$. For both the floating and end bearing cases, and it can be seen that $S_D$/So decreases with increasing $FS_2$, that is, the relative drag settlement reduces as the factor of safety in the stable zone, $FS_2$, increases. A single regression line can be drawn through the points, and as indicated in Figure 7.14, beyond about $FS_2 = 1.25$, there is little further reduction in the relative drag settlement.

From this limited study, it would appear that, from a practical design viewpoint, the use of a factor of safety in the stable zone, $FS_2$, of 1.25 should be adequate to avoid having the piles settle continuously.

### 7.9.6 Cases where soil settlement occurs to considerable depth

In most foundation designs, emphasis is placed on minimising settlements, and this often means supporting the structure on end bearing piles which are founded on rock or on a stiff stratum. However, there may be cases in which such a strategy is neither feasible nor practical, for example, where there is a deep layer of soft clay, most of which is subjected to ground settlements. Such situations are common in certain urban areas (e.g. Bangkok, Mexico City, Houston) because of the pumping of groundwater for water supply. In such cases, it is almost futile to attempt to stop the pile settling as the ground continues to settle. Instead, it seems preferable to accept that continuing settlement of the foundation is inevitable, and to then attempt to have the foundation settle the same amount as the ground. In this way, excessive differential settlements between the structure and the surrounding ground are avoided. In such cases, a proper pile–soil interaction analysis should be carried out to identify the length of piles for which the difference between the pile head settlement and the ground surface settlement is an acceptable value. Poulos (2005c) describes the application of this design philosophy to piled raft foundations.

### 7.9.7 Effects of live load

There is a perception among some engineers that the application of live load can remove the effects of negative skin friction and reduce drag forces. To examine the validity of this concept, the example of an end bearing pile has been examined by Poulos (2008a). The pile, of total length 25 m (with a 5 m embedment into the stable zone), has been subjected to the following history:

1. Dead load of 1.0 MN applied (representing an overall factor of safety of about 3).
2. Application of ground settlement linearly decreasing from 100 mm at the ground surface to zero at 20 m depth.
3. Application of additional (live) loads of increasing magnitude.

Figure 7.15 shows the computed relationship between maximum pile load and the additional live load, while Figure 7.16 shows the corresponding relationship for pile head settlement. These figures show that the maximum load in the pile, and the pile head settlement, continues to increase with increasing live load. When the applied live load is approximately equal to the dead load, the maximum load equals the applied load, that is, the drag force due to the ground settlement is reduced such that the maximum load is now at the pile head. The pile head settlement also becomes similar to the settlement that would have occurred if the ground settlement had not been imposed. From a practical viewpoint, it would appear that, at least in the example considered, the amount of live load that would need to be added to relieve the negative friction effects is far greater than would normally be allowed. Thus, it may be concluded that negative friction effects are unlikely to be completely removed when normal magnitudes of live load are applied.

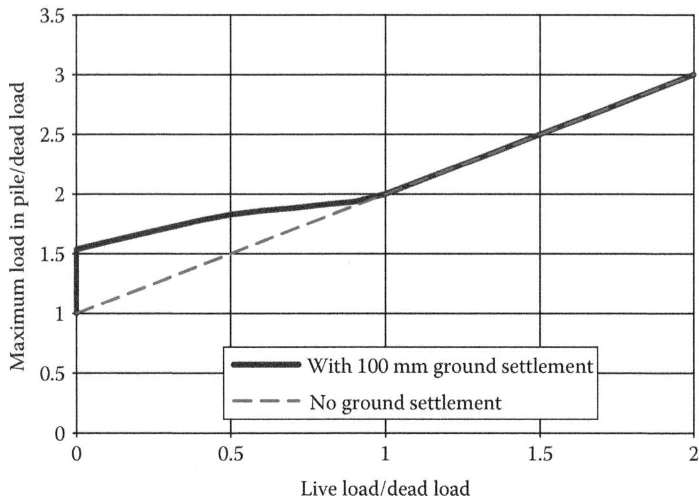


*Figure 7.15* Effect of live load on maximum pile load. (Adapted from Poulos, H.G. 2008a. A practical design approach for piles with negative friction. *Proceedings of the Institution of Civil Engineers, Geotechnical Engineering*, 161(GE1): 19–27. Courtesy of ICE Publishing.)

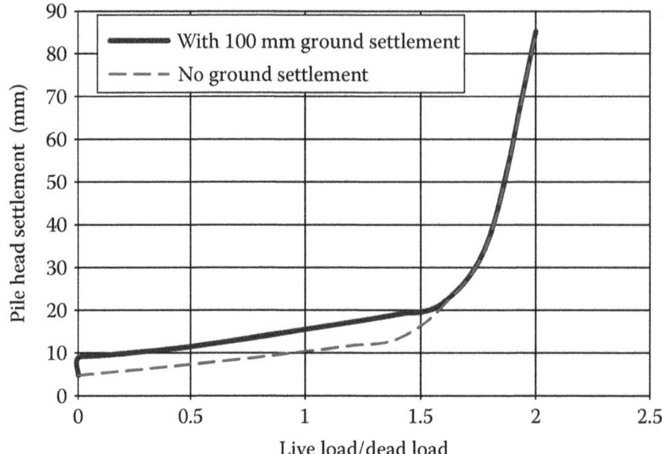


*Figure 7.16* Effect of live load on pile head settlement. (Adapted from Poulos, H.G. 2008a. A practical design approach for piles with negative friction. *Proceedings of the Institution of Civil Engineers, Geotechnical Engineering*, 161(GE1): 19–27. Courtesy of ICE Publishing.)

### 7.9.8 Group effects

It is becoming recognised that group effects may be beneficial in relation to the effects of negative skin friction. To examine the general nature of group effects, Poulos (2008a) has analysed a group of nine end bearing piles, as shown in Figure 7.17, with the ground profile being that of the end bearing case considered previously. Each pile is assumed to have

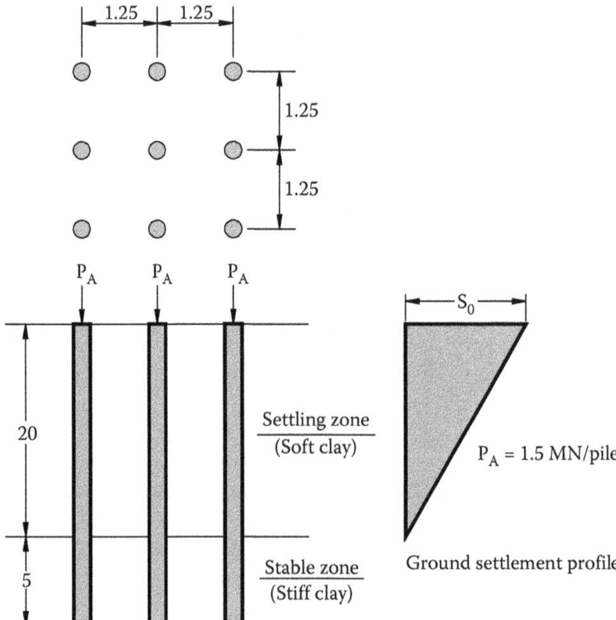


*Figure 7.17* Pile group example. (Adapted from Poulos, H.G. 2008a. A practical design approach for piles with negative friction. *Proceedings of the Institution of Civil Engineers, Geotechnical Engineering*, 161(GE1): 19–27. Courtesy of ICE Publishing.)

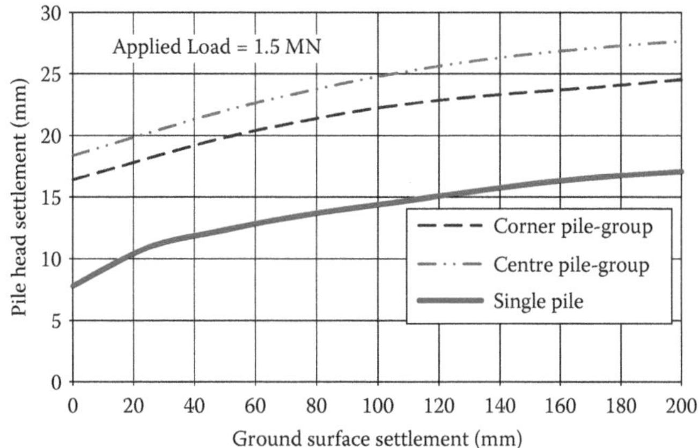


*Figure 7.18* Pile settlement versus ground surface settlement for various piles in group. (Adapted from Poulos, H.G. 2008a. A practical design approach for piles with negative friction. *Proceedings of the Institution of Civil Engineers, Geotechnical Engineering*, 161(GE1): 19–27. Courtesy of ICE Publishing.)

a length of 25 m and to be subjected to a load of 1.5 MN, thus giving an overall factor of safety of about 2 against geotechnical failure. A ground surface settlement is then imposed on the piles of 200 mm, decreasing from a maximum at the surface to zero at 20 m depth. The induced pile loads and settlement are examined for the corner and centre piles of the group, and also for a single isolated pile.

Figure 7.18 shows the computed pile head settlement as a function of the ground surface settlement. It can be seen that

- The pile head settlements increase (but at a diminishing rate) with increasing soil surface settlement
- The centre pile settles more than the corner pile
- Both centre and corner piles in the group settle considerably more than a single isolated pile

It is also found that the maximum pile loads within the group increase with increasing ground settlement, with the load in the centre pile being less than for the corner pile. The rate of increase of load with increasing settlement for both the group piles is however significantly slower than for a single isolated pile. It is not until relatively large ground settlements occur that the loads in the group and single piles become similar. This characteristic is consistent with that found by Kuwabara and Poulos (1989). It can therefore be concluded that group effects may be beneficial in terms of the induced loads in the piles, especially for relatively small magnitudes of ground movement. However, at normal working loads, the pile head settlement is still increased because of group effects.

Briaud and Tucker (1996) have proposed a simple empirical approach for estimating the distribution of drag load in piles within a group in terms of the computed drag force for a single pile, in which the maximum drag load on a pile in a group $P_{Ng}$ is related to that in a single isolated pile, $P_{Nmax}$, as follows:

$$P_{Ng} = \chi \cdot P_{Nmax} \tag{7.58}$$

*Table 7.16* Empirical group factor $\chi$

| | *Empirical group factor* $\chi$ | |
|---|---|---|
| *Pile location* | *Spacing/diameter* = 2.5 | *Spacing/diameter* = 5 |
| Corner | 0.5 | 0.9 |
| Outer side | 0.4 | 0.8 |
| Interior | 0.15 | 0.5 |

where $\chi$ is the empirical group factor.

Values of $\chi$ are shown in Table 7.16 for various ratios of centre-to-centre spacing, s, to diameter, d, within the group, and for piles located at the corner, outer side and interior of the group. The group effect is most marked when the piles are closely spaced, and for piles in the interior of the group which are 'shielded' from the effects of the ground movements by piles nearer the edge.

### 7.9.9 A note on conservatism in design for ground movements

Geotechnical designers have been conditioned to assume that selecting conservative geotechnical parameters for foundation design involves factoring the selected ground strength parameters down to a smaller value. This is indeed relevant when the foundation is subjected to imposed structural loads which have to be supported by the ground. However, when the ground itself becomes the agent of loading, via externally imposed ground movements, then the process of factoring down the strength parameters becomes unconservative, as it places a limit on the actions which can be imposed on the foundation by the ground. On the other hand, if the strength parameters are factored up, then the resistance that the ground can provide against the induced foundations actions can be overestimated.

Since the ground can act, at the same time, as the agent of loading and of resistance, the preferable approach is to *not factor* the geotechnical strength parameters, but to use the best-estimate values in the analysis, and then to factor up the consequent computed actions (axial forces, bending moments and shear forces) to obtain design values.

## 7.10 ASSESSMENT OF PILE LOCATIONS

The arrangement of the piles will depend largely on the distribution of column and core loads, with piles being concentrated in the more heavily loaded areas. In deciding where piles may be required, it is helpful to estimate the maximum column load that can be supported by the raft or basement slab without requiring pile support, and such an estimation can be made using the approach suggested by Poulos (2001b).

The circumstances in which a pile may be needed below the column are as follows:

- If the maximum moment in the raft below the column exceeds the allowable value for the raft
- If the maximum shear in the raft below the column exceeds the allowable value for the raft
- If the maximum contact pressure below the raft exceeds the allowable design value for the soil
- If the local settlement below the column exceeds the allowable value

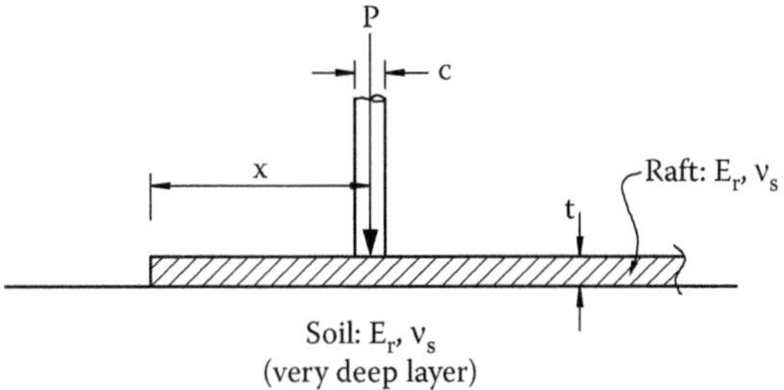


*Figure 7.19* Definition of individual column on a raft or slab.

To estimate the maximum moment, shear, contact pressure and local settlement caused by column loading on the raft, use can be made of the elastic solutions summarised by Selvadurai (1979). These are for the ideal case of a single concentrated load on a semi-infinite elastic raft supported by a homogeneous elastic layer of great depth, but they do at least provide a rational basis for design. It is possible also to transform approximately a more realistic layered soil profile into an equivalent homogeneous soil layer by using the approach described by Fraser and Wardle (1976). Figure 7.19 shows the definition of the problem addressed, and a typical column for which the piling requirements (if any) are being assessed.

1. Maximum moment criterion
   The maximum moments $M_x$ and $M_y$ below a column of radius c acting on a semi-infinite raft are given by the following approximations:

$$M_x = A_x \cdot P \tag{7.59}$$

$$M_y = B_y \cdot P \tag{7.60}$$

where $A_x = [A - 0.0928\ (\ln(c/a))]$, $B_y = [B - 0.0928\ (\ln(c/a))]$, A, B are coefficients depending on $\delta/a$, $\xi$ the distance of the column centre line from the raft edge, a the characteristic length of raft, as defined in Equation 7.61 below, t the raft thickness, $E_r$ the raft Young's modulus, $E_s$ the soil Young's modulus, $\nu_r$ the raft Poisson's ratio, $\nu_s$ the soil Poisson's ratio and P the column load.

The characteristic raft length is defined as

$$a = t \cdot [E_r \cdot (1 - \nu_s^2)/6 \cdot E_s \cdot (1 - \nu_r^2)]^{1/3} \tag{7.61}$$

The coefficients A and B are plotted in Figure 7.20 as a function of the dimensionless distance x/a.

The maximum column load, $P_{c1}$, that can be carried by the raft without exceeding the allowable moment is then given by

$$P_{c1} = M_d/(\text{larger of } A_x \text{ and } B_y) \tag{7.62}$$

where $M_d$ is the design moment capacity of raft.

The estimation of $M_d$ is discussed in Section 7.11.

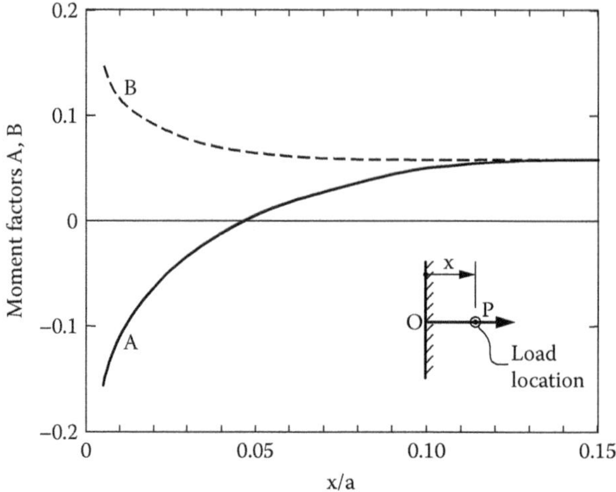


*Figure 7.20* Moment factors A and B for a circular column. (Adapted from Poulos, H.G. 2001b. *Geotechnique*, 51(2): 95–113. Courtesy of ICE Publishing.)

2. Maximum shear criterion
The maximum shear $V_{max}$ below a column can be expressed as

$$V_{max} = (P - q\pi c^2) \cdot C_q / 2\pi c \tag{7.63}$$

where q is the contact pressure below raft, c = the column radius, $C_q$ = the shear factor, plotted in Figure 7.21.

Thus, if the design shear capacity of the raft is $V_d$, the maximum column load, $P_{c2}$, which can be applied to the raft is

$$P_{c2} = V_d \cdot 2\pi c / C_q + q_d \pi c^2 \tag{7.64}$$

where $q_d$ is the design allowable bearing pressure below raft.

The estimation of $V_d$ is discussed in Section 7.11.

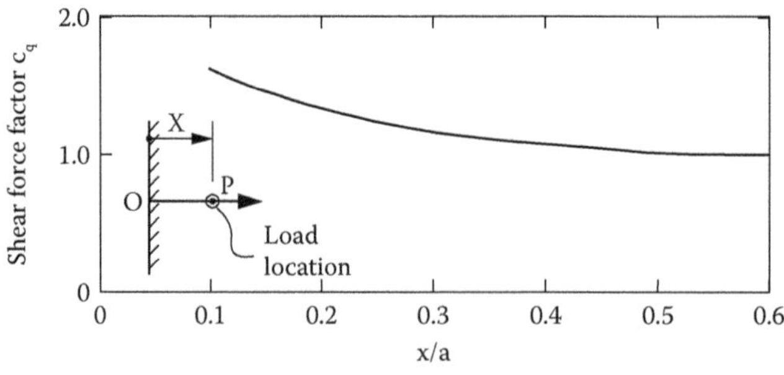


*Figure 7.21* Shear factors $c_q$ for a circular column. (Adapted from Poulos, H.G. 2001b. *Geotechnique*, 51(2): 95–113. Courtesy of ICE Publishing.)

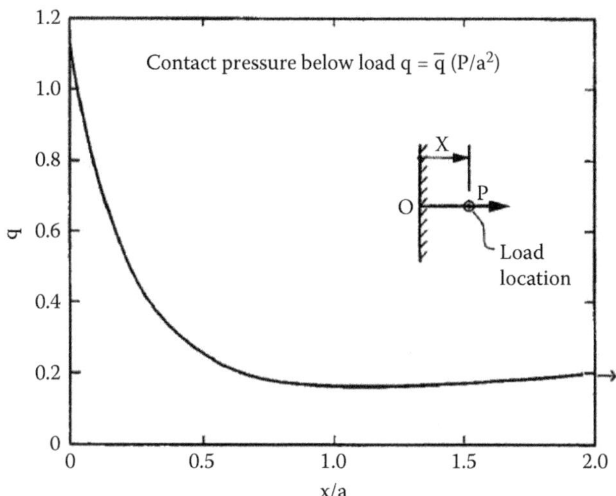


*Figure 7.22* Contact pressure factor q. (Adapted from Poulos, H.G. 2001b. *Geotechnique*, 51(2): 95–113. Courtesy of ICE Publishing.)

3. Maximum contact pressure criterion
   The maximum contact pressure on the base of the raft, $q_{max}$, can be estimated as follows:

$$q_{max} = \bar{q} \cdot P/a^2 \tag{7.65}$$

   where $\bar{q}$ is the factor plotted in Figure 7.22 and a the characteristic length as defined in Equation 7.44.

   The maximum column load, $P_{c3}$, which can be applied without exceeding the allowable contact pressure is then

$$P_{c3} = q_u a^2/(F_s \cdot \bar{q}\ ) \tag{7.66}$$

   where $q_u$ is the ultimate bearing capacity of soil below raft and $F_s$ the factor of safety for contact pressure.

   The estimation of $q_u$ for a raft is discussed in Section 7.7.
4. Local settlement criterion
   The settlement below a column (considered as a concentrated load) is given by

$$S = \omega(1 - \nu_s^{\,2})P/(E_s \cdot a) \tag{7.67}$$

   where $\omega$ is the settlement factor plotted in Figure 7.23.

It should be recognised that this expression does not allow for the effects of adjacent columns on the settlement of the column being considered, and so is a local settlement which is superimposed on a more general settlement 'bowl'.

If the allowable local settlement is $S_a$, then the maximum column load, $P_{c4}$, so as not to exceed this value is then

$$P_{c4} = S_a E_s a/(\omega(1 - \nu_s^{\,2})) \tag{7.68}$$

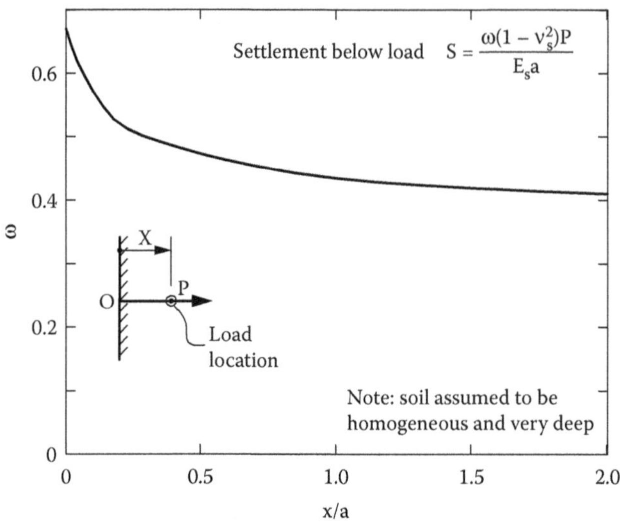


*Figure 7.23* Settlement factor ω. (Adapted from Poulos, H.G. 2001b. *Geotechnique*, 51(2): 95–113. Courtesy of ICE Publishing.)

The estimation of Young's modulus $E_s$ of the ground is discussed in Chapter 6.

If the actual design column load at a particular location is $P_c$, then one or more piles will be required if $P_c$ exceeds the least value of the above four criteria, that is, if

$$P_c > P_{crit} \tag{7.69}$$

where $P_{crit}$ is the minimum of $P_{c1}$, $P_{c2}$, $P_{c3}$ or $P_{c4}$.

The required number of piles below a column can then be estimated via the procedure described by Poulos (2001b). Alternatively, as a very rough indication, for the purposes of preliminary estimation, it may be adequate to estimate the required number of piles, N, as follows:

$$N = (P_c - P_{crit}) \cdot FS/P_u \tag{7.70}$$

where FS is the nominated local factor of safety, for example, 1.5, and $P_u$ the ultimate geotechnical capacity of each pile.

## 7.11 STRUCTURAL DESIGN ASPECTS

### 7.11.1 Introduction

The scope of this book does not extend to details of the structural design of foundations, but it is nevertheless important that the geotechnical designer has a reasonable appreciation of the issues involved in structural design. Accordingly, this section sets out, very briefly, some of these issues, and provides some guidance to the geotechnical designer on the shear and moment capacities of a raft slab and of circular piles. The design and detailing of the necessary reinforcement to achieve these capacities is normally undertaken by the structural designer.

### 7.11.2 Soil–Structure interaction

As discussed by Bull (2012), the structural elements within the foundation and the sub-structure include the raft, the piles, the embedded walls, and possibly shallow pad or strip foundations. The key to developing a sound soil–structure model is to gain a good appreciation of the project on a global basis, and not to focus solely on the geotechnical or structural elements. This requires a close and continued interaction within the project design team, and a cooperative approach between the architect, the structural engineer and the geotechnical engineer.

With most of the commonly employed structural software packages, the most convenient means by which the geotechnical designer can provide input to the structural design is by providing relevant values of the equivalent spring stiffness for piles, and for the soil supporting the raft if a piled raft design is being undertaken. Such springs should take into account the various interactions between the foundation elements and the soil, and also between the superstructure and the foundation system. Suggested methods of obtaining such spring stiffness values are given in Chapter 8, Section 8.10.

### 7.11.3 Material properties

Apart from the supporting soil, the key materials involved in high-rise foundations are concrete and reinforcing steel. From the viewpoint of foundation design and soil–structure interaction, the most important material properties of concrete are strength (primarily compressive strength), modulus and Poisson's ratio. For steel, the key design parameter is its characteristic yield strength in tension.

Concrete strength is usually described in terms of its unconfined compressive strength, with tests carried out either on cylinders or on cube samples. The strength is usually referred to in terms of the strength at 28 days after the sample has been prepared, and to cater for the method of testing, is commonly referred to as, for example, C50/60 grade, where the '50' refers to the 28-day cylinder strength in MPa, and the '60' refers to the 28-day cube strength, in MPa. In super-tall buildings, concrete grades as high as C70/80 are common, with even higher grades for mega-tall structures.

The short-term Young's modulus of uncracked concrete, $E_c$, is generally related to the concrete strength, and for most concretes, the following approximate correlation is commonly used:

$$E_c \approx 4700\sqrt{f_c'}\ \text{MPa} \tag{7.71}$$

where $f'_c$ is the 28-day compressive cylinder strength of concrete, in MPa.

It must however be recognised that

1. Concrete is not a linearly elastic material. If cracks develop, the Young's modulus of the concrete can reduce significantly from the initial value
2. The concrete modulus can also change with time due to such factors as creep and shrinkage
3. Young's modulus can also vary with the type of aggregate used in the mix. For example, Choubane et al. (1996) found that the use of a dense limestone aggregate, with its rough-textured surface and angular shape, produced a concrete with higher strength and stiffness than concretes made with a porous limestone aggregate or a river gravel

Depending on the objectives of the soil–structure interaction analysis, various values of Young's modulus of the concrete may be appropriate. Bull (2012) suggests that a distinction be made in relation to the short-term situation (during construction) and the long-term situation. In the latter case, the concrete stiffness may be affected by creep, shrinkage and the

*Table 7.17* Typical ratios of long-term to short-term Young's modulus values for reinforced concrete

| *Structural element* | *Ratio of long-term to short-term modulus* |
|---|---|
| Raft (>700 mm thick) | 0.5–0.6 |
| Wall | 0.5–0.6 |
| Suspended slab | 0.5–0.6 |
| Pile in compression | 0.8–1.0 |

development of cracks. Table 7.17 provides guidance for the ratio of long-term to short-term Young's modulus of reinforced concrete elements within a foundation system.

Poisson's ratio of concrete is typically within the range 0.1–0.2, but is generally not a critical parameter.

For the steel reinforcement, it is common to use high-strength deformed bars, with a characteristic yield strength of 500 MPa. The steel bars may vary in nominal diameter from 8 to 40 mm.

### 7.11.4 Raft design

The main actions for which the raft has to be designed are bending moment and punching shear. These values are generally obtained from a soil–structure interaction analysis, and are dependent mainly on the magnitude and distribution of the applied column and wall loadings. The raft thickness and amount of reinforcement required will depend on these actions, together with the concrete strength, the tensile strength of the reinforcement and the thickness of the raft.

The ultimate bending moment capacity of a raft slab, $M_u$, may be estimated from the following expression derived from CACA (1991):

$$M_u = K_u \cdot d^2 \text{ MN m} \tag{7.72}$$

where $K_u = f_c' q \cdot (1 - q/1.7)$, $f_c'$ is the compressive strength of concrete (MPa), $q = p \cdot f_{sy}/f_c'$, $p = A_{st}/d$ the proportion of steel reinforcement in unit length section, $A_{st}$ the area of reinforcement per unit length (m$^2$), $f_{sy}$ the steel yield stress (MPa), d the (average) distance from extreme fibre of slab to the centre of the steel reinforcement (m) $= g \cdot t_r$, where $t_r$ the slab thickness and g may be typically 0.8–0.9.

An appropriate reduction factor (typically 0.8) is applied to the above value of $M_u$ to obtain the design moment capacity.

For a typical concrete strength, $f_c'$, of 50 MPa and a steel yield stress, $f_{sy}$, of 500 MPa. Figure 7.24 shows computed values of $M_u$ as a function of the raft thickness, $t_r$, assuming that $d = 0.9t_r$, that is, $g = 0.9$.

Assuming that the load is not applied near an edge of the raft, the punching shear capacity of the raft, $V_u$, can be estimated from the following expression from CACA (1991):

$$V_u = u \cdot d \cdot f_{cv} \tag{7.73}$$

where u is the critical shear perimeter, with $u = \pi(D_c + d)$ for a circular column of diameter $D_c$ or $u = [2(c_1 + c_2) + 4d]$ for a rectangular column with side dimensions $c_1$ and $c_2$ $(c_1 > c_2)$, $f_{cv} = 0.34 f_c'$ for a circular column or $f_{cv} = 0.17(1 + 2c_2/c_1)\sqrt{f_c'} \leq 0.34\sqrt{f_c'}$ for a rectangular column.

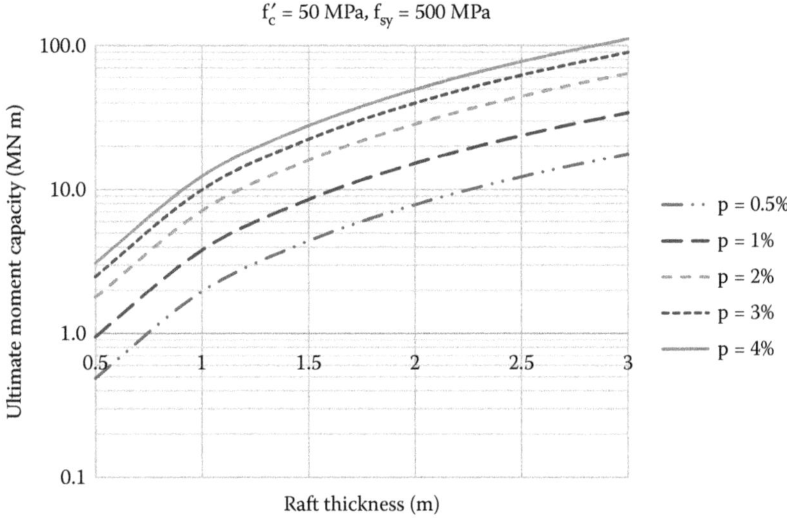


*Figure 7.24* Typical values of design moment capacity for a raft.

Again, an appropriate reduction factor (typically 0.7) needs to be applied to $V_u$ to obtain the design column load that can be sustained by the slab.

Figure 7.25 shows computed values of design punching shear capacity $V_u$ for a circular column of diameter $D_c$, and concrete with $f'_c = 50\,\text{MPa}$.

Preliminary estimates of the moment and shear that are applied to the slab by a column can be obtained from the approximate approach set out in Section 7.10.

## 7.11.5 Pile design

The structural design of the piles will depend on the axial load, the bending moment and the shear force acting on the pile. The amount of reinforcement required will depend on

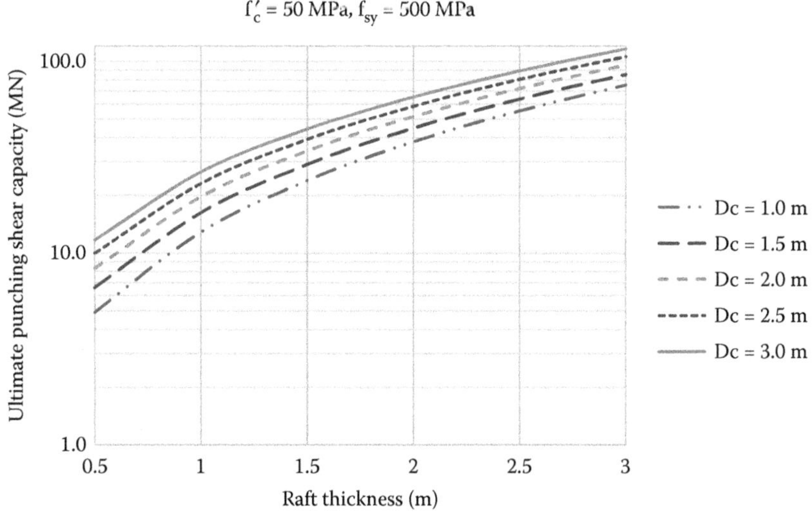


*Figure 7.25* Typical values of design punching shear capacity $V_u$ for a raft ($D_c$ = column diameter).

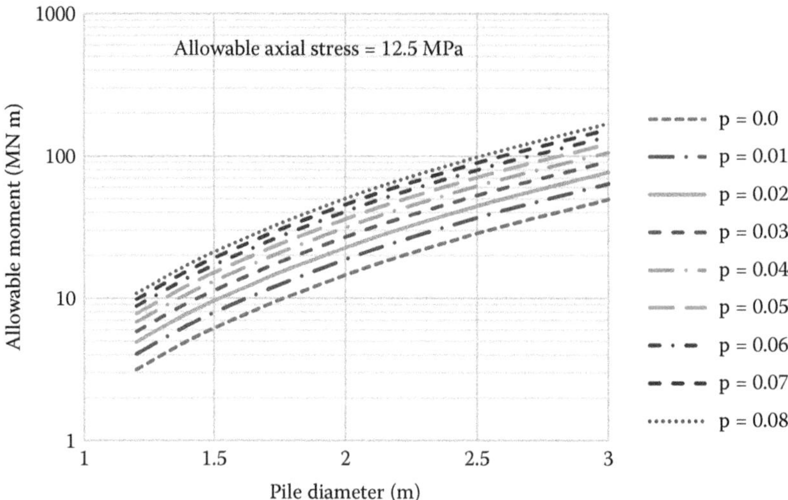


*Figure 7.26* Typical relationship between design moment, pile diameter and steel percentage p.

these actions and on the concrete and steel strengths and the pile diameter. By making some simplifying assumptions, it is possible to obtain approximate relationships for the design bending moment of a pile subjected to axial load. One such series of relationships is shown in Figure 7.26, and has been computed on the following assumptions:

1. The allowable axial stress in the pile is 12.5 MPa and has been applied to the pile. For a concrete strength of $f_c' = 50\,\text{MPa}$, this would imply a strength reduction factor of 0.25.
2. The centreline of the steel reinforcement is located 0.1d from the outside periphery of the pile of diameter d.
3. The yield stress of the steel reinforcement is 500 MPa.

Figure 7.26 should be used only as a rough guide to enable the geotechnical designer to assess whether the proposed pile diameter and reinforcement percentage are likely to be satisfactory.

### 7.11.6 Wall design

The structural design of the basement wall will again depend on the computed bending moments and shears acting on the wall. These will in turn depend on the assessed distributions of earth and water pressure on the wall, together with the details of the supports, which may include anchors, struts or basement floors. Chapter 12 provides a discussion on these aspects of wall design and the methods, both simplified and more detailed, by which design values of wall moments and shears can be obtained. The assessment of the required reinforcement is then the task of the structural designer.

## 7.12 DESIGN FOR DURABILITY

### 7.12.1 Introduction

Durability can be broadly defined as the ability of a material to resist the exposure environment while maintaining its required engineering properties. Durability is not a specific property of a material but relates to the environment to which the material is exposed. For example, a concrete pile with adequate durability for non-aggressive soils might have inadequate durability in a coastal or marine environment.

For civil engineering projects, the construction materials used are required to be durable for the specified design life of the structure. For above-ground structures, periodic inspection, maintenance and repairs are commonly undertaken on accessible components at various times during the design life. However, for footings and other buried elements, such inspection and maintenance programs are generally impractical or economically unviable. As a result, the durability of foundations is more difficult to assess than above-ground structural members, such as beams, columns or slabs, which are more likely to be accessible for maintenance.

The design life of the foundations or other buried structural elements must therefore be equal to or exceed the required design life of the structure, whereas some above-ground components of a structure may be designed for a shorter design life. Durability design must also consider how the various durability requirements will be specified and controlled during construction.

In order to adequately consider durability issues design in foundation design, it is necessary to

- Obtain, through a desk top review and site specific testing, a thorough understanding of the environment which will exist around the foundation during the design life, that is, the exposure conditions
- Assess how the various construction materials used in the foundation will react to the exposure conditions during the nominated design life
- Specify relevant engineering properties for the various construction materials which will be required to achieve the required design life, for example, minimum concrete strength or minimum cover to steel reinforcement

For foundations, the environmental conditions in the surrounding soil, groundwater and seawater (for marine structures) are generally more important than the conditions in the atmosphere, which are generally more relevant to above-ground structures. Nonetheless, some foundations may be exposed to a variety of environmental conditions and it is important to understand these conditions in each exposure zone and design the entire foundation for the most aggressive environment.

### 7.12.2 Durability design considerations

To achieve a durable foundation in an aggressive environment (i.e. one where virtually no maintenance or replacement is required during this design life), specific consideration must be given to the engineering characteristics of the materials used and whether additional protective measures are required.

For reinforced concrete structures, the provision of an adequate thickness of concrete cover to the steel reinforcement, and reducing the permeability of the concrete, are essential to restrict the ingress of chlorides or carbon dioxide. Modifications to the concrete mix to enhance durability include the use of supplementary cementitious material (e.g. fly ash, blast furnace slag or silica fume) and the reduction of the water cement ratio.

Other protective measures for reinforced concrete include protective coatings, cathodic protection, corrosion inhibitors and the use of galvanised or stainless steel reinforcement.

While it is virtually impossible to prevent steel corrosion in reinforced concrete, the corrosion must not be allowed to progress to the point where the strength or serviceability of the structure is impaired.

### 7.12.3 Assessment of the environmental conditions

Laboratory testing to assess the environmental conditions in the ground at a particular site is an important part of the durability design process. For buried foundations, such testing is commonly undertaken as part of the geotechnical investigation for the project.

In the past, geotechnical investigations have focussed mainly on the mechanical properties of the soil (such as shear strength and compressibility) and these properties are of course relevant to ultimate and serviceability limit state design. For durability considerations, the chemical properties of the soil and groundwater are more relevant. Such properties include

- pH, sulphates and chlorides
- Resistivity
- Presence of acid sulphate soils
- Oxygen levels
- Presence of sulphate-reducing bacteria

In situ soil permeability is a relevant physical parameter for durability as it controls the rate of groundwater flow through saturated soils and hence the rate of replenishment of aggressive chemicals.

The required suite of laboratory analyses for a particular project should initially be assessed on the basis of pre-existing information for the site, site reconnaissance, and knowledge of the type of foundation proposed. The suite should be reviewed during, and on completion of, the field investigation, once the subsurface conditions are more reliably known.

### 7.12.4 Durability of concrete piles

Given that the foundations for tall buildings will almost certainly require the use of concrete piles, attention here is focussed on the requirements for such piles. The durability hazards associated with reinforced concrete foundation elements may include the following:

1. Abrasion
2. The effects of the freeze-thaw cycle
3. Aggressive soils, including acid sulphate soils and saline soils
4. Biological attack
5. Chemical attack
6. The effects of fire
7. Construction methods and quality

The concrete may be subjected to various mechanisms of attack, including

1. Carbonation, where the pH of the concrete decreases as the concrete absorbs $CO_2$ from the atmosphere
2. Chloride penetration, whereby chloride ions may reach the reinforcing steel through cracks, or through porous concrete. Contaminated mixing water can also create problems
3. Acid and chemical attack, which may result in actions with acids, ammonium salts, magnesium salts, sulphates and aluminates
4. Alkali-aggregate reactions

The reinforcing steel is also subject to attack via oxidation, resulting in rust. Oxidation requires a low pH environment, depassivation of the region around the steel, an oxygen source and the presence of moisture. In a high pH environment, an oxide layer is formed on the steel, and this tends to inhibit further corrosion.

An example of the durability design requirements for concrete with a design life of 50 and 100 years is presented in Section 6 of AS 2159–2009, '*Piling—Design and Installation*'. The general principles of durability design in this standard are that

- The durability of piles should be assessed from consideration of the aggressivity of the ground and the environmental conditions
- Piles must remain in a safe and serviceable condition to the end of their design life
- Durability of piles also extends to any portion of the pile or pile cap above ground level. The exposure conditions above ground may differ from those below the ground

The standard notes that piles installed in acid sulphate soils require specific durability design considerations to resist acid attack. In particular, the effect of the method of pile construction on the formation of sulphuric acid would need to be considered. For example, driven piles would be preferable in acid sulphate soils as they produce no spoil and expose the surrounding soil to significantly less air than bored piles during construction. If bored piles are used, more stringent corrosion and durability allowances are likely to be required and specific treatment of the excavated soil (e.g. blending with lime) is likely to be required to reduce potential environmental harm. Whether the groundwater in the acid sulphate environment is relatively static, or fluctuates, will also serve as a guide to future ground aggressiveness.

AS 2159-2009 requires durability to be considered in the design of concrete piles by assessing the exposure classification for a particular pile and then for that specific exposure classification specifying:

1. Minimum concrete strength and reinforcement cover
2. Restrictions on content of certain chemicals
3. Limitation on crack width and
4. Selection of concrete aggregates

The exposure classification is selected based on the range of chemical conditions in the soil surrounding the pile. Conditions leading to the most severe aggressive conditions need to be considered as well as likely future changes in groundwater level.

Exposure classifications are provided for concrete piles in water, concrete piles in refuse fill and concrete piles in soil. The latter is reproduced in Table 7.18. Table 7.19 provides

*Table 7.18* Exposure classification for concrete piles in soil

| *Exposure conditions* | | | | *Exposure classification* | |
|---|---|---|---|---|---|
| *Sulphates (expressed as $SO_4$[a])* | | | | | |
| In soil (ppm) | In groundwater (ppm) | *pH* | *Chlorides in groundwater (ppm)* | *Soil conditions A*[b] | [b]*Soil conditions B*[c] |
| <5000 | <1000 | >5.5 | <6000 | Mild | Non-aggressive |
| 5000–10,000 | 1000–3000 | 4.5–5.5 | 6000–12,000 | Moderate | Mild |
| 10,000–20,000 | 3000–10,000 | 4–4.5 | 12,000–30,000 | Severe | Moderate |
| >20,000 | >10,000 | <4 | >30,000 | Very severe | Severe |

Source: AS 2159. 2009. *Piling – Design and Installation*. Standards Australia. Standards Australia.

[a] Approximately 100 ppm $SO_4$ = 80 ppm $SO_3$.
[b] Soil conditions A – high permeability soils (e.g. sands and gravels) which are in groundwater.
[c] Soil conditions B – low permeability soils (e.g. silts and lays) or all soils above groundwater.

*Table 7.19* Concrete strength and reinforcement cover in piles

| | Minimum strength $f_c'$ (MPa) | | Minimum cover to reinforcement | | | |
|---|---|---|---|---|---|---|
| | | | 50 year design life | | 100 year design life | |
| *Exposure classification* | *Precast and prestressed piles* | *Cast in place piles* | *Precast and prestressed piles* | *Cast in place piles* | *Precast and prestressed piles* | *Cast in place piles* |
| Non-aggressive | 50 | 32 | 20 | 45 | 25 | 65 |
| Mild | 50 | 32 | 20 | 60 | 30 | 75 |
| Moderate | 50 | 40 | 25 | 65 | 40 | 85 |
| Severe | 50 | 50 | 40 | 7- | 50 | 100 |
| Very severe[a] | >50 (preferably >60) | >50 (preferably >60) | 40 | 75 | 50 | 120 |

Source: AS 2159. 2009. *Piling – Design and Installation*. Standards Australia.

[a] Consider using an inert liner and/or coating in addition to the specified concrete cover.

requirements for concrete strength and reinforcement cover for the various exposure classifications.

For concrete piles subject to a very severe exposure classification, the particular exposure environment must be taken into account. Consideration must also be given to the suitability of concrete materials, mix proportions, methods of placement, cover and curing and to the possible use of protective surface coatings to the piles or other protective measures.

Chapter 8

# Design for serviceability limit state loadings

## 8.1 INTRODUCTION

As discussed in Section 3.8 of Chapter 3, the critical design criteria for tall building foundations are often the settlement and differential settlement (or angular rotation) of the foundation system, rather than the ultimate capacity. This chapter will discuss design criteria for settlement and angular rotation, and then set out the basic principles of settlement analysis. Category 1 and 2 methods of estimating the settlement of various foundation types will be described in Sections 8.4 through 8.7. Some Category 2 methods of estimating horizontal deflections will then be set out in Sections 8.8 and 8.9.

Category 1 methods, and particularly Category 2 methods, are considered to be valuable tools for making preliminary settlement estimates, and also for checking the results of more complex Category 3 analyses, which will be discussed in Section 8.10. Finally, methods of obtaining the equivalent spring stiffness of foundation elements, for use by the structural designer, will be described in Section 8.11.

## 8.2 DESIGN CRITERIA FOR SERVICEABILITY LIMIT STATE

The design criteria for the serviceability limit state can be stated as follows:

$$\rho_{max} \leq \rho_{all} \quad (8.1)$$

$$\theta_{max} \leq \theta_{all} \quad (8.2)$$

where $\rho_{max}$ is the maximum computed settlement of foundation, $\rho_{all}$ the allowable foundation settlement, $\theta_{max}$ the maximum computed local angular distortion and $\theta_{all}$ is the allowable angular distortion.

Values of $\rho_{all}$ and $\theta_{all}$ depend on the nature of the structure and the supporting soil. Table 8.1 sets out some suggested criteria from work reported by Zhang and Ng (2006). This table also includes values of intolerable settlements and angular distortions. The figures quoted in Table 8.1 are for deep foundations, but Zhang and Ng also consider separately allowable settlements and angular distortions for shallow foundations, different types of structure, different soil types and different building usage. Criteria specifically for very tall buildings do not appear to have been set, but it should be noted that it may be unrealistic to impose very stringent settlement criteria on very tall buildings on clay deposits, as they may not be achievable. For example, experience with tall buildings in Frankfurt, Germany, suggests

*Table 8.1* Serviceability criteria for structures

| *Quantity* | *Value* | *Comments* |
|---|---|---|
| Limiting tolerable settlement (mm) | 106 | Based on 52 cases of deep foundations<br>Std. deviation = 55 mm<br>Factor of safety of 1.5 recommended on this value |
| Observed intolerable settlement (mm) | 349 | Based on 52 cases of deep foundations<br>Std. deviation = 218 mm |
| Limiting tolerable angular distortion (rad) | 1/500 | Based on 57 cases of deep foundations<br>Std. deviation = 1/500 rad |
| Limiting tolerable angular distortion (rad) | 1/250 (H < 24 m)<br>1/330 (24 < H < 60 m)<br>1/500 (60 < H < 100 m)<br>1/1000 (H > 100 m) | From Chinese code<br>H = building height |
| Observed intolerable angular distortion (rad) | 1/125 | Based on 57 cases of deep foundations<br>Std. deviation = 1/90 rad |

Source: Zhang, L. and Ng, A.M.Y. 2006. *Geotech. Special Publication No. 170, Probabilistic Applications in Geotechnical Engineering,* ASCE (on CD Rom).

that total settlements in excess of 100 mm can be tolerated without any apparent impairment of function.

Figure 8.1 shows a suggested approach to the acceptable angular distortion, $\theta_{all}$, of structures, based on Juang et al. (2011). This figure shows that $\theta_{all}$ depends on the lateral strain to which the foundation is subjected, and that the probability of building damage increases significantly as the lateral strain increases Boscardin and Cording (1989). However, for most tall building foundations, the foundation system will be connected to a raft or slab which will largely inhibit lateral strains. A common criterion is $\theta_{all}$ = 1/500 (0.002), and Figure 8.1 shows that, for this value, there is a 20% possibility that damage could occur.

It should also be noted that the allowable angular distortion, and the overall allowable building tilt, reduce with increasing building height, both from a functional and a visual

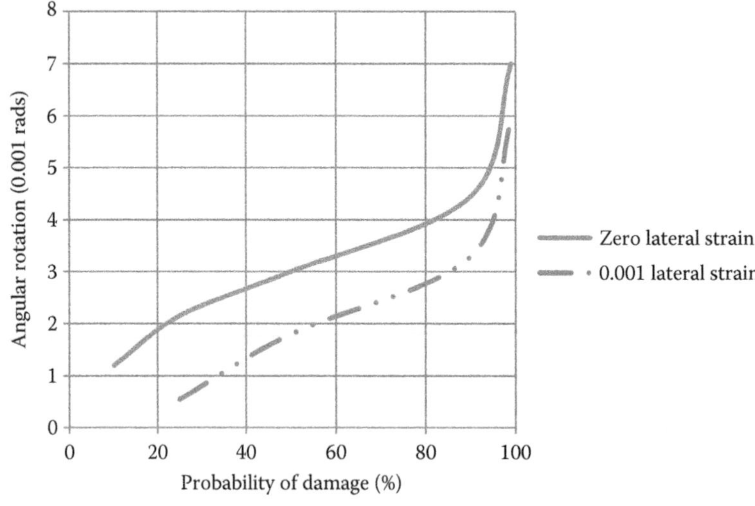


*Figure 8.1* Angular distortion criteria. (Based on Juang, C. et al. 2011. *Journal of Geotechnical and Geoenvironmental Engineering*, 130–139, doi: 10.1061/(ASCE)GT.1943-5606.0000413.)

viewpoint. It can also be noted that, in Hong Kong, the limiting tilt for most public buildings is 1/300 in order for lifts (elevators) to function properly.

Another factor that should be borne in mind is that at least some of the settlement and differential settlement will tend to be 'built out' during the construction process, so that any distortion or tilt when the structure is complete may be considerably less than computed on the implicit assumption of instantaneous construction. This is particularly so for foundations on sand or rock, and in such cases, care should be taken to avoid over-conservative criteria being set for $\theta_{all}$.

## 8.3 PRINCIPLES OF SETTLEMENT ESTIMATION

### 8.3.1 Components of settlement

Under general three-dimensional (3D) conditions, a loaded foundation will experience the following types of settlement:

1. Immediate (or undrained) settlement, which occurs immediately upon application of the load and which, in a saturated soil, arises from shear deformations under constant volume conditions
2. Consolidation settlement, which occurs primarily because of the dissipation of excess pore pressure in the soil and is therefore time-dependent. This component of settlement arises mainly from volumetric deformations, although shear deformations are also involved
3. Creep settlement (frequently termed secondary consolidation), which most frequently manifests itself as a time-dependent settlement after the completion of excess pore pressure dissipation; however, significant creep settlements can also occur under undrained conditions. Creep settlements generally involve both shear and volumetric deformations

The sum of the immediate and consolidation settlements is commonly referred to as the 'primary settlement'.

The total final settlement $S_{TF}$ of a foundation is given by

$$S_{TF} = S_i + S_{CF} + S_{cr} \tag{8.3}$$

where $S_i$ is the immediate settlement, $S_{CF}$ the final consolidation settlement and $S_{cr}$ is the creep settlement.

Leaving aside for the time being the issue of creep settlement, at any time t after the application of the foundation load, the settlement $S_{Tt}$ is

$$S_{T_T} = Si + U_s \cdot S_{CF} \tag{8.4}$$

where $U_s$ is the degree of consolidation settlement.

Thus, the estimation of foundation settlements requires the assessment of three quantities:

1. The immediate settlement, $S_i$
2. The total final settlement, $S_{TF}$
3. The degree of settlement at any time, $U_s$

The general principles involved are described below.

### 8.3.2 Conventional one-dimensional settlement analysis

The classical one-dimensional settlement analysis first developed by Terzaghi (1943) can be classified as a Category 2 method and is applicable primarily to estimating the settlement of shallow foundations and raft foundations. In this method, the assumption is made that vertical load applied to the soil generates only vertical strains, and hence only vertical deformations (settlements). The settlement is usually calculated on the basis of oedometer (consolidometer) tests, and so is often denoted as $S_{oed}$. Depending on the measure of compressibility employed, $S_{oed}$ can be computed as follows:

1. From the coefficient of volume decrease ($m_v$):

$$S_{oed} = \Sigma(m_v \cdot \delta\sigma'_z \cdot \delta h) \tag{8.5}$$

   where $m_v$ is the coefficient of volume decrease, over the relevant range of effective stress, $\delta\sigma'_z$ the increase in vertical effective stress and $\delta h$ is the thickness of layer or sublayer.
2. From the compression index ($C_c$) and recompression index ($C_r$):

$$S_{oed} = \Sigma\left[\left\{C_c \cdot \log\left(\frac{\sigma'_{vf}}{\delta'_{vp}}\right) + C_r \cdot \log\left(\frac{\sigma'_{vp}}{\sigma'_{vo}}\right)\right\}\frac{(\delta h)}{(1+e_o)}\right] \tag{8.6}$$

   where $\sigma'_{vf}$ is the final vertical effective stress, $\sigma'_{vp}$ the vertical preconsolidation pressure, $\sigma'_{vo}$ the initial vertical effective stress, $e_o$ the initial void ratio at centre of layer or sublayer and $\delta h$ is the thickness of layer or sublayer.

The summations in Equations 8.5 and 8.6 are carried out for all layers or sublayers considered within the ground profile. The stress increments in each layer are generally computed via the use of elastic theory.

$S_{oed}$ in the original theory provides an estimate of the total settlement (immediate plus consolidation), but in some subsequent refinements, $S_{oed}$ was taken to be the consolidation settlement and the immediate component of settlement, $S_i$, was computed separately via elastic theory, and added to $S_{oed}$ to provide the final settlement (excluding creep).

### 8.3.3 Application of elastic theory for 3D analyses

In this approach, both the immediate settlement $S_i$ and total final settlement $S_{TF}$ are calculated from elastic theory, using appropriate solutions for the foundation and soil profile characteristics. This theory can provide the vertical displacement directly, or can be used to compute the increments of effective stress which are then used to compute the increments of strain, and hence settlement, within the ground profile. These approaches are discussed in greater detail in Section 8.4 for shallow foundations, and in Sections 8.5 and 8.6 for deep foundations.

For immediate settlement, the relevant deformation parameters to be used in the theory are the undrained Young's modulus, $E_u$ and the undrained Poisson's ratio, $\nu_u$. For a saturated soil, constant volume conditions will occur, and hence $\nu_u = 0.5$.

For total final settlement (immediate plus consolidation), the relevant deformation parameters to be used in the theory are the drained Young's modulus, $E'$ and the drained Poisson's ratio, $\nu'$. $E'$ will be less than $E_u$ and $\nu'$ will generally be less than 0.5. The estimation of these parameters has been discussed in Chapter 6.

### 8.3.4 Approximate allowance for effect of local soil yielding on immediate settlement

An approximate extension to elastic settlement analysis, which allows for non-elastic deformations of shallow foundations on clay, has been described by D'Appolonia et al. (1971) for shallow foundations. This method involves the use of elastic theory to calculate the immediate and final settlements, with a correction factor being applied to the immediate settlement to allow for the effects of local yielding and non-elastic deformations under undrained loading conditions. In this modified elastic method the total final settlement $S_{TF}$ is given by

$$S_{TF} = \frac{S_{ielas}}{F_R} - (S_{Tfelas} - S_{ielas}) \tag{8.7}$$

where $S_{ielas}$ is the immediate settlement calculated from elastic theory, $S_{Tfelas}$ the total final settlement calculated from elastic theory and $F_R$ is the yield settlement factor to account for possible local yield under undrained conditions.

The above method assumes that the magnitude of the consolidation settlement is unaffected by the occurrence of undrained local yield, and is given by the difference between the elastic total final and elastic immediate settlements.

$S_{ielas}$ and $S_{Tfelas}$ may be calculated either by summation of vertical strains beneath the foundation, or directly by the use of elastic displacement theory, as described later in this chapter.

The yield factor $F_R$ has been evaluated for a strip footing on a layer by D'Appolonia et al. (1971), using an elasto-plastic finite element analysis. For shallow foundations, Figure 8.2

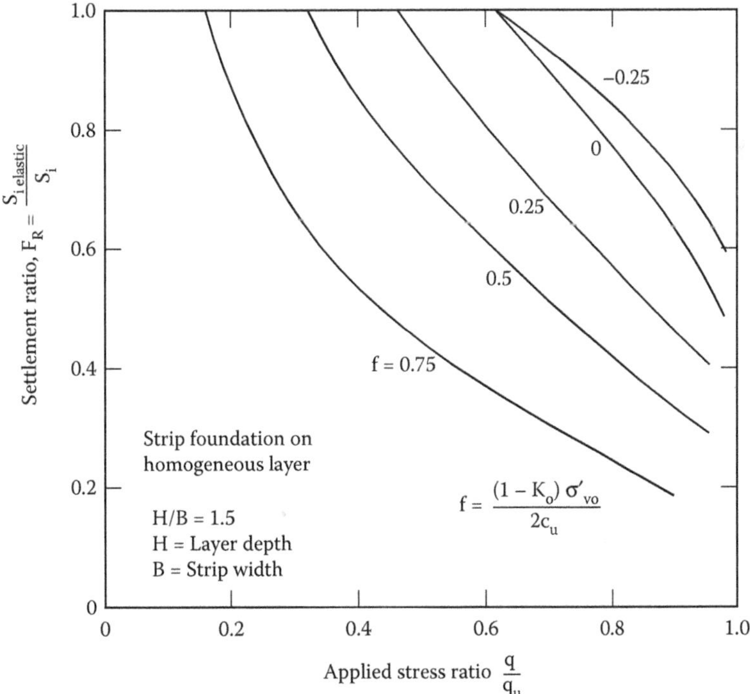


*Figure 8.2* Yield settlement factor, $F_R$, for strip footing. (After D'Appolonia, D.J., Poulos, H.G. and Ladd, C.C. 1971. *JSMFD, ASCE*, 97(SM10): 1359–1397. Courtesy of ASCE.)

shows computed values of $F_R$ as a function of the applied stress ratio $q/q_u$ (the inverse of the factor of safety) and of an initial shear stress ratio f, which is the ratio of the initial in situ geostatic shear stress, to the undrained shear strength of the soil.

### 8.3.5 Estimation of creep settlements

The existence of creep complicates the prediction of both the magnitude and the rate of settlement of foundations on clay soils. Most practical methods of accounting for creep still rely on the early observations of Buisman (1936) that creep is characterised by a linear relationship between settlement and the logarithm of time. The gradient of this relationship is generally represented by the coefficient of secondary compression $C_\alpha$, which is given by

$$C_a = \Delta e/\Delta \log t \tag{8.8}$$

where $\Delta e$ is the change in void ratio and t is the time.

Mesri and Godlewski (1977) have found that $C_\alpha$ is related to the compression index of a soil, as indicated in Table 8.2. It should be noted that, for over-consolidated clays, the ratios in Table 8.2 apply to the recompression index; thus, the creep settlement rate is significantly smaller for an over-consolidated soil than for the same soil in a normally consolidated state.

The difficulty with applying the '$C_\alpha$' concept is that the time at which creep is assumed to commence is not well defined. Considerable controversy exists on this point, with some researchers maintaining that creep only commences at the end of primary consolidation (e.g. Mesri et al. 1994) while others contend that it takes place simultaneously with primary consolidation (e.g. Leroueil, 1996).

Various creep laws can and have been incorporated into consolidation analyses (e.g. Gibson and Lo, 1961; Garlanger, 1972), but it is not common in practice for such analyses to be applied, even for one-dimensional problems.

From a practical viewpoint, the most convenient approach appears to be to add the creep settlement versus time relationship to the conventional time–settlement relationship from consolidation theory, commencing at one of the following times:

- A predetermined time after commencement of loading
- After a predetermined degree of consolidation settlement
- When the gradients of the primary settlement versus log time and the creep settlement versus log time relationships are equal

Overall, it appears that, of all the aspects of settlement analysis, the issue of creep and secondary consolidation is the one in which least progress has been made in terms of fundamental understanding and in the incorporation of research into practice. In the absence of

*Table 8.2* Values of $C_\alpha/C_c$ for geotechnical materials

| *Material* | $C_\alpha/C_c$ |
|---|---|
| Granular soils, including rockfill | $0.02 \pm 0.01$ |
| Shale and mudstone | $0.03 \pm 0.01$ |
| Inorganic clays and silts | $0.04 \pm 0.01$ |
| Organic clays and silts | $0.05 \pm 0.01$ |
| Peat and muskeg | $0.06 \pm 0.01$ |

Source: After Mesri, G., Lo, D.O.K. and Feng, T.W. 1994. *Proceedings of Settlement '94, ASCE Special Pubublication No. 40*, 1: 8–56.

a more satisfactory approach, the method of Buisman, in conjunction with the third option above, appears to be a reasonable means of making an approximate estimate of creep settlements of shallow foundations.

### 8.3.6 Rate of settlement: one dimensional consolidation

The basic equation for one-dimensional consolidation is

$$\frac{\partial u}{\partial t} = c_{vl}\frac{\partial^2 u}{\partial z^2} + \frac{\partial \sigma_z}{\partial t} \tag{8.9}$$

where u is the excess pore pressure, $\sigma_z$ the total vertical stress, $c_{vl}$ the one-dimensional coefficient of consolidation $=k/mv\gamma_w$, k the permeability, $m_v$ the coefficient of volume decrease and $\gamma_w$ is the unit weight of water.

A dimensionless time factor, $T_v$, can be defined as

$$T_v = \frac{c_{vl}t}{H^2} \tag{8.10}$$

where H is the drainage path of soil layer.

For an ideal elastic soil, having Young's modulus $E'$ and Poisson's ratio $\nu'$ for the soil skeleton,

$$c_{vl} = \frac{kE'(1-\nu')}{\gamma_w(1+\nu')(1-2\nu')} \tag{8.11}$$

Figure 8.3 plots solutions for the one-dimensional rate of settlement U versus time factor $T_v$, for various cases of linear stress change with depth, and constant permeability k and coefficient of volume decrease $m_v$. These solutions are relevant to shallow foundations at or near the surface of the soil layer.

### 8.3.7 Rate of settlement: 2D and 3D consolidation

Approximate curves relating degree of settlement $U_s$ to time factor $T_v$ for a shallow circular footing on a layer have been published by Davis and Poulos (1972). In an attempt to provide a more convenient solution, Poulos et al. (2001) have re-expressed these curves in terms of an equivalent coefficient of consolidation that can be used with the one-dimensional rate of settlement curves shown in Figure 8.3. The actual coefficient of consolidation $c_v$ is multiplied by a geometry factor $R_f$ to account for the lateral dissipation. Figure 8.4a and b shows values of $R_f$ for a strip and circular footing respectively.

### 8.3.8 Estimation of horizontal foundation movements

It is possible in principle to estimate horizontal foundation movements due to vertical and/or horizontal loadings. Such estimates are usually carried out using appropriate solutions from the theory of elasticity, for example, as summarised by Poulos and Davis (1974) and Giroud (1973). In employing elastic theory, the following should be borne in mind:

- Accurate estimation of horizontal movements due to vertical loads is difficult, due to the limitations of elastic theory in representing real soil behaviour.

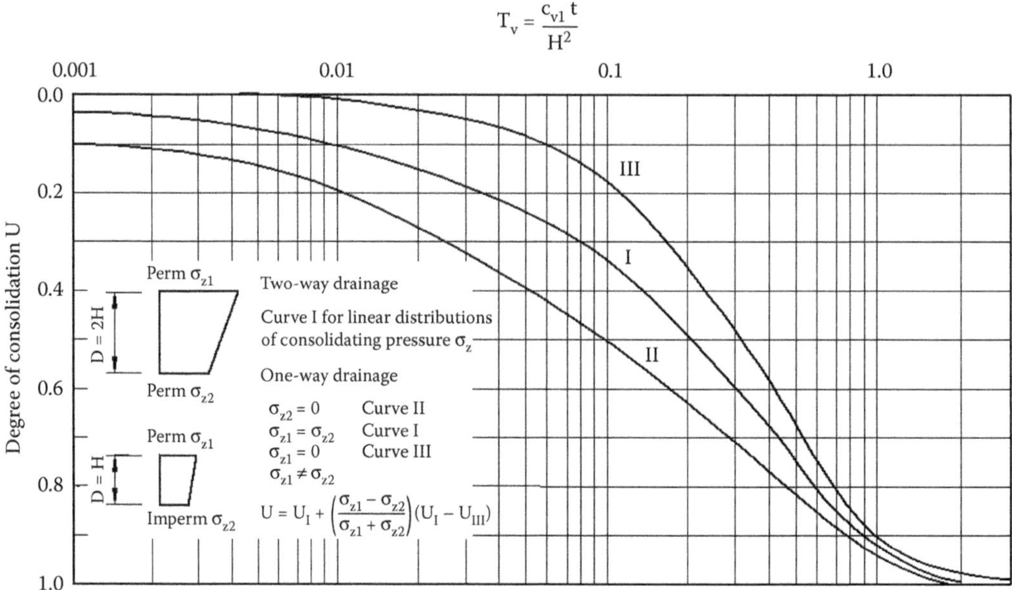


*Figure 8.3* Solutions for one-dimensional rate of settlement.

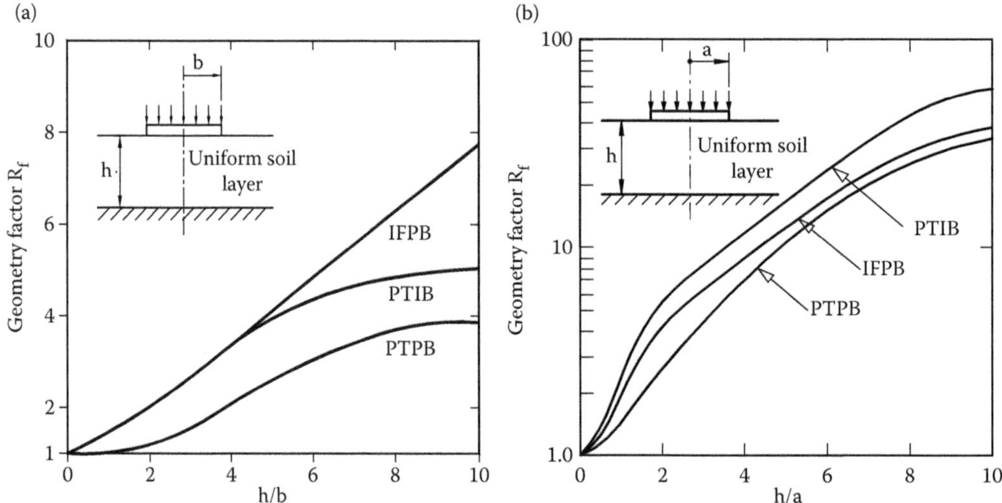


*Figure 8.4* (a) Geometry factor $R_f$ for one-dimensional analysis of rate of consolidation of a strip footing. (b) Geometry factor $R_f$ for one-dimensional analysis of rate of settlement of a circular footing. PT denotes a permeable upper surface, PB a permeable base below soil layer, IF an impermeable footing (but otherwise permeable surface) and IB denotes an impermeable base below soil layer.

- The Young's modulus relevant to horizontal loading may be different from that relevant to vertical loading, because of the different strain levels for each loading type. Thus, care should be taken in selecting modulus values, and the temptation to employ the same values for all types of loading should be tempered by judgement.

## 8.4 ESTIMATION OF SETTLEMENT OF SHALLOW FOUNDATIONS AND RAFTS

### 8.4.1 Category 1 methods

A number of empirical approaches have been put forward for estimating the settlement of vertically loaded shallow foundations. Most are for foundations on sands, and are based on correlations with SPT data, for example, the methods of Burland and Burbridge (1985) and Schultze and Sherif (1973). The basis of the latter method is illustrated in Figure 8.5. The SPT-N value to be used is the average value within a depth of twice the foundation width.

### 8.4.2 Category 2 methods

Category 2 methods include those based on elastic theory, and also methods such as that proposed by Schmertman (1970) which is based on a simplified representation of the strain distribution below a foundation, and a Young's modulus which is correlated with the CPT resistance. In this section, a number of elastic solutions for the settlement of a shallow foundation due to vertical loading will be summarised. Such solutions can be used for raft and piled raft foundations, as well as for shallow isolated footings. Poulos and Davis (1974) provide a wide range of elastic solutions that can be used for foundation settlement estimation.

### 8.4.3 Solutions for shallow foundation settlements from elastic displacement theory

#### *8.4.3.1 Circular footing on a layer*

The following solution may be used to compute the settlement of the centre of a circular footing on an elastic soil layer, whose modulus increases linearly ρ with depth, allowing for footing rigidity and embedment effects (Mayne and Poulos, 1999):

$$\rho = q \cdot d \cdot I_g \cdot I_f \cdot I_E \cdot (1 - v^2)/E_0 \tag{8.12}$$

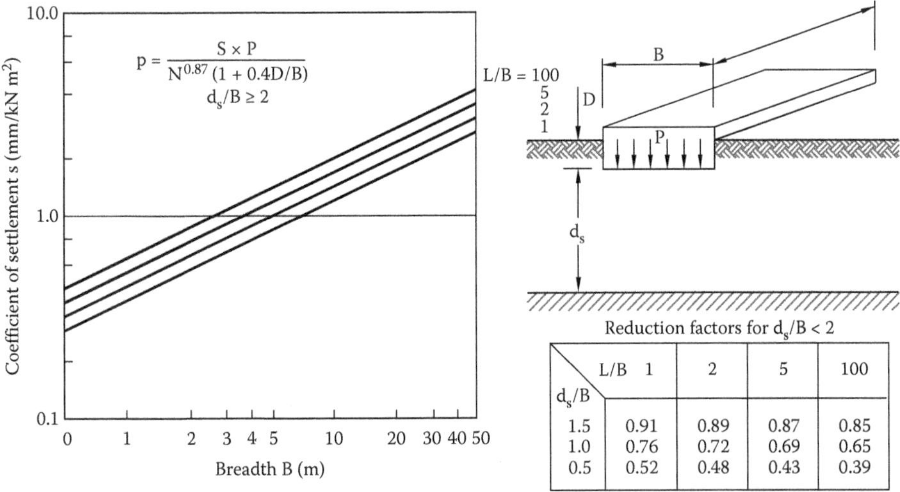


*Figure 8.5* Category 1 method of Schultze and Sherif (1973).

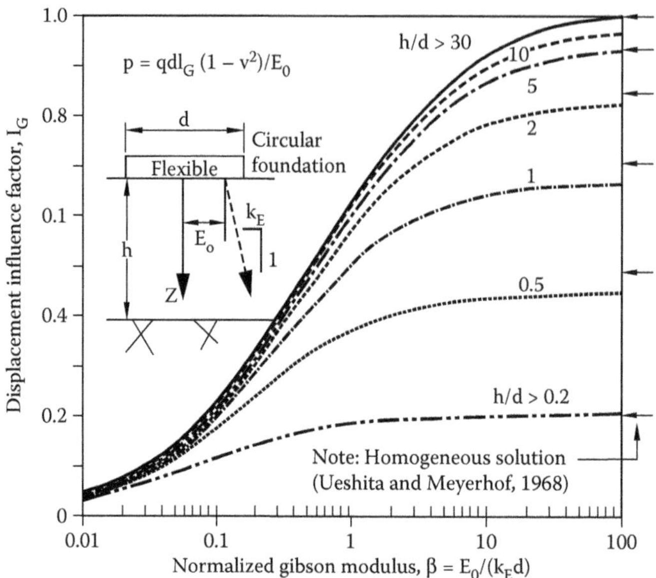


*Figure 8.6* Displacement influence factor $I_G$. (Adapted from Mayne, P.W. and Poulos, H.G. 1999. *Journal of Geotechnical and Geoenvironmental Engineering, ASCE*, 125(6): 453–460. Courtesy of ASCE.)

where q is the average applied loading, d the footing diameter, $I_g$ the displacement influence factor, plotted in Figure 8.6, $I_f$ the foundation flexibility correction factor, $I_E$ the foundation embedment correction factor, $\nu$ the soil Poisson's ratio and $E_0$ is the soil Young's modulus at surface.

$I_f$ is approximately given by

$$I_f \approx \frac{\pi}{4} + \frac{1}{(4 \cdot 6 + 10 \cdot K_F)} \tag{8.13}$$

where $K_F = E_f/E_{sav} \cdot (2t/d)^3$, $E_f$ is the footing Young's modulus, $E_{sav}$ the average soil Young's modulus and t is the footing thickness.

$I_E$ is approximately given by

$$I_E \approx 1 - \frac{1}{[3.5\exp(1.22\nu - 0.4)(d/z_e) + 1.6]} \tag{8.14}$$

where $z_e$ is the depth of embedment of footing base below surface.

#### *8.4.3.2 Uniformly loaded strip on uniform soil*

For a uniform soil mass, the surface settlement of the edge of a uniformly loaded strip on a finite layer is given by

$$S = \frac{ph}{E} \cdot \frac{I_{st}}{\pi} \tag{8.15}$$

Values of $I_{st}$ are plotted in Figure 8.7.

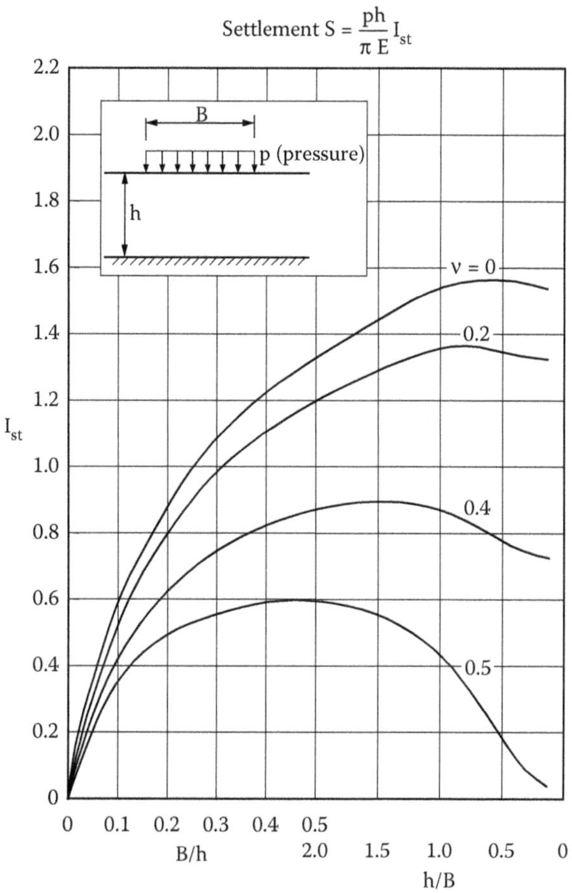


*Figure 8.7* Settlement factor for edge of uniformly loaded flexible atrip on layer. (Adapted from Poulos, H.G. 1967. *Geotechnique*, 17: 378–410. Courtesy of ICE Publishing.)

The settlement of any other point on the soil surface can be obtained by superposition of strips. Thus, for example, the settlement of the centre of a strip of total width B can be calculated as twice the settlement of the edge of a strip of width B/2.

For a rigid strip, the settlement can be approximated with sufficient accuracy as the mean of the centre and edge settlements of a uniformly loaded strip.

#### *8.4.3.3 Circular footing on non-homogeneous Gibson soil*

For a non-homogeneous soil, whose Young's modulus increases linearly with depth according to $E = E_o + mz$ (where z = depth below the surface), the settlement of a uniformly loaded circular area on an infinitely deep layer can be expressed as follows:

$$S = \frac{pa}{E_o} \cdot I \tag{8.16}$$

Values of I are plotted in Figure 8.8.

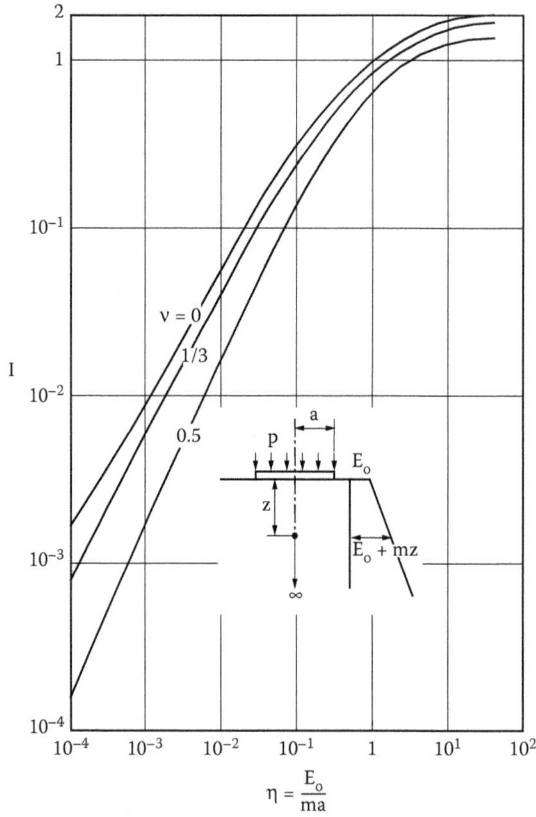


*Figure 8.8* Settlement factor for uniformly loaded circular area on deep non-homogeneous soil.

## 8.5 SETTLEMENT OF SINGLE PILE FOUNDATIONS

### 8.5.1 Introduction

Piles are normally used in groups or as elements of a piled raft foundation. However, the analysis of the settlement of a single pile is an important component of design in that the estimation of the settlement of pile groups or piled rafts very often incorporates the settlement characteristics of a single pile. Moreover, a single pile analysis is necessary when evaluating the performance of test piles. In this section, Category 1 and 2 methods for estimating single pile settlements will be outlined. The settlement of pile groups and piled rafts will then be discussed in the following sections.

### 8.5.2 Category 1 methods

Empirical methods of estimating single pile settlement are not common. Meyerhof (1959) suggested that the settlement $S_1$ of a single pile in sand could be estimated as follows:

$$S_1 = \frac{d_b}{30 \cdot FS} \tag{8.17}$$

where $d_b$ is the diameter of pile base and FS is the factor of safety against axial failure (>3).

For piles in clay, Focht (1967) related the settlement at working load, $S_1$, to the computed free-standing column deformation, $S_{col}$, as

$$S_1 = MR \cdot S_{col} \tag{8.18}$$

where MR is the movement ratio $\approx 0.5$ for highly stressed piles where $S_{col} > 8$ mm, and increasing to $\approx 1$ if $S_{col} < 8$ mm.

### 8.5.3 Category 2 methods

Poulos and Davis (1980) have presented chart solutions for the settlement of a single pile, based on the results of elastic boundary element analyses. Referring to Figure 8.9, the pile head settlement can be expressed as follows:

1. Floating pile

$$S = \frac{P}{dE_s} I_1 R_K R_h R_\nu \tag{8.19}$$

2. End-bearing pile

$$S = \frac{P}{dE_s} I_1 R_K R_b R_\nu \tag{8.20}$$

where P is the applied load, d the pile diameter, $E_s$ the soil modulus ($E_u$ for immediate settlement, $E'$ for total final settlement), $I_1$ the influence factor for rigid pile in semi-infinite mass for $v_s = 0.5$ and $R_K$, $R_h$, $R_b$, $R_v$ are the correction factors for effect of pile compressibility,

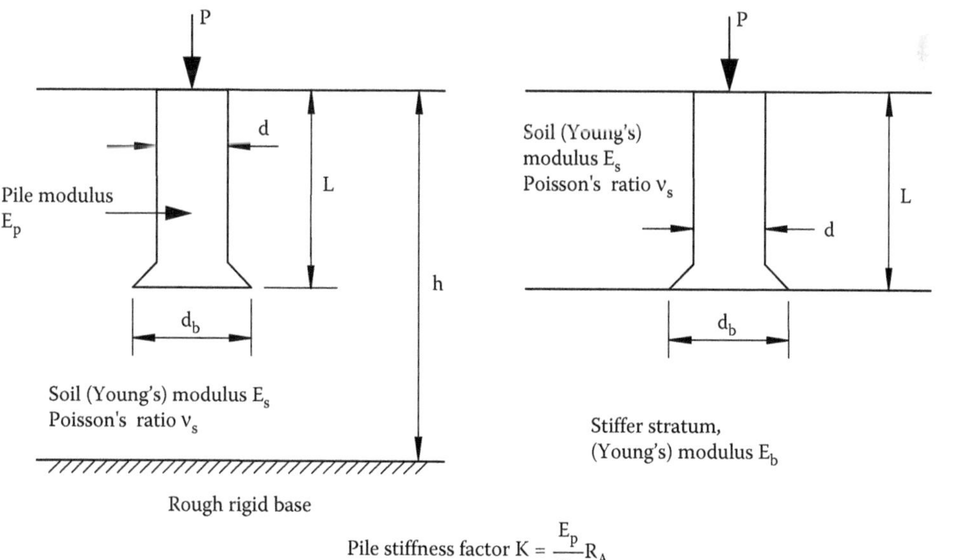


*Figure 8.9* Problem definitions – floating pile, end-bearing pile. (a) Definition of single pile geometry and (b) settlement influence factor $I_1$.

soil depth, bearing stratum and Poisson's ratio. Values of the above factors are plotted in Figures 8.10 through 8.14. Here, the pile stiffness factor, K, is defined as

$$K = \frac{E_p \cdot R_A}{E_s} \tag{8.21}$$

where $E_p$ is the Young's modulus of pile, $R_A$ the area ratio of pile = ratio of area of pile section to gross cross-sectional area. For solid piles, $R_A = 1$.

For layered soil profiles, the solutions for a homogeneous soil can be used approximately if an average soil modulus is used along the length of the pile, and an equivalent modulus is used for the soils on which the pile is founded.

By making use of solutions for the proportion of load carried by the pile base, and estimating the ultimate shaft and base resistances, it is possible to construct a tri-linear load–settlement curve to failure, as described by Poulos and Davis (1980).

Randolph and Wroth (1978) developed a very useful approximate analytical expression for the head settlement of a single pile. They considered a pile in an elastic soil layer with a shear modulus which increases linearly with depth. For an applied load $P_t$, the pile head settlement ($\delta_t$) of a compressible pile is given by the following approximate closed-form solution:

$$\frac{P_t}{\delta_t r_o G_L} = \frac{\left(4\eta_r/(1-\nu_s)\xi\right) + \left((2\pi\rho/\zeta)(\tanh(\mu L)/(\mu L))(L/r_o)\right)}{1 + \left((1/\pi\lambda)(4/(1-\nu_s))(\eta_r/\xi)(\tanh(\mu L)/(\mu L)(L/r_o)\right)} \tag{8.22}$$

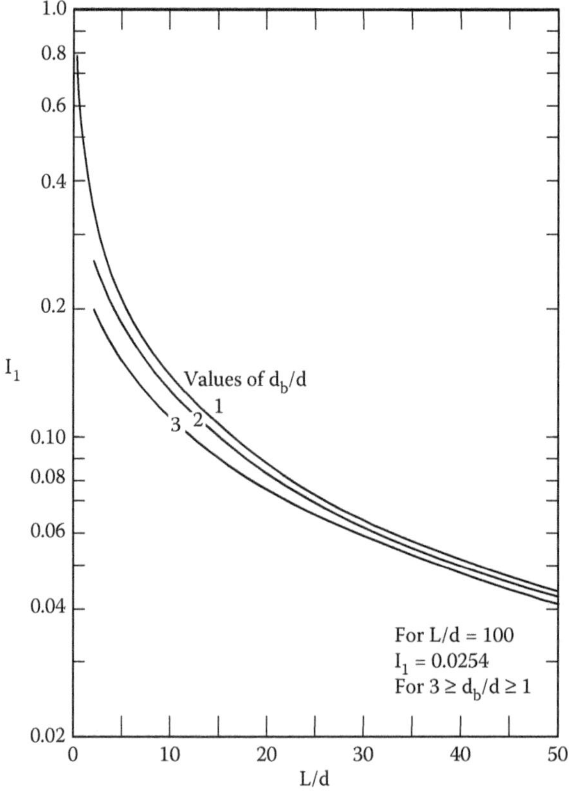


*Figure 8.10* Basic factor $I_1$.

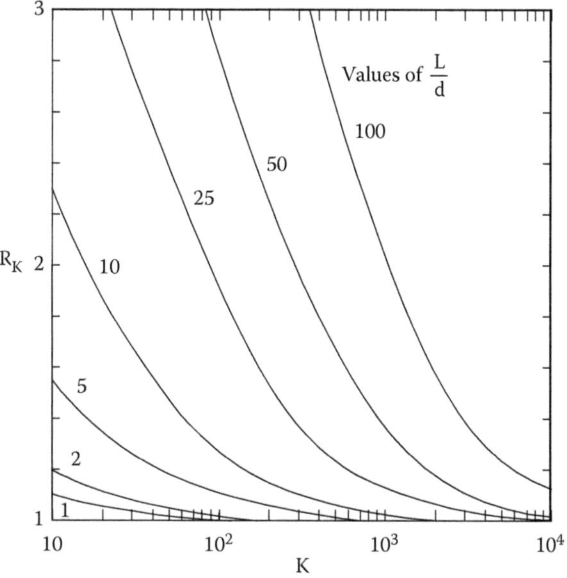


*Figure 8.11* Correction factor $R_K$ for pile compressibility.

where $\eta_r = r_b/r_o$ ($r_o$ is the radius of pile shaft and $r_b$ is radius of pile base), $\xi = G_L/G_b$ ($G_L$ and $G_b$ are shear modulus of soil at depth L and pile base, respectively), $\rho = G_{L/2}/G_L$ (rate of variation of shear modulus of soil with depth), $\lambda = E_p/G_L$ (pile stiffness ratio), $\mu L = [2/\zeta\lambda]^{1/2} L/r_o$, $\zeta = \ln(r_m/r_0)$, $r_m = \{0.25 + \xi(2.5\rho(1 - \nu_s) - 0.25)\} L$, $\nu_s$ = Poisson's ratio of soil.

The settlement profile with depth (z) may be approximated as

$$\delta = \delta_b \cosh(\mu(L - z)) \tag{8.23}$$

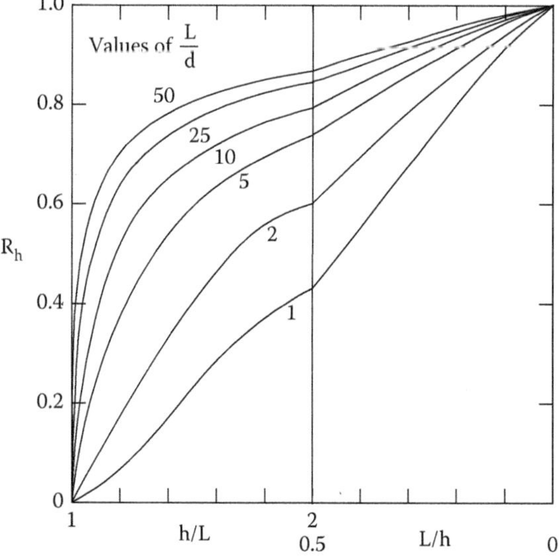


*Figure 8.12* Layer depth correction factor $R_h$.

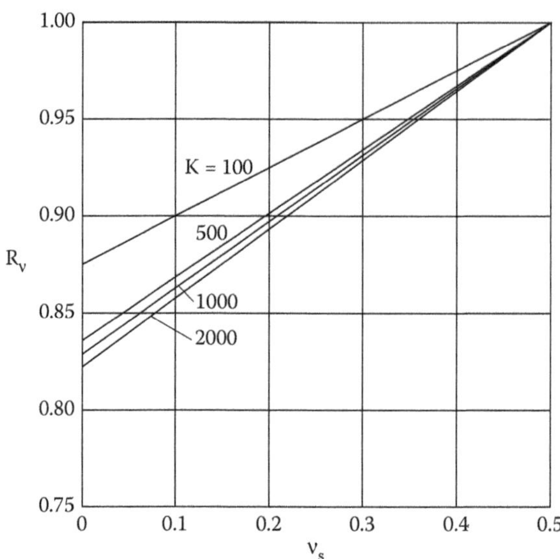


*Figure 8.13* Poisson's ratio correction factor $R_\nu$. Pile compressibility correction factor RK.

where

$$\delta_b = \frac{P_b(1-\nu_s)}{4(r_b G_b)} \tag{8.24}$$

and $P_b$ is the load at pile base.

For a non-circular pile, $E_p$ may be taken as $E_p = (E\ A)_p/(\pi\ r_o^2)$.

When the slenderness ratio L/d is $\leq 0.25(E_p/G_L)^{1/2}$, the pile may be treated as effectively rigid and the pile head stiffness is then derived from the following simplified expression:

$$\frac{P_t}{\delta_t r_o G_L} = \frac{4\eta r}{(1-\nu_s)\xi} + \frac{2\pi\rho L}{r_o} \tag{8.25}$$

When L/d is $\geq 1.5\ (E_p/G_L)^{1/2}$, the pile may be treated as infinitely long. In this case, the effective pile head stiffness is given from

$$\frac{P_t}{(\delta_t r_o G_L)} = \pi\rho(2\lambda/\zeta)^{1/2} \tag{8.26}$$

The Randolph and Wroth approach does have some limitations, namely

- The soil is assumed to be elastic
- The soil stiffness is assumed to increase linearly with depth along the pile shaft
- The pile shaft is of uniform diameter

Subsequent work by Chin and Poulos (1992), Randolph (1994), Guo and Randolph (1997) and Guo (1997) has removed some of the restrictions of the original work by Randolph and Wroth.

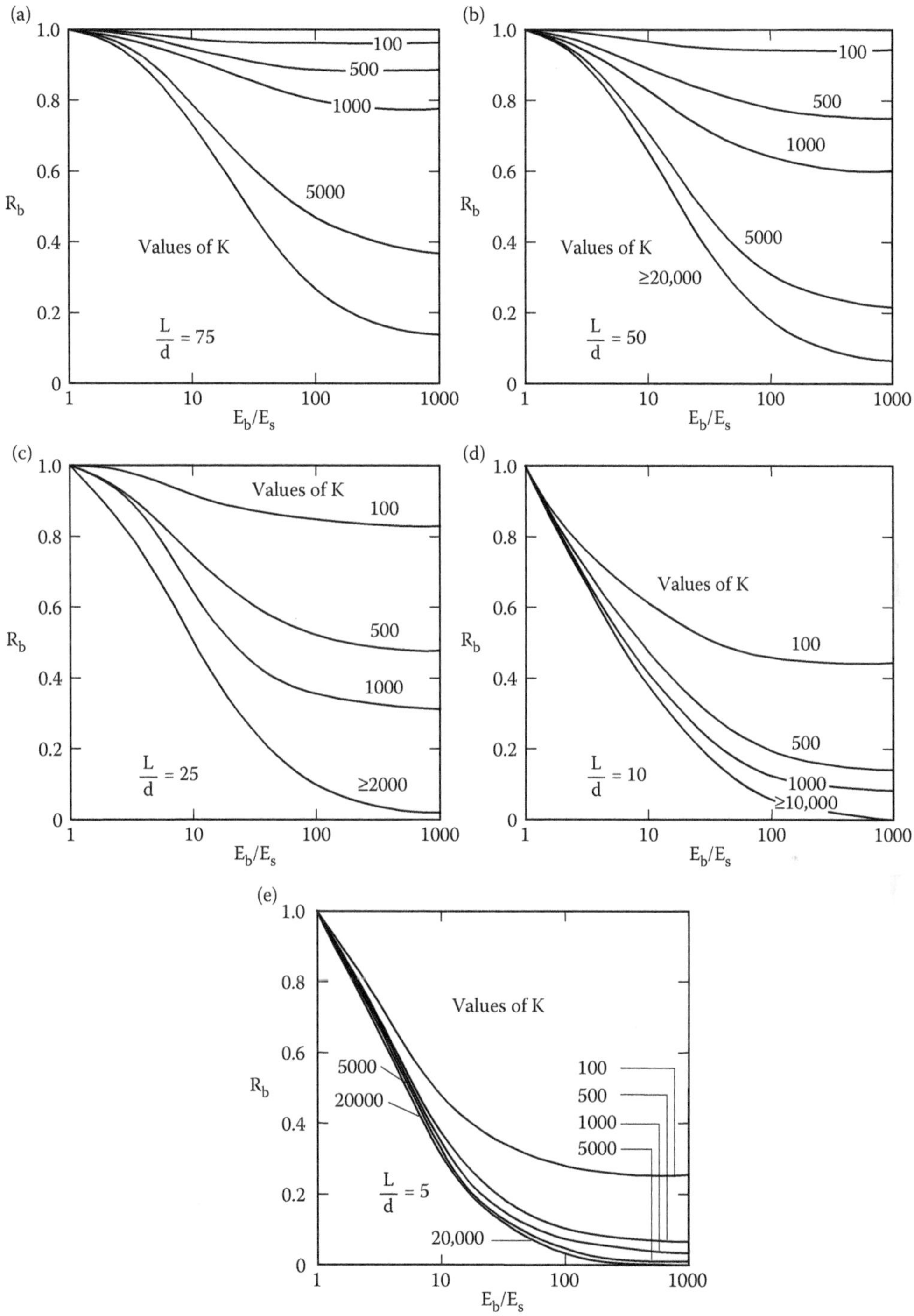


*Figure 8.14* Bearing stratum correction factor $R_b$, for values of L/d of 75, 50, 25, 10, 5.

Randolph (1994) has considered a hyperbolic non-linear response at the pile–soil interface, in which case the load transfer parameter $\zeta$ in Equation 8.22 becomes

$$\zeta = \ln\left[\frac{(r_m/r_0 - \psi)}{(1-\psi)}\right] \tag{8.27}$$

where

$$\psi = \frac{R_f \tau_0}{\tau_s} \tag{8.28}$$

Here $R_f$ is the hyperbolic factor (typically in the range 0.9–1.0), $\tau_0$ the shear stress at pile face and $\tau_s$ is the limiting shaft friction.

Randolph has also suggested that a hyperbolic relationship between base pressure and base displacement can be employed. In this way, load transfer curves can be constructed on a rational basis. Kraft et al. (1981) have developed this approach further.

Poulos (1989) has compared some of the Category 2 and 3 methods (BEM, load transfer, closed-form and finite element) and found that most are capable of giving similar results for single pile settlement, despite differences in the fundamental basis of each analysis. The key to successful settlement prediction for deep foundation systems lies therefore not so much in the method of analysis used, but in the selection of appropriate soil–pile parameters, and in quantifying the relationship between the settlement of a single pile and a pile group. This latter issue is discussed below.

## 8.6 ESTIMATION OF PILE GROUP SETTLEMENTS

### 8.6.1 Introduction

It is now well recognised that the settlement of a pile group can differ significantly from that of a single pile at the same average load level. There are a number of Category 2 and 3 approaches commonly adopted for the estimation of the settlement of pile groups:

- The settlement ratio method, in which the settlement of a single pile at the average load level is multiplied by a group settlement ratio $R_s$, which reflects the effects of group interaction.
- The equivalent raft method, in which the pile group is represented by an equivalent raft acting at some characteristic depth along the piles.
- The equivalent pier method, in which the pile group is represented by a pier containing the piles and the soil between them. The pier is treated as a single pile of equivalent stiffness in order to compute the average settlement of the group.
- Methods which employ the concept of interaction factors and the principle of superposition (e.g. Poulos and Davis, 1980).
- Category 3 numerical methods such as the finite element method and the finite difference method. While earlier work employed two-dimensional (2D) analyses, it is now less uncommon for full 3D analyses to be employed (e.g. Katzenbach et al., 1998).

### 8.6.2 Category 1 methods

Among the few Category 1 methods in the literature for estimating the settlement of a pile group, $S_G$, is the following relationship between group and single pile settlement, via an empirical settlement ratio $R_s$:

$$S_G = R_s \cdot S_1 \tag{8.29}$$

where $S_1$ is the settlement of a single pile at the average load, $R_s$ the settlement ratio, $\approx(4B + 9)^2/(B + 12)^2$, and B is the width of group (in m) (Skempton, 1953), or $R_s \approx s(5 - s/3)/(1 + 1/r)^2$, where s is the ratio of spacing to pile diameter and r is the number of rows in square group (Meyerhof, 1959).

## 8.6.3 Category 2 methods

### *8.6.3.1 Settlement ratio method*

In this method, the pile group settlement, $S_G$, is estimated from the single pile settlement, $S_1$, as per Equation 8.29, but now the group settlement ratio, $R_s$, is estimated on the basis of a Category 2 or 3 pile group analysis. Randolph (1994) has found that $R_s$ can be related approximately to the number of piles within the group, n. For rectangular groups at centre-to-centre spacings of the order of 3–4 diameters, $R_s$ can be approximated as follows:

$$R_s \approx n^{\omega} \tag{8.30}$$

where $\omega$ is the exponent depending on the nature of the soil profile.

Poulos (1989) has found that for friction pile groups in clay, $\omega \approx 0.5$, while for friction pile groups in sand, $\omega \approx 0.33$. For end-bearing groups, lower values of $\omega$ are relevant.

Randolph and Clancy (1993) have related the settlement ratio to the 'aspect ratio' R, where

$$R = \left(\frac{ns}{L}\right)^{0.5} \tag{8.31}$$

with n the number of piles, s the centre-to-centre spacing of piles and L is the pile length.

The settlement ratio is then approximated as follows:

$$R_s = 0.29 \cdot n \cdot R^{-1.33} \tag{8.32}$$

Mandolini et al. (2005) have also approximated the ratio $R_{Dmax}$ of maximum differential settlement to average group settlement as follows:

$$R_{Dmax} = 0.35 \cdot R^{0.35} \tag{8.33}$$

Figures 8.15 and 8.16 plot the relationships between $R_s$ and n, and $R_{Dmax}$ and n, for various ratios of s/L. The settlement ratio approach, while approximate, enables a very rapid estimate of group settlement and differential settlement to be made on the basis of a load test on a single test pile, provided that there are no relatively compressible layers present in the soil profile below the pile tip level.

### *8.6.3.2 Equivalent raft method*

The equivalent raft method has been used extensively for estimating pile group settlements. It relies on the replacement of the pile group by a raft foundation of some equivalent

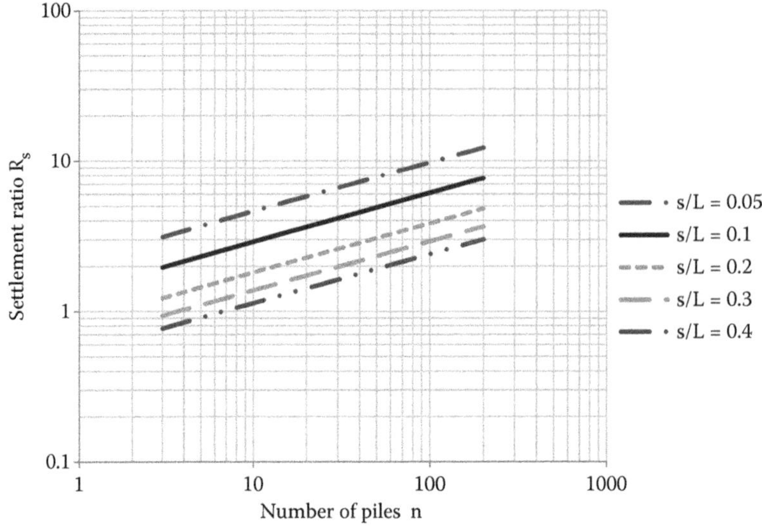


*Figure 8.15* Settlement ratio $R_s$ versus number of piles and relative pile spacing.

dimensions, acting at some representative depth below the surface. There are many variants of this method, but the one suggested by Tomlinson (2001) appears to be a convenient and useful approach. As illustrated in Figure 8.17, the representative depth varies from 2L/3 to L, depending on the assessed founding conditions; the former applies to floating pile groups, while the latter value is for end-bearing groups. The load is spread at an angle which varies from 1 in 4 for friction piles, to zero for end-bearing groups. Once the equivalent raft has been established, the settlement can be computed from normal shallow foundation analysis, taking into account the embedment of the equivalent raft and the compression of the piles above the equivalent raft founding level (Poulos, 1993b).

Much of the success of the equivalent raft method hinges on the selection of the representative depth of the raft and the angle of load spread. Considerable engineering judgment

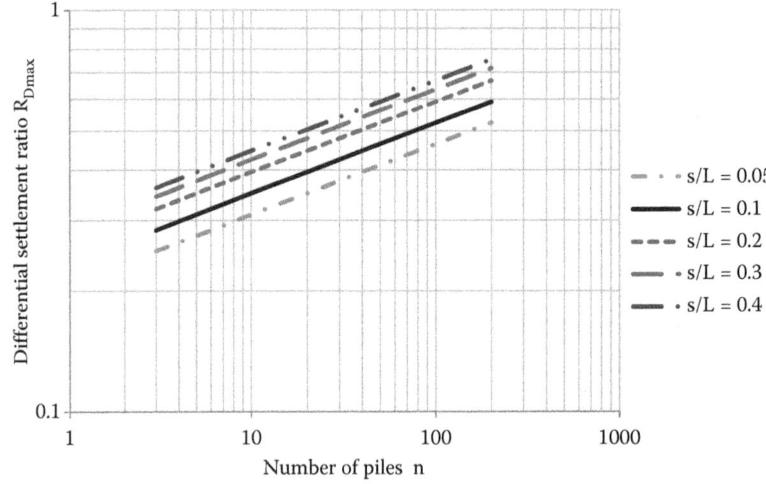


*Figure 8.16* Differential settlement ratio $R_{Dmax}$ versus number of piles and relative pile spacing.

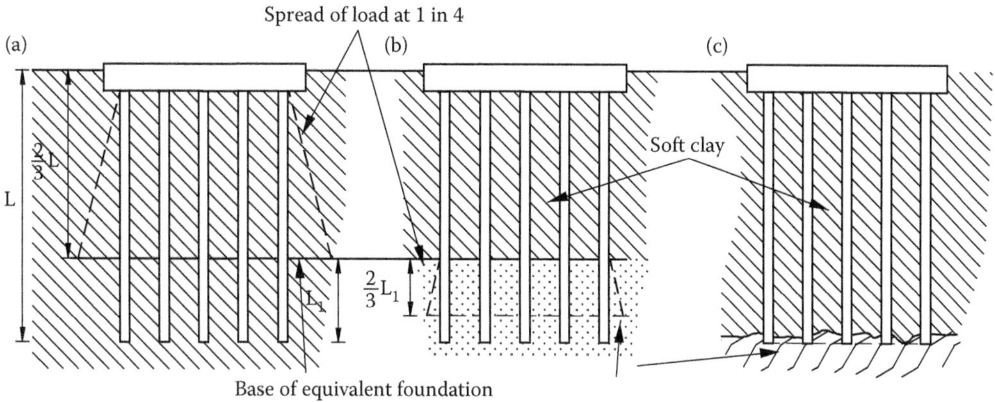


*Figure 8.17* Equivalent raft approach (Tomlinson, 2001): (a) friction pile group, (b) group through soft soil into stiff soil and (c) group bearing on hard layer.

must be exercised here, and firm rules cannot be employed without a proper consideration of the soil stratigraphy.

The applicability of the equivalent raft method has been considered by van Impe (1991) and Randolph (1994). Van Impe (1991) has studied a number of case histories, and related the accuracy of the equivalent raft method to the parameter w, where

$$w = \frac{\text{Sum of pile cross-sectional areas in the group}}{\text{Plan area of pile group}} \tag{8.34}$$

Van Impe has concluded that the equivalent raft method should be limited to cases in which w is greater than about 0.10, that is, the pile cross sections exceed about 10% of the plan area of the group.

Poulos (1993b) has examined the applicability of the equivalent raft method to groups of friction piles and also end-bearing pile groups. He concluded that the equivalent raft method gives a reasonably accurate prediction of the settlement of groups containing more than about 16 piles (at typical spacing of 3 pile diameters centre to centre). This is consistent with the criterion developed by van Impe (1991).

Randolph (1994) has assessed the applicability of the equivalent raft method in terms of the aspect ratio R, defined in Equation 8.31, and has found that the method provides a good analogue for groups having $R > 4$. Viggiani et al. (2012) indicate that the equivalent raft method is to be preferred for large groups and those in which the breadth exceeds the pile length, whereas the equivalent pier method (discussed below) is preferable when the group breadth is less than the pile length.

It may be concluded that the equivalent raft method provides a useful approach to estimating pile group settlements as long as appropriate judgement is exercised in the selection of the equivalent depth (to mirror the actual load transfer mechanisms) and the degree of dispersion along the pile shafts. It also provides a useful check for more complex and complete pile group settlement analyses.

#### *8.6.3.3 Equivalent pier method*

Assessment of the average foundation settlement under working or serviceability loads can be carried out conveniently using the equivalent pier method. In this method, the pile group is replaced by a pier of similar length to the piles in the group, and with an equivalent diameter, $d_e$, estimated as follows (Poulos, 1993b):

$$d_e \cong (1.13 \text{ to } 1.27) \cdot (A_G)^{0.5} \tag{8.35}$$

where $A_G$ is the plan area of pile group.

The lower figure is more relevant to predominantly end-bearing piles, while the larger value is more applicable to predominantly friction or floating piles.

This method utilises elastic solutions for the settlement and proportion of base load of a vertically loaded pier (Poulos, 1994b), provided that the geotechnical profile can be simplified to a soil layer overlying a stiffer layer. Figure 8.18a and b reproduces these solutions, from which simplified load–settlement curves for an equivalent pier containing different numbers of piles can be estimated, using the procedure described by Poulos and Davis (1980). In these figures, the symbol definition is as follows:

Poulos (1993b) and Randolph (1994) have examined the accuracy of the equivalent pier method for predicting group settlements, and have concluded that it generally gives good results. Randolph (1994) has related the accuracy to the aspect ratio R, of the group, as defined in Equation 8.31. The equivalent pier method tends to over-predict stiffness for values of R less than about 3, but the values appear to be within about 20% of those from a more accurate analysis for values of R of 1 or more, provided that the pile spacing is not greater than about 5 diameters. The equivalent pier approach therefore provides a useful tool for preliminary estimates of group settlement.

An attractive feature of the equivalent pier method is the ability to develop a non-linear load–settlement curve, for example, using the simple approach described by Poulos and Davis (1980). It is also possible to estimate the rate of consolidation settlement, using solutions from consolidation theory for a pier within a two-phase poro-elastic soil mass.

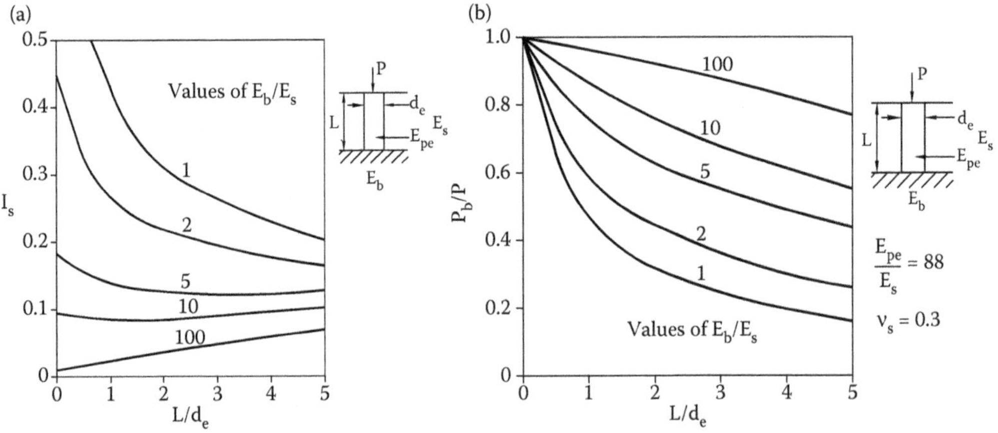


*Figure 8.18* (a) Settlement of equivalent pier in soil layer. (b) Proportion of base load for equivalent pier. Settlement $S = P \cdot I_s/d_e \cdot E_s$. P denotes applied load, $E_s$ the Young's modulus of soil, $E_{pe}$ the Young's modulus of equivalent pier (pile + soil), $d_e$ the diameter of equivalent pier, $I_s$ the settlement influence factor and $P_b$ the load on base of equivalent pier. (Adapted from Poulos, H.G. 1994b. *Special Technical Publication 40, ASCE*, 2: 1629–1649. Courtesy of ASCE.)

#### 8.6.3.4 Interaction factor method

One of the more common means of analyzing pile group behaviour is via the interaction factor method described by Poulos and Davis (1980). The categorisation of this approach borders Category 2 and 3, but in its original form, because of the simplifying assumptions, it may best be classed as a Category 2 approach. In a more comprehensive form, it may be classed as a Category 3 approach, and as such, some computer programs employing it will be described in Section 8.10.

In the interaction factor method, the settlement $w_i$ of a pile i within a group of n piles is given as follows:

$$w_i = \sum_{j=1}^{n} (P_{av} S_1 \alpha_{ij}) \tag{8.36}$$

where $P_{av}$ is the average load on a pile within the group, $S_1$ the settlement of a single pile under unit load (i.e., the pile flexibility) and $\alpha_{ij}$ is the interaction factor for pile i due to any other pile (j) within the group.

In the original approach, the interaction factors were computed from BEM analysis and plotted in graphical form. They were also applied to the total flexibility $S_1$ of the pile, including both elastic and non-elastic components of the single pile settlement.

In subsequent years, significant improvements have been made to the original interaction factor method, among the most important being:

1. The application of the interaction factor to only the elastic component of the single pile flexibility (e.g. Randolph, 1994)
2. The incorporation of non-linearity of single pile response within the interaction factor for the effect of a pile on itself (Mandolini and Viggiani, 1997)
3. The development of simplified or closed-form expressions for the interaction factors, thus enabling a simpler computer analysis of group settlement behaviour to be obtained

In relation to item 1 above, the settlement of a pile i in the group is then given by

$$w_i = \sum_{j=1}^{n} (P_{av} S_{1e} \alpha_{ij}) \tag{8.37}$$

where $S_{1e}$ is the elastic flexibility of the pile.

By assuming that the load–settlement behaviour of the pile is hyperbolic, Mandolini and Viggiani (1997) express the interaction factor, $\alpha_{ii}$, for a pile i due to its own load as

$$\alpha_{ii} = \frac{1}{(1 - R_f P/P_u)^q} \tag{8.38}$$

where $R_f$ is the hyperbolic factor (taken as unity); P the load on pile i; $P_u$ the ultimate load capacity of pile i and q is the analysis exponent =2 for incremental non-linear analysis and 1 for equivalent linear analysis.

#### 8.6.3.5 Estimation of interaction factors

Interaction factors may be computed from BEM or finite element analyses, and many of the Category 3 computer programs incorporate such calculations. However, it is also useful to have alternate means of estimating the interaction factors, and some of these are given below.

Randolph and Wroth (1979) have developed the following closed-form approximation for the interaction factor for a pile in a deep layer of soil whose modulus increases linearly with depth:

$$\alpha_{ij} = \frac{1 - s/(d/\pi + s) + \pi(1-\nu)\rho\Lambda(1/\gamma - 1/\Gamma)}{1 + \pi(1-\nu)\rho\Lambda/\gamma} \tag{8.39}$$

where s is the centre-to-centre spacing between piles i and j, $\rho$ the ratio of soil modulus at mid-length of pile to that at the level of the pile tip (=1 for a constant modulus soil and 0.5 for a 'Gibson' soil); $\gamma = \ln(2r_m/d)$; $\Gamma = \ln(2r_m^2/ds)$; $r_m = 2.5(1-\nu)\rho L$; $\nu$ the soil Poisson's ratio; L the pile length; d the pile diameter and $\Lambda = L/d$.

It has been found that the variation of the interaction factor $\alpha$ with spacing can also be approximated as follows:

$$\alpha = A \cdot \exp\left(-B\left(\frac{s}{d}\right)\right) \tag{8.40}$$

where A, B are empirical factors, s the centre-to-centre spacing of piles and d is the diameter of piles.

Comprehensive parametric studies by Poulos (2008b), using a BEM program, have enabled approximations for the factors A and B to be developed, as functions of the various dimensionless ratios which govern axial pile behaviour. It has been found possible to combine, with acceptable accuracy for practical purposes, the factors which depend on these dimensionless ratios to obtain the following approximations:

$$A = A_l \cdot A_b \cdot A_k \tag{8.41}$$

$$B = B_l \cdot B_b \cdot B_k \tag{8.42}$$

where $A_l$, $B_l$ are factors depending on ratio of length L to diameter d, $A_b$, $B_b$ factors depending on ratio of modulus of bearing stratum to soil along shaft, $A_k$, $B_k$ factors depending on the ratio of pile stiffness to soil stiffness.

Via curve fitting, the following expressions have been derived for the above factors:

$$A_l = 0.376 + 0.0014(L/d) - 0.00002(L/d)^2$$
$$A_b = 1.254 - 0.326 \cdot \ln(E_b/E_s)$$
$$A_k = 0.099 + 0.126 \cdot \ln(K)$$
$$B_l = 0.116 - 0.0164 \cdot \ln(L/d)$$
$$B_b = 0.865 + 0.164 \cdot \ln(E_b/E_s)$$
$$B_k = 1.409 - 0.055 \cdot \ln(K)$$

In the above expressions, L is the pile length, d the pile diameter, $E_b$ the average modulus of bearing stratum below pile tip, $E_s$ the average soil modulus along pile shaft, K the pile stiffness factor $= E_p \cdot R_a/E_s$, $E_p$ the Young's modulus of pile, $R_a$ the area ratio = ratio of area of pile section to total enclosed area (=1 for solid pile).

It is generally assumed that no interaction occurs for spacings greater than a limiting value $s_{max}$, where

$$s_{mas} = \left|0.25 + (2.5\rho(1-\nu) - 0.25)\frac{E_{sL}}{E_b}\right| L + r_g \tag{8.43}$$

where $E_{sL}$ is the soil modulus at mid-length of the pile; $E_b$ the modulus of bearing stratum below pile tip; $r_g$ is a group distance defined by Randolph and Wroth (1979), and the other parameters are defined above.

The presence of a hard layer at the base of a soil layer can substantially reduce the interaction factor and 'damp out' its effect at relatively small pile spacings. The use of solutions for a deep layer may thus lead to significant overestimates of pile interactions and hence, pile group settlements. Mylonakis and Gazetas (1998) and Guo and Randolph (1999) have developed closed-form expressions for the interaction factor in which the important effect of the finite thickness of a soil layer can be taken into account.

Costanzo and Lancellotta (1998) have developed an analytical expression for the interaction factor, taking into account the soil non-linear response. The case of floating piles is considered, with a linear variation of soil shear modulus with radial distance from the pile shaft. McCabe and Sheil (2014) have found that, when using non-linear interaction factors, they should be calculated for the case where the 'receiver' pile is unloaded, otherwise the soil modulus for the receiver pile is degraded excessively, leading to overestimates of the group settlement. With this procedure, they consider that the settlement of groups of up to 200 piles can be estimated adequately.

### 8.6.4 Other issues relevant to pile group settlement

#### *8.6.4.1 Rate of settlement*

Time-dependency of settlement of foundations (under constant loading) usually arises from two sources:

1. Consolidation settlements due to dissipation of excess pore pressures, usually in clay or silty soils
2. Settlements arising from creep of the soil under constant loading; this can occur with all soil types

Figure 8.19 compares solutions for the rate of consolidation settlement of a 25-pile group with that of a rigid impermeable surface footing and a flexible permeable footing of equal plan area. The load is assumed to have been applied instantaneously. The rate of settlement of the pile group is slower than either of the surface footings, and so appropriate caution should be exercised if solutions for the consolidation of a surface footing are applied to a pile group.

In many cases, the rate of settlement during construction, as well as after the completion of construction, will be of considerable interest. In such cases, the solution for an instantaneously applied load can be modified during the construction period by using the following approximate approach developed by Terzaghi (1943):

1. During the construction phase, assuming a linear increase in foundation loading with time, the degree of consolidation U(t), at a time t, is given by

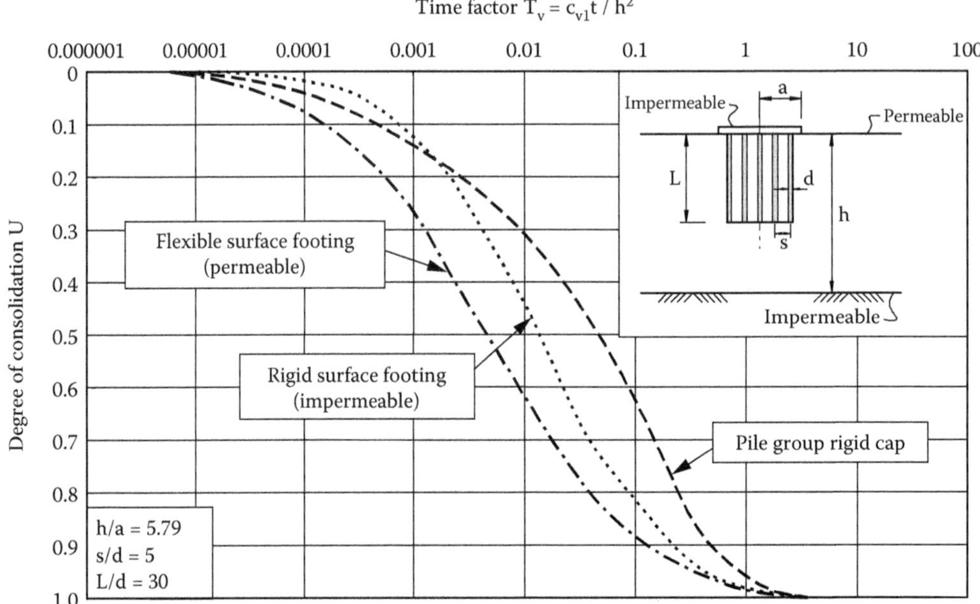


*Figure 8.19* Rate of settlement of 5 × 5 pile group compared to surface footings.

$$U(t) = U_i\left(\frac{t}{2}\right) \cdot \frac{p(t)}{p_f} \tag{8.44}$$

where $U_i(t/2)$ is the degree of consolidation at time t/2 for instantaneous load application, p(t) the applied loading at time t and $p_f$ is the final applied loading.

2. After the completion of construction, which occurs at time $t_c$, the degree of consolidation is

$$U(t) = U_i(t - 0.5t_c) \tag{8.45}$$

where $U_i(t - 0.5t_c)$ is the degree of settlement for instantaneous loading, at a time $(t - 0.5t_c)$.

Settlements due to soil creep are generally not significant at normal working loads, but may become important at load levels of 70% or more of the ultimate. Practical methods for estimating creep settlements are not well developed. Booker and Poulos (1976) have demonstrated that a time-dependent Young's modulus of the soil may be used, but it is not easy to estimate the modulus–time relationship from the data obtained from conventional site characterisation data. Guo (1997, 2000) has developed a closed-form expression for the load transfer characteristics, taking into account both non-linearity and visco-elastic behaviour. His approach shows good agreement with the earlier Booker and Poulos solutions. A method of back-calculating the required creep parameters is also described by Guo.

#### *8.6.4.2 Differential settlements within a group*

Most analyses of pile group settlement make one of the two following extreme assumptions:

1. The pile cap is perfectly rigid so that all piles settle equally (under centric load) and hence there is no differential settlement.
2. The pile cap is flexible, so that the distribution of load onto the piles is known; in this case, the differential settlements within the group can be computed directly.

In reality, the situation is usually between these two extremes. Randolph (1994) has developed useful design guidelines for assessing the differential settlement within a uniformly loaded pile group. For a flexible pile cap, Randolph has related the ratio of differential settlement $\Delta S$ to the average group settlement, $S_{av}$, to the ratio R defined in Equation 8.31, as follows:

$$\frac{\Delta S}{S_{av}} = \frac{fR}{4} \quad \text{for} \quad R \leq 4 \tag{8.46}$$

$$\frac{\Delta S}{S_{av}} = f \quad \text{for} \quad R > 4 \tag{8.47}$$

where $f = 0.3$ for centre-to-midside, and 0.5 for centre-to-corner.

An alternative approximation by Mandolini et al. (2005) has been presented in Equation 8.33 and Figure 8.15.

For pile caps with a finite rigidity, the differential settlements will reduce from the above values (which are for perfectly flexible pile caps), and Randolph suggests that the approach developed by Randolph and Clancy (1993) be adopted. This approach relates the normalised differential settlement to the relative stiffness of the pile cap (considered as a raft). Mayne and Poulos (1999) have developed a closed-form approximation for the ratio of corner to centre settlement of a rectangular foundation, and from this approximation, a rigidity correction factor, $f_R$ can be derived:

$$f_R \approx \frac{1}{(1 + 2.17K_F)} \tag{8.48}$$

where

$$K_F = \left(\frac{E_c}{E_{sav}}\right)\left(\frac{2t}{d}\right)^3 \tag{8.49}$$

where $K_F$ is the foundation flexibility factor; $E_c$ the Young's modulus of pile cap; $E_{sav}$ the representative soil Young's modulus beneath the cap (typically within a depth of about half the equivalent diameter of the cap); t the thickness of pile cap; and d is the equivalent diameter of pile cap (to give equal area with the actual cap). The factor $f_R$ from Equation 8.48 is then applied to the maximum differential settlement estimated from Equations 8.46 and 8.47.

#### *8.6.4.3 Effects of dissimilar or defective piles within a group*

Most of the available methods of pile group settlement analysis assume that all the piles within the group are identical and that the soil profile does not vary over the plan area of the group. In practice, piles are often dissimilar, especially with respect to length, and may also contain structural defects such as necked sections and sections of poor concrete, and/or geotechnical defects such as a soft toe or a section along which the skin friction is reduced because of poor construction practices. The possible consequences of dissimilar or defective piles within a group have been explored by Poulos (1997b), who has found the following indications from theoretical analyses of defects in a single pile:

- Defects within a single pile can reduce the axial stiffness and load capacity of the pile.
- Structural defects such as 'necking' can be characterised by a structural integrity factor, to which the reduction in axial stiffness can be approximately related.
- Geotechnical defects, such as a soft toe, lead to a reduction in pile head stiffness which becomes more severe as the applied load level increases. Failure, or apparent failure, of a pile is more abrupt in piles with structural defects than for piles with geotechnical defects.

For groups containing one or more defective piles, it has been found that the reduction in axial stiffness of a group becomes more marked as the proportion of defective piles, and/or the applied load level, increases. Importantly, the presence of defective piles can result in induced lateral deflection and cap rotation in the group, and additional moments in the piles. This induced lateral response, which can occur under purely axial applied loading, becomes more severe as the location of the defective piles becomes more asymmetric, and can compromise the structural integrity of the sound piles. It is not yet feasible to employ simple methods of calculation to examine the behaviour of groups with defective or dissimilar piles, and even computer methods of group settlement analysis should have the ability to consider both axial and lateral responses, rather than only axial response. In computer programs employing the interaction factor method, modifications need to be made to account for the interaction between dissimilar piles. Such approximations have been explored by Xu (2000) and Wong and Poulos (2006).

#### *8.6.4.4 The effects of compressible underlying layers*

It has been recognised for some time that the presence of soft compressible layers below the pile tips can result in substantial increases in the settlement of a pile group, despite the fact that the settlement of a single pile may be largely unaffected by the compressible layers. Some examples of such experiences include the chimney foundation reported by Golder and Osler (1968) and the 14-storey building described by Peaker (1984).

To emphasise the potential significance of compressible underlying layers, a simple hypothetical problem has been considered. Square pile groups founded in a stiffer layer overlying a softer layer have been analysed, using a BEM program, and assuming that the pile–soil response remains elastic. The settlement of the group is expressed as a proportion of the settlement of the group if the compressible layer was not present, and is related to the number of piles in the group (with the spacing between adjacent piles remaining constant). The results of the analysis are shown in Figure 8.20. It can be seen that, as might be expected, the larger the group (and therefore the width of the pile group), the greater is the effect of the underlying compressible layer on settlement. It is clear that if the presence of such compressible layers is either not identified, or is ignored, the pile group settlements can be several times those which would be predicted for the group bearing on a continuous competent stratum.

#### *8.6.4.5 Significance of non-linearity*

For piles which derive the majority of their resistance from shaft friction, it is found that the load–settlement behaviour at normal design working loads is quasi-linear and dependent

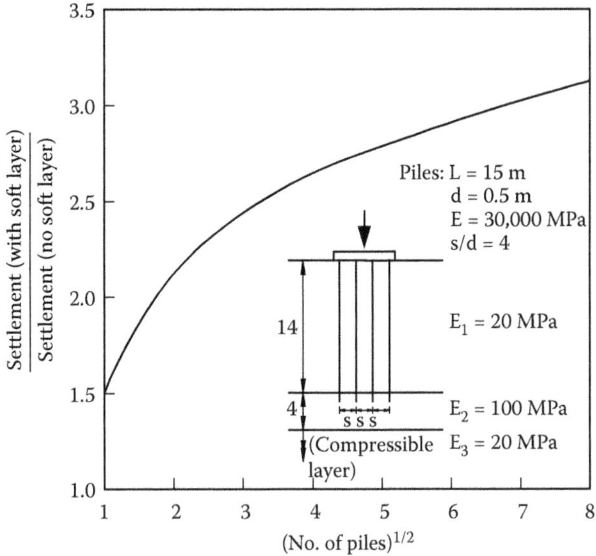


*Figure 8.20* Effect of underlying compressible layer on group settlement.

largely on the stiffness provided by the pile shaft. As a consequence, linear settlement theory is often adequate to predict the settlement. Non-linearity may become more significant under the following circumstances:

- For piles which derive a significant amount of their resistance from the base; in such a case, the full shaft resistance of the pile may be fully mobilised at the working load, and the load–settlement relationship is then dependent on both the shaft stiffness and the base stiffness.
- For piles which are operating at a relatively low overall factor of safety.
- For piles which are slender and relatively compressible; in such cases, pile–soil slip commences near the pile head at relatively low load levels, and progressively works its way down the pile, thus giving rise to a distinctly non-linear load–settlement behaviour, even at normal working load levels.
- For piles in soils which exhibit strain-softening characteristics.

Non-linear pile–soil behaviour can be taken into account in many methods of analysis, and may also be readily incorporated into hand methods of calculation. Such a method has been described by Poulos and Davis (1980), and requires a knowledge only of the elastic stiffness of the pile, the proportion of load carried by the base under elastic conditions, and the ultimate shaft and base capacities. An alternative approach has been developed by Fleming (1992), in which the shaft and base behaviours have been represented by hyperbolic relationships between settlement and load level. In this way, it has been possible to obtain remarkably good agreement between computed and measured load–settlement behaviour of piles, for a very wide range of geometries and soil types.

Poulos (1989) has found that a simple representation of non-linearity is often adequate to model non-linear pile settlement behaviour, and this contention has been supported by the work of Guo and Randolph (1997). They indicated that simple elastic–plastic non-linear behaviour can give a good representation of load–settlement behaviour, and have developed closed-form solutions for the estimation of this behaviour.

Mayne (2001) demonstrated that the use of a stress level-dependent soil Young's modulus such as that in Equation 6.36, in conjunction with elastic solutions for pile settlement, can give load–settlement curves that agree remarkably well with observed responses.

#### 8.6.4.6 Interaction between adjacent groups

It is often assumed that, when a structure is founded on piles, its settlement will be dependent only on the loading applied to that structure. However, if there are other structures nearby, there will be some interaction between the foundations, and as a consequence, the settlement of each of the structures will be greater than that of an isolated structure. This issue may be relevant to developments involving multiple high-rise towers in close proximity.

As an example of the possible consequences of such interaction, an analysis has been carried out for a case involving four identical structures on four identical foundations, each consisting of 25 equally loaded piles which bear on a stiffer stratum and are connected by a flexible cap. Each building has a footprint 36 m square, and the distance between each building is 5 m. The piles are 30 m long and 1.5 m in diameter, and are founded on a layer with a Young's modulus of 100 MPa. The pile shafts pass through a weaker layer with a Young's modulus of 25 MPa. For simplicity, the soil layers are assumed to be linearly elastic. While this case is hypothetical, it is not dissimilar to some high-density housing developments in cities such as Hong Kong.

Figure 8.21 shows the variation of computed settlement of the foundations. The settlement of an isolated foundation is also shown. It will be seen that the settlement taking interaction into account is significantly greater than the settlement of the isolated foundation, with the maximum settlement of the foundations being increased by almost 150% in this case.

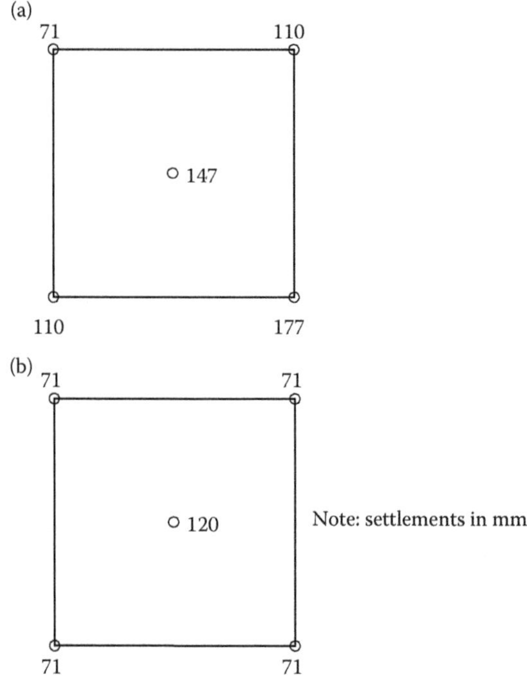


*Figure 8.21* Example of four interacting foundations (only one of the four foundations is shown in (a)): (a) four interacting foundations and (b) isolated foundation.

An important consequence of inter-group interaction is that each of the foundations suffers an induced tilt, despite the fact that the ground conditions are identical beneath all four structures. This tilt can be substantial when the underlying bearing stratum is relatively compressible.

### 8.6.5 A note on soil stiffness for pile settlement calculation

Chapter 6 has dealt in detail with the assessment of soil stiffness for estimates of the deformation of various foundation types. However, some comments specifically related to the estimation of pile settlement are given below.

For estimations of pile settlement, the key geotechnical parameter is the stiffness of the soil. If the analysis is based on elastic continuum theory, the soil stiffness can be expressed by a Young's modulus $E_s$ or shear modulus $G_s$. Both the magnitude and distribution of these moduli are important. It is clear that $E_s$ (or $G_s$) are not constants, but depend on many factors, including soil type, initial stress state, stress history, the method of installation of the pile, the stress system and stress level imposed by the pile and the pile group, and whether short-term or long-term conditions are being considered.

It should also be recognised that, in conventional analyses (including those presented herein), the assumption of lateral homogeneity of the soil is generally made. However, in reality, there are at least four stress regimes operative within the soil surrounding a group of vertically loaded piles, as shown in Figure 8.22, and the following four different values of Young's modulus can be distinguished:

1. The value $E_s$ for the soil in the vicinity of the pile shaft. This value will tend to strongly influence the settlement of a single pile and small pile groups.
2. The value $E_{sb}$ immediately below the pile tip. This value will also tend to influence the settlement of single pile and small pile groups.
3. The small-strain value, $E_{si}$, for the soil between the piles. This will reflect the smaller strains in this region and will affect the settlement interaction between the piles.
4. $E_s$ for the soil well below the pile tips ($E_{sd}$). This value will influence the settlement of a group more significantly as the group size increases.

The first and third values ($E_s$ and $E_{si}$) reflect primarily the response of the soil to shear, while the second and fourth values ($E_{sb}$ and $E_{sd}$) reflect both shear and volumetric strains.

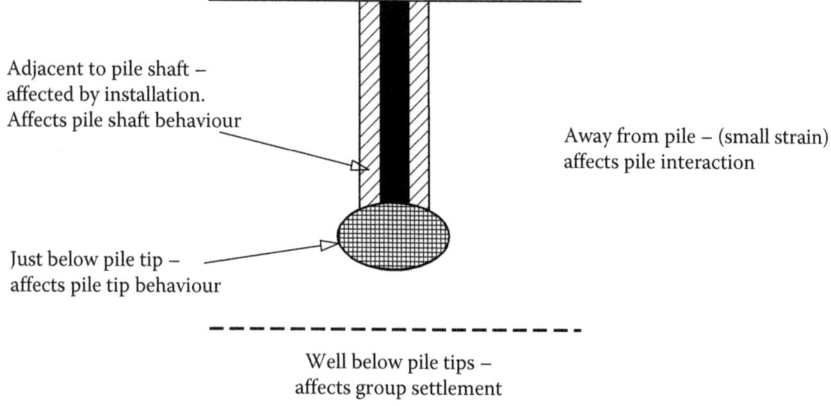


*Figure 8.22* Various values of soil modulus around a pile.

$E_s$ and $E_{sb}$ will both be influenced by the installation process, and would be expected to be different for bored piles and for driven piles. On the other hand, $E_{si}$ and $E_{sd}$ are unlikely to be affected by the installation process, but rather by the initial stress state and the stress history of the soil. As a corollary, the method of installation is likely to have a more significant effect on the settlement of a single pile (which depends largely on $E_s$ and $E_{sb}$) than on the settlement of a pile group, which may depend to a large extent on $E_{si}$ and $E_{sd}$.

The issue of the estimation of the soil modulus values has been discussed at length by Randolph (1994), Poulos (1994b), Mayne (1995), Mandolini and Viggiani (1997), and Yamashita et al. (1994, 1998). Yamashita et al. (1998) suggest that, for a purely elastic analysis, a typical value of modulus of about 0.25–0.3 times the small-strain value can be used, and this suggestion is similar to some of the results presented in Chapter 6.

### 8.6.6 A note on approximate modelling of barrettes

If barrettes are used within the foundation system, analysis via conventional pile analysis programs is generally difficult because most are based on the assumption of circular piles. While, in principle, a three dimensional analysis could be employed, such an analysis may present difficulties, as changes in the size and location of the foundation elements within a group are time consuming to take into account, and also require approximations to be made in order to consider the effects of moment loading. Moreover, some commonly used 3D finite element programs that incorporate piles do not have the capability of defining different stiffness values in the two horizontal directions, although this restriction has been addressed more recently, as described in Section 8.9.3.

Accordingly, for modelling of the foundation system and any diaphragm wall panel elements, it is suggested that the following approximations can be made when using conventional pile analysis programs:

1. Since the main purpose of the analyses is to provide values of head stiffness for each foundation element, for both vertical and lateral loading, separate analyses can be carried out for the following conditions:
   a. Vertical loading on each foundation element
   b. Lateral shear and moment loading in the x-direction on each element
   c. Lateral shear and moment loading in the y-direction on each element
2. For the vertical loading analysis, the barrettes can be transformed into equivalent circular piles via the following means:
   a. An equivalent shaft diameter such that there are equal perimeters for the barrette and the equivalent pile
   b. An equivalent pile base diameter such that there are equal base areas for the barrette and the equivalent pile
   c. An equivalent Young's modulus such that the axial stiffness (EA) of the pile is the same as that of the barrette
3. For the lateral loading analysis in each of the x- and y-direction, the barrettes can be transformed into equivalent piles via the following means:
   a. The pile diameter is taken as the width of the barrette in the direction of loading
   b. The moment of inertia of the pile is taken to be equal to that of the barrette, for the relevant direction of loading
   c. The Young's modulus is kept the same, so that the bending stiffness (EI) of the pile and the barrette are the same for the relevant direction of loading
   d. When assessing the ultimate lateral pile-soil pressure for a non-linear analysis, allowance should be made for the effects of the side shear resistance as well as the

normal lateral bearing resistance. This may be important when considering the capacity of the pile when loaded in the direction of the longer side

e. The effects of the shear stiffness of the two barrette sides is usually neglected, and this will tend to be conservative. Allowance can be made by using the solutions for lateral shear loading on a vertical rectangular area in Poulos and Davis (1974) to estimate the lateral pile–soil shear stiffness and adding this to the pile–soil stiffness for normal loading

4. In those cases where the lateral loading direction is not clearly defined, for example, when estimating the out-of balance soil loads, an average of the properties in the x- and y-direction can be adopted.
5. In those cases where combined axial and lateral loadings are considered simultaneously, the diameter of the shaft and the base can be approximated as the values for axial loading, while the moment of inertia are taken as the average for the x- and y-direction.

## 8.7 ESTIMATION OF PILED RAFT SETTLEMENTS

### 8.7.1 Category 2 methods

A simplified analysis method for piled rafts, denoted as the 'PDR' approach, has been described by Poulos (2001b). In this approach, the simplified equations developed by Randolph (1994) can be used to obtain an approximate estimate of the relationship between average settlement and the number of piles, and between the ultimate load capacity and the number of piles. From these relationships, a first estimate can be made of the number of piles, of a particular length and diameter, to satisfy the design requirements.

The geometry of a pile-cap unit within the piled raft system, as defined by Randolph (1994), is illustrated in Figure 8.23, and using his approach, the stiffness of the piled raft foundation can be estimated as follows:

$$K_{pr} = \frac{[K_p + K_r(1 - a_{cp})]}{(1 - a_{cp}{}^2 K_r / K_p)} \tag{8.50}$$

where $K_{pr}$ is the stiffness of piled raft, $K_p$ the stiffness of the pile group, $K_r$ the stiffness of the raft alone and $\alpha_{cp}$ is the raft–pile interaction factor.

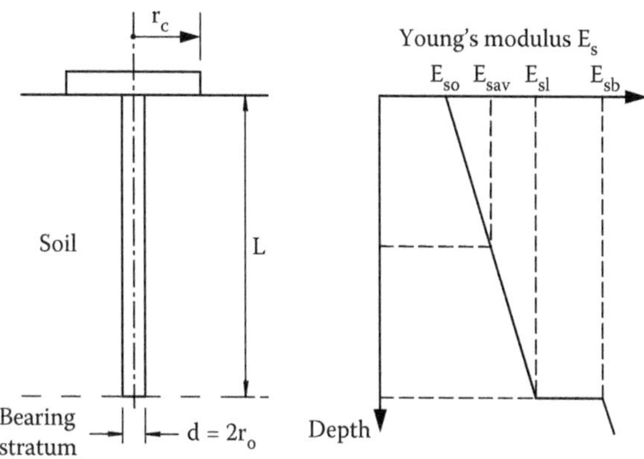


*Figure 8.23* Simplified representation of pile-cap unit.

The raft stiffness $K_r$ can be estimated via elastic theory, for example, as described in Section 8.3. The pile group stiffness can also be estimated from elastic theory, using approaches such as those described in Section 8.6.

The proportion of the total applied load carried by the raft is

$$\frac{P_r}{P_t} = \frac{K_r(1-a_{cp})}{(K_p + K_r(1-a_{cp}))} = X \tag{8.51}$$

where $P_r$ is the load carried by the raft and $P_t$ is the total applied load.

The raft–pile interaction factor $a_{cp}$ can be estimated as follows:

$$\alpha_{cp} = 1 - \ln\frac{(r_c/r_0)}{\zeta} \tag{8.52}$$

where $r_c$ is the average radius of pile cap (corresponding to an area equal to the raft area divided by number of piles), $r_0$ the radius of pile, $\zeta = \ln(r_m/r_0)$, $r_m = \{0.25 + \xi[2.5\,\rho\,(1-\nu)-0.25] * L$, $\xi = E_{sl}/E_{sb}$, $\rho = E_{sav}/E_{sl}$, $\nu$ the Poisson's ratio of soil, L the pile length, $E_{sl}$ the soil Young's modulus at level of pile tip, $E_{sb}$ the soil Young's modulus of bearing stratum below pile tip and $E_{sav}$ is the average soil Young's modulus along pile shaft.

The above equations can be used to develop a tri-linear load–settlement curve as shown in Figure 8.24. First, the stiffness of the piled raft is computed from Equation 8.50 for the number of piles being considered. This stiffness will remain operative until the pile capacity is fully mobilised. Making the simplifying assumption that the pile load mobilisation occurs simultaneously, the total applied load, $P_1$, at which the pile capacity is reached is given by

$$P_1 = \frac{P_{up}}{(1-X)} \tag{8.53}$$

where $P_{up}$ is the ultimate load capacity of the piles in the group and X is the proportion of load carried by the piles (Equation 8.51).

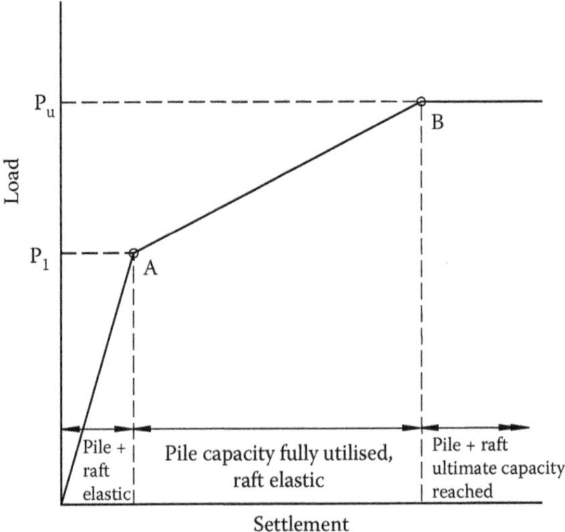


*Figure 8.24* Simplified tri-linear load–settlement curve for piled raft.

Beyond that point (Point A in Figure 8.24), the stiffness of the foundation system is that of the raft alone ($K_r$), and this holds until the ultimate load capacity of the piled raft foundation system is reached (Point B in Figure 8.24). At that stage, the load–settlement relationship becomes horizontal.

The load–settlement curves for a raft with various numbers of piles can be computed with the aid of a computer spreadsheet or a mathematical program such as MATHCAD. In this way, it is simple to compute the relationship between the number of piles and the average settlement of the foundation, and such a relationship is very useful in the concept phase of the foundation design.

### 8.7.2 Category 2 method for compensated piled rafts

The simplified 'PDR' method for piled rafts can be extended to estimate the load–settlement behaviour of a compensated piled raft, and produces a quadri-linear load–settlement curve (Poulos, 2005c). The steps in the approach are as follows:

1. Compute the reduction in vertical pressure, $p_e$ due to the excavation to the basement level at a depth He.
2. Assess the applied average pressure, $p_{ec}$, that needs to be applied to the raft (prior to excavation) to cause virgin loading of the raft. $p_{ec}$ may be estimated approximately as the average difference between the preconsolidation pressure and the in situ effective vertical pressure within the depth of influence of the raft (typically, a depth of about 1.5 times the raft breadth). In a soft normally consolidated soil, $p_{ec}$ would be approximately zero, ignoring any quasi-preconsolidation due to prior creep settlements.
3. Assess the soil modulus within the depth of influence of the raft for reloading, $E_2$. This will typically be 3–5 times greater than the modulus for the soil for the virgin loading state ($E_1$).
4. Compute the incremental stiffness of the raft foundation ($K_r$) for two cases:
   a. For the soil in the virgin loading state ($K_r = K_{rn}$)
   b. For the soil in the reloading state ($K_r = K_{ro}$)
5. The raft stiffness can be computed from a variety of settlement calculation approaches, for example, the elastic method described by Mayne and Poulos (1999) and shown in Figure 8.6. For preliminary purposes, the required geotechnical parameters can be estimated either from in situ testing, field measurements on test piles and footings, or via correlations such as those summarised in Chapter 6.
6. Compute the stiffness of the pile group, $K_p$, for the number of piles being considered. Because of the simplifying assumption made, this value will be the same, whether the soil below the raft is in a virgin loading state or a reloading state. The pile group stiffness can also be computed from elastic theory (Randolph, 1994; Poulos, 1989), as per Section 8.6.
7. Compute the incremental stiffness of the piled raft and the load sharing between the raft and the piles, using Equations 8.50 and 8.51.

   There will be two sets of values for the piled raft stiffness $K_{pr}$ and the load sharing factor X:
   a. The values for the soil below the raft in the virgin loading state, $K_{prn}$ and $X_n$, using the initial loading soil modulus $E_1$
   b. The values for the soil below the raft in the reloading state, $K_{pro}$ and $X_o$, using the unload/reload soil modulus $E_2$
8. Apply the applied load P on the piled raft foundation in a series of relatively small increments. For each increment, while the soil below the raft remains in the reloading state, use $K_{pro}$ to compute the incremental settlement $\Delta S_i$:

$$\Delta S_i = \frac{\Delta P_i}{K_{pro}} \tag{8.54}$$

where $\Delta P_i$ is the increment in applied load in increment i.

9. Compute the average pressure acting on the base of the raft, $p_{ri}$, as follows:

$$p_{ri} = p_{ri-1} + X_o \cdot Dp_i \tag{8.55}$$

where $p_{ri-1}$ is the raft pressure for previous increment, $X_o$ the proportion of raft load for the reloading state and $\Delta p_i$ is the average pressure increment $=\Delta P_i$/raft area.

10. Check for the state of the soil below the raft. While $p_r < p_e + p_{ec}$, the incremental stiffness of the raft is $K_{pro}$ and the load sharing factor is $X_o$. When $p_a > p_e + p_{ec}$, then the remaining increments use the values for the virgin loading state, $K_{prn}$ and $X_n$.
11. Check for the pile capacity being fully mobilised, that is, if $P_{pi} > P_{pu}$, where $P_{pi}$ is the load carried by piles at increment i and $P_{pu}$ is the ultimate capacity of piles. If this condition is met, then, for subsequent increments, only the raft will be able to carry the additional loads. In computing the incremental settlement in this case, the relevant raft stiffness is used, that is, $K_{ro}$ is the soil stiffness below the raft if in the reload state, and $K_{rn}$ if it is in the virgin loading state.
12. Steps 8–11 are repeated until the ultimate capacity of the piled raft system is reached. With the above approximate approach, a quadri-linear or tri-linear load-settlement curve is obtained, as shown in Figure 8.25. There will be three possible cases:
    a. Case 1, in which the soil becomes normally consolidated before the pile capacity is fully mobilised (Figure 8.25a). In this case, up to the load at Point A, the soil remains in the reloading state, and the piled raft stiffness and load sharing are $K_{pro}$ and $X_o$, respectively. At Point A, the stiffness and load sharing of the piled raft system change to $K_{prn}$ and $X_n$, to reflect the reduction in raft stiffness from $K_{ro}$ to $K_{rn}$. At Point B, the capacity of the piles is fully mobilised, and beyond that point, the incremental settlement is governed by the raft stiffness for the virgin loading state, $K_{rn}$. This situation holds until the ultimate capacity of the piled raft system is reached.

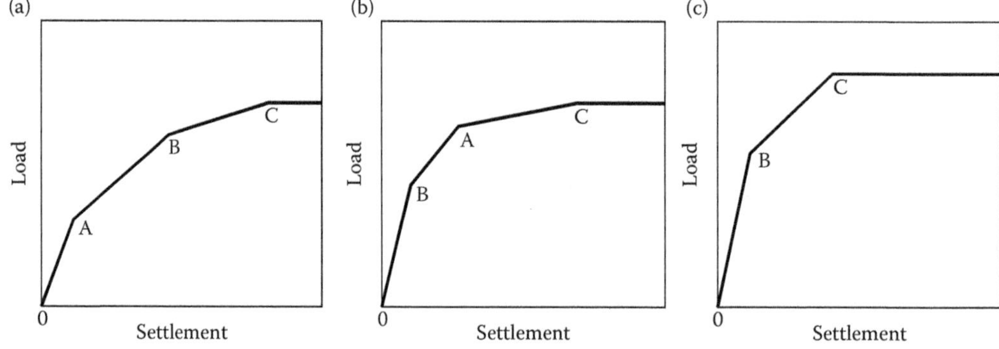


*Figure 8.25* Simplified load–settlement curves for compensated piled raft foundation. (a) Case 1 – Pile capacity mobilised (at Point B) after virgin loading state reached (Point A). (b) Case 2 – Pile capacity mobilised (at Point B) before virgin loading state reached (Point A). (c) Case 3 – Pile capacity mobilised (at Point B) and virgin loading state not reached before failure occurs (Point C).

b. Case 2, in which the pile capacity is fully mobilised while the soil below the raft remains in the reloading state (Figure 8.25b). In this case, the Point B, represents full mobilisation of the pile capacity, and Point A represents the transition from the reloading to the virgin loading state below the raft. The incremental settlements from Point A to Point C (where the ultimate capacity is reached) are governed by the raft stiffness $K_{rn}$.
c. Case 3, in which the soil conditions below the raft remain in the reloading state when the ultimate capacity of the piled raft is reached (Figure 8.25c), In this case, the pile capacity is fully mobilised at Point B, and the incremental settlements from Point B to Point C are governed by the raft stiffness $K_{ro}$. In this case, the load–settlement curve is tri-linear.

The above computational process can be readily evaluated via a spreadsheet.

## 8.8 ESTIMATION OF LATERAL PILE RESPONSE

### 8.8.1 Introduction

The lateral response of piles can be an important consideration in the design of foundations subjected to horizontal forces and overturning moments, especially for tall buildings subjected to wind and seismic loadings. As with vertical loading, consideration must be given in design to both the ultimate lateral resistance of piles, and the lateral deflections under the design serviceability loadings. It is not common for the ultimate lateral resistance to be the governing factor in design unless the piles or piers are relatively short or have a low flexural strength. The issue of ultimate lateral capacity has been discussed in Chapter 7, but for proper consideration of lateral deflections, it is important to consider also the ultimate lateral resistance of the soil, since the latter is an important component of a non-linear analysis of lateral response.

### 8.8.2 Lateral load–Deflection prediction: Linear analyses

Methods of estimating the lateral deflection and rotation of a laterally loaded pile usually rely on either the theory of subgrade reaction (Broms, 1964a,b) or on elastic continuum theory (Poulos, 1971a,b; Randolph, 1981; Budhu and Davies, 1987, 1988). For piles which are 'flexible' (i.e., their embedded length is equal to or longer than the critical length), comparisons show that the various solutions from elastic continuum theory generally agree well.

A very convenient form of the elastic solutions is provided by Poulos and Davis (1980) and by Krishnan et al. (1983). For a soil having a constant modulus with depth, the linear solutions from the latter source for pile groundline deflection, ρ, and rotation, θ, for a free-headed pile may be expressed as follows:

$$\rho = \frac{I_{uH} \cdot H}{(E_s d)} + \frac{I_{uM} \cdot M}{(E_s d^2)} \tag{8.56}$$

$$\theta = \frac{I_{\theta H} \cdot H}{(E_s d^2)} + \frac{I_{\theta M} \cdot M}{(E_s d^3)} \tag{8.57}$$

where H is the applied load at groundline, M the applied moment at groundline, $E_s$ the soil Young's modulus, d the pile diameter of width, $I_{uH}$, $I_{uM}$, $I_{\theta H}$ and $I_{\theta M}$ are influence factors which depend on the ratio K of pile modulus to soil modulus ($E_p/E_s$) and the relative length of the pile.

For a fixed-head pile, the corresponding expression for head deflection, $\rho_F$, is

$$\rho_F = \frac{I_{uF} \cdot H}{(E_s d)} \tag{8.58}$$

where $I_{uF}$ is the displacement influence factor for fixed head pile.

Solutions for the various influence factors are shown in Table 8.3 for piles that can be considered to be 'flexible', that is, their length L exceeds the critical length $L_c$. This value is approximated by Krishnan et al., 1983 as follows:

$$L_c \approx 1.45d\left(\frac{E_p}{E_s}\right)^{0.21} \tag{8.59}$$

where $E_p$ is the pile Young's modulus.

Alternative estimates of $L_c$ have been provided by Randolph (1981) and Poulos and Hull (1989), as follows:

$$L_c \approx 1.31d\left(\frac{E_p}{E_s}\right)^{0.286} \quad \text{(Randolph, 1981, for } v_s = 0.3\text{)} \tag{8.60}$$

$$L_c \approx 1.11d\left(\frac{E_p}{E_s}\right)^{0.25} \quad \text{(Poulos and Hull, 1989)} \tag{8.61}$$

If the soil profile is not homogeneous, then an equivalent soil modulus value needs to be used in Equations 8.56 through 8.58. For a soil profile in which the modulus increases linearly with depth, the equivalent soil Young's modulus can be estimated as the value at an equivalent depth below the surface. These equivalent depths are shown in Table 8.4. An iterative procedure is required as the $E_s$ value appears in the equations.

It is found that for many cases, the equivalent depth is only between 1 and 3 diameters below the soil surface. These results indicate that the lateral response of piles depends critically on the near-surface soil characteristics.

*Table 8.3* Factors for lateral response of a flexible pile ($L \geq L_c$)

| *Pile Head Condition* | *Factor* | *Expression* |
|---|---|---|
| Free head | $I_{uH}$ | $2.50(E_p/E_s)^{-0.31}$ |
| | $I_{uM} = I_{\theta H}$ | $8.80(E_p/E_s)^{-0.73}$ |
| | $I_{\theta M}$ | $2.75(E_p/E_s)^{-0.50}$ |
| Fixed head | $I_{uF}$ | $1.70(E_p/E_s)^{-0.36}$ |

Source: Krishnan, R., Gazetas, G. and Velez, A. 1983. *Geotechnique*, 23(3): 307–325.

*Table 8.4* Equivalent depth for a non-homogeneous soil profile

| *Pile head condition* | *Factor* | *Expression* |
|---|---|---|
| Free head | $z_{uH}$ | $0.38d\ (E_p/E_s)^{0.17}$ |
| | $z_{uM} = z_{\theta H}$ | $0.16d\ (E_p/E_s)^{0.20}$ |
| | $z_{\theta M}$ | $0.34d\ (E_p/E_s)^{0.14}$ |
| Fixed head | $z_{uF}$ | $0.48d\ (E_p/E_s)^{0.20}$ |

Source: Krishnan, R., Gazetas, G. and Velez, A. 1983. *Geotechnique*, 23(3): 307–325.

### 8.8.3 Non-linear analyses

Soil non-linearity can have a significant effect on the lateral response of piles. It leads to increased lateral deflection and pile head rotation, and these increases may become great (compared to the initial linear response) at relatively modest applied load levels. This effect was recognised early in the pioneering work of Reese and his co-workers, who developed the 'p–y' method. The 'p–y' curves in their method are essentially non-linear spring characteristics, and the p–y method itself can be considered as a non-linear subgrade reaction approach. On the basis of carefully instrumented field pile tests, a series of p–y curves was developed for various types of soil, and these form the basis of much common practice today. Details of these empirical curves are summarised by Sullivan et al. (1980) while Murchison and O'Neill (1984) and Bransby (1999) have developed alternative approaches to the development of p–y curves via the use of analytical solutions. An alternative procedure to account for non-linearity has been proposed by Prakash and Kumar (1996) who have assumed that the modulus of subgrade reaction at any depth is dependent on the level of shear strain adjacent to the pile at that depth.

It is also possible to develop non-linear lateral response solutions from elastic continuum theory, by imposing the condition that the lateral pile–soil pressure cannot exceed the ultimate lateral pile–soil pressure, $p_y$. The results of such theory can be expressed in terms of correction factors to the elastic solutions (such as those in Equations 8.56 through 8.58, for example, Poulos and Davis, 1980; Budhu and Davies, 1987, 1988; Poulos and Hull, 1989). For the case of a fixed head pile in a cohesive soil whose properties are constant with depth, the pile head deflection $\rho_F$ can be estimated as follows:

$$\rho_F = \frac{\rho_{el}}{F_u} \tag{8.62}$$

where $\rho_{el}$ is the pile head displacement from elastic theory and $F_u$ is the yield correction factor for deflection, depending on load level and relative flexibility of the pile.

The corresponding pile head fixing moment $M_F$ is expressed as

$$M_F = \frac{M_{FE}}{F_M} \tag{8.63}$$

where $M_{FE}$ is the pile head moment from linear elastic theory and $F_M$ is the yield correction factor for head fixing moment.

Figure 8.26 shows solutions for $F_u$ and $F_M$ for a fixed head pile in a uniform clay.

It has been shown by Poulos (1982a) that similar results can be obtained with this approach, and with the use of simple representations of p–y curves (e.g. hyperbolic or elastic–plastic), as well as the relatively more complex curves commonly used.

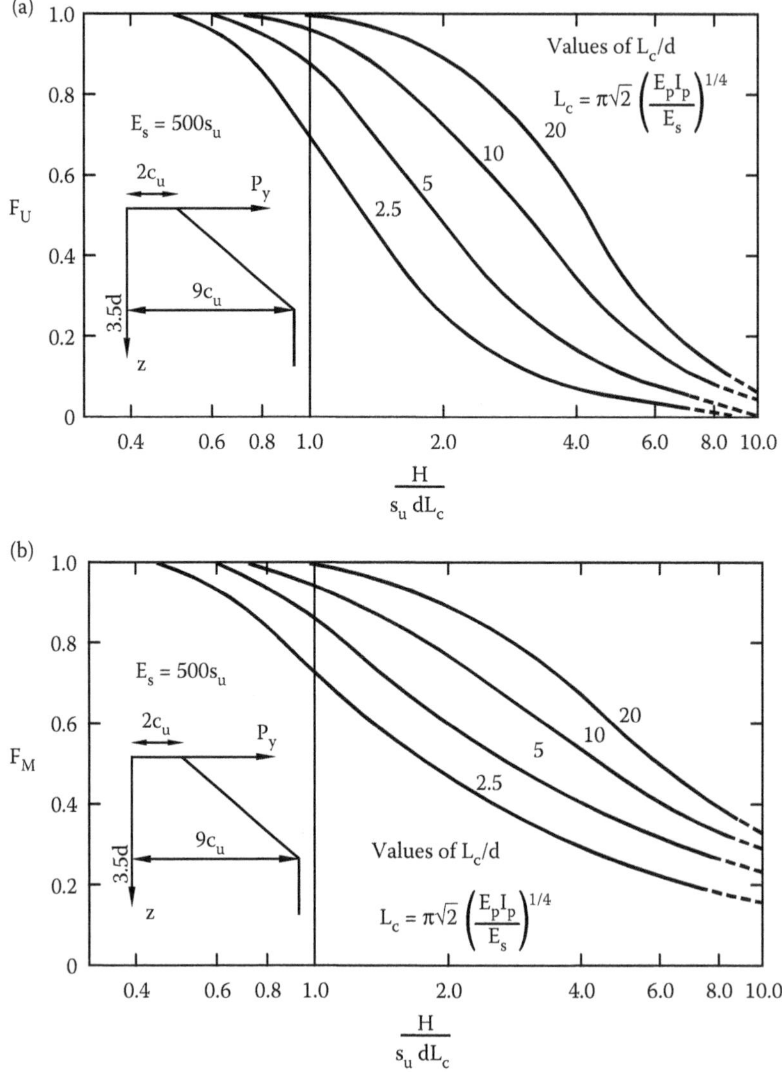


*Figure 8.26* Non-linear correction factors for fixed head pile in uniform clay: (a) deflection factor $F_u$ and (b) head moment factor $F_M$. (Adapted from Poulos, H.G. and Hull, T.S. 1989. *Foundation Engineering – Principles & Practices*. Evanston, IL, ASCE, Vol. 2, pp. 1578–1606. Courtesy of ASCE.)

The research on the behaviour of single laterally loaded piles leads to the following conclusions:

1. There are some differences between the linear solutions from elastic continuum theory and subgrade reaction theory, but these differences can be accommodated in practice via calibration of the analysis with field data and the use of appropriate values of the relevant soil deformation parameters
2. Non-linear pile-soil response is a most important aspect of behaviour. Failure to allow for this behaviour may lead to grossly inaccurate (and unconservative) predictions of lateral deflection and rotation

3. It is not necessary to employ complex representations of non-linear soil behaviour in order to obtain reasonable predictions of lateral pile response. Even a simple elastic-plastic or hyperbolic response may often be adequate to capture the main non-linear effects
4. As with most foundation problems, the key to successful prediction is more the ability to choose appropriate geotechnical parameters rather than the details of the analysis employed

### 8.8.4 A practical procedure for load–deflection estimation

A useful practical procedure for estimating the load–deflection behaviour of single piles was presented by Duncan et al. (1994) and Brettman and Duncan (1996). They introduced the concept of the 'characteristic load', and used dimensional analysis to characterise the non-linear behaviour of piles via relationships between dimensionless variables. These variables are defined as follows:

Characteristic load:

1. For clay:

$$H_c = 7.34d^2(E_pR_1)\cdot\left[\frac{s_u}{(E_pR_1)}\right]^{0.68} \tag{8.64}$$

2. For sand:

$$H_c = 1.57d^2(E_pR_1)\cdot\left[\frac{\gamma' d\phi' K_p}{(E_pR_1)}\right]^{0.57} \tag{8.65}$$

Characteristic moment:

1. For clay:

$$M_c = 3.86d^3(E_pR_1)\cdot\left[\frac{s_u}{(E_pR_1)}\right]^{0.46} \tag{8.66}$$

2. For sand:

$$M_c = 1.33d^3(E_pR_1)\cdot\left[\frac{\gamma' d\phi' K_p}{(E_pR_1)}\right]^{0.40} \tag{8.67}$$

where $H_c$ is the characteristic load, $M_c$ the characteristic moment, d the pile diameter or width, $E_p$ the pile modulus, $R_1$ the ratio of moment of inertia of pile section to that of a solid circular cross section (=1 for a solid circular pile), $s_u$ the undrained shear strength of clay, $\gamma'$ the effective unit weight of sand, $\phi'$ the effective stress friction angle for sand (degrees) and $K_p$ is the Rankine passive pressure coefficient.

For applied horizontal loading, the lateral load–deflection relationship can then be approximated as follows (Brettman and Duncan, 1996):

$$\left(\frac{\rho_h}{d}\right) = a_h \left(\frac{H}{H_c}\right)^{b_h} \tag{8.68}$$

For applied moment loading, the corresponding relationship is

$$\left(\frac{\rho_h}{d}\right) = a_m \left(\frac{M}{M_c}\right)^{b_m} \tag{8.69}$$

The relationship between maximum moment induced in the pile and the applied horizontal loading can similarly be expressed as

$$\left(\frac{H}{H_c}\right) = a_x \left(\frac{M_{max}}{M_c}\right)^{b_x} \tag{8.70}$$

In the above equations, $\rho_h$ is the groundline deflection, $a_h$, $b_h$ are constants for applied horizontal loading, $a_m$, $b_m$ constants for applied moment loading, $a_x$, $b_x$ constants for maximum pile moment, H the applied lateral load at top of pile, M the applied moment at top of pile, $H_c$ the characteristic load (Equations 8.64 and 8.65) and $M_c$ is the characteristic moment (Equations 8.66 and 8.67).

Values of the various constants in the above equations are given in Table 8.5.

When both horizontal load and moment are applied simultaneously, the following procedure is followed:

1. Compute the deflections which would be caused by the load acting alone ($\rho_{hh}$) and the moment acting alone ($\rho_{hm}$).
2. Compute the value of load ($H_m$) that would cause the same deflection as the moment and the value of moment ($M_h$) that would cause the same deflection as the load.
3. Compute the ground-line deflection, $\rho_{h1}$, that would be caused by the sum of the real load and the equivalent load ($H + H_m$), and the deflection, $\rho_{h2}$, that would be caused by the sum of the real moment and the equivalent moment ($M + M_h$).

*Table 8.5* Constants for lateral load–deflection estimation

| | *Clay-* | | *Sand-* | |
|---|---|---|---|---|
| *Constant* | *Free head* | *Fixed head* | *Free head* | *Fixed head* |
| $a_h$ | 50.0 | 14.0 | 119.0 | 28.8 |
| $b_h$ | 1.822 | 1.846 | 1.523 | 1.500 |
| $a_m$ | 21.0 | – | 36.0 | – |
| $b_m$ | 1.412 | – | 1.308 | – |
| $a_x$ | 0.85 | 0.78 | 4.28 | 2.64 |
| $b_x$ | 1.288 | 1.249 | 1.384 | 1.300 |

Source: Brettman, T. and Duncan, J.M. 1996. *Journal of Geotechnical Engineering, ASCE*, 122(6): 496–498.

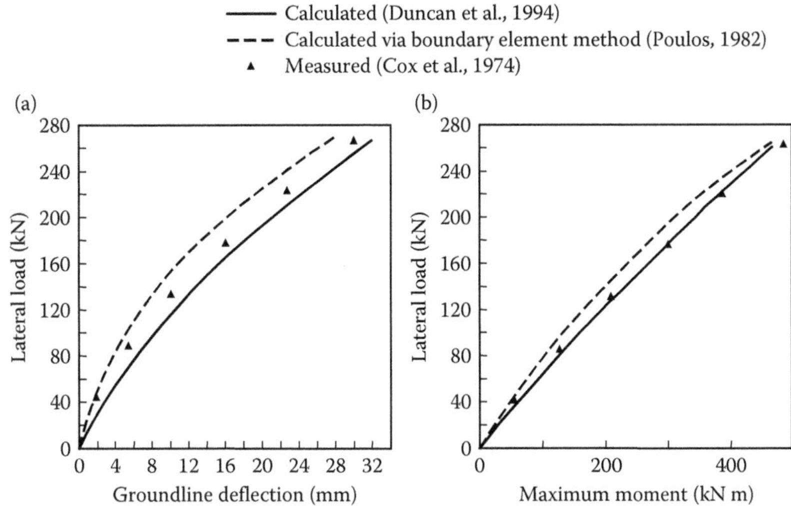


*Figure 8.27* Comparison between solutions for load–deflection response of a single pile.

4. The estimated value of deflection due to both load and moment is taken as the average of the two values computed above, that is,

$$\rho_h = 0.5(\rho_{h1} + \rho_{h2}) \tag{8.71}$$

As pointed out by Duncan et al. (1994), the characteristic load method (CLM) has some limitations. It applies only to 'long' piles that have a length greater than the critical or active length, and it applies only to uniform soils which are sand or clay along the critical length. However, only the soil within the critical depth (usually the upper 8 diameters or so) is important for estimating lateral response, and where ground conditions vary, average properties of the ground profile within this depth can be assumed for the analysis.

To compare the predictions from this approach with that from other methods, the field tests for a pipe pile in sand (Cox et al., 1974) have been analysed. Figure 8.27 shows the comparison for both the head load–deflection relationships and the applied load versus maximum pile moment relationship. Also shown are the relationships predicted from a non-linear BEM analysis by Poulos (1982a). It can be seen that both the Duncan et al. approach and the BEM analysis give results which are comparable and in reasonable agreement with the test data.

## 8.9 ESTIMATION OF GROUP EFFECTS

A group of piles will generally deflect more than a single pile under the same load per pile, due largely to the effects of pile–soil–pile interaction. At the same time, the restraining effects of the pile cap connecting the piles may considerably modify the pile behaviour as compared to a single free-headed pile. Therefore, considerable caution must be exercised in applying the theory for a single pile to pile groups.

Various procedures have been developed to estimate the lateral deflection of pile groups, and these generally fall into the following categories:

- Interaction factor approaches: these were introduced by Poulos (1971b) and involve the consideration of the additional lateral deflections and rotations caused by a loaded pile on adjacent piles. The soil mass is taken to be an elastic continuum, and use is made of the classical equations of Mindlin to compute the various interaction factors. Randolph (1981) has developed very useful approximations for these interaction factors, for soils whose stiffness is either constant or else increases linearly with depth.

- Hybrid approach, combining the p–y method for single piles with elastic continuum analysis to estimate interaction effects (Focht and Koch, 1973; O'Neill et al., 1977).
- Group deflection ratio method. This been used by Poulos (1987b), and uses elasticity theory to derive group factors which are applied to the response of a single pile to allow for group effects. This approach is analogous to the use of a group settlement ratio for estimating pile group settlements.
- Equivalent pier approach, in which the group is represented by an equivalent single pier (Bogard and Matlock, 1983; Poulos, 1975).
- Group reduction factor method. A version of this approach was developed by Davisson (1970), based on model pile tests by Prakash (1962). This approach reduced the subgrade reaction modulus to account for group interaction.
- Elastic continuum BEM analysis, such as used by Banerjee and Driscoll (1976) and Xu (2000).
- Finite element analysis, using either a plane strain model (e.g. Desai, 1974) or a full 3D model (Kimura et al., 1995).

Some of the above approaches for estimating group effects will be discussed below.

### 8.9.1 Interaction factor method

Poulos (1971b) has extended the interaction factor method used for group settlement estimation to consider the interaction among laterally loaded pile groups. However, the application of the interaction factor approach is less straightforward than for settlements because of the following characteristics:

- For free-head piles, there are four different interaction factors to consider: the effect of lateral load on lateral deflection, the effect of lateral load on head rotation, the effect of moment on head deflection, and the effect of moment on head rotation. If elastic pile–soil behaviour is assumed, the effect of lateral load on rotation is the same as the effect of moment on lateral deflection, and so there are then three independent interaction factors.
- The interaction between two piles will depend not only on the spacing between the piles, but also on the orientation of the piles in relation to the direction of loading. The largest interaction occurs if the pile orientation is in the same direction as the direction of loading, and the least interaction occurs when the pile orientation is normal to the direction of loading.

If the piles have a fixed head, the situation is simplified since there is then only one interaction factor, expressing the effect of lateral loading on lateral deflection, but the dependence on pile and loading orientation remains.

Poulos and Davis (1980) have presented a series of plots showing the lateral interaction factors as a function of spacing, orientation, and relative pile flexibility. However, Randolph (1981) has developed more convenient approximate closed-form expressions for these interaction factors, and for the case of fixed head piles equal to or longer than the critical length $L_c$, this interaction factor, $\alpha_{\rho F}$, can be approximated as follows:

$$a_{\rho F} = 0.6\rho_c \left(\frac{E_p}{G_c}\right)^{1/7} \left(\frac{r_0}{s}\right)(1+\cos^2\beta) \qquad (8.72)$$

where $G_c$ is the the average value of $G^* = G(1 + 0.75\nu)$ over the critical length $L_c$ of the pile, $\rho_c = 0.5(G_{c0.25}/G_c)$, $G_{c0.25}$ the value of $G^*$ at a depth of $L_c/4$, $r_0$ the pile radius, s the centre-to-centre spacing between the piles, $\beta$ the departure angle = angle between the line joining the pile centres and the direction of lateral loading and $\nu$ is the soil Poisson's ratio.

The critical length $L_c$ is given by

$$L_c = 2r_0 \left(\frac{E_p}{G_c}\right)^{2/7} \qquad (8.73)$$

For a group of piles, the lateral deflection $\rho_i$ of any pile i in the group can be written as follows:

$$\rho_i = \rho_{F1} \sum(\alpha_{rF\,i,j}) \qquad (8.74)$$

where $\rho_{F1}$ is the head deflection of a single pile at the average load in the group and $\alpha_{rF\,i,j}$ is the lateral interaction factor for the spacing and orientation between pile i and another pile j.

If the distribution of lateral load within the fixed head group is known, Equation 8.74 can be used directly to calculate the lateral deflection of each pile in the group. If the pile cap is assumed to be rigid, then Equation 8.74 is written for each pile in the group, and then, together with the horizontal equilibrium equation, the resulting series of $n + 1$ equations (where $n$ = number of piles in the group) can be solved to give the common lateral deflection of all n piles and the lateral load on each of the piles.

The computer programs DEFPIG and PIGLET use such an approach to analyse groups subjected to lateral, as well as vertical, loadings.

### 8.9.2 Group factors via the hybrid method

A useful practical procedure has been developed by Ooi and Duncan (1994), based on the results of extensive parametric studies using the method of Focht and Koch (1973). Their approach, the 'group amplification procedure', can be summarised as follows:

1. The group deflection, $\rho_g$, is given by

$$\rho_g = C_y \rho_s \qquad (8.75)$$

2. The maximum bending moment, $M_g$, in a pile within a group is given by

$$M_g = C_m M_s \qquad (8.76)$$

where $\rho_s$ is the single pile deflection under the same load per pile; $M_s$ the maximum moment in a single pile under the same load; $C_y$ the deflection amplification factor ($\geq 1$); $C_m$ is the moment amplification factor ($\geq 1$).

In applying Equation 8.75, the pile head condition for the single pile should reflect the conditions of restraint at the pile cap. For a cap which provides little or no restraint, $\rho_s$ is computed for a free-head pile, while for pile caps that provide restraint, it is appropriate to compute $\rho_s$ for a fixed head pile. The latter case would normally be relevant for tall building foundations.

The following expressions were derived from parametric studies by Ooi and Duncan:

$$C_y = \frac{(A + N_{pile})}{(B(s/d + P_s/CP_N)^{0.5})} \tag{8.77}$$

$$C_m = (C_y)^n \tag{8.78}$$

where A = 16 for clay, and 9 for sand; $N_{pile}$ is the number of piles in group; B = 5.5 for clay and 3.0 for sand; s the average spacing of piles; d the diameter of single pile; $P_s$ the average lateral load on pile in group; C = 3 for clay and 16 for sand; $P_N = (s_u d^2)$ for clay and $(K_p \gamma d^3)$ for sand; $\gamma$ the average total unit weight of sand over the top 8 diameters; $K_p$ the Rankine passive pressure; $s_u$ the average undrained shear strength within the top 8 diameters; $n = (P_s/150P_N) + 0.25$ for clay, and $(P_s/300P_N) + 0.30$ for sand.

Ooi and Duncan (1994) have found satisfactory agreement between their approach and the results of a number of field measurements. However, they point out that their method has a number of limitations, including the following:

- It has been developed for uniformly spaced piles which are vertical (not raked)
- The load distribution within the group cannot be obtained
- The results do not depend on the arrangement of the piles in the group
- The method is restricted to piles whose embedded length exceeds the critical length

Figure 8.28 compares load–deflection curves for a typical pile group in clay, computed from Ooi and Duncan's approach, and also via a BEM program DEFPIG (see Section 8.10.2). For the free-headed pile group, the agreement is good, but for a fixed head group, Ooi and Duncan's approach predicts a stiffer response than the elasticity-based DEFPIG analysis.

### 8.9.3 Group deflection ratio approach

A simple approach has been suggested by Poulos (2001a) in which the group lateral deflection $\rho_g$ is estimated as follows:

$$\rho_g = R_\rho \rho_s \tag{8.79}$$

where $R_\rho$ is the group deflection ratio $=(N_{pile})^{\omega l}$; $\rho_s$ the deflection of single pile at the same lateral load; $N_{pile}$ the number of piles; $\omega_l$ the exponent depending on the critical length of the pile, $L_{cr}$, and the pile spacing. Typical values of the exponent $\omega_l$ are plotted in Figure 8.29. From Equation 8.75, it can be seen that Ooi and Duncan's factor $C_y$ has the same meaning as the group deflection ratio $R_\rho$.

Papadopoulou and Comodromos (2010), and Comodromos and Papadopoulou (2012), for sands and clays respectively, have compared the lateral responses of a pile group and a single pile and have identified the influence of the number of piles, the spacing, and the deflection level on the group response on the group deflection ratio, and have developed useful relationships for predicting this ratio. Comodromos et al. (2016) have considered a piled raft subjected to combined vertical and lateral loadings, and have presented expressions for

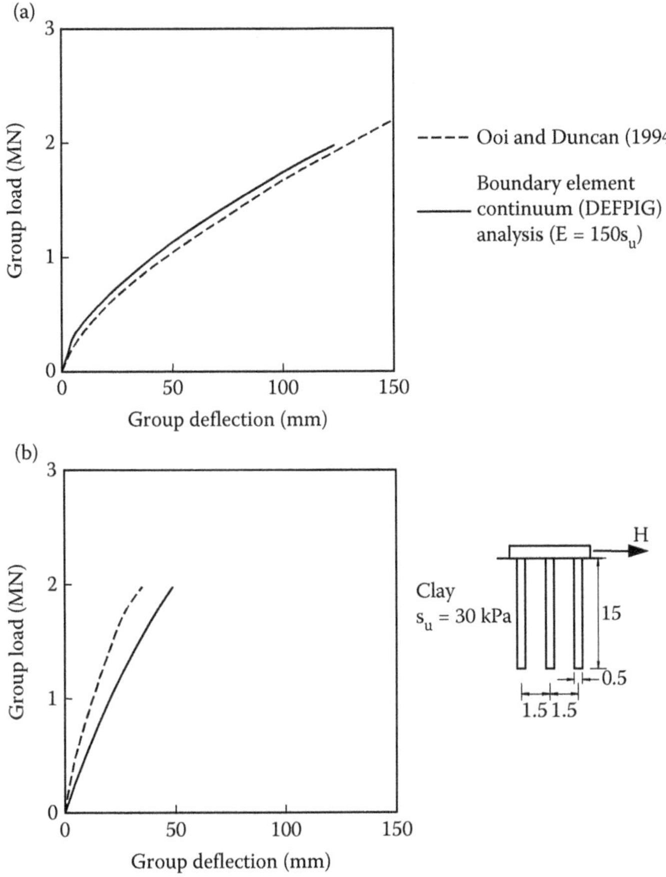


*Figure 8.28* Comparison of load–deflection curves for a nine-pile group: (a) free-head piles and (b) fixed-head piles.

the group deflection ratio of piled rafts, taking into account the effects of group size and configuration, and non-linear effects.

### 8.9.4 Equivalent pier analysis

For preliminary assessments of the lateral response of pile groups, it may be convenient to simplify the group as an equivalent free-headed pier, and then use elastic solutions for the pier head deflection and rotation.

Carter and Kulhawy (1992) have indicated that a pier in a homogeneous layer may be considered to be essentially rigid when the following condition is satisfied:

$$L \leq 0.07d\left(\frac{E_p}{E_s}\right)^{0.5} \tag{8.80}$$

where L is the pier length, d the pier diameter, $E_p$ the pile Young's modulus and E is the soil Young's modulus.

For a practical range of values of $E_p/E_s$, this implies a range of L/d of between 0.5 for stiff to hard soils to about 4 for soft soils.

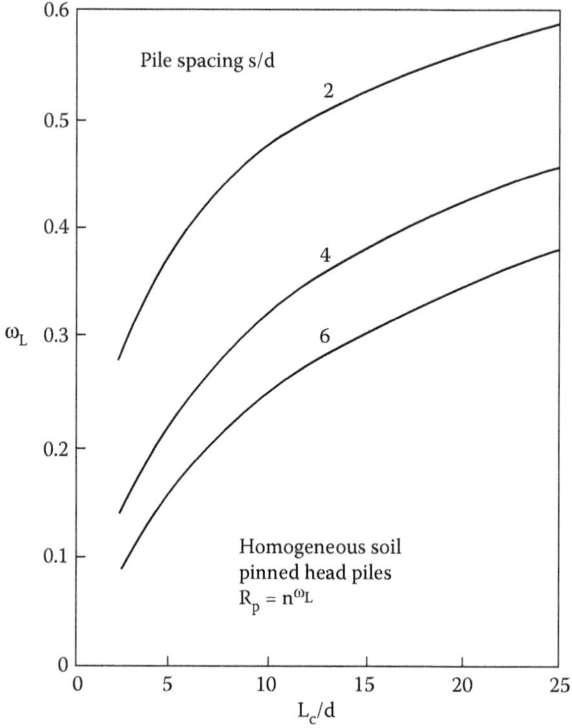


*Figure 8.29* Group factor exponent $\omega_L$.

From their solutions, the following approximate solutions for the pier head influence factors can be derived and substituted into Equations 8.56 and 8.57:

$$I_{uH} = 0.7\left(\frac{L}{d}\right)^{-0.33} \tag{8.81}$$

$$I_{uM} = I_{\theta H} = 0.4\left(\frac{L}{d}\right)^{-0.88} \tag{8.82}$$

$$I_{\theta M} = 0.6\left(\frac{L}{d}\right)^{-1.67} \tag{8.83}$$

For a fixed head pier, the head displacement influence factor is given as follows:

$$f_F = f_{uH} - f_{uM^2}/f_{\theta M} \tag{8.84}$$

where e = M/H is the ratio of moment to applied horizontal load.

When applying the equivalent pier approach to estimating group lateral deflections, it may be prudent to adopt the lesser of the actual pile length (L) and the critical length of a single pile $L_c$ in Equations 8.81 through 8.83.

## 8.10 CATEGORY 3 ANALYSIS METHODS

### 8.10.1 Desirable analysis characteristics

While the preliminary stage of design can generally be undertaken with relatively simple and straightforward techniques to assess both ultimate capacity and overall settlement performance, for the detailed and final design stages, more refined Category 3 techniques are generally appropriate. For these stages, the programs used should ideally have the capabilities listed below.

For overall stability, the program should be able to consider:

- Non-homogeneous and layered soil profiles
- Non-linearity of pile and, if appropriate, raft behaviour
- Geotechnical and structural failure of the piles (and the raft)
- Vertical, lateral and moment loading (in both lateral directions), including torsion
- Piles having different characteristics within the same group

For serviceability analysis, the above characteristics are again desirable, and in addition, the program should have the ability to consider:

- Pile–pile interaction, and if appropriate, raft–pile and pile–raft interaction
- Flexibility of the raft or pile cap
- Some means by which the stiffness of the supported structure can be taken into account

### 8.10.2 Pile analysis programs

There do not appear to be any commercially available pile analysis software packages that have all of the above desirable characteristics. The commercially available programs DEFPIG, PIGLET and REPUTE are based on BEM analyses, and have some of the requirements, but fall short of a number of critical aspects, particularly in their inability to include raft–soil contact and raft flexibility. A brief description of these programs is given below.

#### *8.10.2.1 DEFPIG*

DEFPIG (Poulos, 1990) is a FORTRAN program for determining the deformations and load distribution within a group of piles subjected to vertical, horizontal and moment loading. The program considers a group of identical elastic piles having axial and lateral stiffness that are constant with depth. The piles are supported by a linear elastic medium, but the program allows for the possibility of slippage between the piles and the soil under axial loading and for the development of yield of the soil adjacent to the pile due to lateral loading. The stress distributions are computed from Mindlin's solutions for an isotropic homogeneous linear elastic medium, but non-homogeneity of the soil along the pile length can be approximately taken into account. The piles are assumed to be attached to a rigid pile cap, so that all undergo equal lateral deflection and rotation, or alternatively, the piles can be subjected to specified loads or deflections. Group effects are taken into account via the interaction factor method for both vertical and lateral responses. Raking or battered piled may be present in the group, but only battered in the direction of the lateral loading. Load–deformation relationships to failure may be obtained if desired.

#### 8.10.2.2 PIGLET

PIGLET (Randolph, 1996) employs simplified expressions for the response of single piles to axial, lateral and torsional loading, with interaction among piles in a group being considered via the use of interaction factors which are also expressed in a simplified closed form. The soil is assumed to be an elastic continuum in which the elastic modulus can vary linearly with depth. A stiffer stratum at the level of the pile toes can be specified. The program provides results for the three components of deflection and the three components of rotation, together with the pile head loads, bending moments and torsional moment. The distribution of axial load, settlement, bending moment and deflection along each pile can also be obtained.

#### 8.10.2.3 REPUTE

REPUTE (GeoCentrix, 2013) undertakes the following analysis tasks:

- Various types of single pile, using current and historical design standards (such as Eurocode 7 and BS 8004).
- Pile groups under generalised 3D loading, using linear or non-linear soil models.

Repute considers single pile response using a variety of calculation methods for ultimate and serviceability limit states. Both traditional lumped factors-of-safety and partial factors can be applied in these calculations.

Repute considers pile group behaviour using the BEM, and provides a complete 3D non-linear BEM solution of the soil continuum, which overcomes limitations of traditional interaction-factor methods and gives more realistic predictions of deformations and the load distribution between piles.

There are also two other programs that have been described in the literature but that are not commercially available:

1. geotechnical analysis of raft with piles (GARP), Small and Poulos (2007)
2. Non-linear analysis of piled rafts (NAPRA), Russo (1998)

GARP uses a simplified BEM analysis to compute the behaviour of a piled raft when subjected to applied uniform or concentrated vertical loading, moment loading, and free-field vertical soil movements. The raft is represented by a thin elastic plate and is discretised via the finite element method, using 8-noded elements. The soil is modelled as a layered elastic continuum, and the piles are represented by elastic–plastic or hyperbolic springs, which can interact with each other and with the raft. Pile–pile interactions are incorporated via interaction factors (Poulos and Davis, 1980). Simplifying approximations are utilised for the raft–pile and pile–raft interactions. Beneath the raft, limiting values of contact pressure in compression and tension can be specified so that some allowance can be made for non-linear raft behaviour. The output of GARP includes the settlement at all nodes of the raft; the transverse, longitudinal and twisting bending moments at each Gauss point in the raft; the contact pressures below the raft; and the vertical loads on each pile. It should be noted that analyses that extend the GARP analysis to consider both vertical and horizontal loading have been described by Zhang and Small (1994, 2000).

NAPRA computes the behaviour of a raft subjected to any combination of vertical, distributed or concentrated loading and moment loading. The raft is modelled as a 2D elastic body using thin plate theory via the finite element method, using a 4- or 9-noded rectangular element. The piles and the soil are modelled as interacting linear or non-linear springs. The interaction between the raft and the soil (the piles) is purely vertical; accordingly, only the axial stiffness of the springs is required. The soil is assumed to be a layered elastic continuum. The Boussinesq solution for a rectangular uniformly loaded area at the surface of an elastic half-space is used to calculate the soil displacements due to the contact pressure developed at the raft–soil interface. A layered soil profile is solved by means of the Steinbrenner approximation, in which the stress distribution within an elastic layer is identical with the Boussinesq distribution for an homogeneous half-space. The interaction factor method models pile to pile interaction and a separate BEM analysis gives the two-pile interaction factors at various spacings. Interaction between axially loaded piles beneath the raft and the raft elements is accounted for via pile–soil interaction factors computed from the separate BEM analysis. The reciprocal theorem is used to ensure that the soil–pile interaction factor is equal to the pile–soil interaction factor. A stepwise incremental procedure is used to simulate the non-linear load–settlement relationship of a single pile. The diagonal terms of the pile–soil flexibility matrix are updated at each step, and the nodal reaction vector is computed at each step to check for tensile forces between raft and soil; an iterative procedure is used to make them equal to zero if they are negative. The program outputs the distribution of the nodal displacements of the raft and the pile–soil system, the load sharing among the piles in the group and the soil, and the raft bending moments for each load increment.

Abagnara et al. (2012) and Russo et al. (2013) have compared these latter two programs, first for a simple problem involving a rectangular raft with a relatively small number of piles, and have found reasonable agreement between the computed behaviour from both programs. The programs have then been used for the settlement assessment of the Burj Khalifa Tower in Dubai. The computed values of average and differential settlements for the piled raft from GARP and NAPRA are found to agree well, and to be in reasonable agreement with measured data on settlements taken near the end of construction of the tower.

### 8.10.3 Finite element analyses

Finite element analyses provide a valuable tool for the detailed numerical analysis of piles and pile groups. Typical of such programs are PLAXIS3D, ABAQUS and MIDAS, while the finite difference program FLAC3D has also been widely used. Examples of 3D analyses include Comodromos and Bareka (2005, 2008), Comodromos and Pitilakis (2009) and Comodromos et al. (2016).

2D plane strain analyses have also been employed, but experience has shown that the results of such analyses may be misleading. For example, Prakoso and Kulhawy (2001) have compared 2D and 3D analyses for a piled raft foundation, and found that the 2D analysis tends to overestimate the maximum settlement and underestimate the differential settlement, as shown in Figure 8.30. Their findings may be summarised as follows:

- The 2D model tends to overestimate maximum and average displacements and, for most cases considered, the overestimation is on the order of 5%–25%.
- The plane strain differential displacement is about 2/3 of the centre-corner differential displacement of other models.
- The bending moments are similar to those across the raft centre in a 3D model.
- The plane strain pile tip load is about the average of the pile butt loads in the represented row of 3D models.

Poulos et al. (2001) have also found that a 2D analysis overestimates the maximum settlement, as shown in Figure 8.31. This figure also demonstrates that simple Category 2 analyses such as the PDR method, and custom programs such as GARP and GASP (a piled strip BEM program) compare reasonably well with a 3D finite element analysis.

If the foundation configuration is reasonably regular, then a 2D axi-symmetric analysis, rather than a plane strain analysis, may give a satisfactory estimate of foundation settlements.

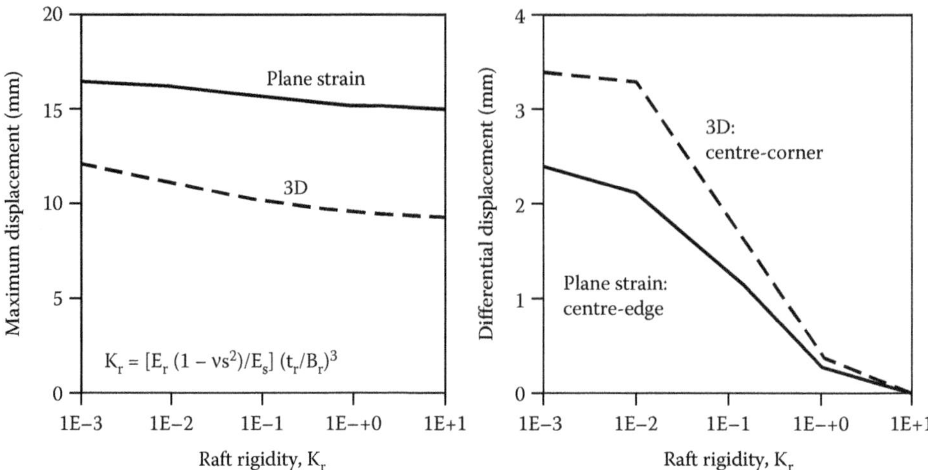


*Figure 8.30* Comparison between 2D and 3D analyses for a uniformly loaded piled raft. (Prakoso, W.A. and Kulhawy, F.H. 2001. *Journal of Geotechnical and Geoenvironmental Engineering, ASCE*, 127(1): 17–24. Courtesy of ASCE.)

Poulos and Bunce (2008) found that, for the Burj Khalifa in Dubai, such an analysis gave a very similar settlement prediction to a 3D analysis using the MIDAS program.

A very useful feature of 3D analyses is that they can model the raft flexibility at the same time as accommodating combined loadings. The deformations of the foundation system can then be illustrated for clarity of interpretation. An example of such an illustration is shown in Figure 8.32 for the Incheon Tower in South Korea (see also Chapter 15). The deformation of the raft is clearly seen in this figure. However, one of the limitations of programs such as PLAXIS3D is that applied bending moments and torsional loads must be imposed as equal and opposite eccentric vertical loads (for moments) and eccentric horizontal loads (for torsion).

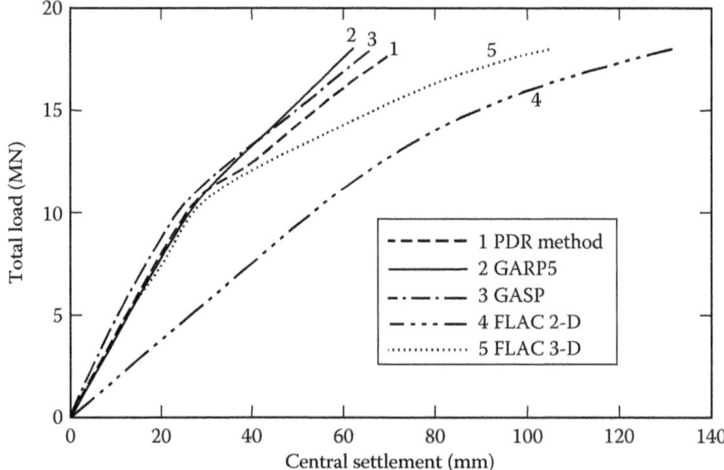


*Figure 8.31* Comparison between 2D, 3D and Category 2 analyses for a uniformly loaded piled raft. (Adapted from Poulos, H.G., Carter, J.C. and Small, J.C. 2001. Foundations and retaining structures – Research and practice. *Theme Lecture, Proceedings of the 15th International Conference on Soil Mechanics and Geotechnical Engineering*, Istanbul, Balkema, Vol. 4, pp. 2527–2606.)

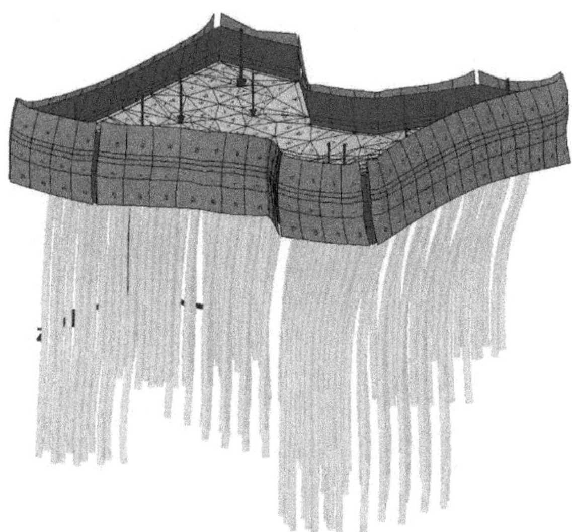

*Figure 8.32* Example of deformations of a piled raft subjected to vertical and lateral loading – Incheon Tower, South Korea. (Adapted from Poulos, H.G., Small, J.C. and Chow, H. 2011. *Geotechnical Engineering, SEAGS*, 42(2): 78–84. Courtesy of SEAGS.)

### *8.10.3.1 The modelling process*

One of the great advantages of modern Category 3 analyses is their ability to model the entire construction process, rather than simply considering the loading of the foundation system itself in isolation from the other construction processes. As an example of such an analysis process, the case described by Tschuchnigg and Schweiger (2013) is set out below. This case involved application of a 3D finite element analysis to twin high-rise towers in Vienna. The following stages in the construction of the foundation were modelled:

1. Generation of initial stresses
2. Activation of the sheet pile wall for supporting excavation
3. Excavation and groundwater lowering
4. Activation of piles (in this case, barrettes) that were 'wished in place'
5. Activation of slabs
6. Full loads of tower I and loads from basement floors of tower II
7. Closing of settlement joint – tower I
8. Full loads of tower II
9. Closing of settlement joint – tower II
10. End of ground water lowering

To reduce the complexity of the twin tower model, the barrettes of only one tower were modelled in full detail by Tschuchnigg and Schweiger and the foundation system of the other tower was modelled as a homogenised block, with the zones of the subsoil in which panels were installed being defined with smeared properties. With this approach, the global settlement behaviour of the entire structure was calculated taking interaction of the towers into account. This modelling assumption was validated by undertaking an analysis where both foundations were explicitly modelled. It was also found that very similar settlements were computed whether the piles were modelled as volume elements or via embedded elements (see below).

For the generation of the initial stress state it was emphasised that it was important to take the over-consolidation of the soil into account. A hardening soil model was adopted, and the analysis was carried out under assumed drained conditions so that the computed settlements were final settlements (excluding creep).

### 8.10.3.2 Embedded piles

In the standard finite element approach, piles are discretised by means of volume elements, which can lead to very large finite element models which become unattractive for many practical engineering projects where time and financial restraints are imposed. Another option for modelling piles is the concept of the so-called embedded pile element. This special element consists of a beam element which can be placed in an arbitrary direction in the subsoil, embedded interface elements to model the interaction of the structure and the surrounding soil, and embedded non-linear spring elements at the pile tip to account for base resistance. Additional nodes are automatically generated inside the existing finite element mesh and the pile–soil interaction behaviour is linked to the relative displacements between the pile nodes and the existing soil nodes. Figure 8.33 illustrates the embedded pile concept (Tschuchnigg and Schweiger, 2013).

The effectiveness of embedded piles is related to their connection to the surrounding soil, which is usually implemented through local node-to-node springs. In representing piles via the embedded pile model, the pile is assumed to be a series of slender beam elements which are virtually connected to the soil by means of the shaft and base interfaces. These elements may have arbitrary inclination and cross the soil elements at any arbitrary position. The interaction between the pile and the soil along the shaft is modelled by means of line-to-volume interface elements and the interaction at the pile base by means of point-to-volume interface elements. Spring elements, either elastic or inelastic, provide force–displacement relationships at the soil–pile interface, governing the static interaction between solid soil elements and embedded pile elements, and the possible occurrence of interface sliding or detachment. Unlike piles represented by fully solid elements, the employment of embedded

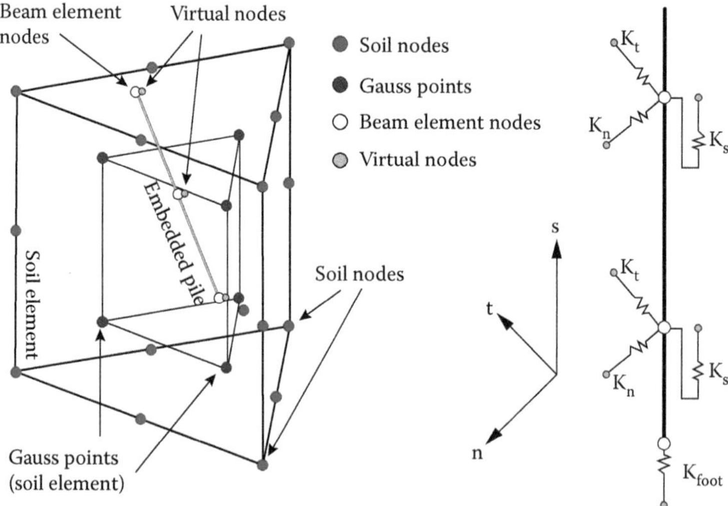


*Figure 8.33* Concept of embedded piles. (After Tschuchnigg, F. and Schweiger, H.F. 2013. *Geotechnical Engineering, SEAGS*, 44(3): 40–46.)

piles does not affect the spatial discretisation of the soil domain, because they are inserted after the global mesh generation and do not require any nodal compatibility between soil and beam elements. As a consequence, the increase in degrees of freedom close to the pile is avoided, and significant computational time can be saved. The big advantage of the embedded pile concept is that different pile lengths, spacing and orientation can be studied without regenerating the entire finite element mesh. If a large number of piles has to be considered, the number of elements in the system is significantly reduced as compared to finite element models with volume piles.

While different values of the moment of inertia for the two bending directions can be specified for embedded elements in the PLAXIS 3D program, it is not clear whether other aspects of barrette behaviour can be considered, as the program appears to consider only circular or square pile sections, and computes an equivalent diameter for the latter case.

Engin et al. (2008) have used the PLAXIS 3D program, with embedded piles, to obtain the load–settlement behaviour of both pile groups and piled rafts. They have demonstrated that their results are in good agreement with those from other analysis methods, including the finite element analyses of Ta and Small (1996). Similar levels of agreement have been reported by Oliveira and Wong (2014). Tradigo et al. (2016) have extended the use of embedded pile elements to consider piled rafts in which the raft and the pile heads are disconnected.

Despite the computational advantages of embedded piles, and the good agreement found between the settlement estimates from volume pile and embedded pile representations, accurate internal forces within individual piles cannot always be obtained.

#### *8.10.3.3 Effect of raft embedment, raft base shear resistance and basement walls*

Chow et al. (2010) have considered the effects of a cap or raft in resisting lateral loading of a group of piles. As would be expected, consideration of the passive resistance of the raft and the frictional resistance of the raft base leads to a reduction in the lateral deflections. This effect is dependent on load level, and becomes more pronounced as the load level increases. In the example they considered, with a limiting shear stress of only 20 kPa at the raft–soil contact, the lateral deflection was reduced by about 50% at a load level approaching that which would cause failure of a pile group with no raft contact. The bending moments at the pile heads were also reduced, although the effects were relatively modest.

A 3D numerical analysis also enables the basement walls to be incorporated into the analysis for the final stages of design. In this case, as would be expected, there was some reduction in lateral deflection under lateral load, but the effect was relatively small, as illustrated in Figure 8.34. Similarly, the effect of raft contact with the underlying ground was also relatively small in this case.

#### *8.10.3.4 Allowance for structure stiffness*

Melis and Rodriguez Ortiz (2001) have demonstrated that the stiffness of common buildings can be evaluated in a relatively simple way. An estimate can be made of the stiffness of the structure by summation of the stiffness of the floor slabs, the basement and the bearing and partition walls. This computed stiffness can then be easily incorporated into the finite element model via an equivalent additional thickness of the raft, as set out below.

The moment of inertia (second moment of area) of the floor slabs is taken relative to the neutral axis of the structure, close to the mid-height of the effective height of the structure, usually neglecting the moment of inertia of each slab relative to its own middle plane. For

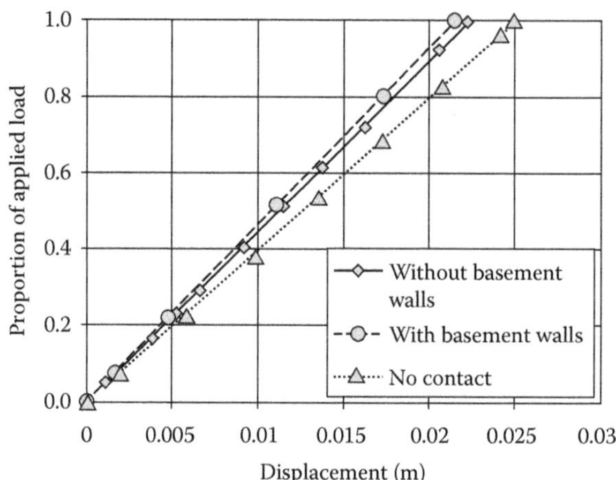


*Figure 8.34* Effect of raft contact and basement walls on lateral load–deflection behaviour – Incheon Tower, South Korea. (Adapted from Poulos, H.G., Small, J.C. and Chow, H. 2011. *Geotechnical Engineering, SEAGS*, 42(2): 78–84. Courtesy of SEAGS.)

this effective height of structure, it would appear appropriate to consider only a limited number of storeys, for example, 2 or 3, since the higher floors will have a relatively minor influence on the combined bending action of the structure–foundation system. This is somewhat analogous to the concept of critical length for laterally loaded piles.

The stiffness of the basement includes the walls with their continuous footings (or diaphragm walls) or the foundation slab. Their moments of inertia are taken relative to the neutral axis of the basement. This can appear as a conservative approach when the basement walls are the lower part of common bearing walls, because the inertia should be computed for the total height of the building.

The stiffness of the walls is considered in the case of bearing walls and in concrete framed structures. The stiffness of the columns in framed structures is usually neglected. Conservatively, only external walls are usually considered, but the contribution of partition walls can be also significant in some cases. The effects of the presence of opening (doors or windows) can be accounted for by means of a reduction factor (on the bending stiffness EI), as shown in Table 8.6.

After computing the $(EI)_b$ value, a modulus $E_m$ is chosen for modelling the building in the finite element mesh (most conveniently, the same modulus as the raft slab itself), and the corresponding additional thickness of the slab, $\Delta t$, is then given as follows:

$$\Delta t = \left[\frac{12(EI)_b}{(B \cdot E_m)}\right]^{1/3} \tag{8.85}$$

where $(EI)_b$ is the computed equivalent stiffness of building, B the width of raft and $E_m$ is the modulus of raft.

This approach 'smears' the additional thickness of the raft over the whole area. However, it may desirable to isolate areas of the raft, especially below walls, where the raft thickness may be thickened in accordance with the wall stiffness, while the other parts of the raft are thickened only in accordance with the stiffness of the floor and the basement.

*Table 8.6* Structural stiffness reduction factors for walls with openings

| *% of openings in wall* | *Length < wall height* | *Length > 2wall height* |
|---|---|---|
| 0 | 1.0 | 1.0 |
| 0–15 | 0.7 | 0.9 |
| 15–25 | 0.4 | 0.6 |
| 25–40 | 0.1 | 0.15 |
| >40 | 0 | 0 |

Source: After Melis, M. and Rodriguez Ortiz, J.M. 2001. *Proceedings of the International Conference on Response of Buildings to Excavation-Induced Ground Movements.* London, CIRIA SP201, pp. 387–394.

## 8.10.4 Compensated piled rafts

A proper consideration of the behaviour of a compensated piled raft requires that the process of excavation, pile installation and raft installation be modelled in addition to the gradual application of the structural loads. While this can now be carried out using finite element analysis, a brief discussion of the simulation of these stages, via an alternative approach developed by Sales et al. (2010), is presented below.

### *8.10.4.1 Modelling excavation and pile installation*

The excavation and pile installation process must be selected to suit each case (see Figure 8.35). In some buildings, with shallow excavations, the piles can be executed before the excavation, from the ground level. In others, where greater depth must be achieved, part or

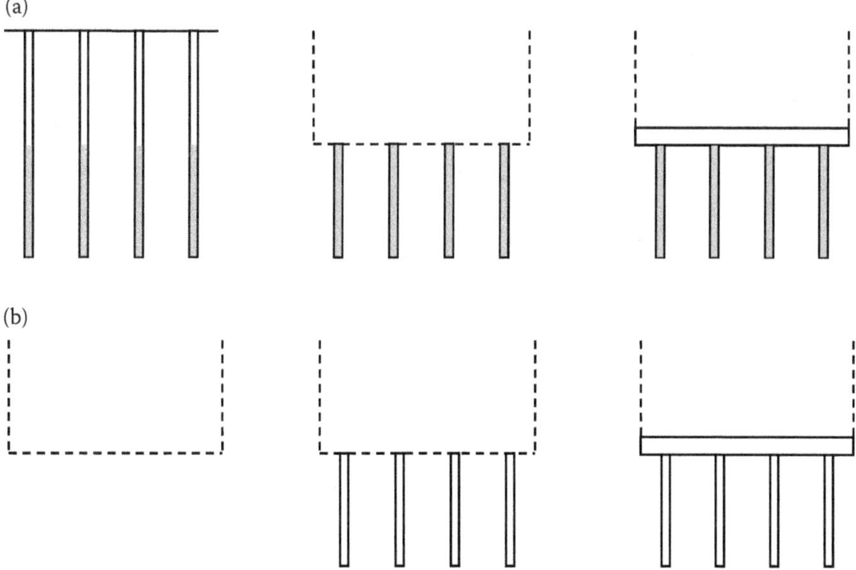


*Figure 8.35* Different construction sequences for compensated piled rafts: (a) the piles are cast in place before the excavation and (b) the soil is firstly excavated.

the whole excavation is carried out first and the piles are installed once excavation is complete. The presence of groundwater will also influence the construction process.

When the piles are constructed in advance of the excavation (Figure 8.35a), the piles will act as anchors, reducing the tendency for bottom soil heave. The upward soil movement will generate tensile stresses in the piles. Sommer (1993) reported 'locked in stresses' for the piles of the Messeturm Building, in Frankfurt, of about 1.5 MN after excavation.

In clayey soils, depending on how much time has passed between the excavation and the pile installation, the piles can still suffer tensile stresses, but this is usually less than for the process shown in Figure 8.35a.

#### *8.10.4.2 Modelling raft installation*

At the time of raft concreting, the soil underneath the raft is no longer experiencing the original stresses but has been unloaded, and so it will behave as an over-consolidated soil. Any in situ tests carried out before excavation may not be sufficient for accurate settlement predictions as the reloading parameters are required. Laboratory tests, or at least previous experience, will be necessary to describe the soil behaviour at this construction stage.

Applying a 'wet load' to the soil, corresponding to the weight of wet concrete in the raft, will result in an initial settlement of the base of the excavation and a small load will be induced into the piles. Figure 8.36 presents the case of a hypothetical raft concreting simulation with different soil/pile stiffnesses. The example involves a 9-piled raft constructed in a 100 m thick homogeneous soil. The square raft is considered to be 8 × 8 m in plan and the pile spacing is 3 m centre-to-centre. The different elastic-moduli (E) and pile diameters (D) adopted are presented in the figure. The shorter the piles or the stiffer the soil, the lower will be the induced pile load.

The resulting settlement of sandy soils due to raft concreting tends to be completed before raft hardening. On the other hand, for clayey soils the consolidation process will take place after raft hardening and the initial load sharing, shown in Figure 8.36, will change.

The downward soil movement will induce compression stresses into the upper part of the piles. Combining with the possible locked in tensile stress due to the excavation process, the

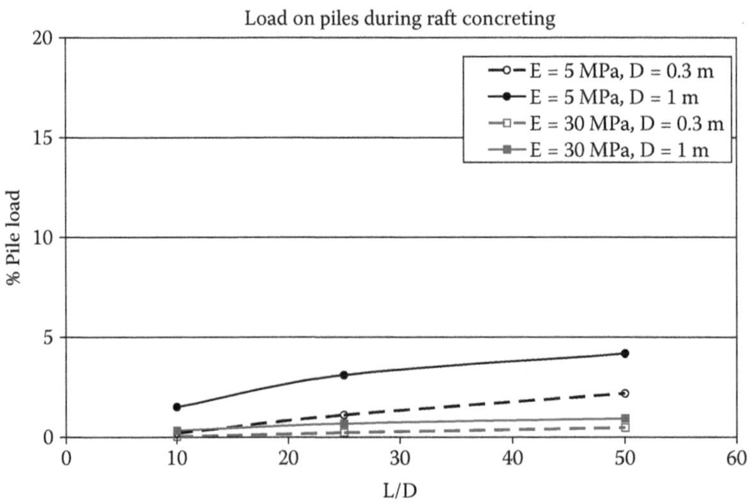


*Figure 8.36* Pile load induced by raft concreting ('wet load').

resulting distribution of stresses along the pile shaft will be complex and will change with elapsed time.

The stresses (locked in tensile or perhaps downward compressive stresses) will normally not change the final bearing capacity of the piles. The pile stiffness, however, will change since the initial levels of load in the soil have changed. As the total piled-raft stiffness is directly related to the pile stiffness, the overall behaviour of a 'compensated piled raft' will be affected by the excavation sequence.

If the raft weight is lower than the effective excavation weight, the soil will still behave as an over-consolidated soil during the first stage of raising the building structure.

Other important points required for a prediction of behaviour of compensated piled rafts are the excavation sequence and also what response is to be measured. The settlement distribution and load sharing is highly dependent on factors such as

- Excavation process
- Time between end of excavation and raft concreting
- Drawdown of the groundwater level
- Instruments to be used to monitor settlements, stresses and loads
- Time of construction

In some case histories, the settlement measurements began after raft concreting, in others, before. The difference in settlement predictions in both cases can be of the order of several millimetres. The load measuring instruments, usually with strain gauges inside the piles, load cells on the tip and/or the top of the piles and earth pressure cells underneath the raft, are normally installed before raft concreting. Thus, in many cases, the load measurements can register the raft installation, but the initial settlement is lost.

The presence (or absence) of buoyant force during the construction period is a key factor in predicting the load sharing. Some tall buildings have the piled raft installed 10 or more metres below the groundwater table. The magnitude of buoyant force in this kind of building can represent a not insignificant percentage of the total weight and the final raft load will be strongly dependent on this component of load. A classical example of this point is the Messeturm, where the groundwater table was drawn down twice after the beginning of construction (Sommer et al., 1985; Reul and Randolph, 2003). The measurements clearly show an increase of the pile loads during the groundwater lowering process and the inverse tendency after stopping the pumping of water.

Increasing costs and tighter deadlines normally define a very short period for raising the superstructure of tall buildings. In clayey soils, such as Frankfurt or London Clays, this means the soil will undergo undrained behaviour during the construction period followed by consolidation, and perhaps creep, settlement after the building completion (Franke et al., 2000).

Salas et al. (2010) set out the following three-stage approach for predicting the settlement of a compensated piled raft foundation:

- Stage 1, when the soil is overconsolidated and the applied load is less than the effective weight of the excavated soil
- Stage 2, in which the settlement is a consequence of the effective net load that exceeds the effective weight of the excavated soil
- Stage 3, which represents the long-term state after primary consolidation is completed. The settlement and load distribution are calculated using the drained deformation parameters

## 8.11 ASSESSMENT OF PILE SPRING STIFFNESS VALUES

### 8.11.1 Introduction

Structural design models for tall buildings generally represent the foundation system as a raft slab supported by springs representing the piles. Such a representation is inevitably a simplified one, but it can be reasonably satisfactory if the spring stiffness values are assessed in an appropriate manner.

In principle, the stiffness of these springs should be assessed by the geotechnical designer, taking into account the stiffness of individual piles and how interaction among the pile group influences (and generally reduces) the stiffness of the piles. However, in past practice, there has been a tendency for the pile stiffness values to be assessed without consideration of the interaction effects, and such a practice can lead to underestimation of the settlements and differential settlements of the foundation and the supported structure.

Alternative approaches to rational estimation of pile spring stiffness values are discussed below. Attention is focussed on the vertical spring stiffness values, but the same principles can be applied to the estimation of lateral and rotational spring stiffness values.

### *8.11.2 Alternative approaches*

There are a number of alternative procedures that can be adopted to assess the stiffness of piles within a group. Because the load–settlement behaviour of piles is non-linear, there will also be alternative assumptions available for both the nature and the magnitude of the applied loadings. The alternative procedures include the following:

1. Using a pile group analysis in which all piles are loaded with the same average load.
2. Using a pile group analysis in which the pile cap or raft is assumed to be rigid, and the total load is applied to the pile cap.

The applied load may be either:

- The working or serviceability load, if the main objective is to estimate the structural and foundation performance under serviceability loading conditions, or
- The ultimate limit state design loading, if the main objective is to estimate the structural response and foundation loadings under the ultimate limit state loadings

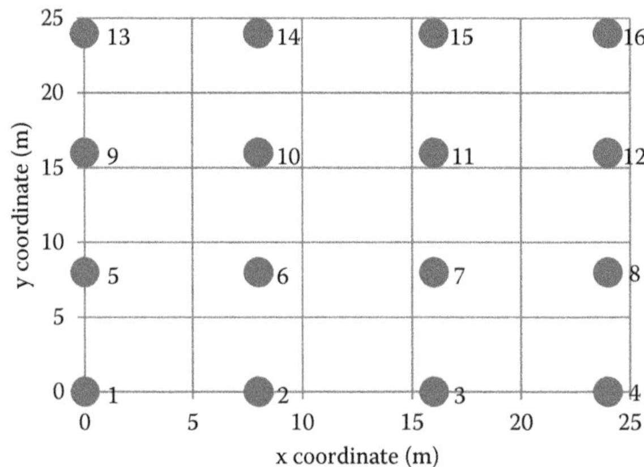


*Figure 8.37* Plan of pile group for example.

*Table 8.7* Details of ground profile for example

| *Depth range (m)* | *Young's modulus Esv (MPa)* | *Ultimate shaft friction (MPa)* | *Ultimate end bearing (MPa)* |
|---|---|---|---|
| 0–20 | 20 | 0.040 | – |
| 20–40 | 50 | 0.060 | – |
| 40+ | 200 | 0.100 | 6.0 |

A simple hypothetical example of a 16-pile group has been analysed, as shown in Figure 8.37. The piles are assumed to be 45 m long and 2.5 m in diameter, and are situated in a ground profile as defined in Table 8.7. It is assumed that the average serviceability load is 20 MN/pile, that is, the total load on the group is 320 MN, while the ultimate limit state vertical load is 448 MN, that is, an average 28 MN/pile.

Figure 8.38 shows, for the serviceability case, the computed vertical pile head stiffness values for each of the above assumptions. Figure 8.39 shows the corresponding stiffness values for the ultimate load case. From these figures, the following observations can be made:

- The stiffness values for the rigid cap assumption show a greater variability than those from the equal load assumption.
- The group stiffness values are significantly lower than the single pile stiffness value.
- The stiffness values for the ultimate limit state case are significantly lower than those for the serviceability case, as would be expected because of the non-linear behaviour of the piles.

It is recommended that the assumption of equal loads in each pile be used in preference to the rigid cap. Moreover, it is essential that pile-pile interaction be allowed for to avoid overestimating the pile stiffness values, and hence underestimating the foundation settlements.

A similar approach can be adopted for assessing the lateral and rotational stiffness of piles within a group. However, it should be borne in mind that, in general, applied lateral loading will cause both a lateral deflection and a rotation, and similarly an applied moment will cause a rotation and a lateral deflection. Accordingly, the pile head stiffness values are not

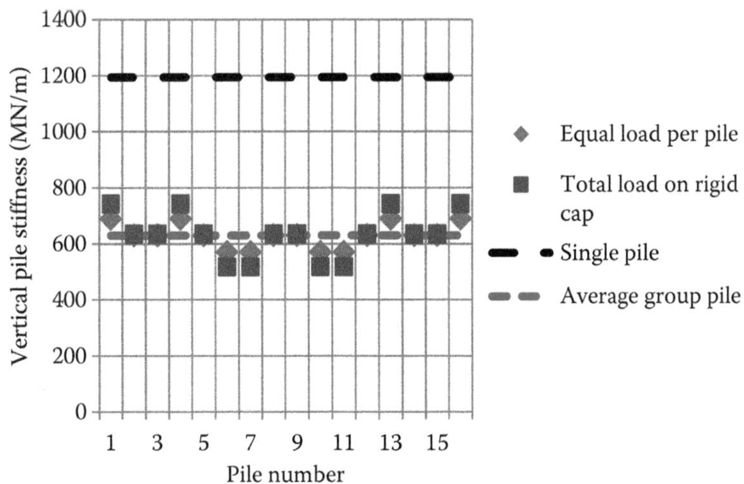


*Figure 8.38* Computed pile head stiffness values from alternative analyses – serviceability case.

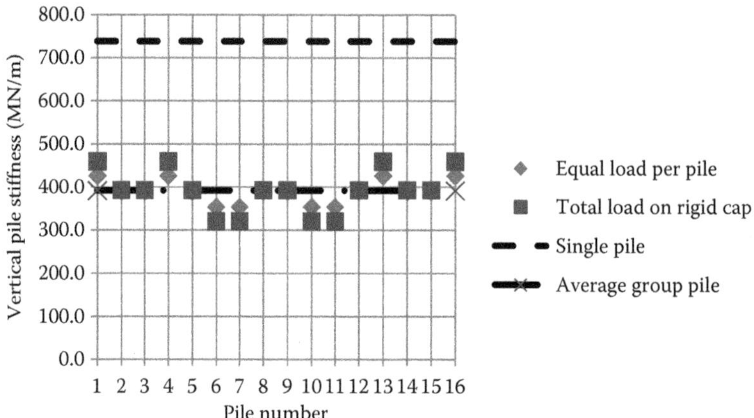


*Figure 8.39* Computed pile head stiffness values from alternative analyses – ultimate load case.

unique but will depend on the ratio of lateral load to moment. If the piles are assumed to be fixed into the raft or basement slab, then only the horizontal stiffness of the fixed head piles is relevant, and the computational process is simplified. The fixed head condition appears to be a reasonable assumption for many tall building foundation systems.

### *8.11.3 Equivalent modulus of subgrade reaction*

For the structural design of piled rafts, many structurally-oriented programs require the input of a modulus of subgrade reaction to characterise the raft-soil contact behaviour. Meaningful values of this pseudo-parameter for vertical loading can be estimated from the Category 3 geotechnical analysis of the piled raft. As part of such an analysis, values of raft contact pressure and settlement are obtained. There are then three options for obtaining the equivalent vertical modulus of subgrade reaction, k:

1. Compute k for each raft element as the ratio of the contact pressure to the settlement of that element
2. Select a number of zones of elements within the raft footprint, and compute an average value of k for each zone, as the ratio of average contact pressure within that one to the average settlement within the zone
3. Obtain an average value of k for the entire raft by computing the ratio of the average raft contact pressure and the average raft settlement

The first approach may imply an undue precision of the analysis, while the third approach may be too coarse an approximation for many cases, unless the raft carries only a small amount of the load. The second approach may then be the most reasonable to adopt, although it requires a level of engineering judgement to select appropriate zones over which to obtain average values of k.

Chapter 9

# Design for ground movements

## 9.1 INTRODUCTION: SOURCES OF GROUND MOVEMENT

There are many circumstances in which pile foundations may be subjected to loadings arising from vertical and/or lateral movements of the surrounding ground. Figure 9.1 illustrates a number of these circumstances. In such cases, at least two important aspects of pile foundation design must be considered:

1. The movements of the piles caused by the ground movements.
2. The additional forces and/or bending moments induced in the piles by the ground movements, and their effect on the structural integrity of the piles.

Problems involving the effects of ground movements on piles may be analysed in at least two ways:

1. Via a complete single analysis (generally numerical) involving modelling of the pile, the soil and the source of the ground movements. This will give a complete solution for the behaviour of both the soil and the pile.
2. Via a simplified approach involving initial separation of the soil and the pile ('substructuring') so that the soil movements are first computed and then imposed on the pile. In this approach, the focus is generally placed on the behaviour of the pile.

This chapter summarises a consistent theoretical approach to the analysis of ground movement effects on piles, for both vertical and horizontal movements, which falls into the second category. Two distinct stages are involved in this analysis:

1. Estimation of the 'free-field' soil movements which would occur if the pile was not present.
2. Calculation of the response of the pile to these computed ground movements.

Some specific cases of ground movement that may be relevant to foundations for tall buildings are then considered, and in each case, a discussion is given of the general features of pile behaviour revealed by the theory, and means by which ground movement effects on the foundations can be estimated.

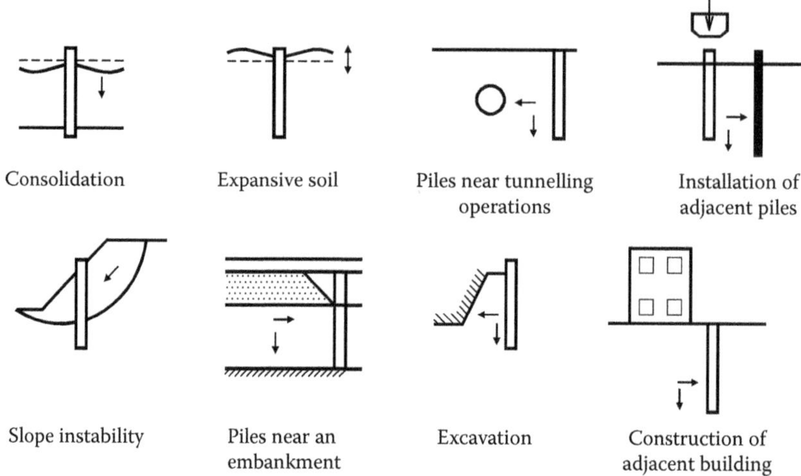


*Figure 9.1* Some sources of ground movements.

## 9.2 ANALYSIS OF THE EFFECTS OF GROUND MOVEMENTS

### 9.2.1 Vertical ground movements

The analysis used for the response of a pile to vertical ground movements has been described by Poulos and Davis (1980) and has been used to analyse problems of negative friction of piles in consolidating soil, and of tension and uplift of piles in expansive soil. It employs a simplified form of boundary-element analysis, in which the pile is modelled as an elastic column and the surrounding soil as an elastic continuum.

The pile is divided into a series of cylindrical elements. The vertical movement of each element is related to the applied load, the pile–soil interaction stresses, the pile compressibility, and the pile tip movement. The vertical movement of each supporting soil element depends on the pile–soil interaction stresses, the modulus or stiffness of the soil, and also on any free-field movements that may be imposed on the pile. To simulate real pile response more closely, allowance may be made for slip at the pile–soil interface, that is, the pile–soil interaction stresses cannot exceed the limiting pile–soil skin friction.

The analysis of axial pile response requires knowledge of the pile modulus, the distribution of soil modulus and limiting pile–soil skin friction with depth, and the free-field vertical soil movements. The assessment of the pile–soil parameters (in particular the soil modulus and limiting pile–soil skin friction) has been discussed in Chapter 6.

It is important to emphasise that, as set out in Section 7.9, vertical ground movements do not affect the geotechnical axial capacity of a pile, since the pile will need to move past the soil if failure is to occur, and for this to happen, any negative friction will alter to positive friction (Poulos, 2008a; Fellenius, 2016). The only exception may be for cases in which there is a significant strain softening of the interface friction during the development of negative friction.

### 9.2.2 Horizontal ground movements

Details of the lateral-response analysis have been given by Poulos and Davis (1980), and this analysis also relies on the use of a simplified boundary-element analysis. In this case, the pile is modelled as a simple elastic beam, and the soil as an elastic continuum. The lateral displacement of each element of the pile can be related to the pile bending stiffness and the horizontal pile–soil interaction stresses. The lateral displacement of the corresponding soil elements is related to the soil modulus or stiffness, the pile–soil interaction stresses, and the free-field horizontal soil movements. A limiting lateral pile–soil stress can be specified so that local failure of the soil can be allowed for, thus enabling a non-linear response to be obtained.

### 9.2.3 Group effects

The analysis of ground movement effects on groups of piles has been reported by several authors, for example, Kuwabara and Poulos (1989), Teh and Wong (1995), Chow et al. (1990), Xu and Poulos (2001). All these authors have found that under purely elastic conditions, group effects tend to be beneficial to the pile response as compared to single isolated piles, that is, group effects tend to reduce the pile movement and the forces and moments induced in the piles. This is especially so for the inner piles within a group, which, because of the pile–soil–pile interaction are, in effect, 'shielded' from the soil movements by the outer piles.

From the viewpoint of design, at least in the first instance, it is generally both convenient and conservative to ignore group effects and analyse a pile as if it were isolated.

### 9.2.4 Loading via ground movements versus direct applied loading

There is a widespread misconception that the effects of externally imposed ground movements on piles can be simulated by the application of equivalent loadings at the pile head. To illustrate the consequences of this procedure, the case in Figure 9.2 has been analysed.

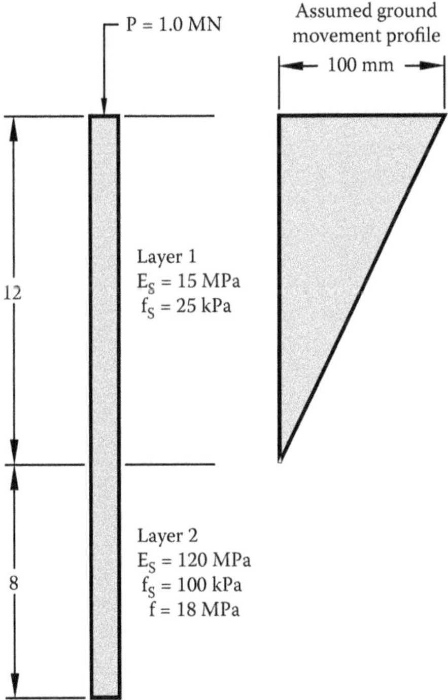


*Figure 9.2* Typical problem analysed. (Adapted from Poulos, H.G. 2006b. Ground Movements – A hidden source of loading on deep foundations. 2006 John Mitchell Lecture, In: *Proceedings of the 10th International Conference on Piling and Deep Foundations*. DFI, Amsterdam, pp. 2–19. Courtesy of DFI.)

A single pile in a two-layer soil profile is considered, and the pile is subjected to the following sources of loading:

- An applied vertical load of 1.0 MN
- An applied lateral load of 0.1MN at the pile head
- Vertical ground movement profile which decreases from 100 mm at the ground surface to zero at a depth of 12m
- A lateral ground movement profile which also decreases from 100 mm at the ground surface to zero at 12 m depth

Figure 9.3 shows the computed axial load distributions for the applied load acting alone, the vertical ground movement acting alone and the applied load and the ground movement acting together. It can be seen that the distributions of axial load in the pile due to applied loading are very different from those induced by ground movements. In the latter case, the maximum axial load occurs near the bottom of the upper soil layer which is subjected to movement. It can also be seen that the addition of the two profiles of axial load gives axial loads which are less than those computed for the combined loading and ground movement case.

Figure 9.4 shows the corresponding distributions of moment computed for the lateral response of the pile. Again, it can be seen that the distribution of induced bending moment is very different for applied loading and for lateral ground movement. In the latter case, the

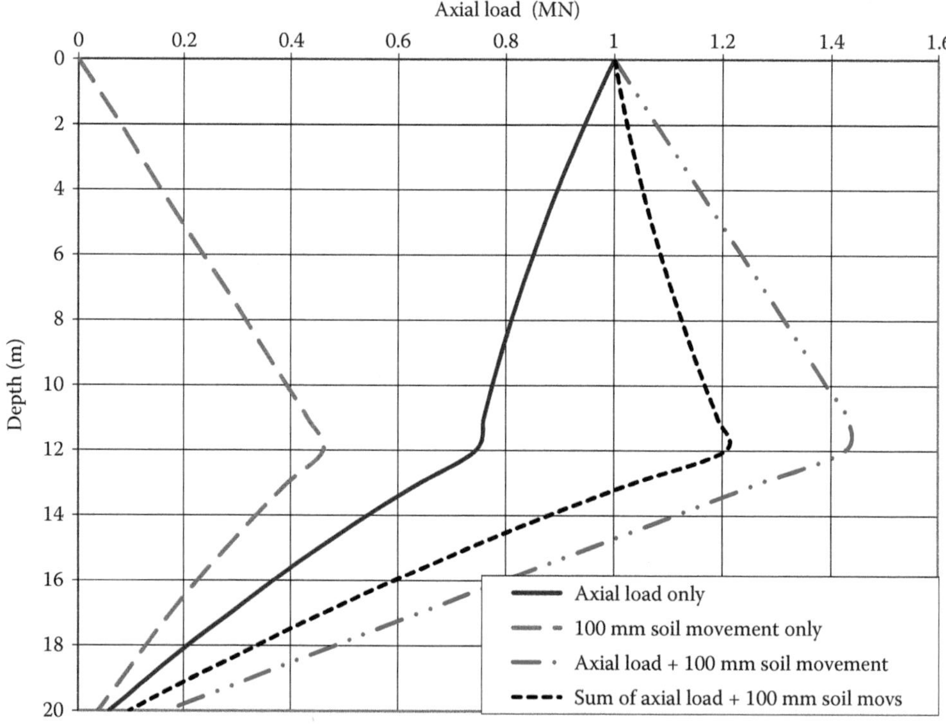


*Figure 9.3* Comparison of axial responses. (Adapted from Poulos, H.G. 2006b. Ground Movements – A hidden source of loading on deep foundations. 2006 John Mitchell Lecture, In: *Proceedings of the 10th International Conference on Piling and Deep Foundations*. DFI, Amsterdam, pp. 2–19. Courtesy of DFI.)

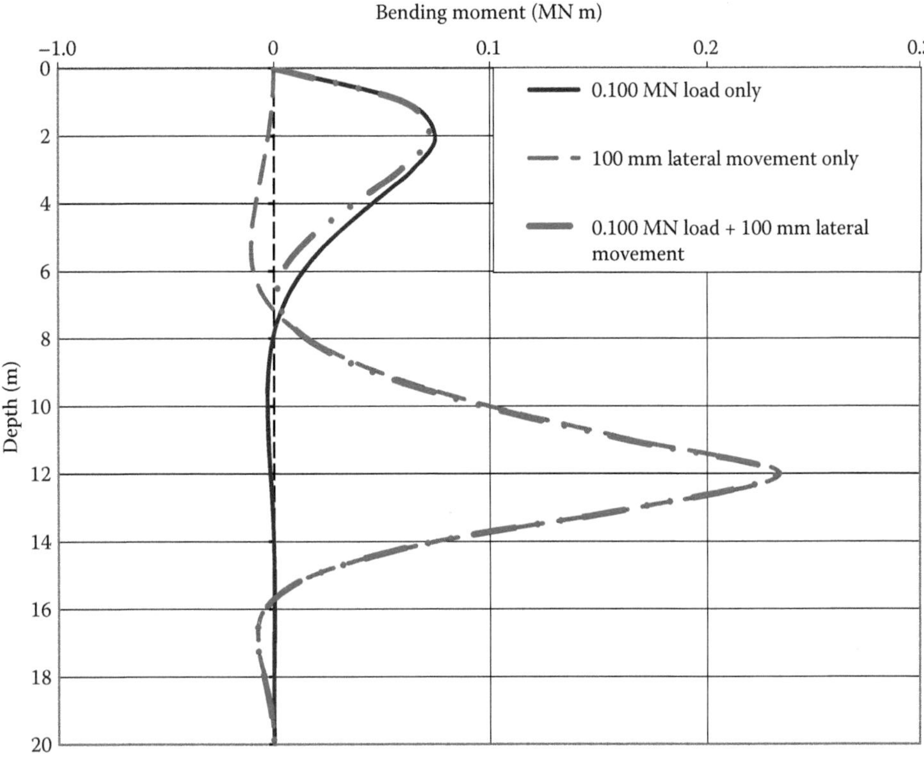


*Figure 9.4* Comparison of lateral responses. (Adapted from Poulos, H.G. 2006b. Ground Movements – A hidden source of loading on deep foundations. 2006 John Mitchell Lecture, In: *Proceedings of the 10th International Conference on Piling and Deep Foundations.* DFI, Amsterdam, pp. 2–19. Courtesy of DFI.)

maximum moment occurs well below the pile head, near the bottom of the zone of ground movement. The maximum bending moment under the combined loadings is also at the latter location, since the moment due to the applied loading is virtually zero where the moment due to the lateral ground movements is largest.

In terms of practical pile design, the above example demonstrates the following important points:

- The effects of ground movements cannot be simulated accurately by the application of a load to the pile head.
- The superposition of axial load distributions due to axial applied loading and vertical ground movements may underestimate the maximum axial load in the pile.
- The maximum moment in a pile subjected to lateral ground movements may occur well below the pile head. In this particular case, having the pile reinforced only to resist applied lateral loading (e.g., in the upper 6 m or so) will be inadequate to resist the ground movement-induced moments. The pile may well fail structurally at a considerable depth below the pile head.

Thus, it is important to consider the possibility of ground movements in pile design, and to allow for reinforcement of the piles to resist deep-seated moments that may be induced by these movements.

## 9.3 GENERIC CATEGORY 2 DESIGN CHARTS

### 9.3.1 Vertical soil movements

The design of piles subjected to negative friction has been discussed in Section 7.9 for the ultimate limit state case. For the important serviceability limit state, estimates need to be made of the settlement of the piles and the induced axial force. For these purposes, Category 2 design charts for the settlement and maximum axial force in a single end bearing pile on rock and subjected to vertical soil movements, have been published by Poulos and Davis (1980). These charts are reproduced in Figures 9.5 and 9.6, and assume a perfectly elastic homogeneous soil mass and pile–soil interface, with a vertical soil movement profile which varies linearly with depth. Similar charts for a single floating or friction pile are shown in Figures 9.7 and 9.8 (Poulos and Davis, 1980). The above charts will tend to give upper bound values of both pile settlement and induced pile force, because there is no limit to the pile–soil shear stress that is developed between the pile and the soil. In reality, the existence of an ultimate skin friction will result in a limit to the axial force and pile movement that can be generated within a pile. The use of elastic solutions therefore tends to be conservative when applied to practical cases. Corrections for pile–soil slip and other practical effects are presented by Poulos and Davis (1980) and Nelson and Miller (1992).

Charts such as those in Figures 9.5 through 9.8 are for a single isolated pile. Since group effects will tend to be beneficial when piles are subjected to ground movements, the settlement of a group due to ground movements will generally be less than that of a single pile

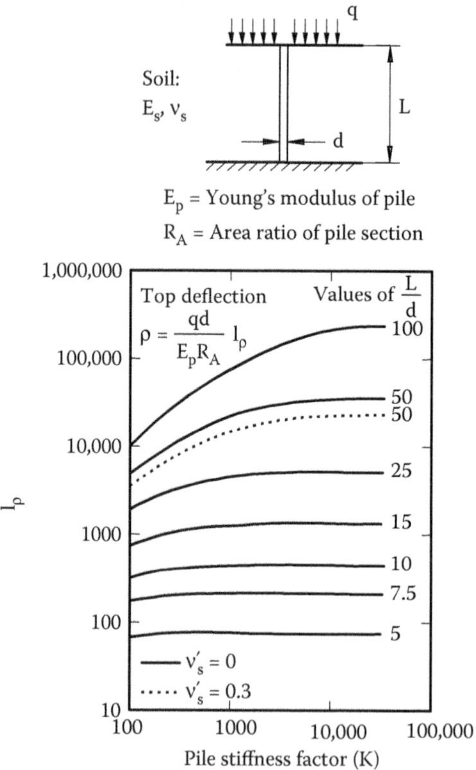


*Figure 9.5* Elastic solutions for top settlement of end bearing pile. (Adapted from Poulos, H.G. and Davis, E.H. 1980. *Pile Foundation Analysis and Design*. John Wiley, New York.)

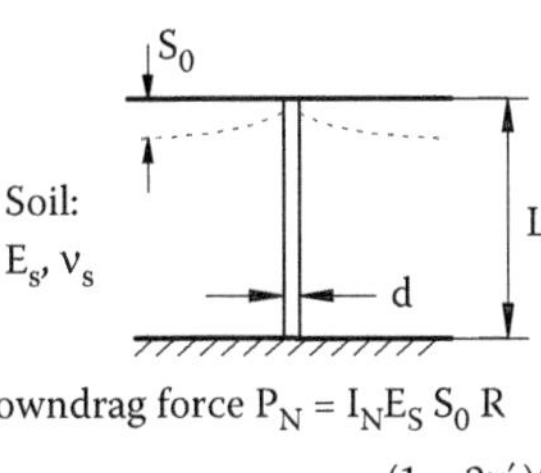


Maximum downdrag force $P_N = I_N E_S S_0 R$

where $R = \dfrac{(1 - 2\nu'_s)(1 + \nu'_s)}{(1 - \nu_s^2)}$

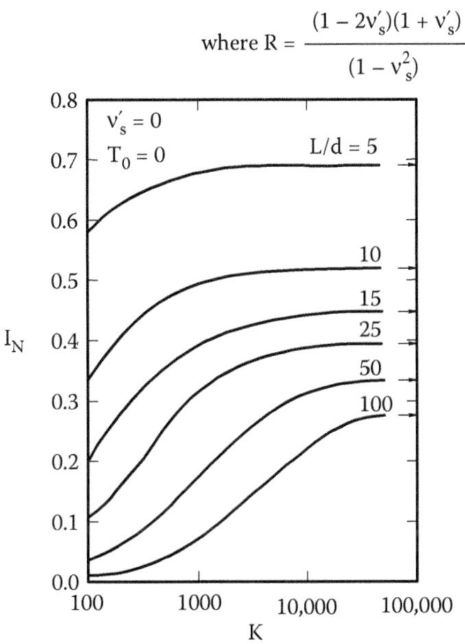


*Figure 9.6* Influence factor for final downdrag force at end bearing pile tip-elastic analysis. (Adapted from Poulos, H.G. and Davis, E.H. 1980. *Pile Foundation Analysis and Design*. John Wiley, New York.)

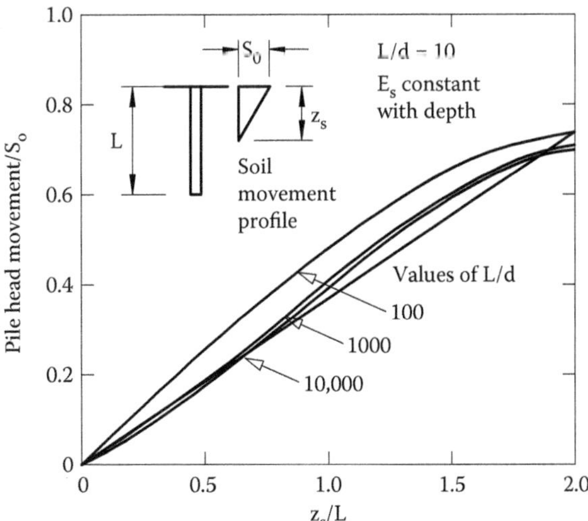


*Figure 9.7* Elastic solutions for floating pile movement in expansive soil. (Adapted from Poulos, H.G. 1989. *Géotechnique*, 39(3): 65–415. Courtesy of ICE Publishing.)

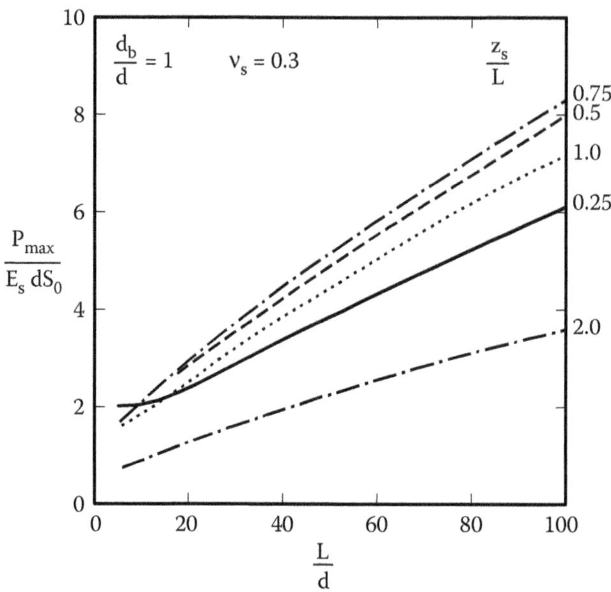


*Figure 9.8* Elastic solutions for maximum pile load–uniform diameter floating pile. (Adapted from Poulos, H.G. and Davis, E.H. 1980. *Pile Foundation Analysis and Design*. John Wiley, New York.)

due to the 'shielding' effect. Accordingly, it is conservative to ignore the group effect in this case.

### 9.3.2 Lateral soil movements

If the distribution with depth of free-field lateral movements can be simplified, it is possible to develop useful design charts to enable approximate assessment of the pile head deflection and the maximum bending moment in the pile. Chen and Poulos (1997, 1999) have presented such charts, both for a pile in soil subjected to a uniform movement with depth (to a depth $z_s$ below the surface), and for a soil in which the horizontal movement decreases linearly with depth, from a maximum at the surface to zero at a depth $z_s$. The first movement profile may be relevant to piles in unstable soil slopes, while the linear profile may be relevant for cases such as piles adjacent to an excavation or an embankment.

For the linear soil movement profile, Figures 9.9 and 9.10 present charts for pile head movement and maximum moment, for a homogeneous (uniform) soil, and a 'Gibson' soil whose modulus increases linearly with depth. The pile head is unrestrained. Figures 9.11 and 9.12 show corresponding charts for the case of a uniform soil movement profile with depth. As discussed by Chen and Poulos (1997), these solutions assume that the soil remains elastic, and they therefore generally give an upper bound estimate of the pile moment and deflection. The extent of the possible overestimation increases with increasing lateral soil movements, due to the progressive departure from elastic conditions which results from the development of plastic flow of the soil past the pile.

The input parameters required for use of the elastic design charts include pile diameter (d), pile length (L), pile bending rigidity ($E_pI_p$), soil Young's modulus ($E_s$, either uniform with depth or $=N_hz$ for 'Gibson' soil, where $N_h$ is a constant), magnitude and distribution of

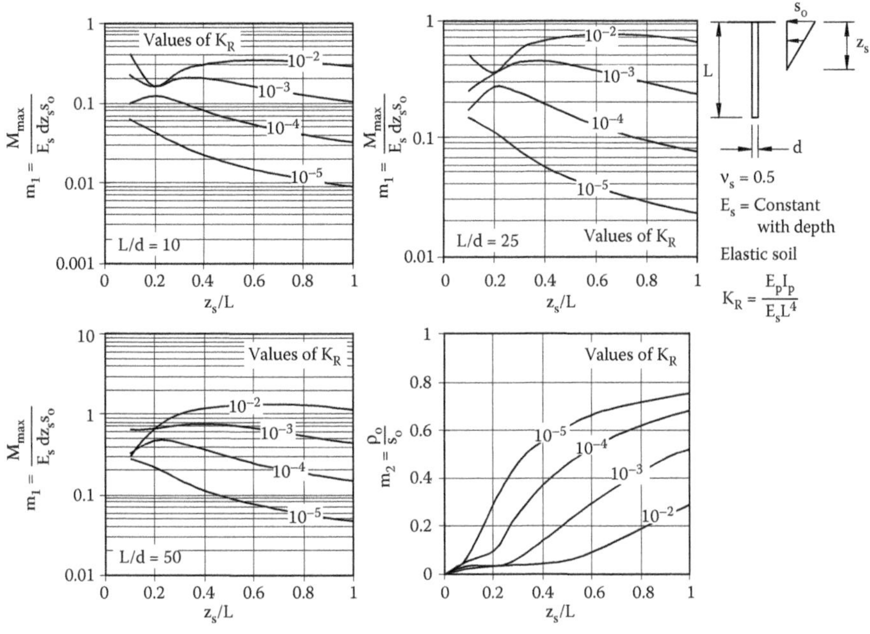


*Figure 9.9* Elastic solutions for pinned-restrained head pile in uniform soil, with linear soil movement profile. (Adapted from Chen, L.T. and Poulos, H.G. 1997. *Journal of Geotechnical and Geoenvironmental Engineering, ASCE*, 123(9): 802–811. Courtesy of ASCE.)

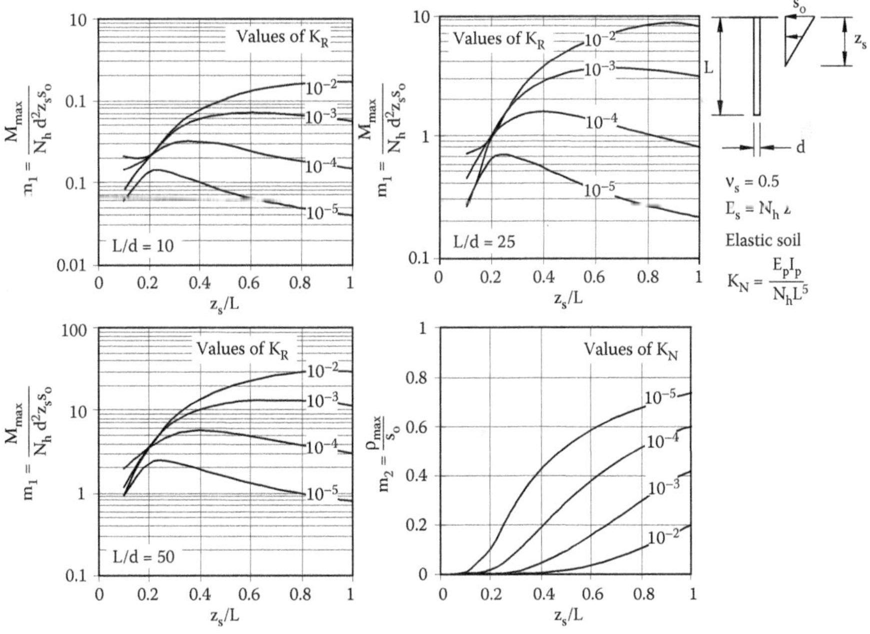


*Figure 9.10* Elastic solutions for pinned-restrained head pile in Gibson soil, with linear soil movement profile. (Adapted from Chen, L.T. and Poulos, H.G. 1997. *Journal of Geotechnical and Geoenvironmental Engineering, ASCE*, 123(9): 802–811. Courtesy of ASCE.)

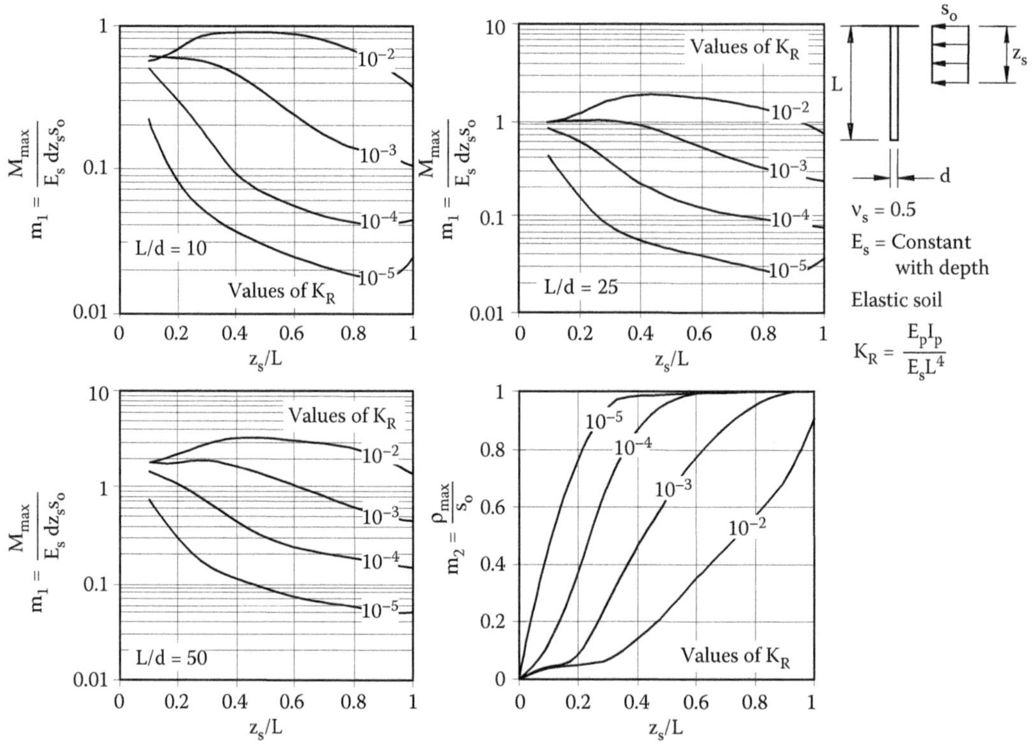


*Figure 9.11* Elastic solutions for pinned-restrained head pile in uniform soil, with uniform soil movement profile. (Adapted from Chen, L.T. and Poulos, H.G. 1997. *Journal of Geotechnical and Geoenvironmental Engineering, ASCE*, 123(9): 802–811. Courtesy of ASCE.)

soil movement at ground surface ($s_o$) and the thickness of moving soil layer ($z_s$). The assessment of the pile–soil parameters frequently involves correlations with laboratory and/or in situ test information such as the SPT and the static CPT, some of which are summarised in Chapter 6.

From a study of several examples, Chen and Poulos (2001) have suggested the following preliminary guidelines for the determination of soil movements in making theoretical predictions via the generic design charts in Figures 9.9 through 9.12, in the absence of measured ground movement data or more accurate estimation by other methods:

- For unstrutted excavations or relatively small slope movements, a linear soil movement profile, with a maximum value at the ground surface and zero at a certain depth below the surface, may be adopted. The maximum value may be estimated from measured ground surface movements or via appropriate empirical approximations which relate movement to the height of the retained soil, for example, Peck (1969), or via the approaches set out in Section 9.4.
- For cases involving unstable sites and relatively large soil movements (e.g., up to about 0.4 pile diameters), a uniform soil movement profile may be adopted.

The above study by Chen and Poulos also showed that the elastic design charts can give reasonably good estimations of the lateral pile response, provided that the ground

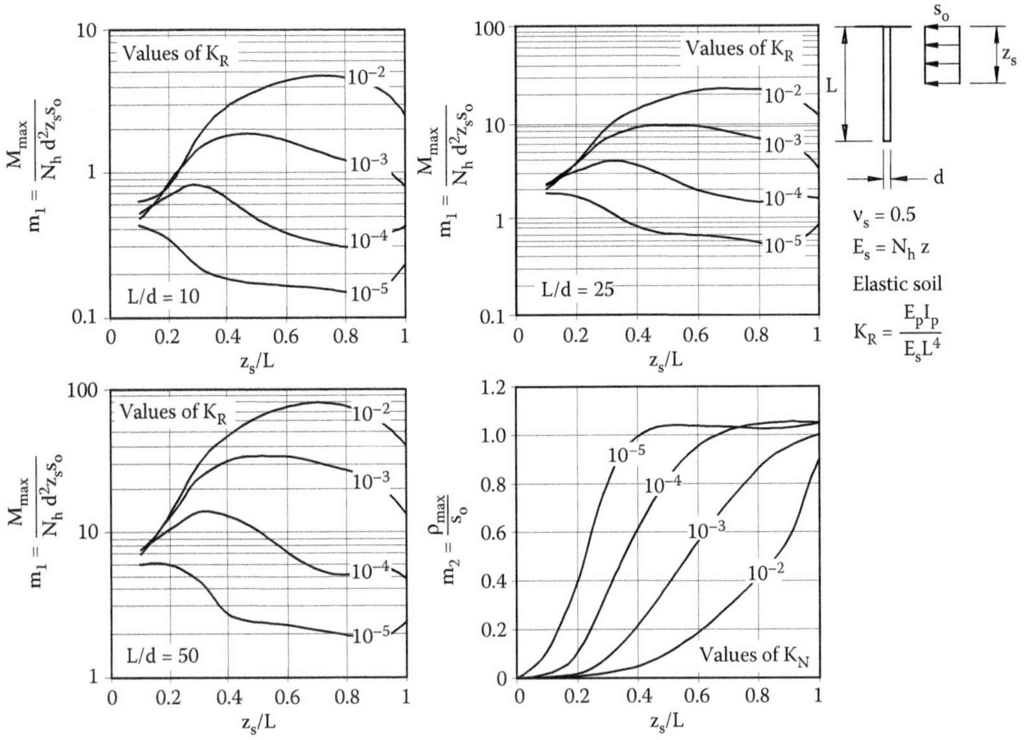


*Figure 9.12* Elastic solutions for pinned-restrained head pile in Gibson soil, with uniform soil movement profile. (Adapted from Chen, L.T. and Poulos, H.G. 1997. *Journal of Geotechnical and Geoenvironmental Engineering, ASCE*, 123(9): 802–811. Courtesy of ASCE.)

movements are not extremely large, for example, less than about 30%–40% of the pile diameter.

## 9.4 EXCAVATION-INDUCED MOVEMENTS AND THEIR EFFECTS

### 9.4.1 Introduction

Excavations for the construction of high-rise buildings in congested urban areas have become increasingly prevalent. They may however cause damage to existing structures because of the soil movements they induce. Examples of such damage have been reported by Finno et al. (1991), Amirsoleymani (1991a,b), Chu (1994) and Poulos (1997c). An excavation will cause both vertical and lateral ground displacements, but the lateral component may be more critical as adjacent piles are not always designed to sustain significant additional lateral loadings. Thus, attention here will be focussed on the lateral response of piles to excavation-induced lateral ground movements.

Poulos and Chen (1996, 1997), Chen and Poulos (1999) and Goh et al. (1996) have outlined an approach to the problem which again utilises the method described herein. The ground movements have been computed via a finite element analysis, while the pile response to these movements has been analysed via a boundary-element computer programme.

## 9.4.2 Estimation of ground movements

### 9.4.2.1 Category 1 method

Peck (1969) developed the plot shown in Figure 9.13, based on observations from field measurements of ground settlements arising from excavations in clays and sands. This figure indicates that settlements less than 1% of the excavation depth should be achieved in sands and clay with average workmanship, but movements in excess of 2% of the excavation depth could occur with marginally stable excavations in soft clays. Figure 9.13 has proved to be a useful practical guide, and has stimulated subsequent studies from which the Category 2 methods set out below have been derived.

### 9.4.2.2 Category 2 methods

Clough and O'Rourke (1990) have presented charts based on a series of finite element analyses for excavation-induced ground movements in terms of a dimensionless lateral movement, $\delta/H$ (where $\delta$ = maximum lateral deflection and H = excavation depth) versus a factor of safety against basal heave, and a dimensionless support stiffness, $EI/\gamma_w h_{av}^4$, where EI is the bending stiffness of excavation support, $\gamma_w$ the unit weight of water and $h_{av}$ the average support spacing. Figure 9.14 shows the relationships obtained.

Clough and O'Rourke (1990) also proposed profiles of surface settlement adjacent to an excavation for various soil types, and these profiles are shown in Figure 9.15.

Kung et al. (2007) suggested that the maximum vertical movement, $\delta_{vm}$, could be related to the maximum lateral movement, $\delta_{hm}$, via a deformation ratio, R, that is:

$$\delta_{vm} = R\delta_{hm} \tag{9.1}$$

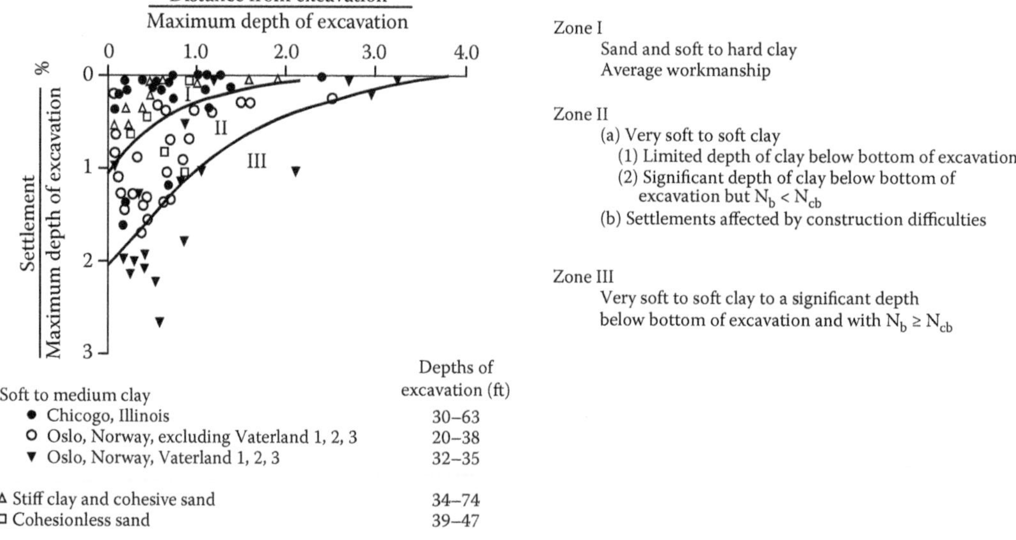


*Figure 9.13* Settlements due to excavation in clays and sands. Note: All data shown are for excavations using standard soldier piles or sheet piles braced with cross-bracing or tiebacks. (After Peck, R.B. 1969. Deep excavations and tunneling in soft ground. *Proceedings of the 7th International Conference on Soil Mechanics and Foundation Engineering*, State-of-the-Art-Report, Mexico City, State-of-the-Art-Volume, pp. 225–290.)

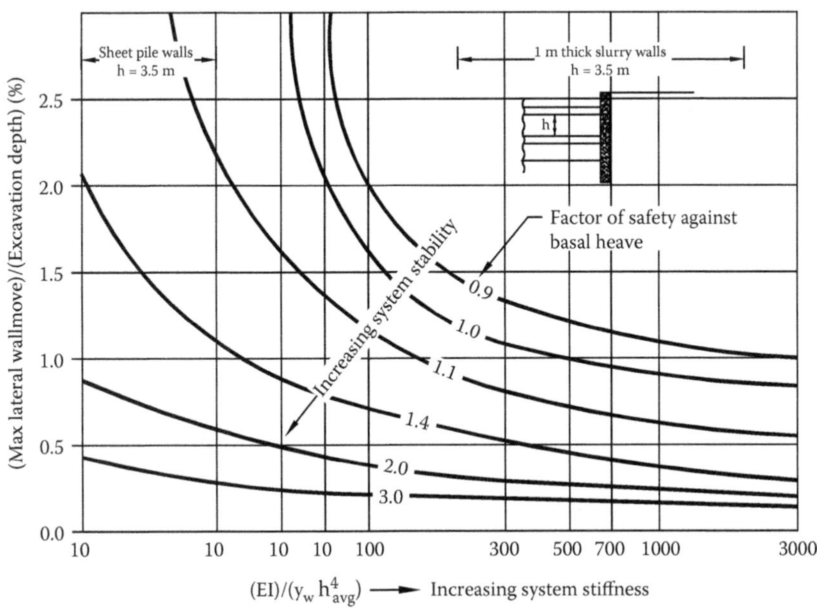


*Figure 9.14* Maximum wall movement. (Adapted from Clough, G.W. and O'Rourke, T.D. 1990. Construction induced movements of in situ walls. *Design and Performance of Earth Retaining Structures, ASCE GSP No.25*, pp. 439–470. Courtesy of ASCE.)

Empirically, the deformation ratio generally fell within the range of 0.5–1.0, but Kung et al. developed the following regression equation for R, based on many finite element analyses:

$$R = c_0 + c_1Y_1 + c_2Y_2 + c_3Y_3 + c_4Y_1Y_2 + c_5Y_1Y_3 + c_6Y_2Y_3 + c_7Y_3^3 + c_8Y_1Y_2Y_3 \quad (9.2)$$

where $Y_1 = \Sigma H_{clay}/H_{wall}$, $Y_2 = s_u/\sigma'_v$, $Y_3 = E_i/1000\sigma'_v$, $\Sigma H_{clay}$ is the thickness of clay layer, $H_{wall}$ the wall length, $s_u$ the undrained shear strength of clay, $\sigma'_v$ the vertical effective stress, $E_i$ the initial Young's modulus of soil, $c_1$ to $c_8$ are coefficients from a least square regression of the element analysis results, with $c_0 = 4.55622$, $c_1 = -3.40151$, $c_2 = -7.37697$, $c_3 = -4.99407$, $c_4 = 7.14106$, $c_5 = 4.60055$, $c_6 = 8.74863$, $c_7 = 0.38092$, $c_8 = -10.58958$.

Another regression equation was developed for the maximum lateral wall deflection, $\delta_{hm}$, within a deep layer, as follows:

$$\delta_{hm}(mm) = b_0 + b_1X_1 + b_2X_2 + b_3X_3 + b_4X_4 + b_5X_5 + b_6X_1X_2 + b_7X_1X_3 + b_8X_1X_5 \quad (9.3)$$

where $X_1 = t(H_e)$, $X_2 = t\left[\ln\left(EI/\gamma_w h_{av}^4\right)\right]$, $X_3 = t(B/2)$, $X_4 = t(s_u/\sigma'_v)$, $X_5 = t(E_i/\sigma'_v)$, t is the wall thickness, $H_e$ the excavation depth, B the excavation width, $b_0$ to $b_8$ are regression coefficients, as follows: $b_0 = -13.41973$, $b_1 = -0.49351$, $b_2 = -0.09872$, $b_3 = 0.06025$, $b_4 = 0.23766$, $b_5 = -0.15406$, $b_6 = 0.00093$, $b_7 = 0.00285$, $b_8 = 0.00198$.

To allow for the possible presence of a hard underlying stratum, the following correction factor, K, can be applied to $\delta_{hm}$:

$$K = 1.5\left(\frac{T}{B}\right) + 0.4, \quad \text{for} \quad \frac{T}{B} \leq 0.4 \quad (9.4)$$

where T is the depth to hard stratum, measured from excavation level.

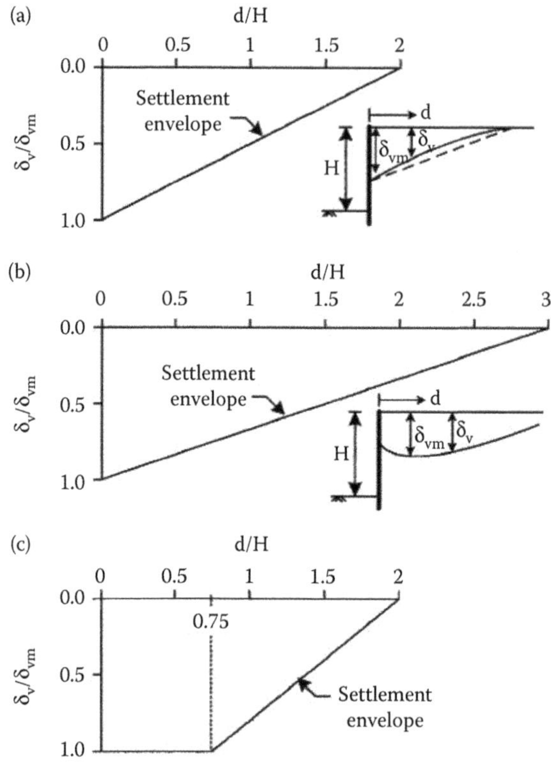


*Figure 9.15* Design charts for surface settlement profiles adjacent to excavations: (a) sand, (b) stiff to very hard clays and (c) soft and medium clays. (Adapted from Clough, G.W. and O'Rourke, T.D. 1990. *Design and Performance of Earth Retaining Structures, ASCE GSP No. 25*, pp. 439–470. Courtesy of ASCE.)

The distributions of movement with depth are difficult to estimate without some form of analysis, as they depend on wall flexibility and excavation support conditions. For excavation with limited support, it may be reasonable to assume a linear distribution with depth.

#### 9.4.2.3 *Category 3 analyses*

It is now common for the ground movements around excavations to be estimated via Category 3 detailed numerical analyses such as the finite element method. While common design practice employs two-dimensional (2D) analyses, three-dimensional (3D) analyses are being used increasingly because of the significant repression of movements near the corners of an excavation. Near the centre of the excavation, 2D analyses can give reasonable soil movement estimates (e.g., Yong et al., 1996).

Category 3 methods also include simplified 'beam on elastic foundation' methods, such as that implemented in the commercially available programme WALLAP (Geosolve, 2013).

### 9.4.3 Some characteristics of pile behaviour

For the case shown in Figure 9.16, a 2D non-linear finite element analysis has been used by Poulos and Chen (1997) to compute the lateral deflection of the ground due to the excavation, and then the lateral response of a single pile has been computed via a boundary-element

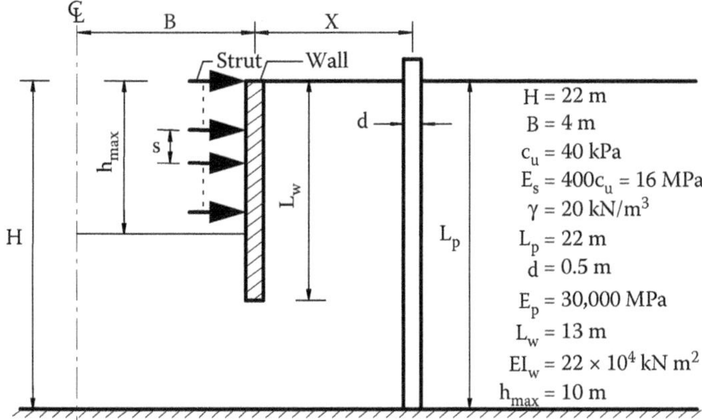


*Figure 9.16* Excavation problem analysed. (Adapted from Poulos, H.G. and Chen L. 1997. *Journal of Geotechnical and Geoenvironmental Engineering, ASCE,* 123(2): 94–99. Courtesy of ASCE.)

analysis. For a pile at a distance of 1 m from the excavation face and wall, Figure 9.17 shows computed distributions of lateral deflection and moment for various values of the stability number $N_c = \gamma H/s_u$, where $\gamma$ is the soil unit weight, H the unsupported height of wall and $s_u$ the undrained shear strength. The pile deflections are very close to the free-field soil movements, reflecting the fact that the pile analysed is relatively flexible. The pile bending moment distributions exhibit a double curvature, with the maximum values increasing with increasing stability number. The rate of increase of bending moments with stability number accelerates at larger stability numbers, when failure of the soil is approached.

### 9.4.4 Design charts for piles near supported excavations

Poulos and Chen (1997) have developed design charts which may be useful for a preliminary assessment of the pile response. The 'standard' problem considered is shown in Figure 9.16, and the analysis techniques outlined above have been used to compute the maximum pile deflection and maximum bending moment for this case, for a single pile in a homogeneous

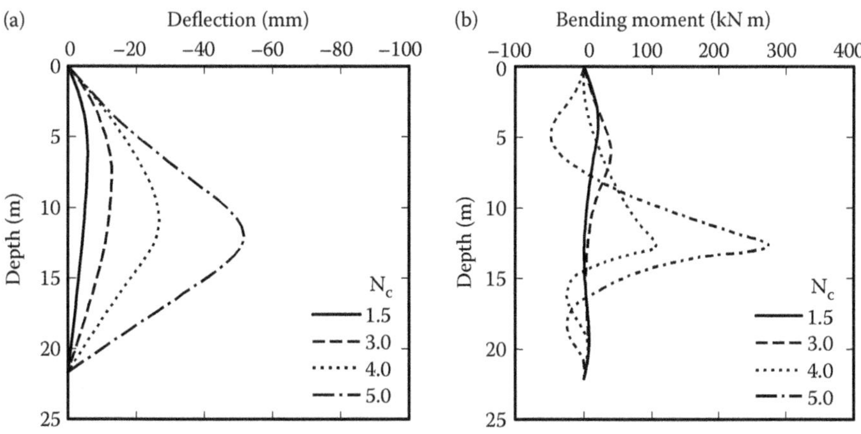


*Figure 9.17* Pile response for basic problem: (a) deflection profile; (b) bending movement profile. (Adapted from Poulos, H.G. and Chen L. 1997. *Journal of Geotechnical and Geoenvironmental Engineering, ASCE,* 123(2): 94–99. Courtesy of ASCE.)

clay layer, at various distances from the excavation. Non-linear behaviour of the pile has been allowed for by imposing limiting lateral pile–soil pressures, assumed to be equal to $9s_u$, where $s_u$ is the undrained shear strength of the clay. A number of parameters have then been varied (one at a time) to obtain to examine the effect of each on the pile deflection and bending moment. The results of these analyses have then been expressed in terms of basic solutions for the 'standard' problem, with correction factors for each of the parameters varied.

The following approximate expressions are derived for the maximum pile deflection, ρ and the maximum bending moment, $M_{max}$:

$$\rho = \rho_b \cdot k'_{cu} \cdot k'_d \cdot k'_{Nc} \cdot k'_{EI} \cdot k'_k \cdot k'_s \tag{9.5}$$

$$M_{max} = M_b \cdot k_{cu} \cdot k_d \cdot k_{Nc} \cdot k_{EI} \cdot k_k \cdot k_s \tag{9.6}$$

where $\rho_b$ is the basic maximum deflection, $M_b$ the basic maximum bending moment, $k_{cu}$, $k'_{cu}$ correction factors for undrained shear strength, $k_d$, $k'_d$ correction factors for pile diameter, $k_{Nc}$, $k'_{Nc}$ correction factors for depth of excavation (depending on $N_c = \gamma h/s_u$, where $\gamma$ is the average soil unit weight, $s_u$ the undrained shear strength, h the excavation height), $k_{EI}$, $k'_{EI}$ correction factors for wall stiffness, $k_k$, $k'_k$ correction factors for strut (or support) stiffness and $k_s$, $k'_s$ correction factors for strut (or support) spacing.

Figures 9.18 through 9.21 plot the various factors, for the case of a free-head unrestrained pile. It should be emphasised that Equations 9.5 and 9.6 give only the additional response of the pile, assuming that the pile has zero initial bending moment. If this is not the case, then the existing bending moment may be decreased or increased due to the excavation, depending on the bending moment and deflection of the pile under the applied loading.

Chen and Poulos (1996) have examined the effects of the pile head boundary condition on the pile response, and have found that it may have a substantial effect. The greater the degree of restraint, the greater the bending moments induced in the pile. Figure 9.22 shows the computed deflection and moment profiles, and the maximum positive and negative bending moments for the standard case in Figure 9.16. It can be seen that, while the maximum

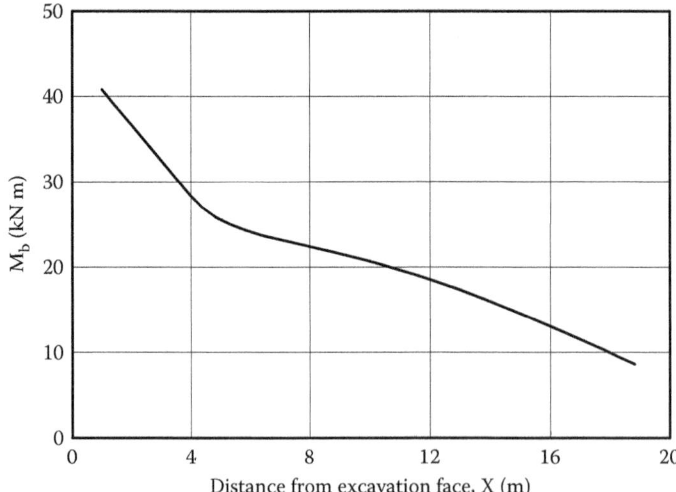


*Figure 9.18* Basic bending movement versus distance from excavation face. (Adapted from Poulos, H.G. and Chen L. 1997. *Journal of Geotechnical and Geoenvironmental Engineering, ASCE*, 123(2): 94–99. Courtesy of ASCE.)

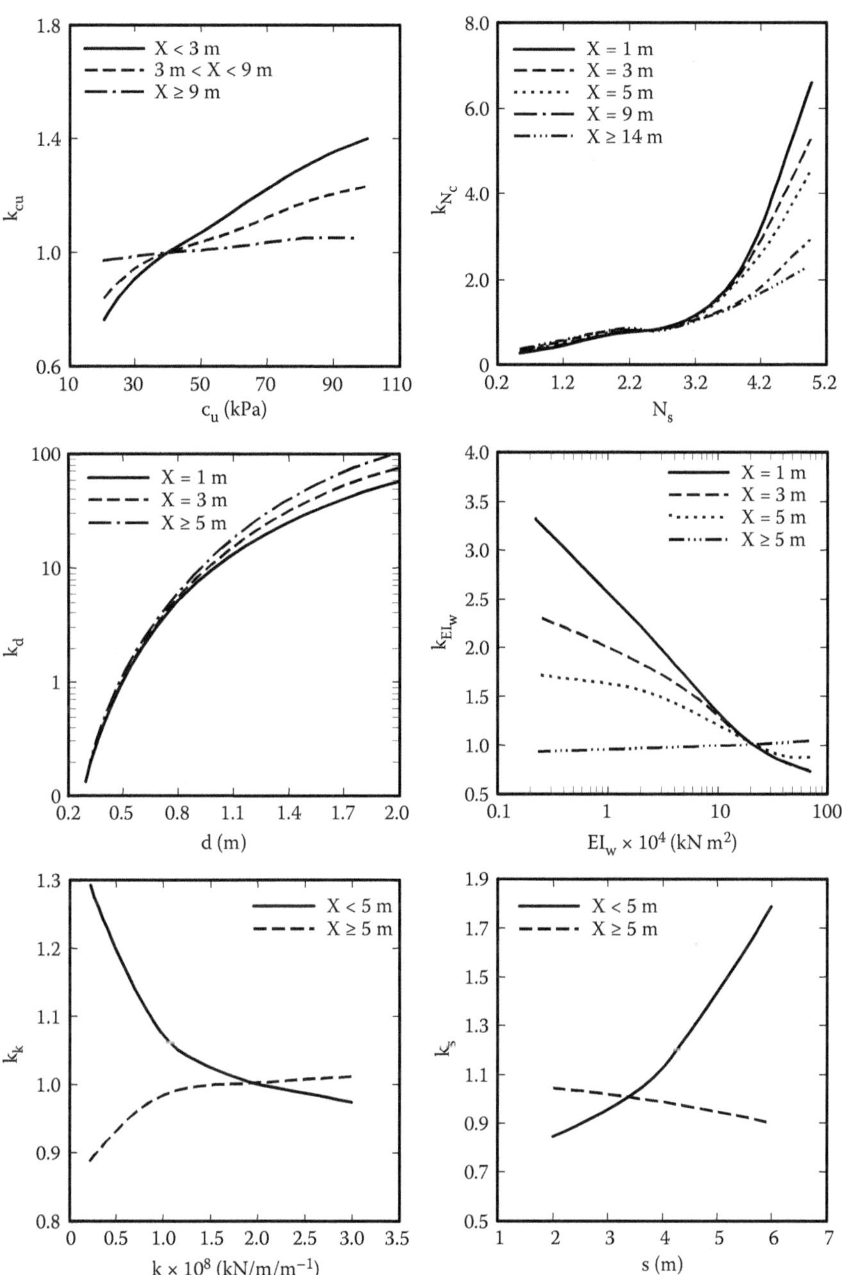


*Figure 9.19* Correction factors for bending moment. (Adapted from Poulos, H.G. and Chen L. 1997. *Journal of Geotechnical and Geoenvironmental Engineering, ASCE*, 123(2): 94–99. Courtesy of ASCE.)

positive bending moment does not vary greatly, a very large negative bending moment is developed when the pile head is restrained against rotation. This negative moment becomes extreme when the pile head is also restrained against translation. In relation to tall buildings, this may be very significant. Piles within an existing large group will normally be supported by a cap or mat, and there will be considerable restraint against both rotation and

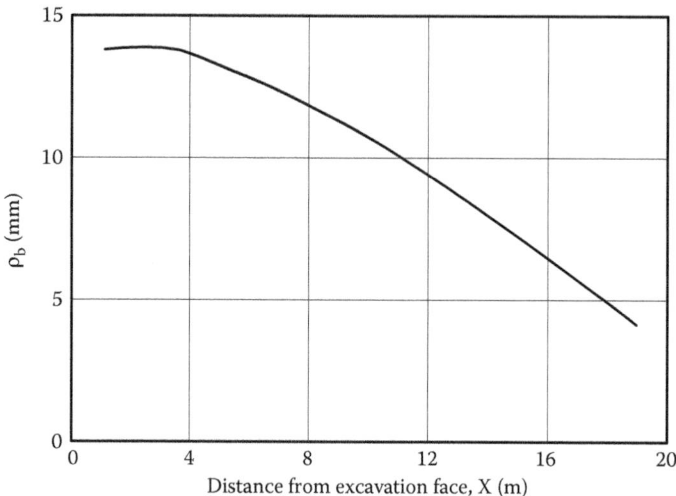


*Figure 9.20* Deflection versus distance from excavation face. (Adapted from Poulos, H.G. and Chen L. 1997. *Journal of Geotechnical and Geoenvironmental Engineering, ASCE*, 123(2): 94–99. Courtesy of ASCE.)

translation of the piles near the excavation because of the cap and the other piles. Thus, the possibility of pile damage due to a nearby excavation may be more severe for piles within a group than for isolated piles without head restraint. In turn, this implies that during construction, there is a reduced risk of damage if the excavation is carried out before any nearby piles are capped.

### 9.4.5 Application of charts to case history

Finno et al. (1991) described a case where a 17.7 m deep tieback excavation was made through primarily granular soils within an existing frame structure, which was supported by groups of step-tapered piles about 21 m long. Although the excavation was provided with temporary support by a tieback sheet–pile wall, the main column pile caps had moved about 64 mm laterally toward the excavation by the time the sheet–pile extraction was about to begin. Measurements of the deflection profile along the pile and the distribution of maximum moment were reported by Finno et al. (1991), and Poulos and Chen (1997) reported reasonable comparisons between these measurements and the computed behaviour from a detailed analysis.

The design charts can also be used to estimate the maximum pile deflection and bending moment, assuming the soil to be the equivalent of a stiff clay layer with $c_u = 100$ kPa. The computations for the fourth stage excavations (to a depth of 15 m) are set out below.

Basic values: for X = 1.5 m, $M_b = 37$ kN m from Figure 9.18, and from Figure 9.20, $\rho_b = 14$ mm. From Figures 9.19 and 9.21, the correction factors are as follows:

For $s_u = 100$ kPa, $k_{cu} = 1.4$, $k'_{cu} = 1.95$
For $N_c = 19*15/100 = 2.9$, $k_{Nc} = 1.0$, $k'_{Nc} = 1.0$
For $d = 0.327$ m, $k_d = 0.2$, $k'_d = 1.0$
For $EI_w = 11*10^4$ kN m$^2$, $k_{EI} = 1.30$, $k'_{EI} = 1.06$
For $k = 1*10^4$ kN/m m$^{-1}$, $k_k = 1.45$, $k'_k = 1.5$
For $s = 5$ m, $k_s = 1.45$, $k'_s = 1.2$

The maximum estimated pile deflection is thus: 14*1.95*1.0*1.0*1.06*1.5*1.2 = 52 mm.
The maximum estimated bending moment is: 37*1.4*1.0*0.2*1.3*1.3*1.45 = 25 kN m.
These estimated values agree adequately with the measured values reported by Finno et al. which were 64 mm and 23 kN m, respectively.

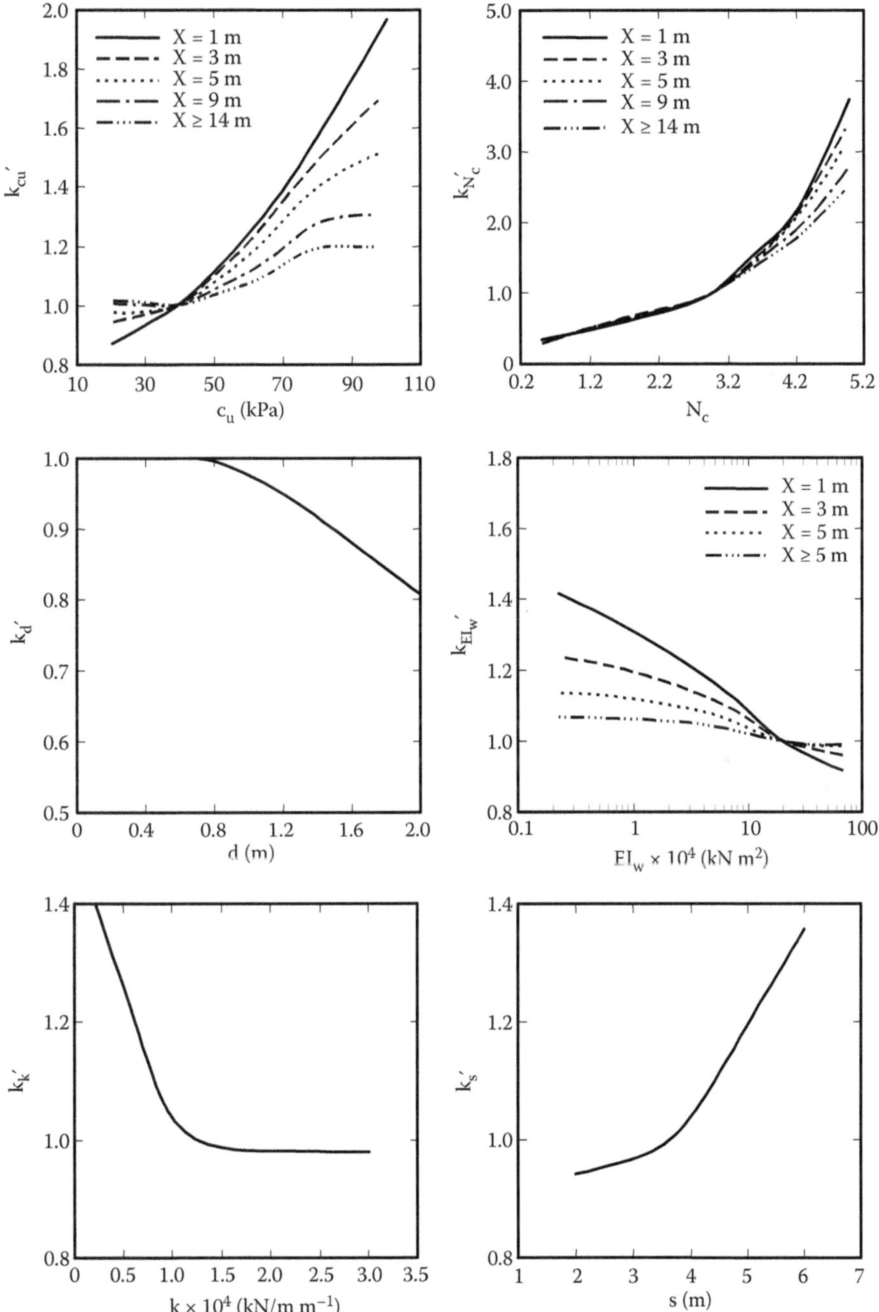


*Figure 9.21* Correction factors for deflection. (Adapted from Poulos, H.G. and Chen L. 1997. *Journal of Geotechnical and Geoenvironmental Engineering, ASCE*, 123(2): 94–99. Courtesy of ASCE.)

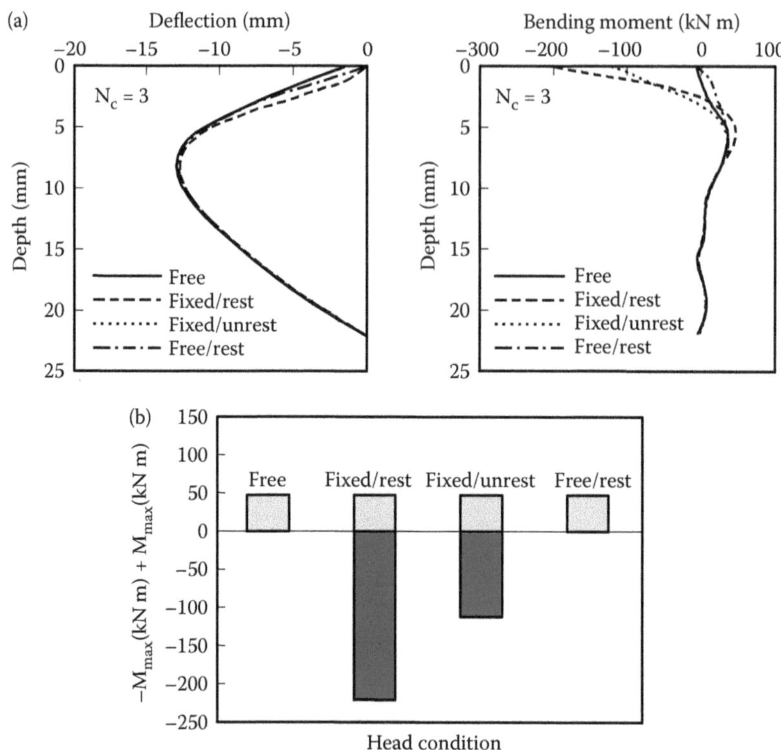


*Figure 9.22* Effect of pile head condition on pile response: (a) deflection and bending moment profile and (b) maximum bending moment. (Adapted from Chen, L.T. and Poulos, H.G. 1996. Some aspects of pile response near an excavation. *Proceedings of 7th Australia and New Zealand Conference on Geomechanics*, Adelaide, pp. 604–609.)

## 9.5 CONSTRUCTION-RELATED MOVEMENTS AND THEIR EFFECTS

### 9.5.1 Introduction

Poulos (2005b) has discussed some of the processes involved in investigation and foundation construction which can lead to 'side effects' on existing foundations, and which, although not always recognised by designers, may have a significant effect on the existing foundation system. Such side effects may also be present when a new project is being constructed adjacent to an existing building or facility. The processes considered included

- The driving of piles
- The installation of piles by jacking
- Carrying out relatively deep excavations which are supported
- Drilling holes in stressed ground for the purposes of investigation or construction
- Carrying out relatively shallow unsupported excavations, for example, for pile cap construction or the installation of services

In each of these cases, the processes involved lead to the generation of additional ground movements, both vertical and horizontal. These movements will interact with existing foundation elements, particularly piles, and induce additional foundation movements, forces and bending moments. Attention will be focussed on the last two processes, as the effects of deeper excavations has been considered in Section 9.4 above, and pile driving and pile jacking are less frequently carried in urban environments in which high-rise structures are constructed.

### 9.5.2 Characteristics of the environment around an existing foundation system

In new foundation construction in 'greenfield' conditions, there is usually relatively unimpeded access to the site and to the areas in which the new foundation system is to be constructed. This contrasts with the environment around or within an existing foundation system which is being investigated and/or upgraded. In this latter case, the following characteristics can be anticipated:

- Access to the area may be very difficult and may limit the range of construction methods that can be employed.
- The ground will often be highly stressed, and thus changes in the stress regime due to investigation or construction may result in larger ground movements than would be the case in a greenfield situation.
- Existing piles will generally be subjected to some measure of restraint from the building which they are supporting, via the attachment to pile caps and the overall foundation system.
- Strict control of investigation and construction processes are likely to be more critical, but more difficult to achieve, than with greenfield situations.
- The consequences of uncontrolled ground movements on the existing structure and foundation system are likely to be more immediate and severe than with a greenfield site.

For these reasons, it is worthwhile to give special attention to the problems of construction and investigation within an existing foundation system, and to examine in some detail the possible consequences of inadequate control of the resulting ground movements.

### 9.5.3 Drilling for investigation and pile construction

The drilling of holes in the ground is a standard process for ground investigation and for the construction of piles, and is so common that little consideration is given to the possible consequences. Of particular concern are the ground movements that can arise because of the removal of the soil during the drilling process. While such movements may be very small in an 'open-field' situation, this may not be the case when the holes are drilled in highly stressed ground, for example, below an existing high-rise building. In this section, an examination is made of the ground movements that can arise from the drilling process, and of the consequences of these ground movements on existing nearby piles.

#### 9.5.3.1 Ground movements

Despite the ubiquitous nature of the hole drilling process, there appear to be few if any studies of the ground movements arising from drilling operations. As a consequence, a numerical analysis employing the commercially available computer programme PLAXIS was carried out by Poulos (2005b) to examine the order of magnitude of such movements, and their dependence on the stress level in the ground. The problem examined is illustrated

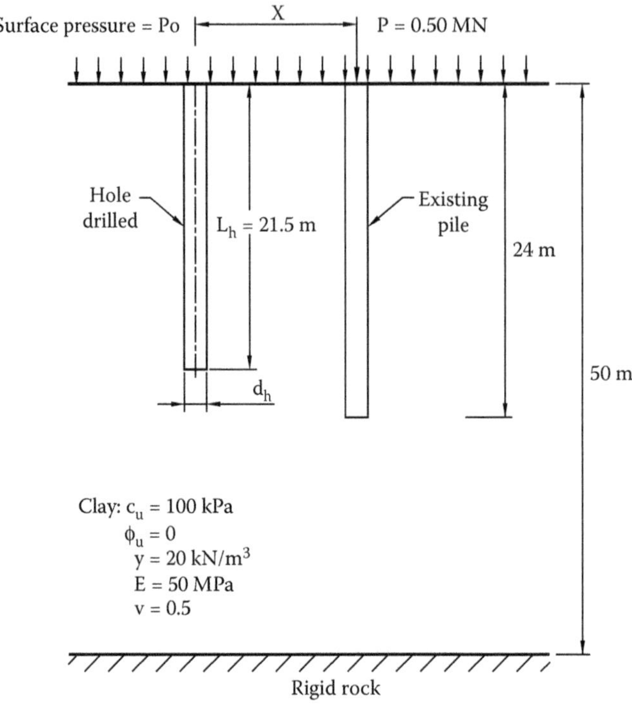


*Figure 9.23* Basic problem of hole drilled near an existing pile. (Adapted from Poulos, H.G. 2005b. *Geotechnical Engineering, SEAGS*, 36(1): 51–67. Courtesy of SEAGS.)

in Figure 9.23, and involved a hole of diameter $d_h$ drilled into a uniform stiff clay profile which was subjected to a surface surcharge loading $p_s$. It was assumed that the drilling was carried out rapidly so that undrained conditions prevailed, and it also assumed that there were no effects arising from differences between the water level inside and outside the hole. Settlements are shown as negative in the following results.

Figure 9.24 shows the computed distribution of vertical movement with depth at a distance of 2 m from the centre of a 0.6 m diameter hole, for values of the surcharge pressure of 0, 100, 200 and 300 kPa. The hole has been drilled to a depth of 21.5 m. The following characteristics may be observed:

- There is generally a settlement above the level of the base of the hole and a heave just above and below this level.
- When there is zero surcharge pressure, the ground movement is small and tends to be a heave of the order of 0.5 mm.
- When the surface surcharge is large, then the ground settlements above the level of the base of the hole become more significant, of the order of 2 mm maximum, with a heave of a similar or greater magnitude near the level of the base of the hole.

Figure 9.25 shows corresponding solutions for the lateral ground movements due to hole drilling. Negative movements are toward the hole. In this case, large lateral movements, of the order of 14 mm, can be generated near the base of the hole when the surcharge pressure is large. Again, in the 'open-field' situation, the lateral movements are much smaller, of the order of 1–2 mm.

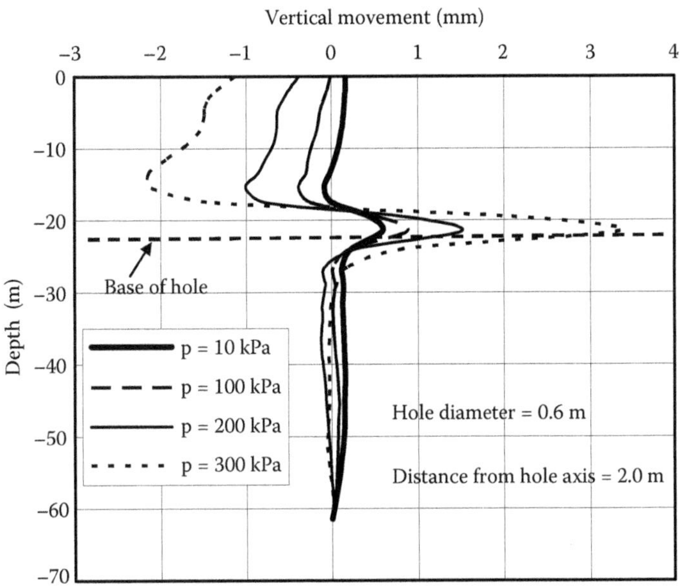


*Figure 9.24* Variation with surface pressure of vertical movement due to hole drilling (settlement is negative). (Adapted from Poulos, H.G. 2005b. *Geotechnical Eng. SEAGS*, 36(1): 51–67. Courtesy of SEAGS.)

Figure 9.26 shows the variation with distance of the horizontal movement at a depth of 17.5 m. The rapid drop-off with increasing distance can be clearly seen.

Figure 9.27 shows the effects of the diameter of the drilled hole on the horizontal movement at a distance of 2 m from the hole, for a surface surcharge pressure of 200 kPa. Clearly, the larger the hole diameter, the larger are the movements.

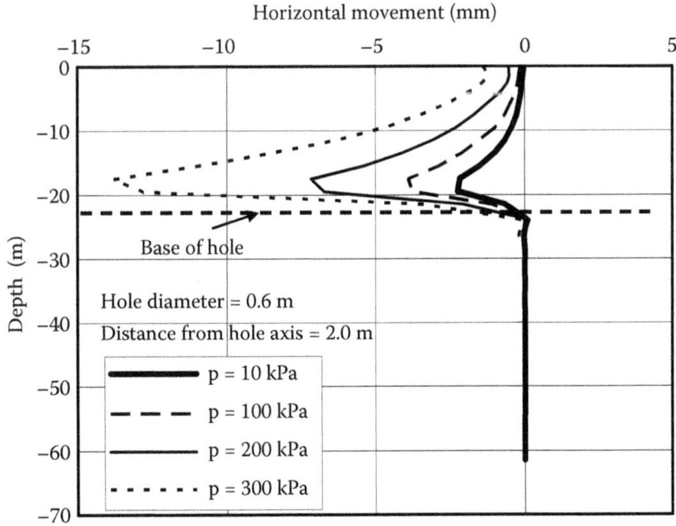


*Figure 9.25* Variation with surface pressure of horizontal movement due to hole drilling (negative toward the hole). (Adapted from Poulos, H.G. 2005b. *Geotechnical Engineering, SEAGS*, 36(1): 51–67. Courtesy of SEAGS.)

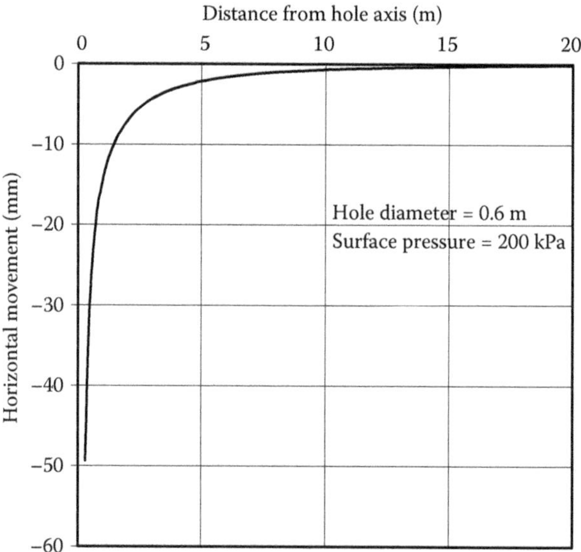


*Figure 9.26* Variation with hole diameter of horizontal movement due to hole drilling (negative towards the hole). (Adapted from Poulos, H.G. 2005b. *Geotechnical Engineering, SEAGS*, 36(1): 51–67. Courtesy of SEAGS.)

As would be expected, the movements are found to decrease with increasing distance from the hole, and reduce to very small values once the distance exceeds about 6–8 hole diameters. However, very close to the hole, significant ground movements can occur, especially in the horizontal direction.

From the above solutions, it can be seen that the drilling of holes may lead to ground movements which are significant, especially at short distances from a hole in ground which is highly stressed, and where the hole diameter is large.

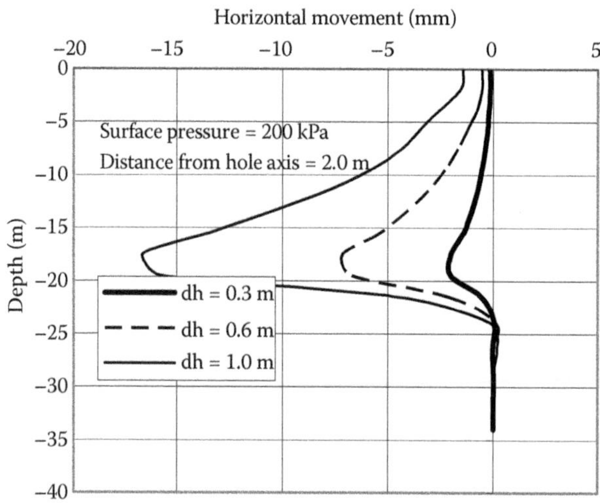


*Figure 9.27* Variation with hole diameter of horizontal movement due to hole drilling (negative toward the hole). (Adapted from Poulos, H.G. 2005b. *Geotechnical Engineering, SEAGS*, 36(1): 51–67. Courtesy of SEAGS.)

#### 9.5.3.2 Pile response to ground movements

As an example of the possible effects of hole drilling on pile response, the case of a single pile 24 m long, in a homogeneous stiff clay, has been considered. The pile head is assumed to be restrained against both translation and rotation (e.g., fixed into a large pile cap). The clay is assumed to have an undrained shear strength of 100 kPa and a Young's modulus of 50 MPa. The axis of the pile is located 2 m from the axis of the newly-drilled hole, which extends to 21.5 m below the surface. The pressure acting on the ground surface is assumed to be 200 kPa.

Figure 9.28 shows the computed maximum induced bending moment in the pile, as a function of the hole diameter, and for three pile diameters. As would be expected, the induced moment increases as the hole diameter increases, and may approach the design moment capacity, especially for the smaller-diameter pile.

Figure 9.29 plots the computed axial movement of the pile head due to the drilling of a hole. In this case, the pile head is assumed to be free to move vertically and is subjected to an applied axial force of about 40% of the ultimate axial capacity (i.e., the factor of safety is 2.5). Again, it can be seen that the pile head movements increase as the hole diameter increases, and may reach the order of 0.8–1.0 mm for a 1 m diameter hole.

If the pile is fixed into a cap which restrains vertical movement, then there will be a tendency for an axial tensile force to be induced in the pile. Figure 9.30 plots the induced axial force in piles of various diameters, as a function of hole diameter. The larger the hole diameter, the larger is the induced force at the pile head. Of course, the net axial force at the pile head will generally still be compressive under normal circumstances, but if the pile has been newly-installed and is yet to carry significant load, the net force may possibly be tensile.

These examples clearly indicate the potential for hole drilling near existing piles to cause additional bending moments and axial force in those piles. The larger the hole diameter, the closer the hole to the pile, or the larger the surface pressure on the soil, the greater are the induced responses. In addition, if multiple holes are drilled near an existing pile, the combined effects may be sufficient to initiate yield of the pile section or excessive additional

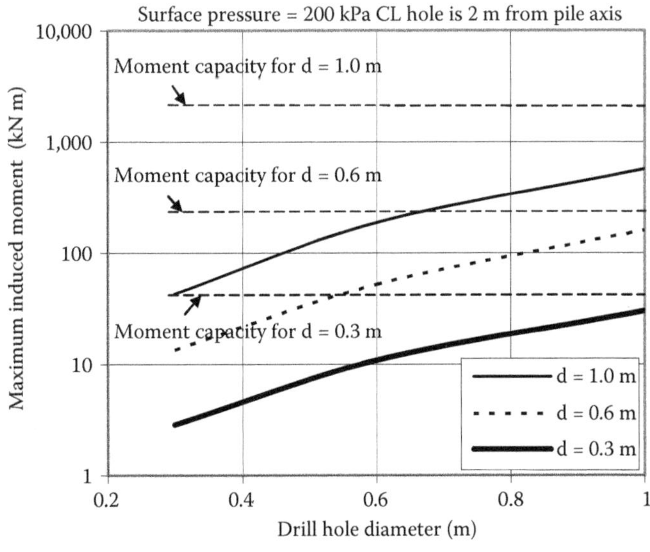


*Figure 9.28* Induced bending moments due to hole drilling. Fixed head restrained pile. (Adapted from Poulos, H.G. 2005b. *Geotechnical Engineering, SEAGS*, 36(1): 51–67. Courtesy of SEAGS.)

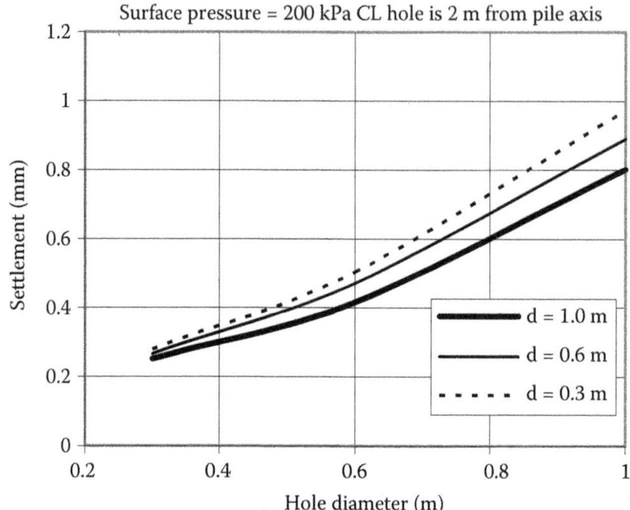


*Figure 9.29* Induced settlement due to drilling. Vertically unrestrained pile head. (Adapted from Poulos, H.G. 2005b. *Geotechnical Engineering, SEAGS*, 36(1): 51–67. Courtesy of SEAGS.)

settlement. A typical field scenario for such a case would be the drilling of holes for new large diameter bored piles adjacent to existing smaller-diameter precast piles.

### 9.5.4 Excavation for a pile cap near existing piles

#### *9.5.4.1 Introduction*

In the context of 'side effects', a common situation is when an excavation is carried out for a new pile cap or raft, in the vicinity of existing piles. In some cases, little or no support may be provided for the excavation, since pile cap thicknesses are typically 1–3 m.

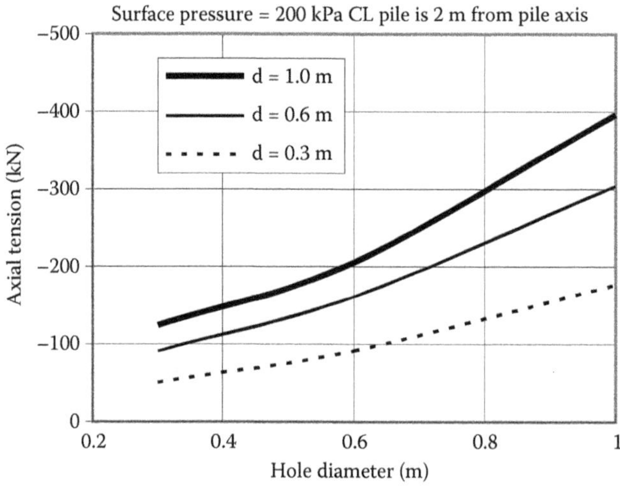


*Figure 9.30* Induced tension due to hole drilling. Vertically restrained pile. (Adapted from Poulos, H.G. 2005b. *Geotechnical Engineering, SEAGS*, 36(1): 51–67. Courtesy of SEAGS.)

However, under conditions in which the ground is highly stressed, even such modest excavations deserve consideration.

### 9.5.4.2 Ground movements

It is now common for the ground movements around excavations to be estimated via detailed numerical analyses such as the finite element method. When numerical analyses cannot be carried out, it is possible to use approximations developed by Clough and his co-workers to estimate vertical and horizontal distributions of ground movements (see Section 9.4).

Common design practice employs 2D analyses, and near the centre of an excavation, 2D analyses can give reasonable soil movement estimates (e.g., Yong et al., 1996). Thus, in the following examples, a 2D analysis, employing the computer programme FLAC, has been used to estimate the ground movements due to excavation for a pile cap. The case examined is shown in Figure 9.31, and involves an excavation in medium-soft clay for a 3 m deep pile cap, 10 m in width, with no lateral support provided for the excavation. It is further assumed that the excavation is carried out relatively rapidly, and that no drop in the level of the water table arises from the excavation.

Figure 9.32a–d shows typical distributions of the vertical and lateral movements with depth, at various distances from the excavation. Two different values of the surface pressure are considered, 0 kPa (a 'greenfield' situation) and 50 kPa, a typical situation that may arise beneath an existing low-rise building. It can be seen that, as would be expected, the movements for the 50 kPa surface pressure are considerably larger than those for zero pressure, and that the movements tend to decrease with increasing distance from the excavation.

### 9.5.4.3 Pile response to ground movements

For the case as shown in Figure 9.31, Figures 9.33 and 9.34 summarise the computed maximum bending moment and shear in an adjacent pile, as a function of the distance from

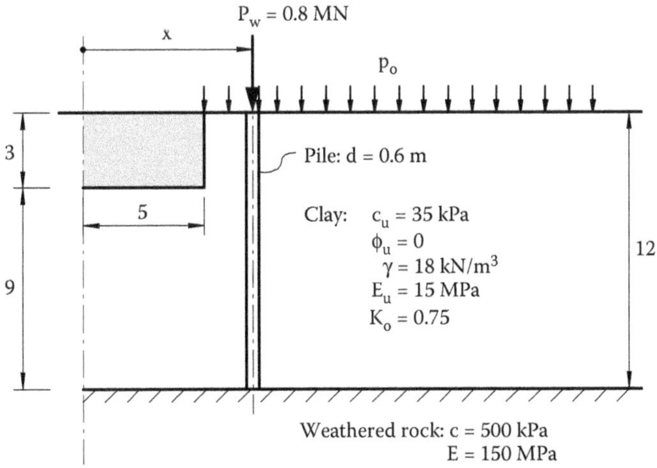


*Figure 9.31* Problem analysed for effect of cap excavation. (Adapted from Poulos, H.G. 2005b. *Geotechnical Engineering, SEAGS*, 36(1): 51–67. Courtesy of SEAGS.)

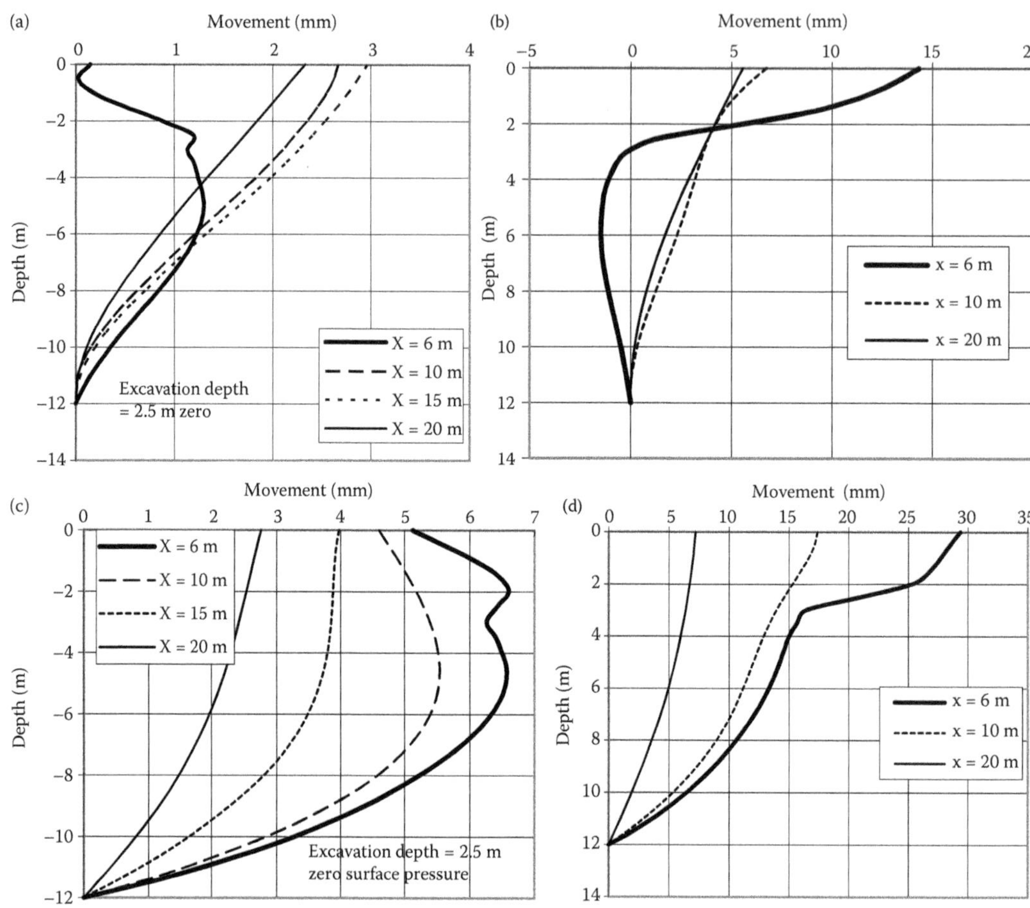


*Figure 9.32* Vertical movement profiles due to cap excavation for (a) zero surface pressure and (b) 50 kPa surface pressure. Horizontal movement profiles due to cap excavation for (c) zero surface pressure and (d) 50 kPa surface pressure. (Adapted from Poulos, H.G. 2005b. *Geotechnical Engineering, SEAGS,* 36(1): 51–67. Courtesy of SEAGS.)

the excavation and the surface pressure. It will be seen that the induced maximum bending moment is very large when the pile is close to the excavation. For a 0.6 m diameter reinforced concrete pile with 1% reinforcement, carrying a working axial load of 800 kN (corresponding to a factor of safety of about 2), the maximum design moment capacity is about 0.56 MN m. Thus, Figure 9.33 implies that piles within about 10 m of the axis of the excavation could have induced moments that exceed the design capacity of the pile, if the surface pressure is as little as 50 kPa.

Figure 9.35 summarises the computed additional movement of an existing pile adjacent to the excavation. In this case, if there is zero surface pressure, the adjacent pile tends to move upward slightly because of the excavation, but it settles if the surface pressure is 50 kPa. In the latter case, the additional axial force induced in the pile by the vertical ground settlement is small, even if the pile is relatively close to the excavation.

Thus, it would appear that the issue that may cause most concern in such situations is the induced bending moment and shear in the pile due to the lateral ground movements.

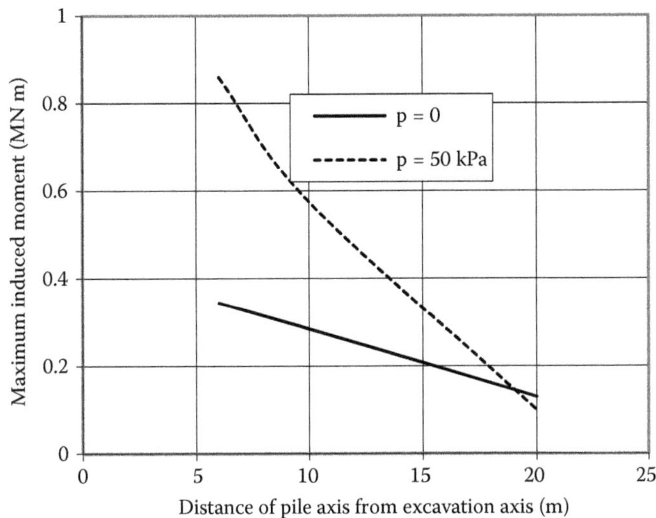


*Figure 9.33* Induced bending moments due to cap excavation. (Adapted from Poulos, H.G. 2005b. *Geotechnical Engineering, SEAGS*, 36(1): 51–67. Courtesy of SEAGS.)

## 9.5.5 Ground movements due to diaphragm wall construction

It is now recognised that the construction of a diaphragm wall will cause movements in the ground nearby. Clough and O'Rourke (1990) reported settlements due to diaphragm wall construction ranging between 5 and 15 mm, while Poh et al. (2001) observed a maximum settlement of about 24 mm and a lateral movement of 45 mm during trench excavation in an instrumented field test on a 55 m deep diaphragm wall in a soft soil deposit.

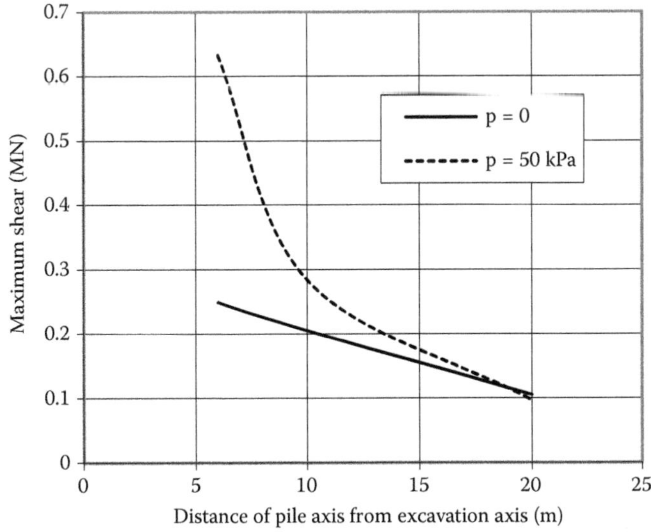


*Figure 9.34* Induced shear in pile due to cap excavation. (Adapted from Poulos, H.G. 2005b. *Geotechnical Engineering, SEAGS*, 36(1): 51–67. Courtesy of SEAGS.)

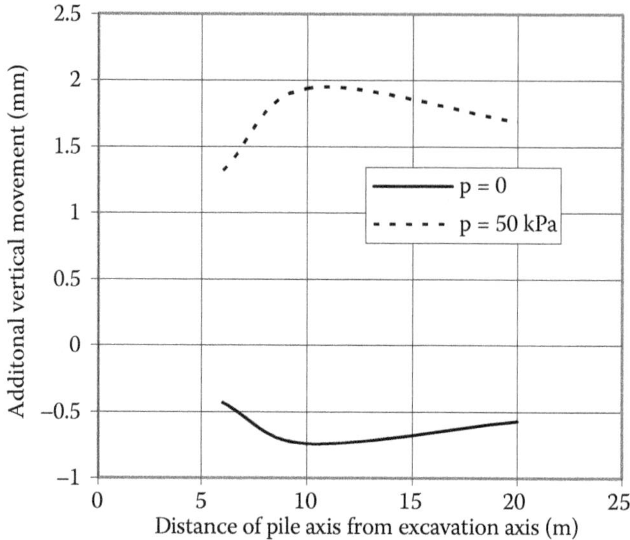


*Figure 9.35* Additional vertical movement of pile due to cap excavation. (Adapted from Poulos, H.G. 2005b. *Geotechnical Engineering, SEAGS*, 36(1): 51–67. Courtesy of SEAGS.)

This problem was examined theoretically by Ng and Lei (2003) who developed an explicit analytical solution for calculating horizontal stress changes and displacements caused by the excavation for a diaphragm wall panel. The theoretical solution was obtained by applying the principle of superposition to model diaphragm wall construction, using a basic elastic solution to the problem of an infinite horizontal plate with a rectangular opening subjected to a uniaxial stress at infinity. Key parameters governing the magnitude of horizontal stress changes and displacements were identified, and computed results were given in a normalised form in terms of aspect ratio (length to width) of a diaphragm wall panel. From the parametric study carried out, charts were developed to allow preliminary designs to be undertaken.

It was found that 3D effects are most important when analysing such ground movements, and that use of a 2D analysis could lead to significant overestimates of such movements. Ng et al. (2004) have presented correction factors to be applied to the results obtained from 2D analyses to allow for 3D effects.

Comdromos and Papadopoulou (2013) described a numerical approach for simulating the effects of installation of diaphragm walls on the surrounding and adjacent buildings. The method employed a 3D non-linear analysis and a constitutive law providing bulk and shear modulus variation, depending on the stress path (loading, unloading or reloading). From the application of the method in a normally to slightly over-consolidated clayey soil, it was found that the panel length was the most important factor affecting ground movements and lateral stress reduction during panel installation. It was found that the effects from the construction of a panel were mainly limited to a zone within a distance of the order of the panel length. The effects on an adjacent building were also investigated by applying a full soil–structure interaction analysis, including the whole building. Settlement profiles and settlements were examined at specific points and, contrary to lateral movements, which mostly take place at the panel under construction, it was found that the effect of settlements covers a larger area leading to a progressive settlement increase. The effect depended greatly on the distance from the panel under construction.

No results are available for the response of piles to ground movements induced by diaphragm wall construction, but a rough estimate could be made by estimating the maximum movement magnitude from the results of Ng and Lei (2003), assuming a linear distribution of ground movement with depth, and then using the generic design charts in Figures 9.5, 9.7, 9.9 and 9.10.

## 9.6 TUNNELLING-INDUCED MOVEMENTS AND THEIR EFFECTS

### 9.6.1 Introduction

In contemporary urban environments, it is not uncommon for tall buildings to be located in the vicinity of future transportation or utility tunnels. Tunnelling will inevitably cause ground movements, and these will in turn impose axial and lateral forces on nearby pile foundations. In such cases, it is prudent to examine the potential effects of ground movements on tall building foundations. Chen et al. (1999) have considered this problem and have studied the basic problem illustrated in Figure 9.36. They developed design charts based on the two-stage analysis involving:

- Estimation of tunnelling-induced ground movements
- Analysis of the response of a pile to these ground movements

These steps are described very briefly below, and then the design charts are reproduced.

### 9.6.2 Ground movements due to tunnelling

#### *9.6.2.1 Category 1 methods*

Existing and well-established Category 1 empirical methods have been used widely for estimating surface settlement. The most commonly used empirical method for estimating surface settlement is that proposed by Peck (1969). This method is based on a number of field measurements and the representation of the surface settlement trough with a probability distribution, or error curve, as shown in the following equation:

$$S = S_{max} \cdot \exp\left(\frac{-x^2}{2i^2}\right) \tag{9.7}$$

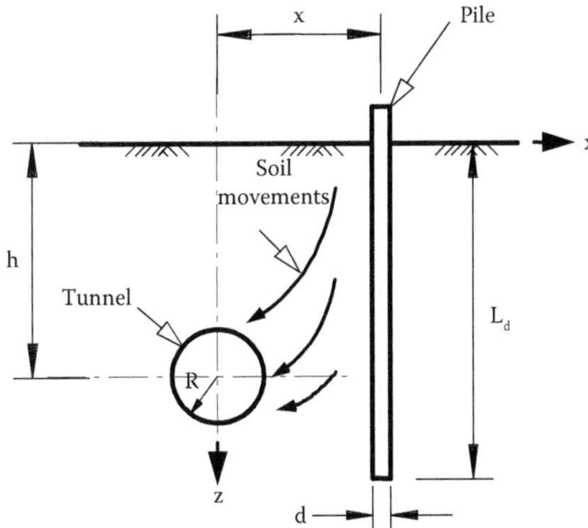


*Figure 9.36* Pile adjacent to tunnelling—the basic problem.

*Table 9.1* Recommended I values in previous studies

| *Study* | *i value* | *Remarks* |
|---|---|---|
| Peck (1969) | $\frac{i}{R}=\left(\frac{z_0}{2R}\right)^n$ : n = 0.8–1.0 | Based on field observations |
| Atkinson and Potts (1977) | $i=0.25(z_0+R)$: loose sand, $i=0.25(1.5z_0+0.5R)$: dense sand and over-consolidated clay | Based on field observations and model tests |
| O'Reilly and New (1982) | $i=0.43z_0+1.1$: cohesive soil $i=0.28z_0-0.1$: granular soil | Based on field observations in UK tunnels |
| Mair (1983) | $i=0.5z_0$ | Based on worldwide field observations and centrifuge tests |
| Attewell and Woodman (1982) | $\frac{i}{R}=\alpha\cdot\left(\frac{z_0}{2R}\right)^n$ : $\alpha=1$ and $n=1$ | Based on field observations in UK tunnels |
| Clough and Schmidt (1981) | $\frac{i}{R}=\alpha\cdot\left(\frac{z_0}{2R}\right)^n$ : $\alpha=1$ and $n=0.8$ | Based on field observations in UK tunnels |

Note: $z_0$ is the depth of tunnel below ground (at tunnel springline) and R is the tunnel radius.

where S is the surface settlement at a transverse distance (x) from the tunnel centreline; $S_{max}$ the maximum settlement at x = 0; and i the trough width parameter (location of maximum settlement gradient or point of inflexion).

A significant amount of research involving field observations and model tests has been devoted to the estimation of $S_{max}$ and the 'i' values for different ground conditions. Various estimates of 'i' values derived in previous studies are given in Table 9.1.

The maximum settlement $S_{max}$ may be estimated as follows:

$$S_{max}=\frac{0.313V_LD^2}{i} \tag{9.8}$$

where $V_L$ is the ground loss volume and D is the diameter of the tunnel.

Figure 9.37 shows a comparison of various predicted surface settlement troughs for a hypothetical 6 m diameter tunnel at a depth of 30 m. The ground loss volume ratio was assumed as 1%. It is observed that the maximum surface settlement predicted using the various methods ranges from 7 to 10 mm. The surface settlement trough width, i, varies from 7.5 to 10.4 m. This shows the variability of the results of the different empirical prediction methods. This variability is a result of the use of differing methods for the derivation of 'i' values, and can cause considerable uncertainty in tunnelling-induced risk assessment.

Relatively few empirical methods are available to predict subsurface settlement profiles. The empirical methods proposed by Mair (1993) and Atkinson and Potts (1977) are widely used in practice. It is often assumed that the shape of subsurface settlement profiles developed during tunnel construction is characterised by a Gaussian distribution, in the same manner as for surface settlement profiles (Mair, 1993).

Mair (1993) proposed the following empirical method to estimate subsurface settlement, $S_z$:

$$S_z=S_{z,max}\exp\left(\frac{-x^2}{2i_z}\right) \tag{9.9}$$

where $i_z=k(z_0-z)$ and $k=(0.175+0.325(1-(z/z_0)))/(1-(z/z_0))$

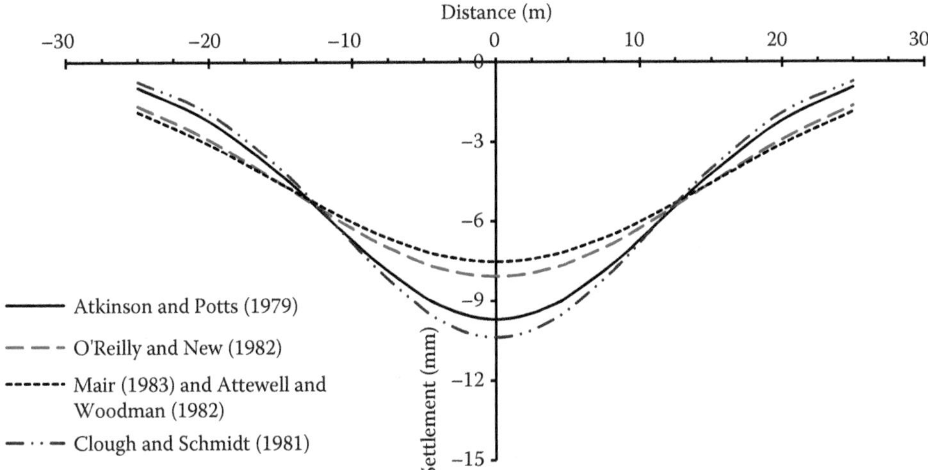


*Figure 9.37* Comparison of various empirical surface settlement troughs.

Therefore,

$$\frac{S_{z,max}}{R} = \frac{1.25V_L}{0.175 + 0.325(1 - (z/z_0))} \cdot \frac{R}{z_0} \tag{9.10}$$

Atkinson and Potts (1977) proposed the following method to estimate subsurface settlement, in shallow tunnels based on model tests:

$$\frac{S_z}{S_{z,max}} = 1.0 - \alpha\left(\frac{z - R}{2R}\right) \tag{9.11}$$

where, $\alpha = 0.57$ for dense sand, 0.40 for loose sand and 0.13 for over-consolidated clays.

$S_z$ is the settlement at depth z, and $S_{z,max}$ the maximum settlement at depth z.

Empirical methods such as those above are subject to certain important limitations in their applicability to different ground conditions and construction techniques, and provide only limited information on subsurface horizontal movements and settlements.

#### 9.6.2.2 Category 2 methods

The ground movements arising from tunnelling can be obtained conveniently from the approximate closed-form solutions published by Loganathan and Poulos (1998), which give the distributions of vertical and horizontal movements with depth and distance as functions of the volume loss and the tunnel geometric parameters. Their generalised, modified, analytical solutions for the estimation of the surface settlement, subsurface settlement and the lateral deformation are given in Equations 9.12 through 9.14, respectively:

$$U_{z=0} = \varepsilon_0 R^2 \cdot \frac{4H(1-\nu)}{H^2 + x^2} \cdot \exp\left\{-\frac{1.38x^2}{(H\cot\beta + R)^2}\right\} \tag{9.12}$$

$$U_z = \varepsilon_0 R^2 \left( -\frac{z-H}{x^2+(z-H)^2} + (3-4\nu)\frac{z+H}{x^2+(z+H)^2} - \frac{2z[x^2-(z+H)^2]}{[x^2+(z+H)^2]^2} \right) \cdot \exp\left\{ -\left[ \frac{1.38x^2}{(H\cot\beta+R)^2} + \frac{0.69z^2}{H^2} \right] \right\} \tag{9.13}$$

$$U_x = -\varepsilon_0 R^2 x \left[ \frac{1}{x^2+(H-z)^2} + \frac{3-4\nu}{x^2+(H+z)^2} - \frac{4z(z+H)}{(x^2+(H+z)^2)^2} \right] \cdot \exp\left\{ -\left[ \frac{1.38x^2}{(H\cot\beta+R)^2} + \frac{0.69z^2}{H^2} \right] \right\} \tag{9.14}$$

where $U_{z=0}$ is the ground surface settlement; $U_z$ the subsurface settlement; $U_x$ the lateral soil movement; R the tunnel radius; z the depth below ground surface; H the depth of tunnel axis level; $\nu$ the Poisson's ratio of soil; $\varepsilon_0$ the average ground loss ratio and x the lateral distance from tunnel centreline.

These equations allow rapid estimation of ground deformation and require only an estimate of the ground loss ratio and Poisson's ratio, $\nu$, of the soil. The ground strength, stiffness and elasto–plastic behaviour of the ground are considered in the estimation of the ground loss values. In most cases, tunnel excavation is carried out within the elastic strain range of the ground. The tunnelling-induced strain around the excavated face is controlled by applying the appropriate face pressure, installing the tunnel support system on time, or by improving the ground around the tunnel.

The applicability of Equations 9.12 through 9.14 has been evaluated with reference to 10 case histories (Loganathan and Poulos, 1998, 1999; Loganathan and Flanagan, 2001), together with the results obtained from three detailed centrifuge tests (Loganathan et al., 2000), and also using numerical analysis using FLAC$^{3D}$ (Loganathan et al., 2002).

Comparisons between predicted ground movements using the above equations and measured ground movements from carefully controlled laboratory centrifuge tests (Loganathan et al., 2000) show reasonable agreement, as indicated in Figures 9.38 and 9.39.

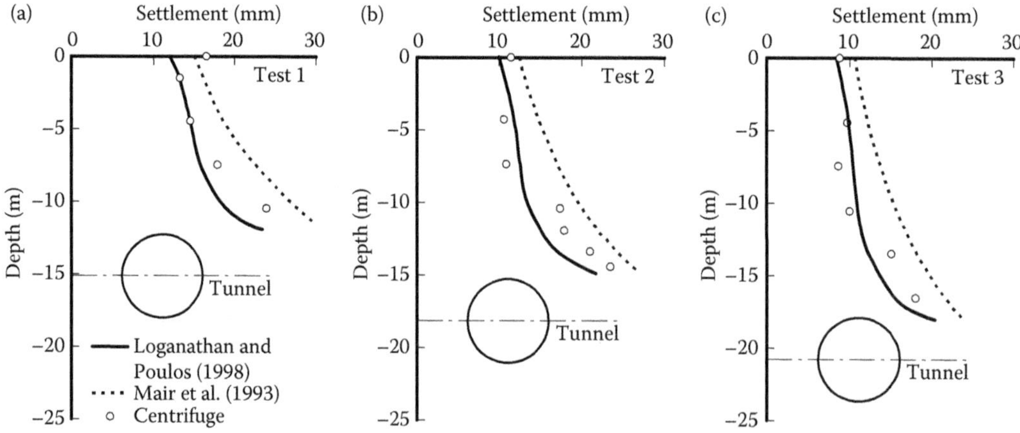


*Figure 9.38* Comparison of subsurface settlement profiles: (a) Test 1, (b) Test 2 and (c) Test 3. (Adapted from Loganathan, N., Poulos, H.G. and Stewart, D.P. 2000. *Geotechnique*, 50(3): 283–294. Courtesy of ICE Publishing.)

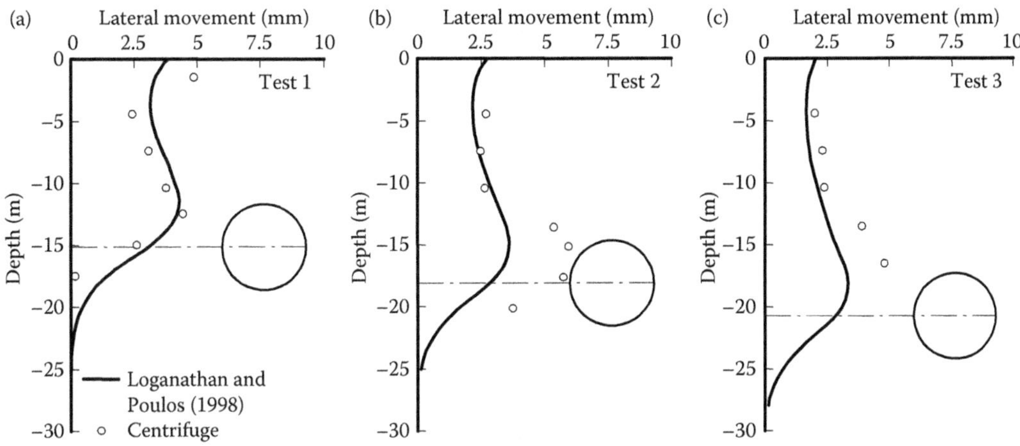


*Figure 9.39* Comparison of subsurface lateral soil movement profiles: (a) Test 1, (b) Test 2 and (c) Test 3. (Adapted from Loganathan, N., Poulos, H.G. and Stewart, D.P. 2000. *Geotechnique*, 50(3): 283–294. Courtesy of ICE Publishing.)

## 9.6.3 Pile–Soil interaction and characteristics of behaviour

### *9.6.3.1 Single piles*

To better understand the general characteristics of behaviour of piles affected by tunnelling, a series of analyses has been carried out by Chen et al. (1999). The response of a pile to vertical and lateral ground movements has been analysed via computer programmes based on a simplified boundary-element analysis. The simple case of a pile in a homogeneous clay layer has been considered, with the following assumptions being made:

- Undrained shear strength = 60 kPa, Young's modulus = 24 MPa.
- Tunnel radius = 3 m, depth of tunnel axis = 20 m.
- Pile diameter = 0.5 m, ultimate shaft friction = 48 kPa, ultimate end bearing pressure = ultimate lateral pile–soil pressure = 540 kPa; pile is precast with 2.5% steel in the top half and 1% in the bottom half.
- Pile lengths of 15 m (short pile), 20 m (medium pile) and 25 m (long pile) have been analysed.
- Ground loss ratios of 1% (a common design value) and 5% (an extreme value) are considered.

Figure 9.40 shows the computed response of the pile to the tunnelling-induced ground movements, together with the computed ground movements themselves. The following points can be noted:

- The lateral pile deflections are similar to the ground movements
- The bending moment profile has a double curvature, with the maximum occurring just above the level of the tunnel axis
- The pile settles relatively uniformly along the whole pile shaft, but the pile head movement is less than the maximum soil movement (which occurs near the crown of the tunnel)
- Both compressive and tensile forces are induced in the pile, with the compressive force in this case being larger

- Both the axial forces and the bending moments are significant in relation to the structural strength of the pile, and for the 5% volume loss, the bending moment exceeds the allowable value for the pile
- In general, the pile response decreases as the distance of the pile from the tunnel decreases, although the axial compressive force reaches a maximum at a distance of about 10 m from the tunnel in the case of 5% volume loss. The effects of tunnelling on axial response are found to extend to a greater distance than is the case with lateral response

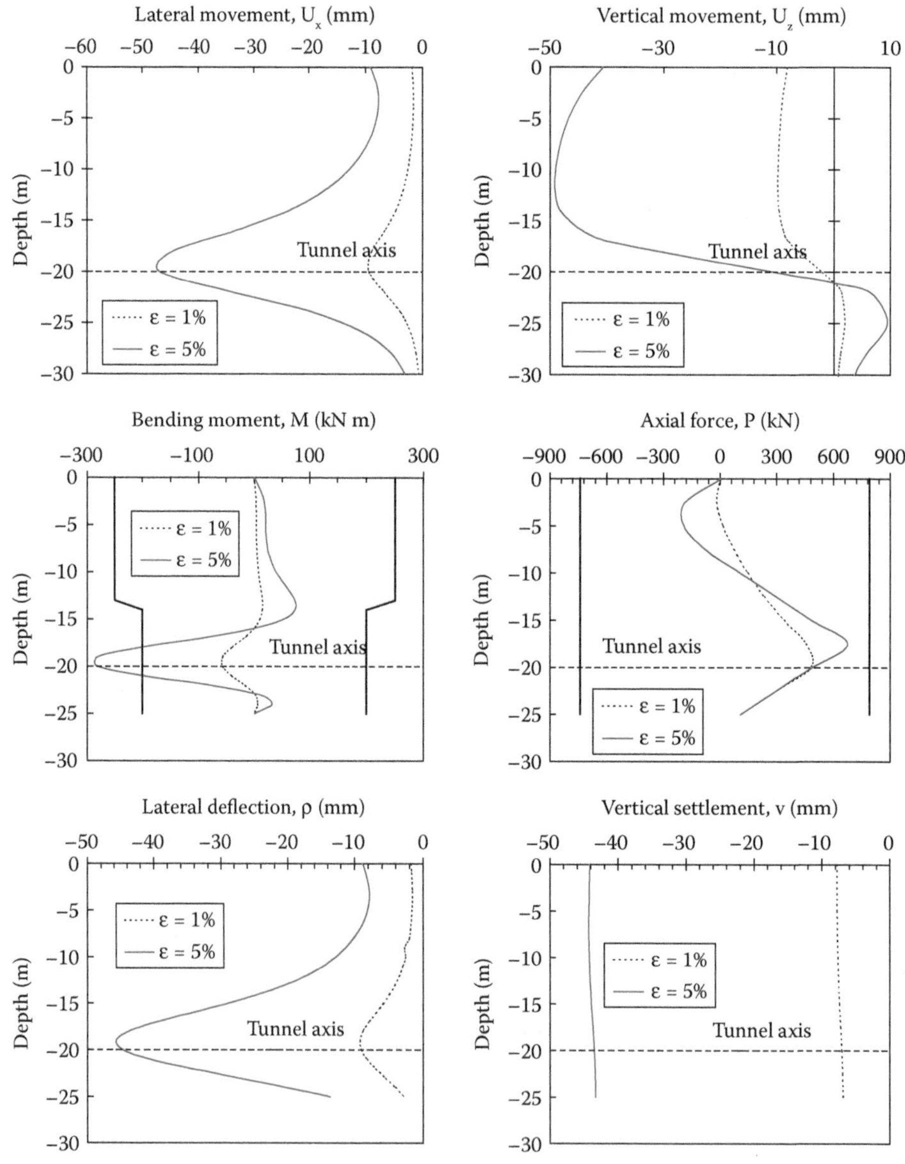


*Figure 9.40* Typical soil movements and pile response at x = 4.5 m for long pile case ($L_P$ = 25 m). (Adapted from Chen, L.T., Poulos, H.G. and Loganathan, N. 1999. *Journal of Geotechnical and Geoenvironmental Engineering, ASCE*, 125(3): 207–215. Courtesy of ASCE.)

#### 9.6.3.2 *Groups versus single piles*

Loganathan et al. (2001) have examined the differences between the response of a single isolated pile and a pile within a group, when they are subjected to tunnelling-induced ground movements. It has been found that, for piles at the same distance from the tunnel, there is relatively little difference between the isolated pile and group pile responses, for both axial and lateral responses. In general, the axial forces and bending moments in the isolated pile are larger than in the group pile, again reflecting the beneficial effects of pile–soil–pile interaction. Some differences however occur near the head of the group pile because of the restraint provided by the pile cap (assumed rigid in these analyses). Nevertheless, it would appear reasonable to consider a single isolated pile when designing piles to resist the effects of tunnelling-induced ground movements.

### 9.6.4 Design charts

As with the case of piles near an excavation, it has been found possible to develop approximate design charts which take into account the main parameters affecting pile response. Within the range of parameters examined by Chen et al. (1999), the following approximate expressions for pile response were derived:

Maximum moment:

$$M_{max} = M_b \cdot K_{cu}^M \cdot K_d^M \cdot K_{Lp}^M \tag{9.15}$$

Maximum lateral deflection:

$$\rho_{max} = \rho_b \cdot K_{cu}^\rho \cdot K_d^\rho \cdot K_{Lp}^\rho \tag{9.16}$$

Maximum compressive axial force:

$$P_{max} = +P_b \cdot K_{cu}^{+P} \cdot K_d^{+P} \cdot K_{Lp}^{+P} \tag{9.17}$$

Maximum tensile force:

$$-P_{max} = -P_b \cdot K_{cu}^{-P} \cdot K_d^{-P} \cdot K_{Lp}^{-P} \tag{9.18}$$

Maximum settlement:

$$v_{max} = v_b \cdot K_{cu}^v \cdot K_d^v \cdot K_{Lp}^v \tag{9.19}$$

where $M_b$, $\rho_b$ etc. are the 'basic' values for the standard set of parameters chosen, plotted in Figure 9.41; $K_{cu}^M$ etc. correction factors for undrained shear strength, plotted in Figure 9.42; $K_d^M$ etc. correction factors for pile diameter, plotted in Figure 9.43; $K_{Lp}^M$ etc. correction factors for relative pile length, plotted in Figure 9.44.

It must be emphasised that the application of the above equations is approximate only, as the superposition of the effects of the various parameters via multiplication is not strictly valid. In addition, these charts give only the effect of the tunnelling, to which must be added any other effects of applied loading. Finally, the charts apply only for single piles, with no account being taken of interaction among piles in a group.

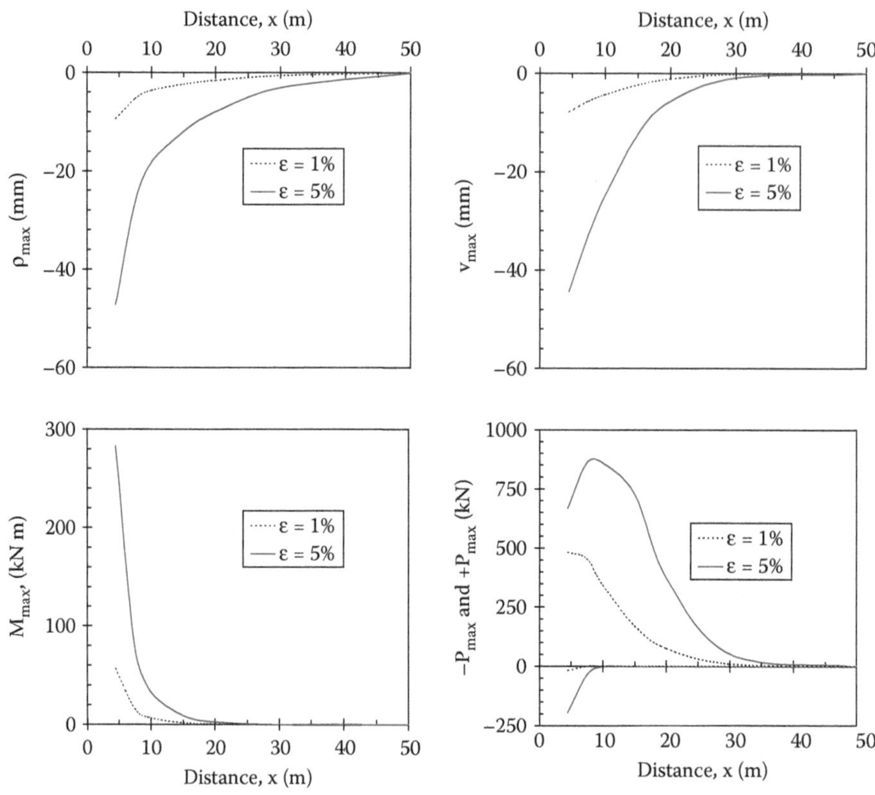


*Figure 9.41* Basic maximum pile responses versus distance x, for long pile ($L_P = 25$ m). (Adapted from Chen, L.T., Poulos, H.G. and Loganathan, N. 1999. *Journal of Geotechnical and Geoenvironmental Engineering, ASCE*, 125(3): 207–215. Courtesy of ASCE.)

### 9.6.5 Example of application: Case history in London

Lee et al. (1994) described a case involving the construction of a tunnel for the Angel Underground Station in London. As shown in Figure 9.45, the tunnel was driven between pile foundations supporting a seven-storey building with a two-storey basement, the tunnel axis line being about 5.7 m from the nearest piles. The tunnel was excavated using hand tools in two stages, the first being a pilot tunnel of 4.5 m diameter and the second an enlargement to 8.25 m diameter. Measured ground loss ratios were approximately 1.5% for the pilot tunnel and 0.5% for the tunnel enlargement (Mair, 1993). The piles were driven through 28 m of London Clay to the underlying Woolwich and Reading beds. Ground investigation data showed that the average undrained shear strength of the London Clay increased linearly from about 50 kPa at the top to about 220 kPa at the bottom. Inclinometers were installed at various locations to measure the lateral soil movement, and within some piles to measure lateral pile deflections. Measured data showed that some of the piles had moved laterally toward the tunnel by about 10 mm when the tunnelling operation was complete.

The computed lateral pile deflection profiles were obtained using a boundary-element programme, in conjunction with ground movements estimated from Equations 9.13 and 9.14, and these are shown in Figure 9.46. Also shown are the measured values and those predicted from a finite element analysis by Lee et al. (1994). The agreement between the

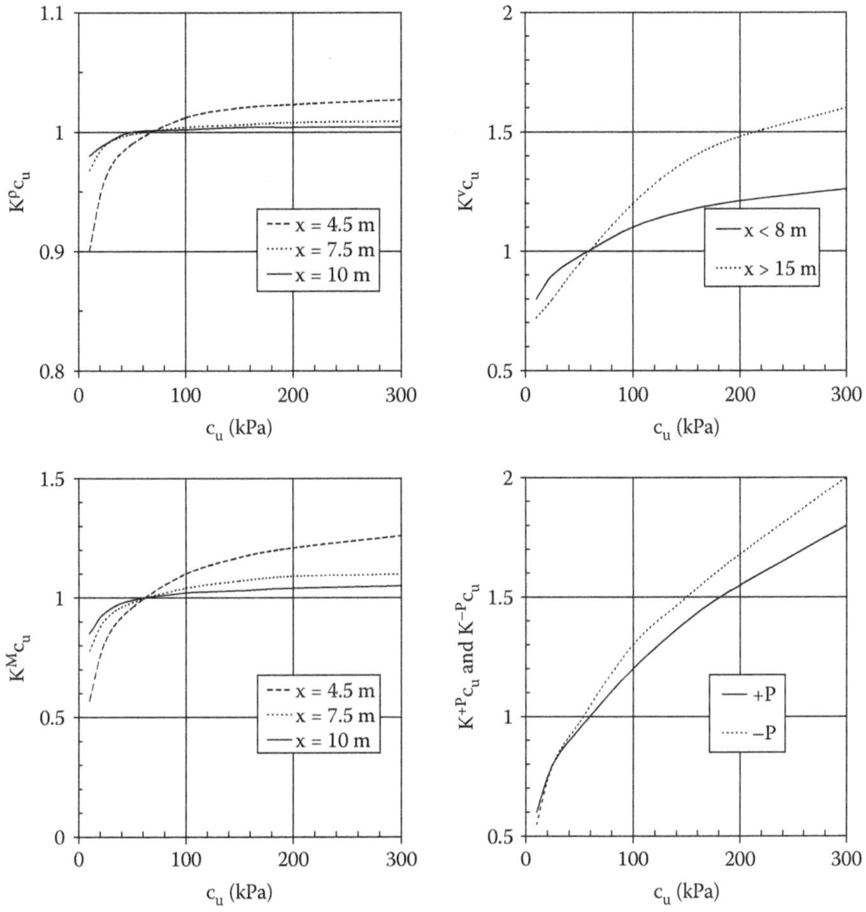


*Figure 9.42* Correction factors for undrained shear strength. (Adapted from Chen, L.T., Poulos, H.G. and Loganathan, N. 1999. *Journal of Geotechnical and Geoenvironmental Engineering, ASCE,* 125(3): 207–215. Courtesy of ASCE.)

computed final profile and the measured profile is good, although the predicted maximum lateral deflection occurred slightly above the location of the measured maximum deflection, which was at the tunnel centreline.

Chen et al. (1999) have also used the design charts shown in Figures 9.41 through 9.44 to estimate the maximum lateral pile deflection. Calculations were carried out for both the pilot tunnel and the enlargement, for the respective ground loss values measured. The predicted maximum lateral deflections were 4.5 mm for the pilot tunnelling stage and 5.5 mm for the enlargement stage, giving a total maximum lateral of 10 mm. This value agrees well with the measured value shown in Figure 9.46.

## 9.7 CATEGORY 3 ANALYSES

The finite element method has been used widely for detailed analyses of the process of tunnelling and the consequent ground movements, for example, Rowe and Lee (1992), Komiya et al. (1999), Moller and Vermeer (2008), Moldovan and Popa (2012), Likitlersuang et al.

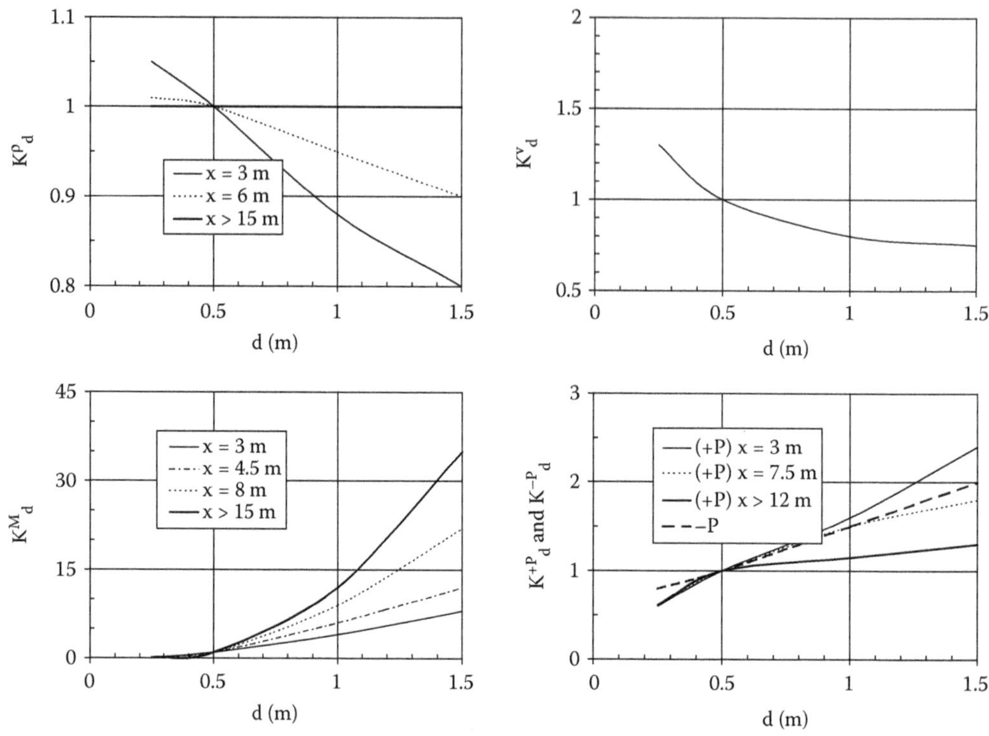


*Figure 9.43* Correction factors for diameter d. (Adapted from Chen, L.T., Poulos, H.G. and Loganathan, N. 1999. *Journal of Geotechnical and Geoenvironmental Engineering, ASCE*, 125(3): 207–215. Courtesy of ASCE.)

(2014). Both 2D and 3D analyses are now used relatively routinely, although 2D analyses continue to be the predominant choice.

One of the decisions that needs to be made in the finite element analyses is the method by which the tunnel excavation and associated ground loss due to tunnelling is to be represented. Among the possible methods are the following, as described by Moldovan and Popa (2012) and Likitlersuang et al. (2014):

1. The gap method (Rowe and Lee, 1992), in which a pre-defined void is introduced into the finite element mesh, representing the total expected ground loss.
2. The convergence method, which is implemented either by gradually reducing the material stiffness inside the periphery of the tunnel lining (stiffness reduction approach) or reducing the pressure within the tunnel periphery.
3. The volume loss control method, in which the volume loss that is assessed to occur on the completion of excavation is prescribed.
4. The progressive softening method, which can be used to model a sequential excavation such as the New Austrian Tunnelling Method (NATM), but with the material stiffness being reduced progressively for each part of the tunnel, the top, the bench and the invert, individually.
5. The stress reduction method, in which an 'unloading factor' β is used to take into account the 3D tunnelling effects within a 2D analysis. Three steps are involved:
   a. The initial support pressure acting on the tunnel periphery is calculated
   b. The periphery pressure is reduced by a factor β to allow the surrounding soil to deform

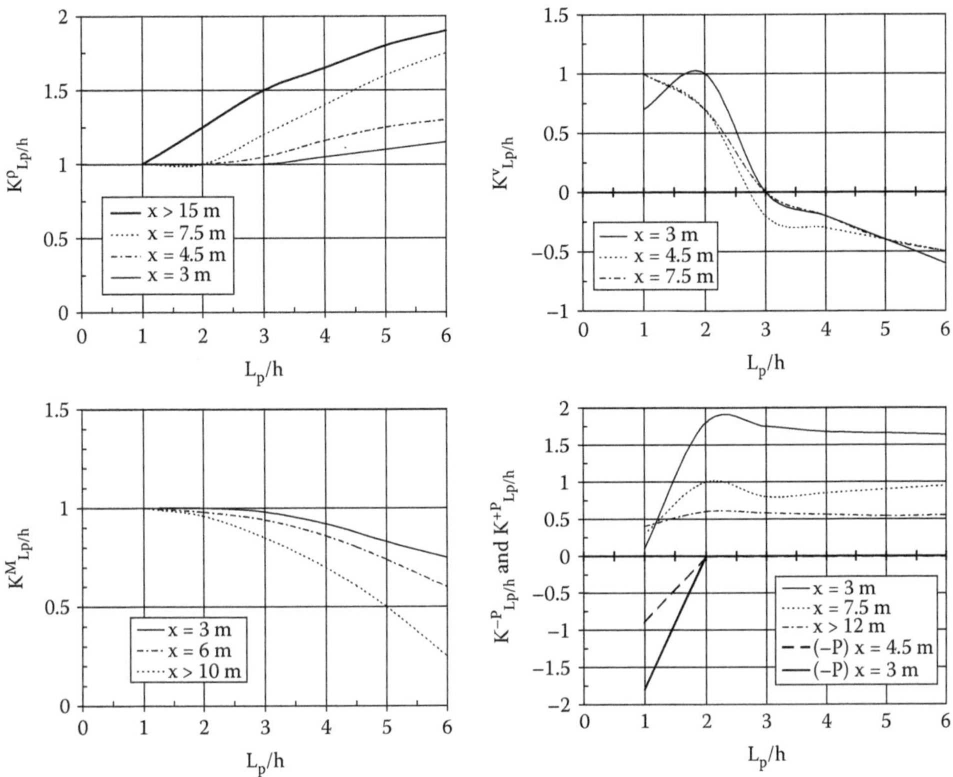


*Figure 9.44* Correction factors for $L_p/h$. (Adapted from Chen, L.T., Poulos, H.G. and Loganathan, N. 1999. *Journal of Geotechnical and Geoenvironmental Engineering, ASCE*, 125(3): 207–215. Courtesy of ASCE.)

   c. The soil elements within the tunnel periphery are deactivated, while the tunnel lining is activated
6. The modified grout pressure method, in which a gap between the tunnel lining and the soil is modelled, and where the following analysis sequence occurs:
   a. The soil elements within the tunnel are deactivated, and at the same time, the face pressure (representing the slurry pressure) is applied to the entire tunnel boring machine (TBM) section
   b. The tunnel lining is activated and the grouting pressure is applied to the physical gap area
   c. The grout pressure is removed, and the physical gap is replaced by the hardened grout material

Likitlersuang et al. (2014) found that the latter three methods gave comparable values of ground movement which matched measured movements reasonably well. They also concluded that simplified 2D modelling can be used to model 3D problems of tunnelling-induced ground movements, with adequate accuracy. Vermeer et al. (2003) have demonstrated how 'smart' 3D analysis can be used to develop appropriate values of the unloading factor β to be used in a 2D analysis.

An important factor in applying a Category 3 finite element method to tunnelling problems is the soil model adopted in the analysis. Simple elastic and elasto-plastic models are

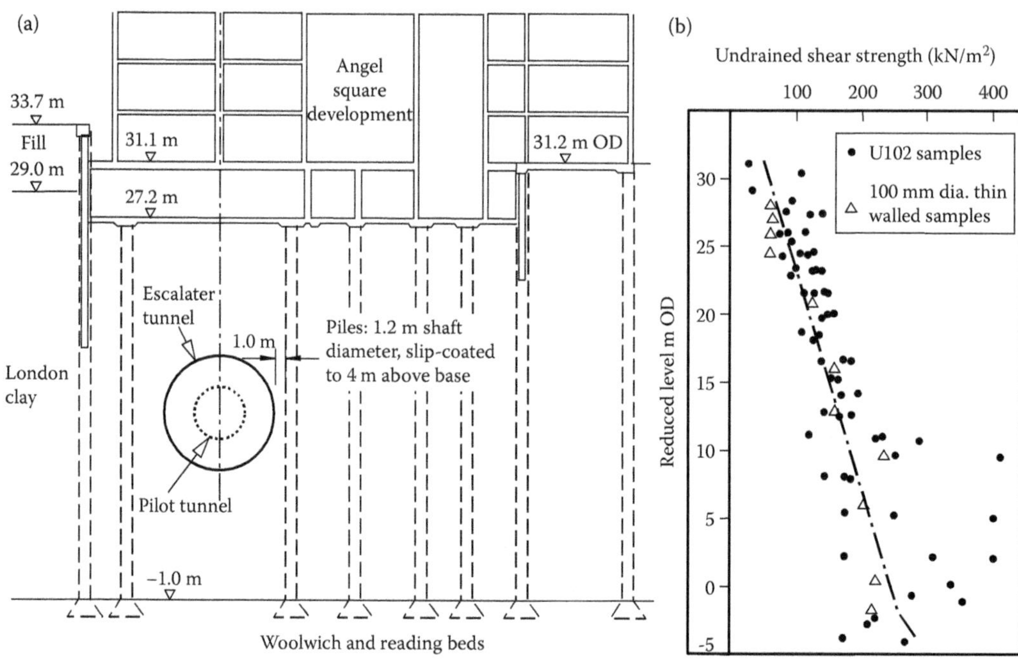


*Figure 9.45* Case history—Angel Islington Station: (a) section through Angel Escalator Tunnel and building foundations and (b) undrained shear strength (After Mair, R.J., Taylor, R.N. and Bracegirdle, A. 1993. *Geotechnique*, 43: 315–320. Lee, R.G., Turner, A.J. and Whitworth, L.J. 1994. Deformations caused by tunneling beneath a piled structure. *Proceedings of the XIII International Conference on Soil Mechanics and Foundation Engineering*, New Delhi, University Press, London, pp. 873–878.)

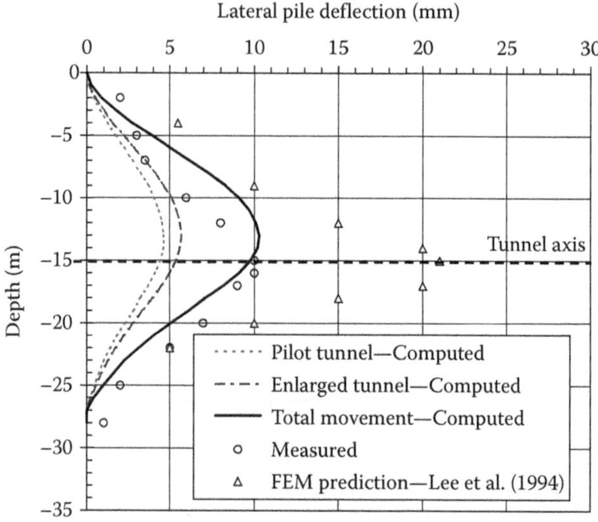


*Figure 9.46* Lateral pile deflection for case history. (Chen, L.T., Poulos, H.G. and Loganathan, N. 1999. *Journal of Geotechnical and Geoenvironmental Engineering, ASCE*, 125(3): 207–215. Courtesy of ASCE.)

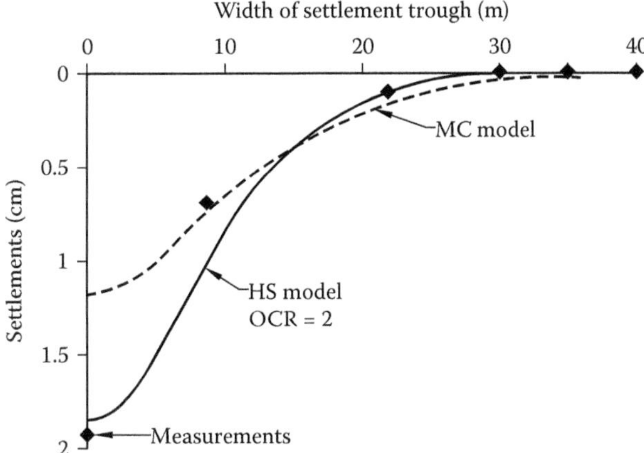


*Figure 9.47* Observed and computed settlement trough of Steinhaldenfeld tunnel. (Adapted from Vermeer, P.A., Möller, S.C. and Ruse, N. 2003. On the application of numerical analysis in tunnelling. *Post Proceedings 12th Asian Regional Conference on Soil Mechanics and Geotechnical Engineering (12 ARC)*, Singapore, 4–8 August, Vol. 2, pp. 1539–1549.)

generally not adequate and more advanced models such as those involving plastic hardening and strain-dependent soil stiffness are required. An example of the difference in results between using a simple Mohr Coulomb (MC) model and a Hardening Soil (HS) model is shown in Figure 9.47 (Vermeer et al., 2003). The HS model gives results for the settlement trough that are in better agreement with measured settlements than are the MC results.

Pile foundations can be incorporated into 2D and 3D analyses, for example, as demonstrated by Mroueh and Shahrour (2002), among several others.

Chapter 10

# Design for dynamic loadings

## 10.1 INTRODUCTION

Dynamic loadings are generally applied to a building via wind and seismic actions. This chapter will deal primarily with dynamic response to wind loadings, while seismic loadings will be considered in Chapter 11. However, the general principles of dynamic foundation response are applicable to both sources of loading.

Wind loadings are dynamic and random phenomena in both time and space. Observations show that the wind speed can be described by a mean value on which random fluctuations (gusts) are superimposed, and that the wind speed increases with height above ground. The response of a building to wind loading depends to a considerable extent on its fundamental natural frequency of vibration (see Chapter 2). The dynamic response of relatively low-rise buildings is not likely to be significant, but high-rise buildings are dynamically sensitive. The ASCE Standard ASCE 7-05 classifies a structure as dynamically sensitive if it is 'flexible' and has a natural frequency $f_n < 1$ Hz, that is, a natural period $T_n > 1.0$s. Thus, tall buildings in excess of about 50 m in height will fall into this category and require consideration of their dynamic response. This means that the natural frequency and mode shapes of the first few modes of vibration will need to be calculated, and these in turn will be influenced by the dynamic characteristics of the foundation system. Thus, the stiffness and damping of the foundation system, which determine the dynamic foundation response, are of considerable importance in tall building design.

This chapter will deal with methods for estimating stiffness and damping for foundation systems. Some basics of dynamic response will be reviewed first, and then most of the remainder of the chapter will deal with Category 2 methods of computing foundation stiffness and damping. Detailed information on foundation dynamics can be obtained from a number of sources, for example, Richart, Hall and Woods (1970), Gazetas (1991), Wolf (1994), El Naggar (2001). A relatively simplified approach will be adopted here, such that geotechnical designers without a deep grounding in engineering dynamics should still be able to undertake the necessary calculations. A brief review of some Category 3 methods will be given at the end of the chapter.

## 10.2 DYNAMIC LOADINGS

There are six modes of vibration that may be excited, as illustrated in Figure 10.1. For tall buildings, given that lateral wind loading provides a major source of dynamic loading, the most important modes of vibration are likely to be the horizontal and rotational modes, including the torsional modes.

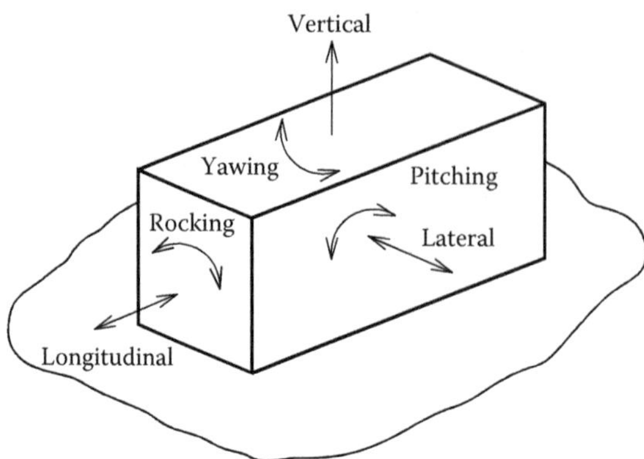


*Figure 10.1* Modes of vibration.

Boggs and Dragovich (2006) point out that the dominant wind gust wavelengths are large compared to the dimensions of most buildings, and as a consequence, it is often necessary to consider only the lowest mode of vibration when considering dynamic response of a structure to wind. In contrast, with earthquakes, the dominant excitation energy is in the frequency range of low-rise buildings or the higher modes of tall buildings, and so a large number of modes must generally be considered when evaluating the dynamic building response to earthquakes.

Seismic activity generates dynamic loading which is transmitted through the ground to the foundation–structure system. This will generate inertial forces in the structure that will be transmitted to the foundation, but also kinematic movements within the supporting ground which will in turn induce additional forces and bending moments in the foundation system. These issues will be addressed in Chapter 11.

## 10.3 SOME BASIC ASPECTS OF DYNAMIC RESPONSE

### 10.3.1 Dynamic response curves

All physical systems composed of material possessing mass and elasticity are capable of vibrating at a characteristic natural frequency. The response of a system to dynamic or vibratory forces depends to a large degree on the relationship between these forces and the natural frequency of the system. For a relatively simple system subjected to relatively simple (generally periodic) forms of loading, the objective of a dynamic analysis is to obtain the relationship between the system response (displacement, velocity or acceleration) and the frequency and magnitude of the loading. Such a relationship is termed a response curve and is illustrated diagrammatically in Figure 10.2. The maximum system response generally occurs when the frequency of the dynamic force is at or near the natural frequency of the system. Thus, for foundation–soil systems, one of the main objectives of a dynamic analysis is to estimate the system stiffness and damping and its response to the imposed loading.

### 10.3.2 Single degree of freedom systems

The simplest mathematical model of a dynamic system is the single degree of freedom (SDOF) system which is illustrated in Figure 10.3. This system consists of a mass, a spring

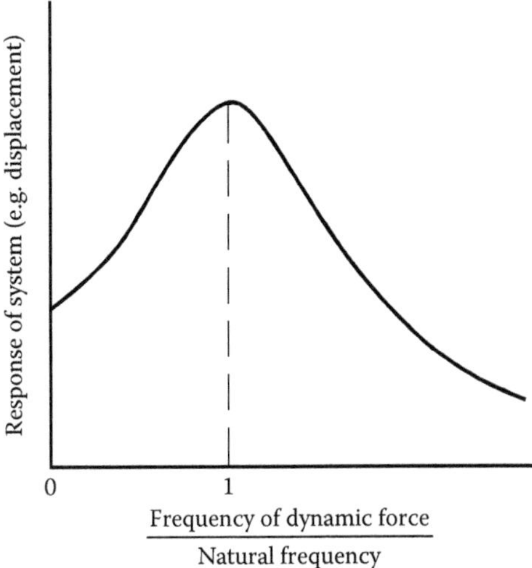


*Figure 10.2* Response curve for SDOF motion.

and a dashpot. When a time-dependent normal force F(t) acts on the system, its response is governed by the following basic differential equation:

$$md^2z/dt^2 + cdz/dt + kz = F(t) \qquad (10.1)$$

where m is the mass, c the damping constant (with units of mass/time), k spring stiffness (with units of force/length or mass/time$^2$), z the displacement and t the time.

The following basic characteristics of the system are as follows:

- Natural frequency:

  $$f_n = (k/m)^{0.5}/2\pi \qquad (10.2)$$

  It can be noted that the circular natural frequency $\omega_n$ is related to $f_n$ as $\omega_n = 2\pi f_n$.

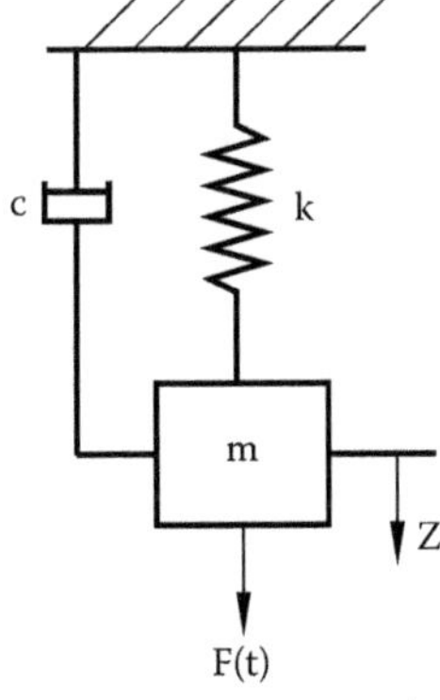


*Figure 10.3* SDOF system.

- Critical damping of system:

$$c_c = 2(km)^{0.5} \tag{10.3}$$

- Damping ratio:

$$\zeta = c/c_c \tag{10.4}$$

- Magnification factor:

$$M = \text{dynamic/static response} \tag{10.5}$$

- For constant force excitation:

$$\text{Maximum magnification factor: } M_{max} = 1/[2\zeta(1-\zeta^2)^{0.5}] \tag{10.6}$$

- Amplitude of motion $= F/[(K - m\omega^2)^2 + \omega^2 c^2]^{0.5}$ (10.7)

  where F is the amplitude of force.

It is also possible to express the damping ratio in a more convenient form which incorporates the stiffness, k, and excludes the mass m, that is,

$$\zeta = c \cdot f_n \pi/k = c \cdot \omega_n/2k \tag{10.8}$$

When a rotational form of loading is applied, the relevant equation of motion is modified to

$$I_m d^2\theta/dt^2 + c_\theta d\theta/dt + k_\theta \theta = M(t) \tag{10.9}$$

where $I_m$ is the mass moment of inertia about the appropriate axis and passing through the centre of gravity, $c_\theta$ the rotational damping constant, $k_\theta$ the rotational stiffness, $\theta$ the rotation and M(t) the time-dependent moment loading.

In this case:

- Rotational natural frequency:

$$f_\theta = (k_\theta/I_m)^{0.5}/2\pi \tag{10.10}$$

- Rotational critical damping:

$$c_{c\theta} = 2(k_\theta I_m)^{0.5} \tag{10.11}$$

- Rotational damping ratio:

$$\zeta_\theta = c_\theta/c_{c\theta} = c_\theta f \pi/k_\theta \tag{10.12}$$

- Amplitude of motion:

$$M/[(k_\theta - I_m\omega^2)^2 + \omega^2 c_\theta{}^2]^{0.5} \tag{10.13}$$

where M is the amplitude of applied moment loading.

Solutions to Equations 10.1 and 10.9 may be obtained analytically or numerically, depending on the nature of F(t). For constant force harmonic loading, $F = F_o \sin(\omega t)$, where $\omega$ is the circular frequency and $F_o$ the force amplitude, the response curves derived from solution of Equation 10.1 are shown in Figure 10.4.

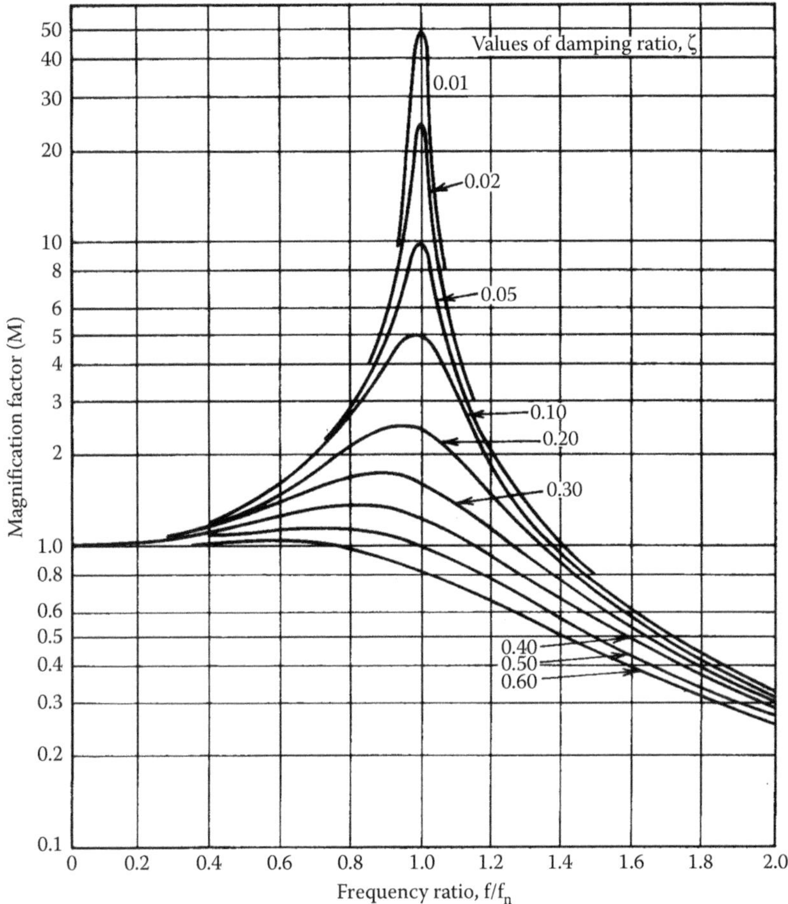


*Figure 10.4* Response curves for SDOF system, with constant amplitude excitation.

The key features of the response curves in Figure 10.4 are

- The magnification factor M can become very large when the frequency of loading is at or near the natural frequency
- If the damping ratio is very small, for example, less than 0.05, the magnification factor can be very large
- The magnification factor decreases markedly as the damping ratio D increases

Thus, to reduce the dynamic response of a system, benefit will ensue from keeping the frequency of loading well away from the natural frequency and increasing the damping ratio of the system.

### 10.3.3 Two-degree of freedom systems

If translational and rocking motions occur simultaneously, for example horizontal wind shear loading together with wind moment loading, two equations of motion need to be considered simultaneously, one describing translational motion (Equation 10.1) and the other describing rotational loading (Equation 10.9). In this case, there are two 'peaks' in

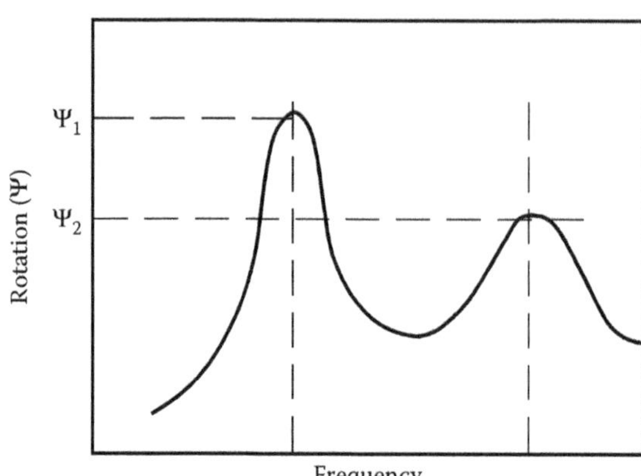


*Figure 10.5* Response curve for two-degree-of-freedom motion (e.g. horizontal and rocking).

the response curve, as shown in Figure 10.5, one each for the translational and rotational modes.

The resonant frequencies (ω) and amplitudes (A) of vibration of the foundation in coupled horizontal and rocking motion can be obtained from the following equations given by Beredugo and Novak (1972) and El Naggar (2001):

$$\omega_i^2 = 0.5(k_h/m + k_\theta/I_m) \pm [0.25(k_h/m - k_\theta/I_m)^2 + k_{h\theta}^2/(mI_m)]^{0.5} \tag{10.14}$$

$$A_i = \left(k_\theta - I_m\omega_i^2\right)/(-k_{h\theta}) \tag{10.15}$$

where i = 1,2 (for horizontal and rotation), $k_h$ is the horizontal stiffness, $k_\theta$ the rotational stiffness, $k_{h\theta}$ the horizontal-rotational cross-stiffness, m the mass and $I_m$ the mass moment of inertia.

### 10.3.4 Lumped parameter models for foundation design

From a practical engineering design viewpoint, it is convenient to replace the foundation system by an equivalent lumped parameter model which is characterised by three properties:

1. Mass (for translational modes of vibration) or inertia (for translational modes)
2. Stiffness
3. Damping

Figure 10.6 illustrates a typical lumped parameter system for vertical, horizontal and rotational vibrations. Each mode will have properties that will generally be different. How the characteristics of the lumped parameter models can be estimated is set out in the remainder of this chapter.

### 10.3.5 Mass and inertia of lumped system

The mass of the models can usually be taken as the mass of the supported structure plus that of the foundation. In some earlier sources, an allowance has also been suggested for an

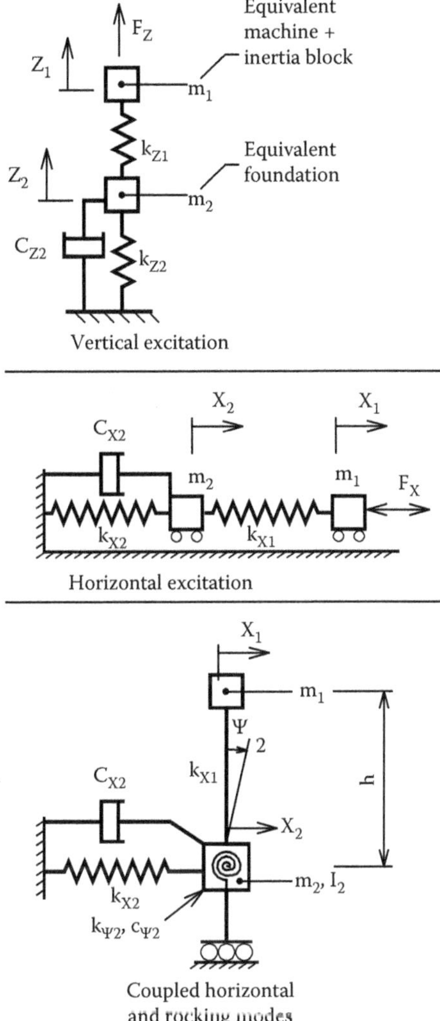


*Figure 10.6* Lumped parameter model of structure and foundation system for various modes of vibration.

additional mass of soil participating in the vibration process, but this may not be significant for relatively heavy structures.

To estimate the inertia, a foundation system in many cases can be approximated by a prism whose shape is representative of the foundation footprint. For convenience, the cases of a rectangular prism and a circular prism are shown in Figure 10.7, together with the expressions for their mass moment of inertia. In each case, m is the mass of the prism.

### 10.3.6 Stiffness

The stiffness of the lumped system will depend on the type of foundation and the properties of the soil in which the foundation is located. Solutions will be presented in Sections 10.5–10.9 for various foundation types.

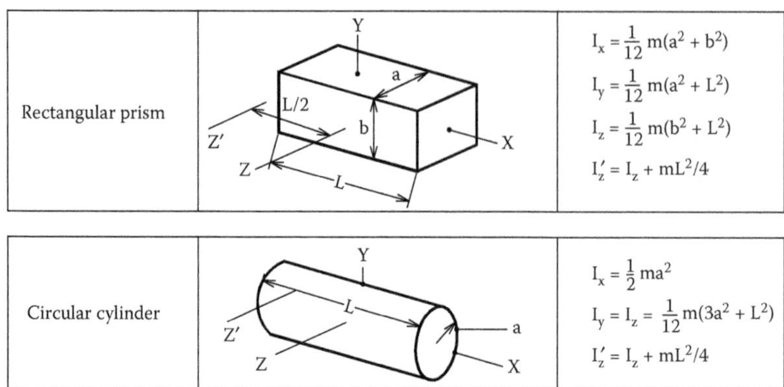


*Figure 10.7* Mass moment of inertia for rectangular and circular prisms.

### 10.3.7 Damping

There are two sources of damping within a lumped parameter system:

- Radiation (or geometric) damping, which is a consequence of the dissipation of energy from the foundation into the surrounding soil. This damping depends on the nature of the foundation, its mass and inertia characteristics, and the type of loading.
- Internal damping of the ground in which the foundation is located. The internal damping is a measure of energy lost as a result of hysteresis effects, and will depend on the nature of the ground and on the level of strain (or stress) which is imposed on the ground by the foundation.

The damping ratio for a lumped system can be taken to be the sum of the damping ratio from each component. Then, internal damping can be incorporated into the overall damping, $C_{tot}$, of the foundation system via the following expression:

$$C_{tot} = C_{rad} + 2K\zeta_i/\omega \tag{10.16}$$

where $C_{rad}$ = radiation damping coefficient (the imaginary component of the dynamic impedance), K = dynamic stiffness (the real component of the dynamic impedance), $\zeta_i$ = internal damping ratio of the soil, $\omega$ = circular frequency of loading.

When the frequency of loading is less than the natural frequency of the soil profile, the radiation damping can be ignored, and only the internal damping contributes to the overall foundation damping.

For translational modes and relatively deep soil layers, the radiation damping will generally dominate, but with rotational modes, the radiation damping ratio may be small and most of the damping may be derived from material damping. The selection of material damping values, and values of soil stiffness, are discussed in Section 10.5.

## 10.4 FOUNDATION RESPONSE TO COMBINED LOADING

The response of a foundation system to combined loading can be expressed by the following expression:

$$\{u\} = [K]^{-1}\{F\} \tag{10.17}$$

where $\{u\} = 1 \times 6$ vector of displacements and rotations; $[K] = 6 \times 6$ stiffness matrix; $\{F\} = 1 \times 6$ matrix of applied forces and moments.

For a foundation system that can be simplified to a circular shape, symmetry enables the stiffness matrix to be reduced to a $4 \times 4$ matrix, and the foundation response can then be expressed via the following equation:

$$\begin{bmatrix} \rho_v \\ \rho_h \\ \theta \\ \psi \end{bmatrix} = \begin{bmatrix} K_z & 0 & 0 & 0 \\ 0 & K_{hh} & K_{hm} & 0 \\ 0 & K_{\theta h} & K_{\theta m} & 0 \\ 0 & 0 & 0 & K_T \end{bmatrix}^{-1} \begin{bmatrix} V \\ H \\ M \\ T \end{bmatrix} \tag{10.18}$$

where $\rho_v$ is the vertical movement, $\rho_h$ the horizontal movement, $\theta$ the rotation (rocking), $\psi$ the torsional rotation, $K_z$ the vertical stiffness for vertical load, $K_{hh}$ the horizontal stiffness for horizontal load, $K_{hm}$ the horizontal stiffness for moment load, $K_{\theta h}$ the rocking rotational stiffness for horizontal load, $K_{\theta m}$ the rocking rotational stiffness for moment load, $K_T$ the torsional stiffness for torsional load, V the applied vertical load, H the applied horizontal load, M the applied moment loading and T the applied torsion loading.

For linear system response, $K_{hm} = K_{\theta h}$, and so only four independent values of stiffness need to be computed.

For static loading, the elements of the stiffness matrix are all real numbers, but for dynamic loading, where damping is also involved, the stiffness matrix elements are complex. For example, for vertical loading in the z-direction, the elements are of the following form:

$$KI = K_z + i\omega C_z \tag{10.19}$$

where KI is the complex stiffness (generally termed the impedance), $K_z$ the dynamic stiffness; $\omega$ the circular frequency, $C_z$ the dashpot coefficient and $i = \sqrt{(-1)}$.

Estimation of the dynamic response requires assessment of the elements of $K_z$ and $C_z$, which are, in general, functions of the frequency. The concept of dynamic impedance is discussed further in Section 10.8.3.

For combined horizontal and moment loading, the overall horizontal and rotational stiffness values can be conveniently expressed in terms of the components $K_{hh}$, $K_{hm}$, $K_{\theta h}$ and $K_{\theta m}$ as follows:

$$K_h = \left[K_{hh} \cdot K_{\theta m} - K_{hm}^2\right] / \left[K_{\theta m} - e \cdot K_{hm}\right] \tag{10.20}$$

$$K_\theta = \left[K_{hh} \cdot K_{\theta m} - K_{hm}^2\right] / \left[K_{hh} - K_{hm}/e\right] \tag{10.21}$$

where $K_h$ is the overall horizontal stiffness matrix, $K_\theta$ the overall rotational stiffness matrix and $e = M/H$ the ratio of applied moment to applied horizontal load in the relevant direction.

This chapter will focus on methods of estimating the various stiffness and radiation damping values.

## 10.5 SELECTION OF SOIL PARAMETERS FOR DYNAMIC FOUNDATION DESIGN

In assessing the dynamic response of foundation systems, the key geotechnical parameters are the stiffness and damping of the soils (and/or rocks) in which the foundations are located.

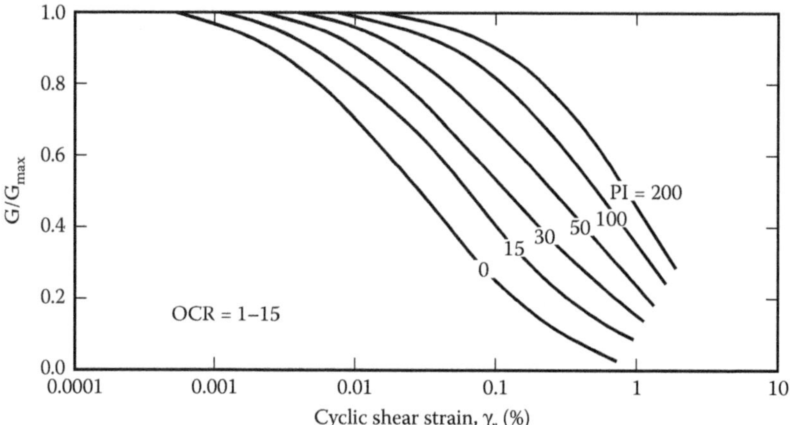


*Figure 10.8* Shear modulus reduction curves for fine-grained soils. (Adapted from Vucetic, M. and Dobry, R. 1991. *Journal of Geotechnical Engineering, ASCE*, 117(1): 89–107. Courtesy of ASCE.)

For such applications, the soil stiffness is generally characterised by the shear modulus while the damping is characterised by the internal or material damping ratio. Both the shear modulus and damping are dependent strongly on the level of shear strain in the soil. Hardin and Drnevich (1972) provided data on the dependency of modulus and damping on strain level, number of cycles of loading, frequency of loading and effective stress. Subsequently, Vucetic and Dobry (1991) have developed design curves, examples of which are shown in Figures 10.8 and 10.9 for fine-grained soils. The shear modulus relationship is expressed in terms of the small-strain shear modulus $G_o$, which can be measured via in situ testing or estimated from correlations with in situ tests (see Chapter 6). It is clear from these figures that, with increasing strain level, the shear modulus decreases, while the damping ratio increases.

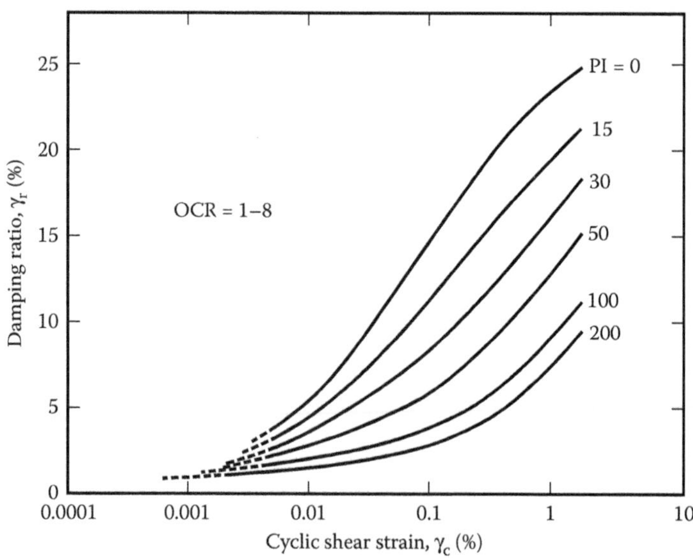


*Figure 10.9* Damping ratio dependence on strain for fine-grained soils. (Adapted from Vucetic, M. and Dobry, R. 1991. *Journal of Geotechnical Engineering*, 117(1): 89–107. Courtesy of ASCE.)

Towhata (2008) quotes the work of Yasuda and Yamaguchi (1985) who express both the shear modulus reduction ratio, $G/G_{max}$, and the damping ratio D for coarser-grained soils as functions of the effective mean principal stress p′ and the average grain size, $D_{50}$, as follows:

$$G/G_{max} = (A_1 + A_2 \log_{10} D_{50}) \cdot p'^{(X1)} \tag{10.22}$$

$$D = (C_1 + C_2 \log_{10} D_{50}) \cdot p'^{(X2)} \tag{10.23}$$

where $X_1 = (B_1 + B_2 \log_{10} D_{50})$, $X_2 = (D_1 + D_2 \log_{10} D_{50})$, $A_1$, $A_2$, $B_1$, $B_2$, $C_1$, $C_2$, $D_1$, $D_2$ are empirical parameters shown in Table 10.1, p′ is the mean effective stress, in kgf/cm$^2$ (1 kgf/cm$^2$ = 98 kPa) and $D_{50}$ is the mean particle size, in mm.

When $D_{50}$ is unknown, Towhata (2008) suggests the following typical values of $D_{50}$ for various soil types:

- Coarse gravelly sand: 0.4 mm
- Medium sand: 0.3 mm
- Fine sand: 0.2 mm
- Silty sand: 0.1 mm
- Sandy silt: 0.04 mm
- Clayey sand: 0.03 mm
- Silt: 0.007 mm
- Clayey silt: 0.005 mm.

Ishibashi and Zhang (1993) have presented expressions for the dynamic shear modulus and damping in terms of the maximum shear modulus, the mean effective stress, the shear strain magnitude and the plasticity index. At very small strains, the data for sands indicate very small values of internal damping ratio $D_0$, typically in the order of 1%. The damping increases with increasing shear strain, and reaches values of around 30% at very large strains.

Senetakis et al. (2013) presented the results of a laboratory investigation of the strain-dependent dynamic properties of volcanic granular soils composed of a rhyolitic crushed rock along with additional experiments on quartz sand through a high-amplitude resonant column testing program. The sands were tested in a dry state in the torsional mode of vibration and the degradation of the normalised shear modulus, and the increase of damping ratio in shear as a function of the shear strain amplitude, were examined. It was found that, for a given mean effective confining pressure and coefficient of uniformity (Cu), the volcanic sands showed higher linearity in comparison to the quartz sands, and that this trend became more pronounced with decreasing mean effective stress and increasing Cu. In contrast to the

*Table 10.1* Empirical parameters for shear modulus reduction and damping

| *Shear strain* | $A_1$ | $A_2$ | $B_1$ | $B_2$ | $C_1$ | $C_2$ | $D_1$ | $D_2$ |
|---|---|---|---|---|---|---|---|---|
| $10^{-4}$ | 0.827 | −0.044 | 0.056 | 0.026 | 0.035 | 0.005 | −0.0559 | −0.258 |
| $3 \times 10^{-4}$ | 0.670 | −0.068 | 0.184 | 0.086 | | | | |
| $10^{-3}$ | 0.387 | −0.099 | 0.277 | 0.130 | 0.136 | 0.036 | −0.375 | −0.173 |
| $3 \times 10^{-3}$ | 0.189 | −0.089 | 0.315 | 0.147 | | | | |
| $10^{-2}$ | 0.061 | −0.054 | 0.365 | 0.167 | 0.234 | 0.037 | 0.000 | 0.000 |
| $3 \times 10^{-2}$ | 0.041 | −0.019 | 0.403 | 0.183 | | | | |

Source: After Towhata, I. 2008. *Geotechnical Earthquake Engineering*. Springer, Berlin.

general trend observed in the quartz soils, the confining pressure and the grain-size characteristics hardly affected the rate of normalised modulus degradation and damping increase in the volcanic sands. These differences were considered to be related to the micro-mechanisms that dominate at particle contacts in the range of small to medium shear strain amplitudes.

Payan et al. (2016) found that the small strain damping ratio was also dependent on the particle shape of the sand, and developed an expression relating $D_0$ to the particle shape roundness and sphericity.

Ideally, both the shear modulus and the damping ratio should be selected for a level of strain that is appropriate for the problem being considered. The strain level relevant to dynamic foundation response is likely to be less than for earthquake response (Chapter 11). However, the effects of strain level are not easy to estimate in the early stages of design, and so it is useful to adopt an approximate, pragmatic and simple-to-use assessment approach. It is suggested that the following procedure may be adopted:

1. Adopt the values of shear modulus for a shear modulus reduction ratio $G/G_{max}$ of 0.7
2. Adopt the value of internal damping ratio corresponding to the strain level for $G/G_{max} = 0.7$

From Figure 10.8, this strain level ranges from about 0.01% to about 0.35%, with higher values being for soils of higher plasticity. However, referring to Figure 10.9, the damping ratio corresponding to these strain levels varies only within a very small range of 5%–7%, for the range of plasticity index of 0–200 shown therein. Thus, in the absence of other information, an internal damping ratio of 5% may be reasonable to adopt as a preliminary value. This value is toward the upper end of the range of 2%–6% suggested by Gazetas (1991).

## 10.6 THE IMPORTANCE OF SITE NATURAL FREQUENCY

The natural frequency of the site (or its inverse, the natural period) is an important determinant of the dynamic response of a foundation. For various simplified profiles of soil shear modulus with depth, Gazetas (1991) gives the following expressions for the fundamental (first) natural site frequency, $f_n$, for horizontal excitation:

- Constant shear modulus with depth:

$$f_n = 0.25V_s/H \tag{10.24}$$

  where $V_s$ is the shear wave velocity within layer and H is the layer depth above bedrock
- Linearly increasing shear modulus with depth:

$$f_n = 0.19V_{sH}/H \tag{10.25}$$

  where $V_{sH}$ is the shear wave velocity at base of layer
- Parabolically increasing shear modulus with depth:

$$f_n = 0.22V_{sH}/H \tag{10.26}$$

For vertical excitation, and a constant shear modulus with depth, the fundamental natural frequency, $f_{nv}$, is as follows:

$$f_{nv} = [0.25V_s/H]\cdot(2(1-\nu)/1-2\nu)^{0.5} \tag{10.27}$$

where $\nu$ is the Poisson's ratio of soil.

For higher modes of vibration, the above fundamental natural frequencies are multiplied by a factor (2n − 1), where n = mode number.

In cases involving layered soils, a weighted average shear wave velocity can be used to estimate $f_n$, via the following approximation:

$$f_n = 0.25\,\Sigma(V_{si}/H_i) \qquad (10.28)$$

where $V_{si}$ is the shear wave velocity in layer i, and $H_i$ is the thickness of layer i, with the summation being carried out for all the layers within the profile.

The shear wave velocity $V_s$ is related to the small strain shear modulus, $G_o$, as follows (see Equation 6.2):

$$V_s = (G/\rho)^{0.5} \qquad (10.29)$$

where ρ is the mass density of soil.

El Sharnouby and Novak (1985) indicate that, for frequencies less than the fundamental natural frequency of a stratum, two important points emerge:

1. The low-frequency dynamic stiffness of a pile group can be assumed to be approximately equal to the static stiffness.
2. The damping can be assumed to stem only from material damping (hysteresis) of the soil and the piles.

As an illustration of the circumstances under which dynamic effects are not likely to be significant, Figure 10.10 plots the natural frequency of layers with shear wave velocities typical of soft, medium, stiff and hard soils, assuming a constant shear wave velocity, $V_s$, with depth. The natural frequency decreases as the layer depth increases or $V_s$ decreases. Also shown in this figure are typical ranges of loading frequency from wind and from earthquake loading, derived from Figure 5.1. From Figure 10.10, the following inferences can be drawn:

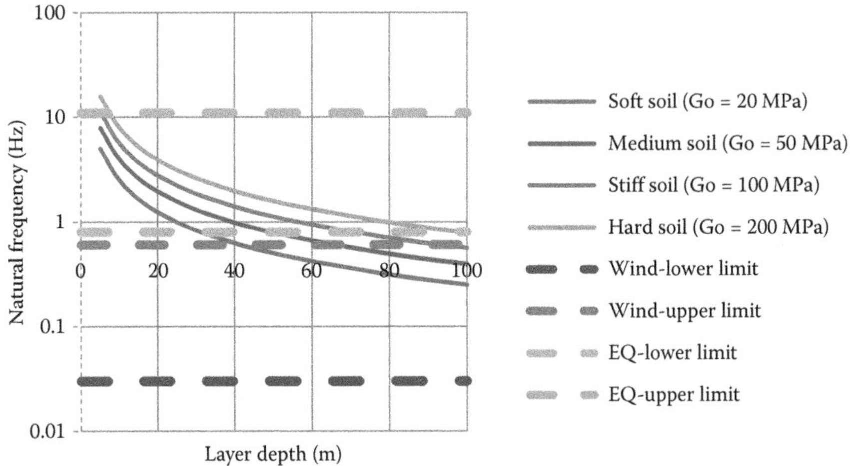


*Figure 10.10* Relationships between natural frequency and layer depth.

1. In many cases involving wind loading, dynamic effects are likely to be small (i.e. the natural frequency of the layer is greater than that of the loading). The dynamic effects may become more significant for soft soil layers more than about 40 m deep.
2. In many cases involving earthquake loading, dynamic effects are likely to become important. Only for relatively shallow soil layers are the natural frequencies greater than those of typical earthquakes.
3. Dynamic effects are more likely to be important for soft soil layers than for stiff soil layers.

## 10.7 DESIGN CRITERIA

Issues related to the effects of dynamic wind loading are generally dealt with by the structural engineer, with geotechnical input being limited to an assessment of the stiffness and damping characteristics of the foundation system. However, it is useful for the geotechnical designer to note the following general principles of design that can be applied to dynamic loadings:

- The natural frequency of the foundation system should be greater than that of the structure it supports, to avoid potential resonance phenomena.
- The amplitude of dynamic motions of the structure–foundation system should be within tolerable limits. The amplitude will depend on the stiffness and damping characteristics of both the foundation and the structure.

The acceptable levels of dynamic motion can be expressed in terms of dynamic amplitude of motion, or velocity, or acceleration. Table 10.2 reproduces guidelines for human perception levels of dynamic motion, expressed in terms of acceleration (Mendis et al., 2007). These are for vibration in the low-frequency range of 0–1 Hz encountered in tall buildings, and incorporate such factors as the occupants' expectancy and experience, their activity, body posture and orientation, and visual and acoustic cues. They apply to both the translational and rotational motions to which the occupant is subjected. The acceleration levels are a function of the frequency of vibration, and decrease as the frequency increases. For example, allowable vibration levels at a frequency of 1 Hz are typically only 40%–50% of those acceptable at a frequency of 0.1 Hz. For a 10-year return period event, with a duration of 10 minutes, American practice typically allows accelerations of between 0.22 and 0.25 $m^2/s$ for office buildings, reducing to 0.10–0.15 $m^2/s$ for residential buildings.

*Table 10.2* Human perception levels of dynamic motion

| *Level of motion* | *Acceleration ($m^2/s$)* | *Effect* |
|---|---|---|
| 1 | <0.05 | Humans cannot perceive motion. |
| 2 | 0.05–0.1 | Sensitive people can perceive motion. Objects may move slightly. |
| 3 | 0.1–0.25 | Most people perceive motion. Level of motion may affect desk work. Long exposure may produce motion sickness. |
| 4 | 0.25–0.4 | Desk work difficult or impossible. Ambulation still possible. |
| 5 | 0.4–0.5 | People strongly perceive motion, and have difficulty in walking. Standing people may lose balance. |
| 6 | 0.5–0.6 | Most people cannot tolerate motion and are unable to walk naturally. |
| 7 | 0.6–0.7 | People cannot walk or tolerate motion. |
| 8 | >0.85 | Objects begin to fall and people may be injured. |

Source: Adopted from Mendis, P. et al. 2007. *EJSE Special Issue, Loading on Structures*, 3: 41–54.

### 10.7.1 Strategies for control of building response

In order to control building response to lateral loading structural engineers may utilise one or more of the following strategies, some of which are referred to in Chapter 2:

1. Increase the stiffness of the system
2. Increase the building weight
3. Increase the density of the structure with fill-ins
4. Use efficient shapes
5. Generate additional damping forces (see Section 2.9)

## 10.8 STIFFNESS AND DAMPING FOR SHALLOW FOUNDATIONS

### 10.8.1 Introduction

Although tall buildings are generally supported by deep foundations, knowledge of the stiffness and damping of shallow foundations is valuable in assessing the response of combined piled raft foundations, for low-rise areas adjacent to high-rise areas, and for cases where the foundation system is simplified to an equivalent footing or raft.

In this section, some Category 2 solutions will be presented for the stiffness and damping of shallow foundations, either on the surface of a soil profile or embedded below the surface. These solutions are derived from the theory of elasticity and hence their application requires representative values of the soil modulus and Poisson's ratio to be assessed.

### 10.8.2 Early solutions for a deep uniform layer

Whitman and Richart (1967) developed simple expressions for the stiffness and radiation damping ratio of shallow foundations resting on an infinitely deep uniform elastic layer. Although very idealised with respect to real soil profiles, these solutions were valuable in demonstrating some of the important aspects of dynamic foundation response. Table 10.3 summarises their solutions for the stiffness and damping ratio of a circular foundation of radius $r_0$ on the surface of a deep elastic layer having a shear modulus G, Poisson's ratio $\nu$ and a mass density $\rho$.

In this table, $I_\psi$ and $I_\theta$ are mass moments of inertia of the foundation about the appropriate axis of rocking or torsion.

As a first approximation, a rectangular foundation can be converted to an equivalent circular footing of radius, $r_{0e}$, as follows:

For vertical or horizontal translation:

$$r_{0e} = [BL/\pi]^{0.5} \quad (10.30)$$

*Table 10.3* Stiffness and damping of a circular footing on a deep layer

| *Mode of vibration* | *Spring stiffness* | *Damping ratio* | *Mass or inertia ratio* |
|---|---|---|---|
| Vertical | $k_z = 4Gr_0/(1-\nu)$ | $D_z = 0.425/\sqrt{B_z}$ | $B_z = 0.25(1-\nu)m/[\rho r_0^3]$ |
| Horizontal sliding | $k_x = 32(1-\nu)Gr_0/(7-8\nu)$ | $D_x = 0.288/\sqrt{B_x}$ | $B_x = (7-8\nu)m/[32\rho r_0^3]$ |
| Rocking | $k_\psi = 8Gr_0^3/[3(1-\nu)]$ | $D_\psi = 0.15/[(1+B_\psi)\sqrt{B_\psi}]$ | $B_\psi = 0.375(1-\nu)I_\psi/[\rho r_0^5]$ |
| Torsion | $k_\theta = 16Gr_0^3/3$ | $D_\theta = 0.50/[1+2B_\theta]$ | $B_\theta = I_\theta/[\rho r_0^5]$ |

Source: After Whitman, R.V. and Richart, F.E., Jr. 1967. *Journal of Soil Mechanics and Foundations Division, ASCE*, 93(6), 169–193.

For rocking:

$$r_{0e} = [BL^3/3\pi]^{0.25} \tag{10.31}$$

For torsion:

$$r_{0e} = [BL(B^2 + L^2)/6\pi]^{0.25} \tag{10.32}$$

where B is the footing width (perpendicular to the direction of rocking) and L the footing length (parallel to the direction of rocking).

Examination of the above solutions indicates the following characteristics of behaviour:

1. The stiffness increases as both the soil shear modulus and the footing size increase.
2. The damping ratio depends on the mass or inertia ratio. The larger this ratio, the smaller is the damping ratio.
3. The mass or inertia ratio increases as the mass (or inertia) increase, but decreases with increase in footing size.
4. Evaluation of the damping ratios reveals that the damping for the translational modes of vibration is generally much greater than the ratios for the rotational modes of rocking or torsion.

#### 10.8.2.1 Effects of embedment

Arya et al. (1979) summarised correction factors which can be applied to the solutions for surface foundations on an elastic half-space to allow for the effects of embedment. For a soil Poisson's ratio of 0.25, these correction factors are plotted in Figure 10.11 for stiffness and in Figure 10.12 for damping, and depend on the ratio of the depth of embedment ($h_e$) to the radius of the footing, $r_0$. It can be seen that, for both stiffness and damping, the correction factors all increase with increasing relative depth of embedment, with the rotational stiffness and damping ratio being most affected.

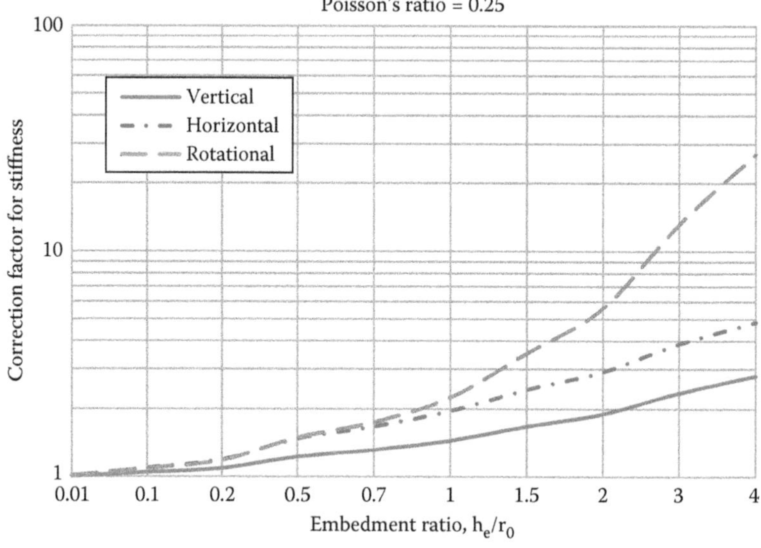


*Figure 10.11* Stiffness correction factors for embedment of footing in a half-space.

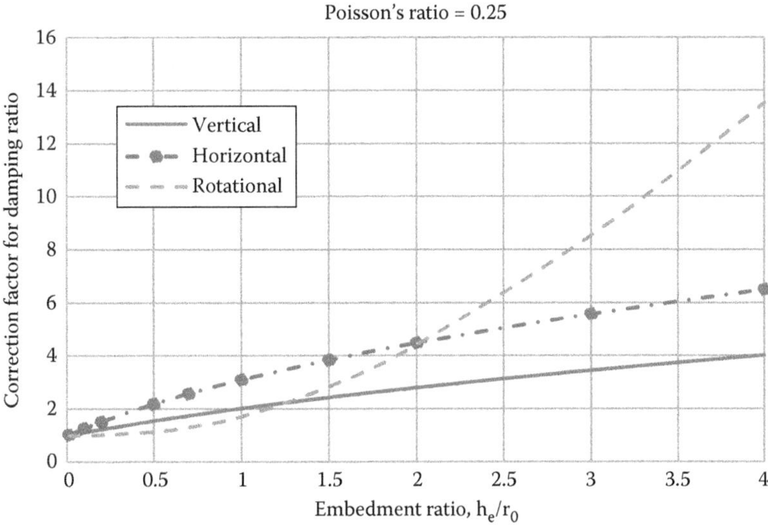


*Figure 10.12* Damping factor correction factors for embedment of footing in half-space.

In principle, these results, in conjunction with the solutions for a surface footing, could be applied, very approximately, to represent a pile group or a piled raft as an equivalent raft, with the level of the raft at some fraction of the pile length (e.g. between 1/2 and 2/3) below the surface. This approximation will be discussed later in this chapter.

The above solutions, while convenient, are independent of the frequency of vibration, but it is known that stiffness and damping are indeed dependent on frequency. This frequency dependence has been captured in the solutions of Gazetas (1991) and El Naggar (2001), which involve the use of the concept of dynamic impedance and which are discussed below.

### 10.8.3 Dynamic impedance

The concept of dynamic impedance has been an important step in the development of the dynamic analysis of foundation behaviour. This concept has been introduced briefly in Section 10.4 and can be defined as the ratio of the steady state force (or moment) and the resulting displacement (or rotation) of a mass-less foundation (Gazetas, 1983). It is a complex number with the real part representing the stiffness and inertia characteristics of the system and the complex part representing the frequency-dependent damping (energy loss) in the system. The impedance, KI, for a particular mode of vibration can be expressed in one of the following ways:

- In terms of a dimensional damping coefficient $c_s$, that is,

  $$KI = K \cdot [k_d + i\omega c_s] \tag{10.33}$$

  where K is the static stiffness, $k_d$ the dynamic correction factor for stiffness, $\omega$ the circular frequency, $c_s$ the damping coefficient and $i = \sqrt{-1}$.
  An alternative form of this expression is

  $$KI = K' + i\omega C \tag{10.34}$$

  where $K'$ is the dynamic stiffness, $C = K \cdot c_s$ the alternative damping coefficient.

- In terms of a dimensionless frequency factor $a_0$:

$$KI = K(k_d + ia_0 \cdot c) \tag{10.35}$$

or

$$KI = K \cdot k_d + ia_0C \tag{10.36}$$

where

$$a_0 = \text{dimensionless frequency} = \omega B/V_s \tag{10.37}$$

Here, B is a suitable dimension of the foundation, for example, width, radius, or diameter, $V_s$ the shear wave velocity, $c = c_sV_s/B$ the dimensionless damping coefficient and $C = c \cdot K$.

In utilising the available solutions for dynamic foundation response, care must be taken to employ the relevant definition of the dynamic impedance KI, the damping term and the dimensionless frequency $a_0$.

The dynamic impedance can be used and manipulated in the same way as is the stiffness for static problems. Indeed, for a static problem, the impedance reduces to the static stiffness of the system.

The damping ratio $\zeta$ can be expressed in terms of the damping coefficient C, the static stiffness K and the natural circular frequency $\omega_n$ as

$$\zeta = C/[2K/\omega_n] \tag{10.38}$$

### 10.8.4 Solutions for surface and subsurface footings

Gazetas (1991) has presented tables and charts for the stiffness and damping of surface and embedded footings of various shapes, including arbitrary shapes. In these solutions, the frequency dependence of stiffness and damping is incorporated via the dimensionless parameter $a_0$. The larger the value of $a_0$, the more significant are likely to be the dynamic effects on footing response.

Gazetas presents results for circular, rectangular and strip footings, and also footings of arbitrary shape, on the surface of a uniform layer of finite depth, on the surface of a layer, embedded footings, and footings on the surface of a soil whose stiffness increases with depth in a linear or parabolic manner.

For the relatively simple case of a rigid square foundation of side 2B on a very deep uniform soil layer, values of the stiffness and the dynamic and embedment correction factors are given in Table 10.4, while corresponding values for the damping coefficient are given in Table 10.5. These expressions may be used for circular footings of equal area (for translational modes) or equal moments of area (for rotational modes). In these solutions, the dynamic impedance is defined via Equation 10.34.

*Table 10.4* Stiffness for embedded shallow foundations

| *Vibration mode* | *Static stiffness $K_s$* | *Dynamic correction factor $k_d$* | *Embedment correction factor $k_e$* |
|---|---|---|---|
| Vertical | $4.54\ GB/(1 - \nu)$ | $1–0.013a_0–0.0884a_0^2$ | $k_{ev} = [1 + 0.11(L/B)] \cdot [1 + 0.2(2h_e/B)^{2/3}]$ |
| Horizontal | $9\ GB/(2 - \nu)$ | 1.0 | $k_{eh} = [1 + 0.5(L/B)^{0.5}] \cdot [1 + 0.52(2h_e/B)^{0.8}]$ |
| Rocking | $3.6\ GB^3/(1 - \nu)$ | $1–0.2a_0$ | $k_{er} = [1 + 1.26(h_e/B)] \cdot [1 + h_e/B(h_e/L)^{-0.2}]$ |
| Torsion | $8.3\ GB^3$ | $1–0.14a_0$ | $K_{et} = 1 + 2.8*(h_e/B)^{0.9}$ |

Source: Derived from Gazetas, G. 1991. *Foundation Engineering Handbook*, 2nd Ed., Chapman & Hall, New York, pp. 563–593.

*Table 10.5* Damping coefficient for embedded shallow foundations

| *Vibration mode* | *Damping coefficient $C_s$ for surface footing* | *Frequency correction factor $c_f$* | *Additional damping coefficient due to embedment $C_e$* |
|---|---|---|---|
| Vertical | $\rho V_{la} A_b$ | $0.90 + 0.05a_0$ | $C_{ev} = 8\rho V_s B h_e$ |
| Horizontal | $\rho V_s A_b$ | 1.0 | $C_{eh} = 4\rho B h_e (V_s + V_{la})$ |
| Rocking | $\rho V_{la} I_{bx}$ | $0.059a_0^2 + 0.226a_0$ | $C_{er} = 1.33\rho V_{la} h_e^3 B \cdot c_{1b} + 1.33\rho V_s B h_e (B^2 + h_e^2) c_{1b} + 4\rho B^3 h_e \cdot c_{ib}$ where $c_{1b} = 0.25 + 0.65(a_0)^{0.5}(h_e/L)^{-a0/2} \cdot (L/B)^{-0.25}$ |
| Torsion | $\rho V_s J_b$ | $-0.067a_0^3 + 0.2a_0^2 + 0.117a_0$ | $2.67\rho V_{la} h_e B^3 \cdot c_{2b} + 8\rho V_s h_e B^3 \cdot c_{2b}$ where $c_{2b} = (h_e/L)^{-0.5}[a_0^2/(a_0^2 + 0.5)]$ |

Source: Derived from Gazetas, G. 1991. *Foundation Engineering Handbook*, 2nd Ed., Chapman & Hall, New York, pp. 563–593.

When computing the stiffness, the static surface value $K_s$ is multiplied by the dynamic correction factor $k_d$ and by the embedment factor $k_e$. For the damping coefficient, the radiation damping value for a surface footing $C_s$ is multiplied by the frequency correction factor $c_f$, and then added to the additional damping coefficient due to embedment, $C_e$.

In these tables, B is the half-width of the footing, L the depth of footing below the surface, $h_e$ the length of contact between sides of footing and the soil, $A_b$ the area of footing base, $I_{bx}$ the moment of area of footing base ($=1.33B^4$), $J_b$ the polar moment of area of footing base ($=2.67B^4$) and $a_0 = \omega B/V_s$.

When these tables are being used to estimate the stiffness and damping of an equivalent raft, the length of contact between the footing and the soil, $h_e$, can be taken as small.

The damping coefficients in Table 10.5 relate to radiation damping. When internal damping is incorporated, the total damping, $C_{tot}$, is then given by

$$C_{tot} = C + 2K_s\zeta_0/\omega \tag{10.39}$$

where C is the radiation damping coefficient, $K_s$ the static stiffness for the relevant mode of vibration, $\zeta_0$ the internal damping ratio and $\omega$ the circular frequency.

Also, if the total damping ratio $\zeta_{TOT}$ is required, it can be obtained as follows:

$$\zeta_{TOT} = C_{tot}\omega/(2K_s) \tag{10.40}$$

### *10.8.4.1 Effect of a finite layer*

Gazetas (1991) has provided solutions for surface and embedded footings on a finite soil layer underlain by a rigid base. For an embedded footing of equivalent radius R, with a depth of embedment L, and a length of sidewall contact $l_s$, on a layer of thickness H, Table 10.6 shows correction factors, $k_H$, for the static stiffness $K_s$ for the various modes of vibration, while Table 10.6 shows corresponding corrections for the damping coefficient.

The dynamic stiffness correction factor $k_d$ for an embedded foundation is approximately the same as that for a foundation on a very deep uniform layer in Table 10.4. For damping, Gazetas indicates that the damping coefficient for an embedded foundation exceeds that for a surface foundation by an amount that depends on the geometry of the sidewall–soil contact surface, and is almost independent of the presence or absence of a rigid base. Table 10.7 summarises the suggestions of Gazetas for the radiation damping coefficients within a layer.

*Table 10.6* Correction factors $k_H$ for stiffness of a rigid footing on a finite layer

| *Mode of vibration* | *Correction factor $k_H$* |
|---|---|
| Vertical | $(1 + 0.55l_s/R)*[1 + (0.85 - 0.28L/R)*L/(H - L)]$ |
| Horizontal | $(1 + l_s/R) \cdot (1 + 1.25L/H)$ |
| Rocking | $(1 + 2l_s/R) \cdot (1 + 0.65L/H)$ |
| Torsion | $(1 + 2.67l_s/R)$ |

Source: Derived from Gazetas, G. 1991. *Foundation Engineering Handbook*, 2nd Ed., Chapman & Hall, New York, pp. 563–593.

*Table 10.7* Radiation damping coefficients for a finite layer

| *Mode of vibration* | *Radiation damping coefficient C* |
|---|---|
| Vertical | For $f \geq 1.5f_c$, $C = 0.8$ times the value for a very deep layer<br>For $f < f_c$, $C = 0$<br>For intermediate f, interpolate linearly between the above |
| Horizontal | For $f > 4/3f_s$, $C = C$ for a very deep layer<br>For $f < 0.75f_s$, $C = 0$<br>For intermediate $f_s$, interpolate linearly between the above |
| Rocking | For $f \geq f_c$, $C = C$ for a very deep layer<br>For $f < f_c$, $C = 0$ |
| Torsion | $C = C$ for a very deep layer |

Source: Derived from Gazetas, G. 1991. *Foundation Engineering Handbook*, 2nd Ed., Chapman & Hall, New York, pp. 563–593.

Note: $f_c = V_{la}/4H$, $f_s = V_s/4H$, $V_{la} = 3.4V_s/[\pi(1 - \nu)]$.

The following characteristics of behaviour have been noted by Gazetas:

- The static stiffness in all modes increases as the layer depth becomes shallower. The effect is most pronounced for vertical stiffness.
- The dynamic stiffness is frequency dependent, but not always in a smooth monotonic manner.
- The dashpot coefficients, and hence the damping ratios, are affected by frequency, and at low frequencies which are below the fundamental natural frequency of the layer, the radiation damping is zero or negligible for all modes of vibration and for all footing shapes.
- At frequencies greater than the fundamental natural frequency of the stratum, the damping fluctuates around the value for an infinitely deep layer (half-space), with the amplitude of these fluctuations tending to decrease as the stratum thickness increases.

### 10.8.5 Solutions for an embedded footing in a backfill layer underlain by a deep layer

Useful solutions have been summarised by El Naggar (2001) for the case shown in Figure 10.13, where an upper ('backfill' layer as denoted by El Naggar) overlies a lower deep layer (half-space). These solutions have been derived from those presented by Novak and Beredugo (1972) and Beredugo and Novak (1972). The various geometric and material parameters are defined in this figure, and have been retained as those used by El Naggar, to avoid confusion in transcription. Stiffness and damping values are shown in Tables 10.8 and

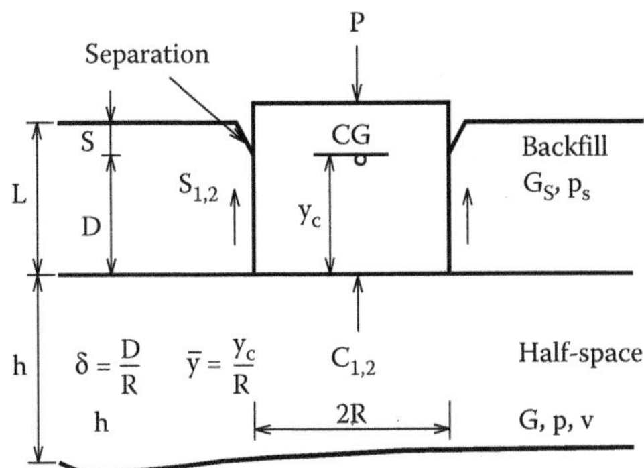


*Figure 10.13* Embedded footing geometry. (Adapted from El Naggar, M.H. 2001. *Geotechnical and Geoenvironmental Engineering Handbook*, Kluwer Academic Publishers, Boston. Courtesy of Kluwer Academic Publishing.)

10.9. The stiffness values are taken to be frequency-independent, as a first approximation, given the other assumptions and uncertainties involved in the analysis.

The complex impedance, $KI_E$, for El Naggar's solutions is defined as follows:

$$KI = K[k_d + i \cdot a_0 \cdot c'] \tag{10.41}$$

where K is the static stiffness, $k_d$ the dynamic factor which depends on the dimensionless frequency $a_0$, which is defined as $\omega R/Vs$, where R is the foundation radius and $c'$ the damping coefficient, also dependent on $a_0$.

*Table 10.8* Stiffness and damping values for embedded footings

| *Motion* | *Stiffness constant, k* | *Damping coefficient, C* |
|---|---|---|
| Vertical | $GH[C_{v1} + (G_s/G)\delta S_{v1}]$ | $R^2(\rho G)^{0.5}[C_{v2} + S_{v2}\delta\{\rho_s G_s/\rho G\}^{0.5}]$ |
| Torsional | $GR^3[C_{\eta 1} + (G_s/G)\delta S_{\eta 1}]$ | $R^4(\rho G)^{0.5}[C_{\eta 2} + S_{\eta 2}\delta\{\rho_s G_s/\rho G\}^{0.5}]$ |
| Horizontal | $GR[C_{u1} + (G_s/G)\delta S_{u1}]$ | $R^2(\rho G)^{0.5}[C_{u2} + S_{u2}\delta\{\rho_s G_s/\rho G\}^{0.5}]$ |
| Rocking | $GR^3[\bar{Y}^2 C_{u1} + (G_s/G)\delta(\delta^2/3 + \bar{Y}^2 - \delta\bar{Y})S_{u1}] + GR^3(C_{\psi 1} + (G_s/G)\ \delta S_{\psi 1})$ | $R^4(\rho G)^{0.5}[\bar{Y}^2 C_{u2} + (\rho_s G_s/G\rho)^{0.5}\delta\ (\delta^2/3 + \bar{Y}^2 - \delta\bar{Y})S_{u2}] + R^4(\rho G)^{0.5}(C_{\psi 2} + (\rho_s G_s/G\rho)^{0.5}\delta S_{\psi 2})$ |
| Coupling | $-GR[y_c C_{u1} + (G_s/G)\delta(y_c - D/2)S_{u1}]$ | $-R^2(\rho G)^{0.5}[y_c C_{u2} + (\rho_s G_s/G\rho)^{0.5}\delta\ (y_c - D/2)]S_{u2}$ |

Source: El Naggar, M.H. 2001. *Geotechnical and Geoenvironmental Engineering Handbook*, Kluwer Academic Publishers, Boston. Courtesy of Kluwer Academic Publishing.

*Table 10.9* Stiffness and damping parameters for embedded footings ($\zeta_\iota = 0$)

| *Motion* | *Soil* | *Side layer* | | *Half-space* | |
|---|---|---|---|---|---|
| Vertical | Cohesive | $S_{v1} = 2.7$ | $S_{v2} = 6.7$ | $C_{v1} = 7.5$ | $C_{v2} = 6.8$ |
| | Granular | $S_{v1} = 2.7$ | $S_{v2} = 6.7$ | $C_{v1} = 5.2$ | $C_{v2} = 5.0$ |
| Horizontal | Cohesive | $S_{\upsilon 1} = 4.1$ | $S_{u2} = 10.6$ | $C_{u1} = 5.1$ | $C_{u2} = 3.2$ |
| | Granular | $S_{\upsilon 1} = 4.0$ | $S_{u2} = 9.1$ | $C_{u1} = 4.7$ | $C_{u2} = 2.8$ |
| Rocking | Cohesive | $S_{\psi 1} = 2.5$ | $S_{\psi 2} = 1.8$ | $C_{\psi 1} = 4.3$ | $C_{\psi 2} = 0.7$ |
| | Granular | $S_{\psi 1} = 2.5$ | $S_{\psi 2} = 1.8$ | $C_{\psi 1} = 3.3$ | $C_{\psi 2} = 0.5$ |
| Torsion | Cohesive and granular | $S_{\eta 1} = 10.2$ | $S_{\psi 2} = 5.4$ | $C_{\psi 1} = 4.3$ | $C_{\psi 2} = 0.7$ |

Source: El Naggar, M.H. 2001. *Geotechnical and Geoenvironmental Engineering Handbook*, Kluwer Academic Publishers, Boston. Courtesy of Kluwer Academic Publishing.

*Table 10.10* Correction factors for stiffness due to finite soil layer depth H

| *Motion* | *Symbol* | *Correction factor for finite depth* |
|---|---|---|
| Vertical | $k_v''$ | $(1 + 1.28R/H)(1 + 0.47\delta) \times [1 + 2(0.85 - 0.28\delta)(D/H)/(1 - D/H)$ |
| Horizontal | $k_u''$ | $(1 + 0.5R/H)(1 + 2\delta/3)(1 + 1.25D/H)$ |
| Rocking | $k_\phi''$ | $(1 + R/6H)(1 + 2d)(1 + 0.7D/H)$ |
| Cross-rocking/horizontal[a] | $k_{u\phi}''$ | $(0.4\delta - 0.03)R \cdot k_u''$ |
| Torsion | $k_t''$ | $(1 + 2.67\delta)$ |

Source: El Naggar, M.H. 2001. *Geotechnical and Geoenvironmental Engineering Handbook*, Kluwer Academic Publishers, Boston. Courtesy of Kluwer Academic Publishing.

Note: Valid for $\delta = D/R \leq 1.5$, $D/H \leq 0.75$ and $R/H \leq 0.5$.

[a] This correction factor is applied to the lateral stiffness for the deep layer.

When the definitions of $a_0$ and $c'$ used by El Naggar are applied in Equation 10.41, it is found that the dynamic impedance is of the same form as Equation 10.34, with the damping coefficient being equal to C in that equation.

It will also be noticed that the horizontal–moment coupling values are negative, indicating that, for example, the presence of a moment in addition to a horizontal load will lead to a reduction in horizontal stiffness.

#### *10.8.5.1 Effect of a finite layer*

If there is a rigid base below the base of the foundation, such that the thickness of the soil layer below the backfill is H, it is possible to derive approximate correction factors for the effect of the finite layer on stiffness, based on the expressions quoted by El Naggar (2001). Table 10.10 shows these stiffness correction factors.

Some differences in computed stiffness values have been noted from alternative analyses, for example, the solutions of Gazetas for an embedded footing tend to give smaller vertical stiffness values than those of El Naggar. It may thus be prudent to check the static stiffness values obtained from these solutions with solutions for a rigid pier given in Chapter 8.

As mentioned previously, at frequencies lower than the natural frequency of the layer, the only source of damping is the material damping, in which case the damping parameters are given by the following expressions:

$$S_{u2} = 2\zeta_{s\iota} S_{u1}/a_0 \qquad (10.42)$$

$$S_{v2} = 2\zeta_{s\iota} S_{v1}/a_0 \qquad (10.43)$$

where $\zeta_{s\iota}$ is the internal material damping ratio.

## 10.9 STIFFNESS AND DAMPING OF CYLINDRICAL PIERS

### 10.9.1 Introduction

As set out in Chapter 8, in the preliminary assessment and design of tall building foundations, it is often convenient to represent the piled or piled raft foundation system as an equivalent pier. In this section, solutions for the vertical and lateral response of a cylindrical pier will be presented.

### 10.9.2 Vertical response

A solution for the vertical stiffness of an equivalent pier within a finite layer can be obtained as the ratio of applied load to settlement from the results presented previously in Figure 8.18 of Chapter 8.

Another solution for the vertical static stiffness of a vertically loaded rigid pier of length L and diameter D, bearing on a stiffer layer has been derived by Kulhawy and Carter (1992), as follows:

$$K_v = E_b D/\left(1-\nu_b{}^2\right) + \pi/\zeta^* \cdot E_s L/(1+\nu_s)] \tag{10.44}$$

where $\zeta = \ln[5(1-\nu_s)L/D]$, $E_s$, $E_b$ are Young's modulus values for the upper and lower layers and $\nu_s$ and $\nu_b$ are the Poisson's ratios for the upper and lower layers.

No solution for damping appears to be available, and so it is suggested that the damping ratio be assessed as for an embedded foundation in Section 10.8.

### 10.9.3 Lateral response

Varun (2006) and Varun et al. (2009) have developed expressions for the horizontal and rocking stiffness and damping of an embedded cylindrical pier. The geometry of the pier is shown in Figure 10.14.

Varun's original results are for a homogeneous deep layer, but have been adapted for a two-layer case in which the pier of diameter D is located within a soil layer of depth L, having Young's modulus $E_s$, and resting on a stiffer layer with Young's modulus $E_b$.

For dynamic loading, the complex impedance KI is given by

$$KI = K_{stat} \cdot k'(a_0) + ia_0 C(a_0) \tag{10.45}$$

where $K_{stat}$ is the static stiffness, $a_0 = \omega D/V_s$ the dimensionless frequency, $\omega$ the circular frequency, $V_s$ the soil shear wave velocity, D the pier diameter, $k'(a_0)$ the frequency-dependent stiffness coefficient and $C(a_0)$ the frequency-dependent radiation damping coefficient.

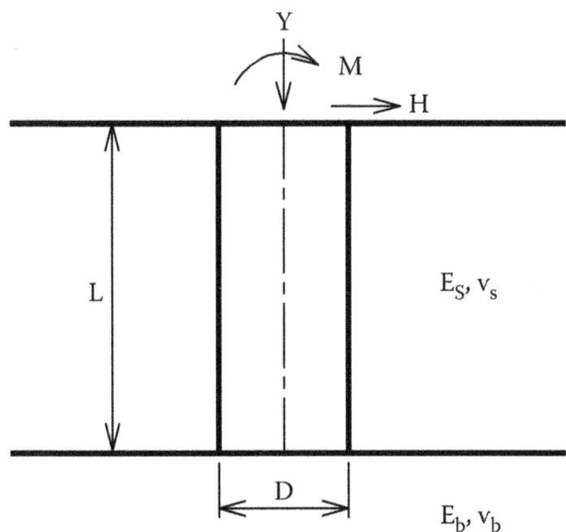


*Figure 10.14* Geometry of cylindrical pier.

The dynamic response to combined horizontal and moment loading is given by the following equation:

$$\begin{vmatrix} H^* \\ \\ M^* \end{vmatrix} = \begin{vmatrix} KI_{xx} & KI_{xr} \\ \\ KI_{rx} & KI_{rr} \end{vmatrix} \cdot \begin{vmatrix} u^* \\ \\ \theta^* \end{vmatrix} \tag{10.46}$$

where H* and M* are the dynamic forcing functions (lateral load and moment), $KI_{xx}$ the impedance for horizontal deflection due to lateral load, $KI_{xr}$ the impedance for horizontal deflection due to moment, $KI_{rx}$ the impedance for rotation due to lateral load (=$KI_{xr}$ for an elastic soil), $KI_{rr}$ the impedance for rotation due to moment, u* the dynamic lateral deflection and θ* the dynamic rotation.

For the two-layer system shown in Figure 10.14, the (real) horizontal stiffness values are expressed as follows:

$$K_{xx} = k_x L + k_{bx} \tag{10.47}$$

$$K_{xr} = K_{rx} = -(k_x L^2/2 + k_{bx} L) \tag{10.48}$$

$$K_{\theta m} = k_x L^3/3 + k_{bx} L^2 + k_\theta L + k_{b\theta} \tag{10.49}$$

where L is the embedded length of pier, $k_x$ the stiffness of translational springs, $k_\theta$ the stiffness of rotational spring, $k_{bx}$ the base translational spring and $k_{b\theta}$ the base rotational spring.

The complex component of stiffness (damping coefficients) can be combined in a similar fashion.

The above equations can be extended to multiple layers along the pile shaft, by using the following replacement terms and summing over these layers:

1. L is replaced by $\Sigma(z_i - z_{i-1})$
2. $L^2$ is replaced by $\Sigma\left(z_i^2 - z_{i-1}^2\right)$
3. $L^3$ is replaced by $\Sigma\left(z_i^3 - z_{i-1}^3\right)$

where $z_i$ is the depth of the base of layer i below the surface. Varun's solutions for the translational spring stiffness values and the corresponding damping coefficients are shown in Table 10.11. Note that the damping values are for the case $E_s = E_b$, that is, a homogeneous

*Table 10.11* Solutions for stiffness and damping of cylindrical pier (L/D = 2–6)

| *Stiffness parameters* | *Dynamic stiffness coefficient* $k'(a_0)$ | *Damping coefficients (for* $E_s = E_b$*)* |
|---|---|---|
| $k_x = 1.828\ Es\ (L/D)^{-0.15}$ | $(1.0 - 0.1a_0)$ | $1.85a_0$ for $a_0 < 1$<br>1.85 for $a_0 > 1$ |
| $k_{bx} = E_b D[0.669 + 0.129(L/D)]$ | 1.0 | $0.6a_0$ for $a_0 < 0.6$<br>0.36 for $a_0 > 0.6$ |
| $k_\theta = E_s D^2[1.106 + 0.227(L/D)]$ | $(1.0–0.225a_0)$ | $-0.21(L/D)a_0$ for $a_0 < 1$<br>$-0.21(L/D)(2 - a_0)$ for $1 < a_0 < 2$<br>0 for $a_0 > 2$ |
| $k_{b\theta}$ can be ignored for L/D > 1 | – | – |

Source: Varun, Assimaki, D. and Gazetas, G. 2009. *Soil Dynamics and Earthquake Engineering*, 29: 268–291.

deep layer. If $E_b > E_s$, then the radiation damping will be reduced and should be disregarded if $a_0$ is less than the value corresponding to the natural frequency of the upper layer.

The expressions in Table 10.11 can be used to derive the impedance functions in Equations 10.47 through 10.49, which can then, via use of Equation 10.45, be inserted into Equation 10.46. For given values of the dynamic lateral load and moment, the lateral deflection and rotation can then be obtained by inversion of the impedance matrix.

The following recommendations have been made by Varun in relation to the horizontal stiffness values:

1. For $L/D < 2$, use the solutions for a shallow embedded foundation.
2. For L/D in the range 2–6, use the solutions shown in Table 10.11.
3. For $L/D > 6$, use the solutions for a single pile (see Section 10.10).

## 10.10 STIFFNESS AND DAMPING OF SINGLE PILES

### 10.10.1 Chart solutions

Pioneering research in this area was carried out by Novak and his associates (e.g. Novak, 1974; Novak and El Sharnouby, 1983). Results were presented in chart and table form for the stiffness and damping constants for single piles subjected to vertical, horizontal, rocking and torsional loadings. The main variables affecting the response were found to be the ratio of pile to soil modulus, the nature of the soil profile (i.e. the variation of soil stiffness with depth) and, for vertical motions, the bearing conditions at the pile tip.

In the solutions provided by Novak and El Sharnouby (1983), the dynamic impedance, in their notation, is defined as follows:

$$KI = k + i\omega c \tag{10.50}$$

where k is the dynamic stiffness (=K′ in Equation 10.34), $\omega$ the circular frequency and c the damping coefficient (=C in Equation 10.34).

The stiffness and damping values are defined in Table 10.12, and the 'f' factors in this table are dimensionless coefficients which depend on the following factors:

- The relative stiffness of pile and soil, expressed as $E_p/G$, where $E_p$ is the pile modulus, G the soil shear modulus
- The dimensionless frequency $a_0 = \omega R/V_s$, where $\omega$ is the circular frequency, R the pile radius and $V_s$ the soil shear wave velocity
- Slenderness ratio L/R, where L is the pile length
- The material damping of both the soil and the pile
- The distribution of soil stiffness with depth, and the founding conditions at the pile tip

*Table 10.12* Stiffness and damping values for single piles, for $a_0 = 0.3$

| *Vertical* | *Horizontal* | *Rocking* | *Coupling* | *Torsion* |
|---|---|---|---|---|
| $k_v = f_{v1} \cdot E_pA/R$ | $k_u = f_{u1} \cdot E_pI/R^3$ | $k_\psi = f_{\psi 1} \cdot E_p.I/R$ | $k_c = f_{c1} \cdot E_pI/R^2$ | $k_\eta = f_{\eta 1} \cdot G_pJ/V_s$ |
| $c_v = f_{v2} \cdot E_pA/V_s$ | $c_u = f_{u2} \cdot E_pI/(R^2V_s)$ | $c_\psi = f_\psi \cdot E_pI/V_s$ | $c_c = f_{c2} \cdot E_pI/(RV_s)$ | $c_\eta = f_{\eta 2} \cdot G_pJ/V_s$ |

Source: After Novak, M. and El Sharnouby, B. 1983. *Journal of Geotechnical Engineering, ASCE* 109(7): 961–974. Courtesy of National Research Council, Canada.

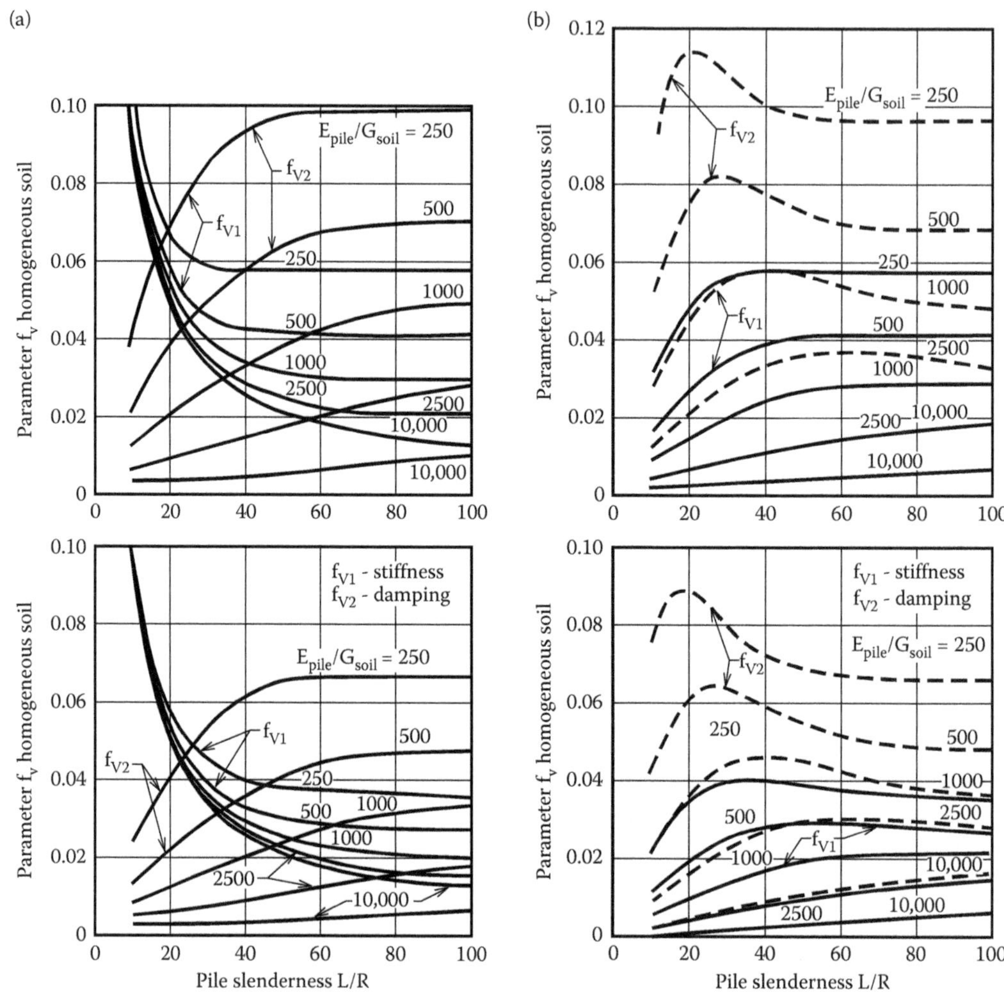


*Figure 10.15* Stiffness and damping values for vertically excited single piles. (a) End bearing piles. (b) Floating piles. (Adapted from Novak, M. and El Sharnouby, B. 1983. *Journal of Geotechnical Engineering*, 109(7): 961–974. Courtesy of National Research Council, Canada.)

In Table 10.12, I and J are the moment of inertia and polar moment of inertia of the pile, respectively, and $G_p$ is the shear modulus of the pile.

Figure 10.15 shows factors for vertical excitation, for both end bearing piles and floating piles, and for a dimensionless frequency of $a_0 = 0.3$. Table 10.13 gives the relevant factors for horizontal response.

## 10.10.2 Approximate closed-form solutions

Gazetas (1991) has presented simplified expressions for the stiffness and damping constants for single piles. For lateral loading, these expressions are for 'long' piles, that is, piles whose length exceeds the critical length discussed in Chapter 8. Three variations of soil shear modulus with depth are considered: linearly increasing, parabolically increasing and constant.

*Table 10.13* Stiffness and damping values for horizontal excitation of single pile

| *Soil profile* | $\nu$ | *Ep/G* | $f_{\psi 1}$ | $f_{c1}$ | $f_{u1}$ | $f_{\eta 1}$ | $f_{\psi 2}$ | $f_{c2}$ | $f_{u2}$ | $f_{\eta 2}$ |
|---|---|---|---|---|---|---|---|---|---|---|
| G constant with depth | 0.25 | 10,000 | 0.2135 | −0.023 | 0.0042 | 0.0021 | 0.1577 | −0.033 | 0.0107 | 0.0054 |
| | | 2500 | 0.2998 | −0.043 | 0.0199 | 0.0061 | 0.2152 | −0.065 | 0.0297 | 0.0154 |
| | | 1000 | 0.3741 | −0.067 | 0.0236 | 0.0123 | 0.2598 | −0.099 | 0.0579 | 0.0306 |
| | | 500 | 0.4411 | −0.093 | 0.0395 | 0.0210 | 0.2953 | −0.134 | 0.0953 | 0.0514 |
| | | 250 | 0.5186 | −0.128 | 0.0659 | 0.0358 | 0.3299 | −0.177 | 0.1556 | 0.0864 |
| | 0.40 | 10,000 | 0.2207 | −0.023 | 0.0047 | 0.0024 | 0.1634 | −0.036 | 0.0119 | 0.0060 |
| | | 2500 | 0.3097 | −0.046 | 0.0132 | 0.0068 | 0.2224 | −0.069 | 0.0329 | 0.0171 |
| | | 1000 | 0.3860 | −0.071 | 0.0261 | 0.013 | 0.2677 | −0.105 | 0.0641 | 0.0339 |
| | | 500 | 0.4547 | −0.099 | 0.0436 | 0.0231 | 0.3034 | −0.142 | 0.1054 | 0.0570 |
| | | 250 | 0.5336 | −0.136 | 0.0726 | 0.0394 | 0.3377 | −0.190 | 0.1717 | 0.0957 |
| G parabolic with depth | 0.25 | 10,000 | 0.1800 | −0.014 | 0.0019 | 0.0008 | 0.1450 | −0.025 | 0.0060 | 0.0028 |
| | | 2500 | 0.2452 | −0.027 | 0.0047 | 0.0020 | 0.2025 | −0.048 | 0.0159 | 0.0076 |
| | | 1000 | 0.3000 | −0.040 | 0.0086 | 0.0037 | 0.2499 | −0.074 | 0.0303 | 0.0147 |
| | | 500 | 0.3489 | −0.054 | 0.0136 | 0.0059 | 0.2910 | −0.101 | 0.0491 | 0.0241 |
| | | 250 | 0.4049 | −0.073 | 0.0215 | 0.0094 | 0.3361 | −0.137 | 0.09793 | 0.0398 |
| | 0.40 | 10,000 | 0.1857 | −0.015 | 0.0020 | 0.0009 | 0.1508 | −0.027 | 0.0067 | 0.0031 |
| | | 2500 | 0.2529 | −0.028 | 0.0051 | 0.0022 | 0.2101 | −0.052 | 0.0177 | 0.0084 |
| | | 1000 | 0.3094 | −0.043 | 0.0094 | 0.0041 | 0.2589 | −0.079 | 0.0336 | 0.0163 |
| | | 500 | 0.3596 | −0.058 | 0.0149 | 0.0065 | 0.3009 | −0.108 | 0.0544 | 0.0269 |
| | | 250 | 0.4170 | −0.078 | 0.0236 | 0.0103 | 0.3468 | −0.146 | 0.0880 | 0.0443 |

Source: After Novak, M. and El Sharnouby, B. 1983. *Journal of Geotechnical Engineering*, 109(7): 961–974. Courtesy of National Research Council, Canada.

For the case of a uniform soil profile, the solutions for stiffness are summarised in Table 10.14 while Table 10.15 shows the solutions for damping. In these solutions, the dynamic impedance is defined as

$$KI = K_s \cdot k_d + i\omega C \tag{10.51}$$

In these solutions, $V_s$ is the shear wave velocity, L the pile length, d the pile diameter, $E_s$ the soil Young's modulus, $E_p$ the pile Young's modulus, $V_{la}$ the wave velocity defined in Table 10.7 and f the frequency of loading.

*Table 10.14* Stiffness of a single pile in a uniform soil mass

| *Mode of vibration* | *Static stiffness $K_s$* | *Dynamic factor $k_d$* | *Remarks* |
|---|---|---|---|
| Vertical $K_v$ | $1.9E_s\, d\, (L/d)^{2/3}(E_p/E_s)^{-(L/d)(Es/Ep)}$ | 1.0 for $L/d < 15$<br>$1+\sqrt{a_0}$ for $L/d \geq 50$<br>Interpolate in between | For $a_0 = \omega d/V_s < 1$ |
| Lateral $K_h$ | $dE_s\, (E_p/E_s)^{0.21}$ | 1.0 | Solution applies for $L \geq 2d(E_p/E_s)^{0.25}$ |
| Rocking $K_r$ | $0.15d^3\, E_s\, (E_p/E_s)^{0.17}/V_s$ | 1.0 | Solution applies for $L \geq 2d(E_p/E_s)^{0.25}$ |
| Lateral-rocking coupling $K_{hr}$ | $-0.22d^2\, E_s\, (E_p/E_s)^{0.50}$ | 1.0 | Solution applies for $L \geq 2d(E_p/E_s)^{0.25}$ |

*Table 10.15* Damping for a single pile in a uniform soil mass

| *Mode of vibration* | *Damping coefficient C or damping ratio D* | *Remarks* |
|---|---|---|
| Vertical | $C_z = a_0^{-0.2}\rho V_s \pi d L r_d$ for $f > f_r$<br>$C_z = 0$ for $f \le f_r$<br>Linearly interpolate for $f_r < f < 1.5f_r$ | $f_r \approx f_c = V_{la}/4H$<br>$r_d = 1 - e^{-X}$<br>$H$ = thickness of soil layer<br>$X = (E_p/E_s)(L/d)^{-2}$<br>Add internal damping contribution $(2K_v\zeta_{sl}/\omega)$ |
| Lateral | $D_h = 0.80\beta + 1.10fd(E_p/E_s)^{0.17}/V_s$ for $f > f_s$<br>or $0.80\beta$ for $f \le f_s$ | Solution applies for<br>$L \ge 2d(E_p/E_s)^{0.25}$<br>$f_s = 0.25V_s/H$<br>$C_h = 2K_hD_h/\omega$ |
| Rocking | $D_r = 0.35\beta + 0.35fd(E_p/E_s)^{0.20}/V_s$ for $f > f_s$<br>or $0.25\beta$ for $f \le f_s$ | Solution applies for<br>$L \ge 2d(E_p/E_s)^{0.25}$<br>$f_s = 0.25V_s/H$<br>$C_r = 2K_rD_r/\omega$ |
| Lateral-rocking coupling | $D_{hr} = 0.80\beta + 0.85fd(E_p/E_s)^{0.18}/V_s$ for $f > f_s$<br>or $0.50\beta$ for $f \le f_s$ | Solution applies for<br>$L \ge 2d(E_p/E_s)^{0.25}$<br>$f_s = 0.25V_s/H$<br>$C_{hr} = 2K_{hr}D_{hr}/\omega$ |

## 10.11 STIFFNESS AND DAMPING OF PILE GROUPS

### 10.11.1 Two-pile interaction

The method of pile group analysis via superposition of interaction factors has been described by Poulos and Davis (1980) for static loading cases and is discussed in Chapter 8. A similar approach has been found to be possible for dynamic loading, and closed form approximations are presented by Gazetas (1991) for the dynamic interaction factors between piles in a uniform deep soil layer, for both vertical and horizontal loading. These solutions are summarised in Table 10.16.

*Table 10.16* Dynamic interaction factors for two piles

| *Mode of vibration* | *Interaction factor* $\alpha$ | *Remarks* |
|---|---|---|
| Vertical | $\alpha_z \approx (2s/d)^{-0.5}e^{-Y1} \cdot e^{-Y2}$ | $Y_1 = \beta\omega s/V_s$<br>$Y_2 = i\omega s/V_s$<br>s = pile centre-to-centre spacing,<br>$\beta$ = internal damping ratio for soil<br>Note that $Y_2$ is a complex number |
| Lateral | $\alpha_h \approx 0.75\alpha_z$ (for $\theta = 90°$)<br>$\alpha_h \approx 0.75 \cdot (s/d) - 0.5e^{-X1}e^{-X2}$ (for $\theta = 0°$)<br>$\alpha h(\theta°) = \alpha_h(0°)\cos^2\theta + \alpha_h(90°)\sin^2\theta$ | Solution applies for<br>$L \ge 2d(E_p/E_s)^{0.25}$<br>$X_1 = \beta\omega s/V_{la}$<br>$X_2 = i\omega s/V_{la}$<br>$\theta$ = angle between line joining piles and direction of lateral loading<br>Note that $X_2$ is a complex number |
| Rocking | $\alpha_r = 0$ | |
| Lateral-rocking coupling | $\alpha_{hr} = 0$ | |

Source: After Gazetas, G. 1991. *Foundation Engineering Handbook*, 2nd Ed., Chapman & Hall, New York, pp. 563–593.

Difficulties arise because the interaction factors for dynamic loading are complex, and do not monotonically decrease with increasing spacing between the two piles. Rather, the interaction factors can oscillate, depending on the pile spacing and the loading frequency. Consequently, the effects of pile–soil–pile interaction on stiffness will differ from those on damping.

### 10.11.2 Superposition for pile groups

In using interaction factors to obtain stiffness and damping values for a group of piles, it is necessary to apply them to the single pile impedance, which incorporates both the stiffness and damping of the single pile. For a group of piles subjected to dynamic vertical loading, the impedance function $KI_{vG}$ for a group will be given as follows (El Naggar, 2001):

Vertical impedance:

$$KI_{vG} = kI_{v1}\sum\sum \varepsilon v_{i,j} \tag{10.52}$$

where $kI_{v1}$ is the vertical impedance of a single pile and $\varepsilon v_{i,j} = [\alpha v]^{-1}$ the inverted matrix of the complex vertical interaction factors $\alpha v_{i,j}$ between two piles i and j within the group.

Similar expressions hold for the lateral and rotational responses of the group are set out below (Mitwally and Novak, 1987).

Horizontal impedance:

$$KI_{vG} = kI_{h1}\sum\sum \varepsilon h_{i,j} \tag{10.53}$$

where $kI_{h1}$ is the horizontal impedance of a single pile and $\varepsilon h_{i,j} = [\alpha h]^{-1}$ the inverted matrix of the complex horizontal interaction factors $\alpha h_{i,j}$ between two piles i and j within the group.

Rotational impedance: This value is derived from two components, the moments required to produce unit rotations at the pile heads, and the moments resulting from the vertical pile forces.

$$KI_{rG} = kI_{r1}\sum\sum \varepsilon r_{i,j} + kI_{v1}\sum\sum \varepsilon v_{i,j}x_i x_j \tag{10.54}$$

where $kI_{v1}$ = is the rotational impedance of a single pile, and $\varepsilon r_{i,j} = [\alpha r]^{-1}$ = the inverted matrix of the complex interaction factors $\alpha r_{i,j}$ between two piles i and j within the group, $kI_{v1}$ = the vertical impedance of a single pile, $\varepsilon v_{i,j} = [\alpha v]^{-1}$ = the inverted matrix of the complex vertical interaction factors $\alpha v_{i,j}$ between two piles i and j within the group, and $x_i$ and $x_j$ are the distances from piles i and j from the axis of rotation for the direction of loading.

Horizontal-rocking cross impedance:

$$KI_{hrG} = kI_{hr1}\sum\sum \varepsilon hr_{i,j} \tag{10.55}$$

where $kI_{hr1}$ is the horizontal-rocking cross impedance of a single pile and $\varepsilon hr_{i,j} = [\alpha v]^{-1}$ the inverted matrix of the complex horizontal-rocking interaction factors $\alpha hr_{i,j}$ between two piles i and j within the group.

In all cases, the double summations in the above expressions are carried out for all the piles within the group.

### 10.11.3 Use of dynamic group factors

A useful approximate approach has been developed by Gazetas et al. (1993) in which dynamic group factors for stiffness and damping are developed and applied to the single pile values. These group factors are given for various pile groups within a deep soil layer, with configurations ranging from a $1 \times 2$ group to a $6 \times 6$ square group. Results are presented for vertical, horizontal and rotational impedances, and in each case, a dynamic stiffness group factor and a damping group factor are applied to the real and imaginary components of the single pile dynamic impedance respectively. The effects of a 5% internal damping ratio are also included on their results.

The group factors are functions of the following variables:

1. Pile centre-to-centre spacing
2. Dimensionless frequency of applied loading
3. Pile group size and configuration
4. Relative stiffness of pile to soil
5. Pile length to diameter ratio
6. The distribution of soil modulus with depth

Unfortunately, the results provided are for a limited range of parameters which may not be relevant to many high-rise situations.

El Naggar and El Naggar (2007) have presented a similar approach to the Gazetas et al. approach, with computed values of a complex group factor that can be applied to the values of impedance for a single pile. For both vertical and lateral loading, the group impedance, $KI_G$, is expressed as follows:

$$KI_G = KI_s n/\alpha_G \tag{10.56}$$

where $KI_s$ is the impedance of single pile, $\alpha_G$ the complex group factor and n the number of piles in the group.

The complex group factor $\alpha_G$ can be expressed as follows:

$$\alpha_G = \alpha_1 + i\alpha_2 \tag{10.57}$$

where $\alpha_1$ and $\alpha_2$ are real and imaginary components of the dynamic group factor.

El Naggar and El Naggar (2007) have obtained solutions for the group factors $\alpha_G$ for a limited number of cases involving square $2 \times 2$, $3 \times 3$ and $4 \times 4$ groups, for the following parameters: $L/d = 20$, internal damping ratio $= 0.05$, soil Poisson's ratio $= 0.4$, spacing ratio $s/d = 2$, 5 and 10, and for various ratios of the relative pile stiffness $E_p/G_s$, where $E_p$ = pile Young's modulus, $G_s$ = soil shear modulus. The soil is assumed to be a uniform semi-infinite elastic mass. Only group factors for vertical and horizontal motions are provided.

The effect of $E_p/G_s$ is relatively small, and Figures 10.14 through 10.16 reproduce their results for a typical value of $E_p/G_s = 1000$. In these figures, the dimensionless frequency, $a_0$, is shown on the horizontal axis, and is defined as

$$a_0 = \omega d/V_s \tag{10.58}$$

where $\omega$ is the circular frequency of loading ($=2\pi f$), f the frequency of loading, d the pile diameter and $V_s$ the shear wave velocity of soil.

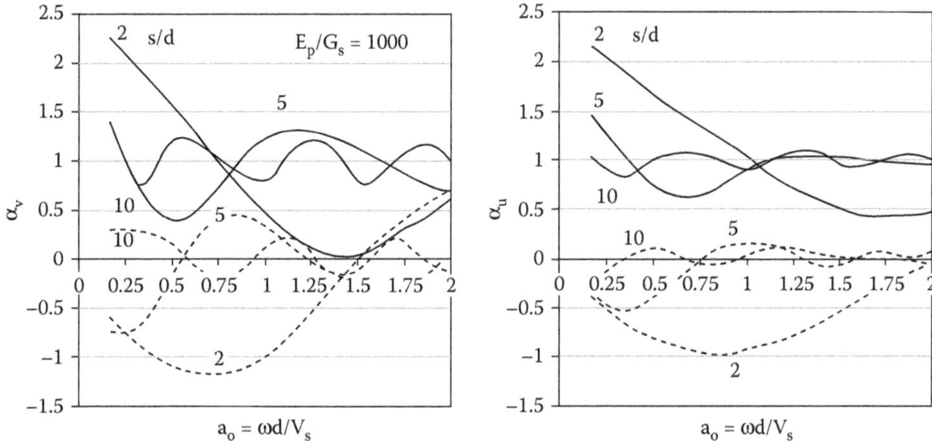


*Figure 10.16* Vertical and horizontal interaction factors for 2 × 2 group. Full lines are for stiffness, dashed lines are for damping. (Adapted from El Naggar, H. and El Naggar, M.H. 2007. Simplified approximate approach to group effect in pile dynamics. *Proceedings of the 4th International Conference on Earthquake Geotechnical Engineering*, Paper No. 1325, Thessaloniki, ISSMGE.)

The following points should be noted:

1. The group factors tend to oscillate depending on the pile spacing and the frequency of loading.
2. The group factors shown in Figure 10.16 through 10.18 are for piles within a deep uniform layer. Accordingly, the interaction effects indicated by these solutions may well be overestimated for piles within a layer or a profile of limited depth. In particular, if the frequency of loading is less than the natural frequency of the soil profile, then radiation damping may be negligible and thus only internal damping of the soil can be relied upon.

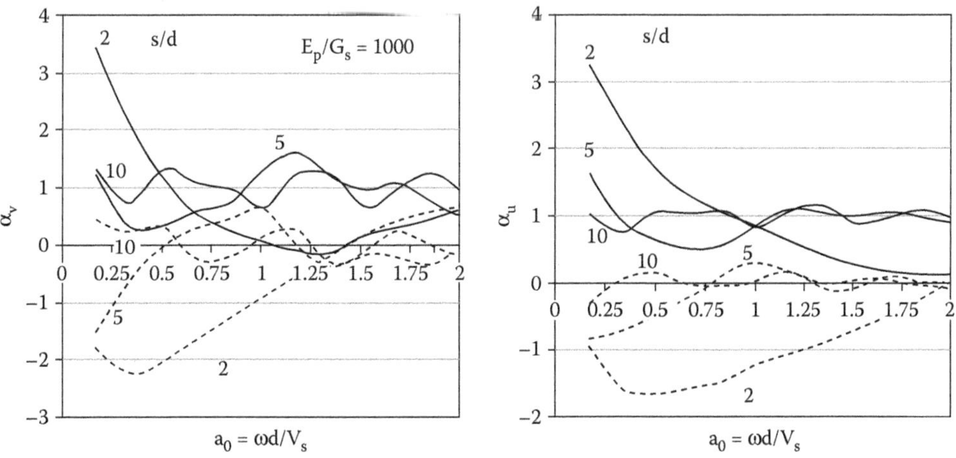


*Figure 10.17* Vertical and horizontal interaction factors for 3 × 3 group. Full lines are for stiffness, dashed lines are for damping. (Adapted from El Naggar, H. and El Naggar, M.H. 2007. Simplified approximate approach to group effect in pile dynamics. *Proceedings of the 4th International Conference on Earthquake Geotechnical Engineering*, Paper No. 1325, Thessaloniki, ISSMGE.)

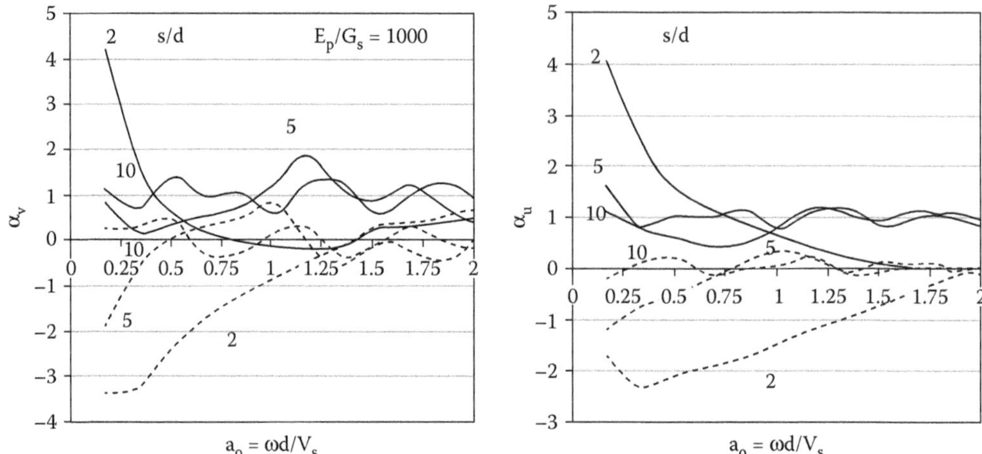


*Figure 10.18* Vertical and horizontal interaction factors for 4 × 4 group. Full lines are for stiffness, dashed lines are for damping. (Adapted from El Naggar, H. and El Naggar, M.H. 2007. Simplified approximate approach to group effect in pile dynamics. *Proceedings of the 4th International Conference on Earthquake Geotechnical Engineering*, Paper No. 1325, Thessaloniki, ISSMGE.)

The approach developed by El Naggar and El Naggar may be summarised as follows:

1. Obtain the single pile stiffness and damping values from the results in Section 10.9.
2. Obtain the equivalent group interaction factors from the nearest square group to the width of the group being analysed. Figure 10.19 shows an example for an 8 × 16 group, where the factors for a 4 × 4 group are used. Piles that are spaced more than 20d away will have an insignificant effect on the piles being considered in this equivalent group, and this can help to assess the most appropriate smaller square group to be used.
3. Calculate the overall group stiffness and damping using Equations 10.56 and 10.57. It is implicit that the number of piles, n, is the total number of piles in the group.

For low loading frequencies, for example with $a_0 < 0.1$, it may be reasonable to use the static interaction factors in place of the dynamic values. In such cases, the group effects will

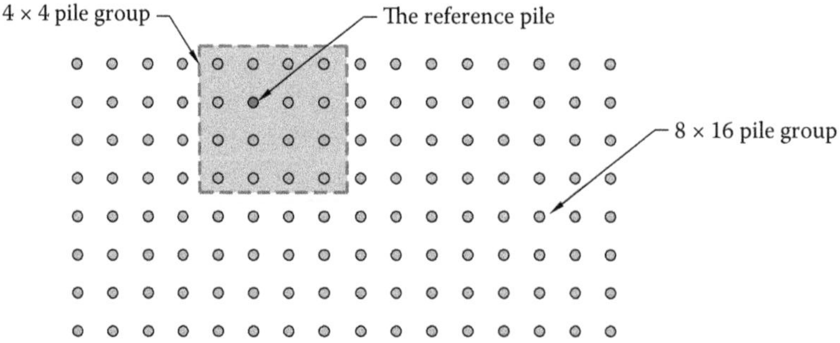


*Figure 10.19* Example of use of group interaction factors. (Adapted from El Naggar, H. and El Naggar, M.H. 2007. Simplified approximate approach to group effect in pile dynamics. *Proceedings of the 4th International Conference on Earthquake Geotechnical Engineering*, Paper No. 1325, Thessaloniki, ISSMGE.)

be similar to those for static loading, and the approximations given in Chapter 8 can then be employed for the group stiffness values.

### 10.11.4 Some observations on dynamic pile group response

Kaynia and Kausel (1982) have demonstrated some of the unusual dynamic interaction characteristics of piles within a group, as contrasted to the more familiar static interaction effects. They present the dynamic stiffness components for a 4 × 4 square pile group as a function of dimensionless frequency, $a_0$, for three values of centre-to-centre spacing, 2d, 5d and 10d, where d is the pile diameter. The group stiffness and damping values are presented as a proportion of the stiffness and damping of 16 individual piles without any interaction. For some combinations of spacing and frequency, the stiffness can become negative, that is, the motion is in the opposite direction to that of the loading.

Dobry and Gazetas (1988) and Gazetas et al. (1993) have also made some useful observations on the characteristics of pile group behaviour under dynamic loading, including the following:

1. As might be expected, group interaction becomes more pronounced as the number of piles in the group increases.
2. There is a radical change of group stiffness and damping when there are two piles rather than one. In a linear group (piles in a line), increasing the number of piles beyond two or three has only a relatively small effect on the dynamic stiffness and damping factors.
3. The interaction effect becomes more pronounced as the number of rows within the group increases, and 'peaks' and 'valleys' appear in the relationships between stiffness versus frequency and damping versus frequency. Thus, at certain frequencies, there can be a dynamic group efficiency greater than unity.
4. The frequency dependence of group response appears to be more marked for the translational modes of vibration than for the rotational modes.
5. In some cases, at relatively high frequencies, the dynamic vertical stiffness can become negative, especially in groups with closely spaced piles.
6. The use of static interaction factors cannot properly predict the dynamic response of pile groups, except at very low frequencies. Certainly, the use of static interaction factors cannot replicate dynamic group efficiencies greater than unity.
7. As mentioned previously, when the frequency is less than a 'cut-off' frequency (approximately equal to the natural frequency of the soil profile), radiation damping becomes very small and can be ignored. This 'cut-off' circular frequency is $\omega_s = \pi V_s/2H_s$ for horizontal vibrations, and $\omega_c = 3.4\omega_s/\pi(1 - \nu_s)$ for vertical vibrations, where $V_s$ is the average shear wave velocity, $\nu_s$ the soil Poisson's ratio and $H_s$ is the thickness of soil layer.
8. For very close spacings, for example, s/d = 2, the stiffness and damping exhibit a smooth variation with frequency, with the pile group behaving in a similar manner to an isolated embedded foundation. The soil mass between the piles tends to vibrate in phase with the piles and so the pile group-soil system responds as a 'block'. Thus, in such cases, the use of a dynamic equivalent pier or buried footing should provide a good approximation to the group behaviour.
9. For larger spacings, for example, s/d = 5, groups exhibit a more complicated behaviour, with both the stiffness and damping having 'peaks' and 'valleys' which depend on the group size and the pile spacing.

Caution should be exercised in applying the dynamic interaction factor approach to large groups of piles. As pointed out by Dobry and Gazetas (1988), the interaction between two piles is likely to be overestimated by the theoretical expressions, as scattering of waves and interference by the in-between piles will occur. It must also be borne in mind that the available solutions for dynamic interaction factors are for infinitely deep soil layers, and thus will tend to overestimate interaction effects within a layer (or layers) of limited thickness. Accordingly, it seems prudent to undertake checks with a simple equivalent pier, equivalent buried footing or equivalent raft method to guard against the possibility that the more detailed interaction analysis may give unrealistic or unreliable results.

### 10.11.5 Comparative example

It is of interest to compare the results for analysing the dynamic response of a pile group, using both equivalent pier approaches and approaches considering dynamic pile interaction. For this purpose, an idealised case has been considered in which a group of 9 identical piles is located in a semi-infinite uniform soil mass. For the piles, the length is 15 m, the diameter is 1.0 m, Young's modulus is 30,000 MPa, and the mass density is 2.4 t/m$^3$. The piles are in a square 3 × 3 configuration, with a centre-to-centre spacing of 2 m for the first case, and 5 m for the second case. For the soil, Young's modulus of the soil is 30 MPa, Poisson's ratio is 0.40 and the mass density is 1.7 t/m$^3$. The assumed frequency of loading is 2 Hz.

Calculations have been carried out using the following approaches:

1. The solutions from a dynamic pile group analysis provided by Dobry and Gazetas (1988).
2. A static analysis of the pile group, using the program DEFPIG.
3. A static analysis of the pile group, using the program PIGLET.
4. A static analysis of an equivalent pier, using the solution for vertical and lateral loading on a rigid pier by Kulhawy and Carter (1992).
5. An equivalent raft analysis using the solutions of Gazetas (1991) for an embedded footing in a half-space (Section 10.8.2), and assuming various equivalent depths for the raft.
6. A dynamic analysis of an equivalent buried footing by Gazetas (1991).
7. A dynamic analysis of an equivalent buried footing by El Naggar (2001).
8. A dynamic analysis of an equivalent pier using the results of Varun et al. (2009).
9. The simplified group factor analysis of El Naggar and El Naggar (2007).

Attention is concentrated on the stiffness and damping values for vertical, horizontal and moment loadings.

Table 10.17 shows the computed values from the various methods for the case of spacing/diameter = 2, while Table 10.18 shows the corresponding results for spacing/diameter = 5. The values of damping coefficient have been expressed in terms of the value C, as defined in Equation 10.34, and the internal damping ratio is taken as 0.05.

The following observations can be made from Tables 10.17 and 10.18:

1. Most of the approaches considered give comparable values of vertical stiffness. However, the equivalent buried footing approach of El Naggar (2001) gives somewhat higher values of the vertical stiffness.
2. The static DEFPIG and PIGLET analyses appear to underestimate the lateral group stiffness as compared with the dynamic analyses.
3. The values of dynamic lateral and rotational stiffness for a buried footing and an equivalent pier are all significantly greater than the values from the static DEFPIG and PIGLET analyses.

*Table 10.17* Computed values of stiffness and damping for 3 × 3 pile group by various methods[a]

| *Approach* | *Reference* | $K_z$ *(MN/m)* | $C_z$ *(MN s/m)* | $K_h$ *(MN/m)* | $C_h$ *(MN s/m)* | $K_r$ *(MNm/rad)* | $C_r$ *(MN ms/rad)* |
|---|---|---|---|---|---|---|---|
| Dynamic pile group analysis | Gazetas et al. (1993) | 464 | 62 | 276 | 14 | $1.4 \times 10^4$ | 196 |
| Static pile group analysis DEFPIG | Poulos and Davis (1980) | 511 | – | 262 | – | $1.04 \times 10^4$ | – |
| Static pile group analysis PIGLET | Randolph (1996) | 556 | – | 270 | – | $1.13 \times 10^4$ | – |
| Equivalent pier-static | Kulhawy and Carter (1992) | 605 | – | – | – | – | – |
| Dynamic Equivalent raft via Gazetas (1991) | Gazetas (1991) $h_e = 2L/3$ | 502 | 16 | 661 | 15 | $1.74 \times 10^4$ | 143 |
| | Gazetas (1991) $h_e = L/2$ | 426 | 14 | 533 | 13 | $0.91 \times 10^4$ | 77 |
| Dynamic buried footing | El Naggar (2001) | 606 | 54 | 878 | 75 | $4.67 \times 10^4$ | 2955 |
| Dynamic buried footing | Gazetas (1991) | 601 | 47 | 794 | 59 | $2.51 \times 10^4$ | 2880 |
| Equivalent rigid pier-dynamic | Varun et al. (2009) | – | – | 817 | 56 | $6.31 \times 10^4$ | 1472 |
| Dynamic group factor | El Naggar and El Naggar (2007) | 541 | 41 | 294 | 14 | – | – |

s/d = 2.

[a] Analysis includes 5% internal damping.

*Table 10.18* Computed values of stiffness and damping for 3 × 3 pile group by various methods[a]

| *Approach* | *Reference* | $K_z$ *(MN/m)* | $C_z$ *(MN s/m)* | $K_h$ *(MN/m)* | $C_h$ *(MN s/m)* | $K_r$*(MN m/rad)* | $C_r$ *(MN ms/rad)* |
|---|---|---|---|---|---|---|---|
| Dynamic pile group analysis | Gazetas et al. (1993) | 464 | 155 | 449 | 30 | $4.1 \times 10^4$ | 2760 |
| Static pile group analysis DEFPIG | Poulos and Davis (1980) | 747 | – | 426 | – | $4.33 \times 10^4$ | – |
| Static pile group analysis PIGLET | Randolph (1996) | 556 | – | 309 | – | $1.27 \times 10^4$ | – |
| Equivalent pier-static | Carter and Kulhawy (1991) | 767 | – | 955 | – | $5.39 \times 10^4$ | |
| Dynamic Equivalent raft via Gazetas (1991) | Gazetas (1991) $h_e = 2L/3$ | 767 | 75 | 955 | 54 | $5.39 \times 10^4$ | 583 |
| | Gazetas (1991) $h_e = L/2$ | 687 | 46 | 842 | 36 | $3.49 \times 10^4$ | 583 |
| Dynamic buried footing | El Naggar (2001) | 443 | 463 | 332 | 406 | 1178 | $1.2 \times 10^4$ |
| Dynamic buried footing | Gazetas (1991) | 774 | 118 | 1040 | 150 | $6.13 \times 10^4$ | 7660 |
| Equivalent rigid pier-dynamic | Varun et al. (2009) | – | – | 949 | 296 | $12.8 \times 10^4$ | 5980 |
| Dynamic group factor | El Naggar and El Naggar (2007) | 630 | 113 | 504 | 36 | – | – |

s/d = 5.

[a] Analysis includes 5% internal damping.

4. The values of lateral and rotational damping from the buried footing and equivalent pier are all significantly greater than the values given by the other dynamic approaches.
5. The damping coefficient for vertical motion is similar from the Gazetas and El Naggar (2001) solutions, but the value from the El Naggar and El Naggar (2007) approach is somewhat higher.
6. There is a disparity between the computed values of damping for lateral response, although the values for rotational response are similar.

It should be noted that the numerical results from the El Naggar and El Naggar (2007) approach are very sensitive to the values of the dynamic interaction factors interpolated from Figure 10.18. While this approach is inherently attractive, it would appear that its accuracy relies on using accurate values of the interaction factors, and such values are difficult to obtain from the published figures. It should further be noted that the group results from the Gazetas et al. solutions will depend on the method of calculating the single pile impedance values.

The results for this competitive example suggest that considerable caution needs to be exercised in employing the equivalent pier approach to assess the dynamic response of a pile group. In particular, there appears to be a strong tendency for the lateral stiffness and damping values to be significantly larger than the values derived from a static analysis. It would therefore appear prudent to use a static pile group analysis as a guide to the dynamic stiffness values, while the internal damping can be considered as a lower limit to the overall damping.

## 10.12 CATEGORY 3 ANALYSES

Category 3 analyses of the dynamic response of pile foundations include the following approaches:

1. Methods in which the stiffness matrix for a pile group is assembled from the single pile stiffness values and the corresponding dynamic interaction factors and the ensuing equations solved for the specified loadings. El Naggar and Novak (1995) describe the development of the matrices and their solution, while Mitwally and Novak (1987) demonstrate how superstructure stiffness can also be incorporated. Typical of programs of this type is DYNAPILE (Ensoft, 2014) and DYNA6.1 (El Naggar, 2014). DYNA6.1 provides values of foundation stiffness and damping which can be used in soil–structure interaction analyses. Rigid footings, flexible mats (with or without piles) and pile groups can be considered. For rigid footings, all six degrees of freedom are considered as coupled. Transient, shock, harmonic, pulse and random loadings can be analysed.
2. Finite element programs with a dynamic analysis capability. DYNAFLOW, PILE-3D, ANSYS, ABAQUS and PLAXIS are examples of such programs.
3. Finite difference programs with a dynamic analysis capability, such as FLAC3D (Itasca, 2006).

Such methods are inherently powerful and can take into account, at least to some extent, the complexities of subsurface geometry, pile configuration, load application and real soil behaviour. On the other hand, they also may have limitations in their practical use, in terms of the assignment of geotechnical parameters, and uncertainties in the nature of the dynamic loading. As with all other aspects of foundation design, if a Category 3 analysis is employed, it should be checked with a simpler Category 2 method to avoid the possibility of spurious or erroneous results.

Chapter 11

# Design for seismic events

## 11.1 INTRODUCTION

Consideration of the effects of earthquakes and seismic loadings is an increasingly important aspect of modern foundation design, and most contemporary standards have a mandatory requirement for such consideration. For example, the Australian Piling Code, AS 2159-2009, states that 'a pile shall be designed for adequate strength, stiffness and ductility under load combinations including earthquake design actions'. The designer must then consider the following issues:

1. 'Inertial' effects, via loads applied to the pile by the supported structure
2. 'Kinematic' effects, via ground movements generated by the earthquake acting on the pile
3. Possible loss of soil support during the earthquake due to liquefaction or partial loss of soil strength

However, the methods by which such considerations can be undertaken are generally not set out in the standards, and a wide range of approaches have been utilised, ranging from very simplistic methods to extremely complex computer analyses. Accordingly, there appears to be scope for an approach that is soundly based but which is neither too simplistic nor too complex.

As set out by Villaverde (2009), earthquake-resistant design requires the participation of various professionals, including architects, seismologists, geologists, geotechnical and foundation engineers and structural engineers, and will involve many of the following steps:

1. Identification of likely future earthquake sources
2. Assessment of the probable size of future earthquakes on the basis of the attributes of the identified sources
3. Assessment of the distance and orientation of each seismic source with respect to the location of the structure
4. Establishment of correlations between ground motion characteristics and earthquake size, orientation and distance
5. Dynamic analysis of the soil deposits at the site to quantify possible amplification of bedrock motions
6. Selection of the structure and its components to best resist earthquake effects
7. Dynamic analysis of the structure and its components to estimate the maximum internal forces and displacements that may be generated
8. Design and detailing of the structural members and connections in accordance with the computed internal forces and deflections
9. Analysis of the foundation soil profile to assess its susceptibility to earthquake effects, such as liquefaction

10. Analysis of the foundation system to estimate the forces and bending moments induced by earthquake actions
11. Assessment of the need to improve the foundation soil properties to reduce the susceptibility of the site to earthquake effects

The objective of this chapter is to set out a systematic approach by which the foundation designer can address the relevant aspects of those mentioned above, including assessment of the relevant earthquake characteristics, quantification of the relevant geotechnical parameters and the undertaking of relevant calculations to satisfy the foundation design requirements for seismic regions.

The following matters are dealt with:

1. A summary of the main effects of earthquakes
2. Assessment of seismic hazard
3. Characteristics of the design earthquake
4. Structural response spectra
5. Site response analyses
6. Assessment of liquefaction potential
7. Pile design for cases where liquefaction does not occur
8. Pile design for cases in which liquefaction does occur
9. Measures to mitigate against liquefaction effects

Emphasis is placed herein on Category 2 methods that do not require 'black box' software or which employ complex soil models in which the physical meaning of the parameters may be unclear.

## 11.2 THE KEY EFFECTS OF EARTHQUAKES RELATED TO FOUNDATION BEHAVIOUR

Some of the key effects of earthquakes relevant to foundation design may be summarised as follows:

1. Motion of the causative fault causes an excitation of the bedrock below the near-surface ground profile, and this is manifested as a time-dependent acceleration at the surface of the bedrock.
2. There is a transmission of waves (with shear waves generally being predominant) from bedrock through soil profile to the surface. This gives rise to ground movements which are in turn time-dependent, and which will act on the foundations. There may be an amplification of ground motions during the wave transmission, particularly if the predominant period of the earthquake is similar to the natural period of the ground profile above bedrock. This ground motion amplification is often quantified by an amplification factor or a 'site factor' which relates the peak acceleration at (or below) the ground surface to the peak acceleration at bedrock level.
3. Structures and their supporting foundations will respond to the ground motions, and there may be a further amplification of motions if the natural period of the structure is similar to the natural period of the ground profile above bedrock. The response of the structure is generally represented by a response spectrum, which relates the maximum acceleration (or velocity, or displacement) of the structure to its natural period. Such a relationship forms one means of assessing the earthquake-induced inertial forces which the structure will impose on the foundation system.

Therefore, seismic foundation design will require the assessment of the relevant earthquake characteristics and the effects of the transmission of waves from the bedrock through the ground to the surface. The following sections summarise a simplified but practical engineering approach to the assessment of earthquake characteristics.

## 11.3 ESTIMATION OF SEISMIC HAZARD AND EARTHQUAKE CHARACTERISTICS

### 11.3.1 Introduction

The assessment of the seismic hazard due to future earthquakes involves the three main steps that are illustrated in Figure 11.1:

1. The development of a seismicity model for the location and size of future earthquakes in the region
2. The development of a ground motion model to predict the expected levels of seismic shaking at the site, arising from any of these earthquake scenarios
3. The integration of these two models to give the expected levels of shaking at the site

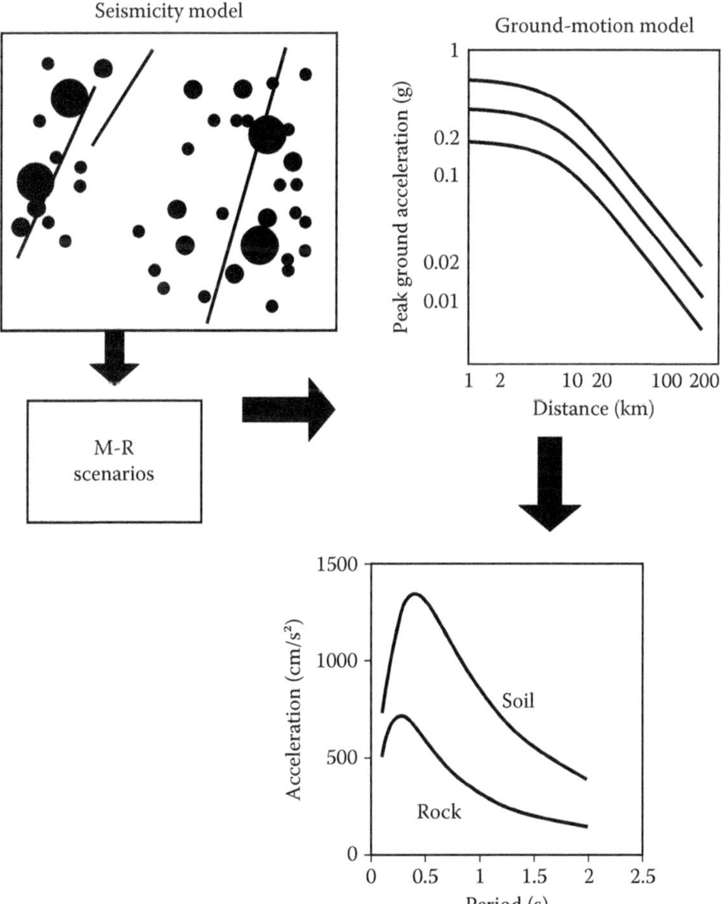


*Figure 11.1* Process of seismic hazard analysis.

It is desirable that this assessment be carried out by an experienced seismologist, but it is nevertheless important for the foundation designer to have some general appreciation of the processes involved.

The seismicity model considers scenarios of earthquakes of magnitude M at a distance R from the site. The ground motion model predicts the shaking parameter of interest (e.g., ground acceleration) for this combination of magnitude and distance. The results are usually expressed in terms of response spectra relating a response (acceleration, velocity and/or displacement) to the natural period of a structure at the site.

More details of the process are given by Bommer and Stafford (2009).

### 11.3.2 Characteristics of the design earthquake

The major characteristics of the design earthquake that need to be assessed are

1. The earthquake size or magnitude
2. The seismicity rate
3. The maximum bedrock acceleration, and its attenuation with distance from the source fault
4. The duration
5. The predominant period
6. The time–acceleration relationship at bedrock level

### 11.3.3 Earthquake size or magnitude

There are two common measures of the size of an earthquake:

1. The *earthquake intensity*, which is a qualitative measure of the effects of an earthquake at a particular location. The most common measure of intensity is the modified Mercalli intensity (MMI). Table 11.1 reproduces the significant parts of the MMI scale in terms of the effects to structures and facilities.
2. The *earthquake magnitude*, which is a quantitative measure of the size of the earthquake. The traditional measure of earthquake magnitude is the Richter magnitude ($M_L$), now known as the local magnitude. This is defined as the logarithm (base 10)

*Table 11.1* Modified Mercalli scale of earthquake intensity

| *MMI* | *Description of effects* |
|---|---|
| VI | Felt by all, many frightened. Some heavy furniture moved. A few instances of fallen plaster. Damage slight. |
| VII | Damage negligible in buildings of good design and construction; slight to moderate in well-built ordinary structures. Considerable in poorly built structures. Some chimneys broken. |
| VIII | Damage slight in specially design structures; considerable in ordinary substantial buildings with partial collapse; great in poorly built structures. Fall of chimneys, factory stacks, columns, walls. Heavy furniture overturned. |
| IX | Damage considerable in specially designed structures; well-designed frame structures thrown out of plumb. Damage great in substantial buildings, with partial collapse. Buildings shifted off foundations. |
| X | Some well-built wooden structures destroyed; most masonry and frame structures with foundations destroyed. Rails bent. |
| XI | Few, if any, masonry structures remain standing. Bridges destroyed. Rails bent greatly. |
| XII | Damage total. Lines of sight and level are distorted. Objects thrown into the air. |

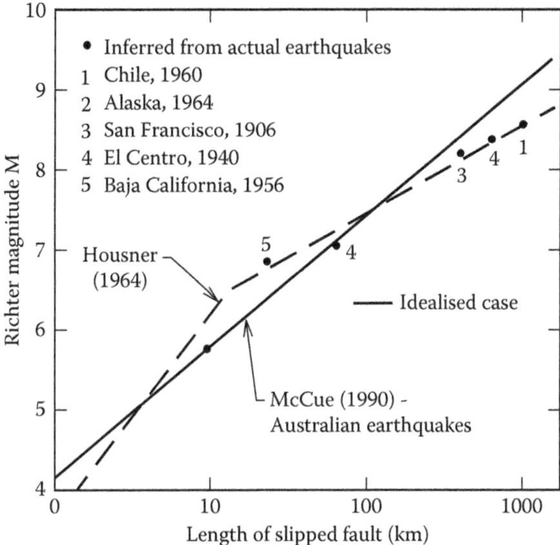


*Figure 11.2* Relationship between length of fault and Richter magnitude.

of the maximum trace amplitude (in μm) of a seismometer located 100 km from the epicentre of the earthquake. For larger earthquakes, a more commonly used measure is the moment magnitude ($M_w$), which is related to the seismic moment. Up to about $M_w = 6.5$, $M_L$ and $M_w$ are very similar, but for larger events, $M_w$ becomes the most appropriate measure.

The Richter magnitude $M_L$ has been related to the length of the rupture fault (see Figure 11.2). This figure includes data from Australia for intra-plate events, for which the relationship is similar to that for inter-plate earthquakes.

Approximate relationships between MMI and $M_L$ have been proposed, for example, by Lam et al. (2003) for two areas in Australia:

$$\mathrm{MMI} = 1.5\,M_L - 3.2\log R + 2.2 \text{ (Western Australia)} \tag{11.1}$$

$$\mathrm{MMI} = 1.5\,M_L - 3.9\log R + 3.9 \text{ (South-Eastern Australia)} \tag{11.2}$$

where R is the distance between the source and the site.

### 11.3.4 Seismicity rate

The seismicity rate refers to the frequency of occurrence of an earthquake of a particular magnitude within the region of interest. It is usually described by the following relationship developed by Gutenberg and Richter (1954):

$$\mathrm{Log}(N) = A - b \cdot M_L \tag{11.3}$$

where
N is the number of earthquakes within a time interval, with magnitudes greater or equal to $M_L$,
A the number of earthquakes exceeding magnitude zero and
b the recurrence parameter

*Table 11.2* Seismicity rate parameters

| *Area* | A | b |
|---|---|---|
| East of Perth, Western Australia | 3.66 | 0.94 |
| SE Queensland | 2.10 | 0.66 |

Source: Gaull, B.A., Michael-Leiba, M.O. and Rynn, J.M.W. 1990. *Australian Journal of Earth Sciences*, 37: 169–187.

Values of A and b for two areas in Australia are given in Table 11.2 (Gaull et al., 1990).

### 11.3.5 Maximum bedrock acceleration

The maximum bedrock acceleration is related to the magnitude of the earthquake and the distance from the source. A great number of attenuation curves (peak acceleration vs. distance) have been proposed, many of which have been summarised by Douglas (2002). For Australian conditions, attenuation curves have been proposed by Rynn (1986) and Gaull et al. (1990), and in each case, the following empirical relationship has been used:

$$a_{max} = \frac{c_1 \cdot \exp(c_2 M_L)}{R^{c_3}} \tag{11.4}$$

where
- $a_{max}$ is the peak ground acceleration (PGA) (m/s$^2$),
- $M_L$ the Richter magnitude,
- R the hypocentral distance (km) $= (D^2 + h^2)^{0.5}$,
- D the epicentral distance (km),
- h the depth of fault below surface (km) and
- $c_1$, $c_2$, $c_3$ are attenuation constants.

For four Australian/Pacific regions, values of the attenuation constants are given in Table 11.3.

It should be noted that Equation 11.4 can also be applied to other earthquake characteristics such as maximum velocity and intensity (MMI). Appropriate constants for these components are given by Gaull et al. (1990).

Gaull et al. (1990) have also provided contours of PGA for Australia. These are presented in probabilistic terms, as values of the 'hazard factor' Z, the proportion of gravitational acceleration which has a 10% chance of being exceeded in a 50-year period. These

*Table 11.3* Acceleration attenuation coefficients

| | *Attenuation coefficients* | | |
|---|---|---|---|
| *Region* | $c_1$ | $c_2$ | $c_3$ |
| Western Australia | 0.025 | 1.10 | 1.03 |
| SE Australia | 0.088 | 1.10 | 1.20 |
| NE Australia | 0.060 | 1.04 | 1.08 |
| Indonesia (for Indonesian Zone, R > 300 km) | 37.6 | 1.11 | 2.08 |

Source: Gaull, B.A., Michael-Leiba, M.O. and Rynn, J.M.W. 1990. *Australian Journal of Earth Sciences*, 37: 169–187.

*Table 11.4* Probability factor $k_p$ – AS 1170.4-2007

| *Annual probability of exceedance* | *Probability factor $k_p$* |
|---|---|
| 1/2500 | 1.8 |
| 1/2000 | 1.7 |
| 1/1500 | 1.5 |
| 1/1000 | 1.3 |
| 1/800 | 1.25 |
| 1/500 | 1.0 |
| 1/250 | 0.75 |
| 1/200 | 0.7 |
| 1/100 | 0.5 |
| 1/50 | 0.35 |
| 1/25 | 0.25 |
| 1/20 | 0.20 |

hazard factors range from 0.03 in Hobart to 0.11 in Newcastle. The Australian Standard, AS1170.4-2007 also includes a 'probability factor' $k_p$ to account for a wide range of annual probabilities of exceedance.

Table 11.4 reproduces values of the probability factor $k_p$. The use of Z and $k_p$ is discussed further in Section 11.5.

Approximate relationships have also been proposed between PGA and MMI. An example of such relationships, for various parts of the world, is shown in Figure 11.3 (Linkimer, 2008). It will be noted that there is a considerable spread in the values of PGA for a particular MMI, depending on region.

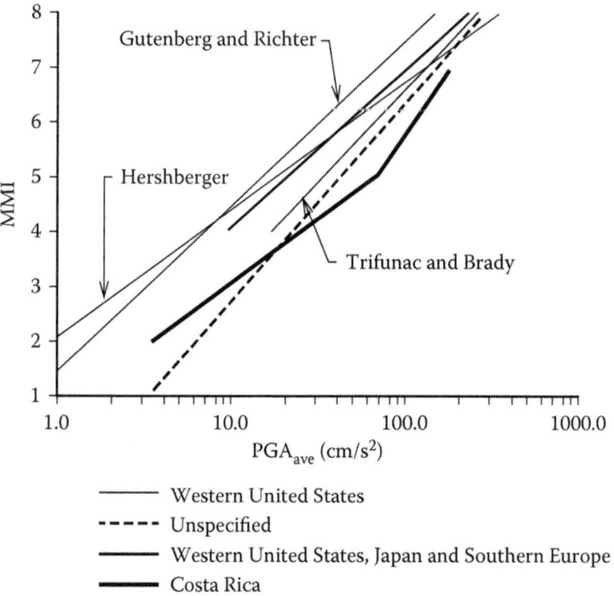


*Figure 11.3* Relationships between PGA and MMI. (Adapted from Linkimer, L. 2008. *Revista Geologica de America Central*, 38: 81–94.)

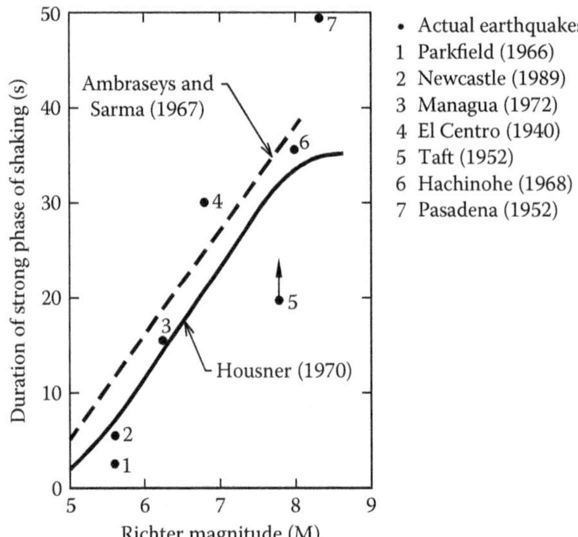


*Figure 11.4* Relationship between earthquake duration and Richter magnitude. (Adapted from Seed, H.B. and Idriss, I.M. 1969. *Journal of Soil Mechanics and Foundations Division, ASCE*, 95: 99–137. Courtesy of ASCE.)

### 11.3.6 Duration

The duration of an earthquake is related primarily to its magnitude, with longer durations being associated with larger magnitudes. Figure 11.4 shows two suggested relationships and some selected observations from some earthquakes. The duration shown in this figure is that of the strong phase of motion defined by Ambraseys and Sarma (1967) as accelerations which remain above 0.03 g.

For distances less than 10 km from the causative fault, an alternative approximation has been suggested by Bommer and Martinez-Pereira (1999) as follows:

$$\log(D_E) = 0.69\,M_w - 3.70 \tag{11.5}$$

where
$D_E$ is the effective duration (s) and
$M_w$ is the moment magnitude of earthquake.

### 11.3.7 Predominant period

There is usually a considerable range of frequency components associated with an earthquake but it is generally possible to define an average predominant or characteristic period of an earthquake. This period tends to increase as either the earthquake magnitude increases or as the distance from the causative fault increases. Relationships developed by Seed et al. (1969) are shown in Figure 11.5. This data relates largely to inter-plate events, and it can be seen that typical predominant periods range between about 0.25 and 0.4 s, for distances of up to about 50 km from the source.

For intra-plate earthquakes, such as those experienced in Australia, there is a tendency for the predominant period to be smaller than for inter-plate events, typically between 0.1s and 0.25s.

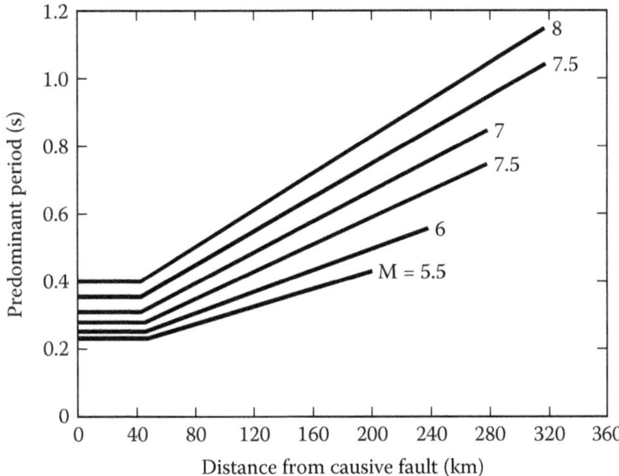


*Figure 11.5* Predominant period for maximum rock acceleration. (Adapted from Seed, H.B. and Idriss, I.M. 1969. *Journal of Soil Mechanics and Foundations Division, ASCE*, 95: 99–137. Courtesy of ASCE.)

### 11.3.8 Time–Acceleration histories

If analyses are to be carried out to compute the site response to a seismic excitation at bedrock level, it is necessary to use appropriate time–bedrock acceleration relationships. The use of existing earthquake records was discussed by Seed et al. (1969), who suggested the following steps:

1. Select earthquake records with a similar predominant period to that assessed for the site.
2. Scale the record to obtain an appropriate peak acceleration, as estimated from Section 11.3.5.
3. Modify the length of the record to obtain the approximate duration of strong motion. If this duration (say X seconds) is less than the duration of the selected record, then the first X seconds of the record is used. If the duration, X, is greater than the duration of the record, then appropriate parts of the record can be repeated to give a total of X seconds.

Two main procedures have been used for modifying ground motions:

- Direct scaling, where the amplitude is changed by a constant scale factor.
- Spectral matching, in which the frequency content of the accelerogram is adjusted until the response spectrum is within accepted limits of a target response spectrum over a defined period range.

In recent years, the procedure for selection of earthquake records has been refined, and it is now more common to select records that give response spectra that are similar to those computed from a site-specific analysis, or else those that are specified in the relevant code or standard. A large number of records are now accessible on the Pacific Earthquake Engineering Research Centre (PEER) website http://peer.berkeley.edu/smcat/.

*Table 11.5* Typical ground motion characteristics

| Event | Record | *Peak ground acceleration (g)*[a] | *Approximate number of cycles,* $N_c$ | *Range of predominant periods,* $T_p$ *(s)*[b] |
|---|---|---|---|---|
| EC8, S1 soil | Artificial | 0.13 | >20 | 0.10–0.50 |
| Northridge (1994) | Pac_down, ch. 1 | 0.43 | 2–3 | 0.15–0.50 |
| Pyrgos (1993) | Pyrrtran | 0.46 | 1 | 0.12–0.45 |
| Whittier (1987) | La16th, ch.1 | 0.39 | 4–5 | 0.10–0.25 |
| | Pacoima, ch.1 | 0.16 | 3–4 | 0.10–0.30 |
| | Tarzana, ch. 3 | 0.40 | 10 | 0.30–0.40 |
| Loma Prieta (1989) | Anderson, downstream | 0.25 | 6–7 | 0.15–0.30 |
| Kobe (1995) | Kobe JMA, NS | 0.83 | 4 | 0.30–0.90 |
| Mexico (1985) | La Villita, 0° | 0.12 | 3 | 0.50–0.60 |

Source: Nikolaou, S. et al. 2001. *Geotechnique*, 51(4): 425–440.

[a] Before normalisation.
[b] Based on the 5% damped acceleration spectrum.

Nikolaou et al. (2001) have reported characteristics of a number of earthquakes, including the PGA, the range of predominant periods and the approximate number of cycles. Table 11.5 reproduces this information.

## 11.4 ASSESSMENT OF GROUND CONDITIONS

### 11.4.1 Introduction

It is critical to develop an appropriate ground model, both for conventional foundation design and for the proper assessment of seismic effects on the foundations and the supported structures. From the viewpoint of seismic response, three of the most important parameters are the stiffness, damping and mass density of the various soil strata. The density is a standard parameter that is generally obtained during the ground investigation stage, but the stiffness and damping are less frequently measured or deduced. Some of the more common methods of obtaining these parameters have been discussed in Chapters 6 and 10 but are further summarised below.

### 11.4.2 Soil stiffness: Small-strain shear modulus

In geotechnical earthquake engineering, the soil stiffness is commonly characterised via the shear modulus, G. The value of G for very small strains, $G_{max}$, can be derived from in situ or laboratory measurements of the shear wave velocity, $V_s$, via the fundamental relationship given in Equation 10.27. In cases where direct in situ measurements of $V_s$ are not available, correlations between $V_s$ and SPT-N, such as those discussed in Chapter 6, can be used.

### 11.4.3 Shear modulus degradation

It is well recognised that the shear modulus decreases with increasing cyclic shear strain level, and typical relationships between the shear modulus ratio ($G/G_{max}$) and cyclic shear strain have been shown in Figure 10.8.

*Table 11.6* Typical values of reference strain $\gamma_r$

| *Soil type* | $\gamma_r$ |
|---|---|
| Gravelly soils (relative density ≈ 80%) | $1.3 \times 10^{-4}$ |
| Quartz sands | $3.7 \times 10^{-4}$ |
| **Clays:** | |
| PI = 5–10 | $4.0 \times 10^{-4}$ |
| PI = 10–20 | $7.0 \times 10^{-4}$ |
| PI = 20–40 | $1.1 \times 10^{-3}$ |
| PI = 40–80 | $2.0 \times 10^{-3}$ |
| PI > 80 | $3.6 \times 10^{-3}$ |

Source: Hardin, B.O. and Drnevich, V.P. 1972. *Journal of Soil Mechanics and Foundations Division, ASCE*, 98(SM7): 6678–6692.

Hardin and Drnevich (1972) proposed that the relationship between the degradation of shear modulus and increasing shear strain could be represented by the following expression:

$$G/G_{max} = \frac{1}{(1+\gamma/\gamma_r)} \tag{11.6}$$

where

$\gamma$ is the cyclic shear strain and
$\gamma_r$ is the reference cyclic shear strain.

Typical values of $\gamma_r$ are given in Table 11.6.

### 11.4.4 Soil damping

In almost all soils, internal damping at low strain levels is small, but it increases significantly as the cyclic shear strain increases, as shown in Figure 10.9. The relationship between damping ratio and cyclic shear strain can be approximated as follows:

$$D = Do + Dl \cdot \left(\frac{\gamma/\gamma_r}{(1+(\gamma/\gamma_r))}\right) \tag{11.7}$$

where

D is the damping ratio.
Do the damping ratio for very small strains,
Dl the damping ratio for very large strains,
$\gamma$ the cyclic shear strain and
$\gamma_r$ is the reference strain.

For a wide range of soils, values of Do are typically about 0.005 while Dl is about 0.26.

### 11.4.5 Soil parameters related to acceleration levels

McKenzie and Pender (1996) have carried out analyses to relate the shear modulus ratio G/Go (where Go ≡ $G_{max}$) and the damping ratio to the PGA. Their results are shown in

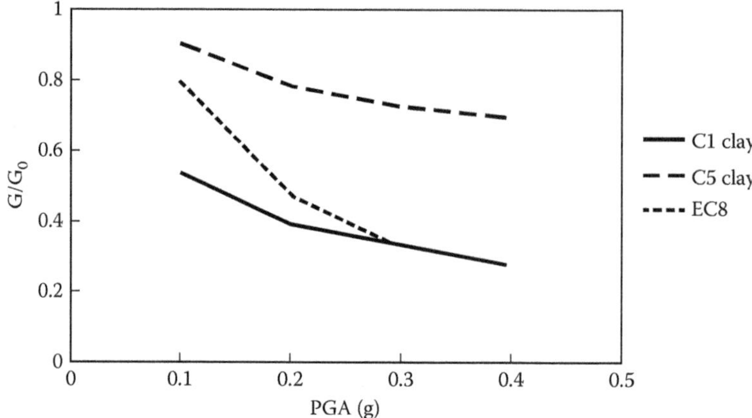


*Figure 11.6* Shear modulus ratio related to PGA. (Adapted from McKenzie, N.P. and Pender, M.J. 1996. Representative shear modulus for shallow foundation seismic soil-structure interaction. *Proceedings of the 11th World Conference on Earthquake Engineering*, Acapulco, Mexico, Paper 931.)

Figures 11.6 and 11.7 for two clay types: C1, a low plasticity clay (PI 0-10) and C5, a high plasticity clay (PI > 80). The modulus degradation is more rapid for the low plasticity clay, but the damping ratio is similar for both clays. Also shown in Figure 11.6 are draft recommendations from Eurocode EC8, which lies between the relationships derived by McKenzie and Pender (1996).

### 11.4.6 Estimation of natural site period Ts

The natural period Ts of the site is conventionally estimated as the inverse of the expressions for natural site frequency given in Section 10.6.

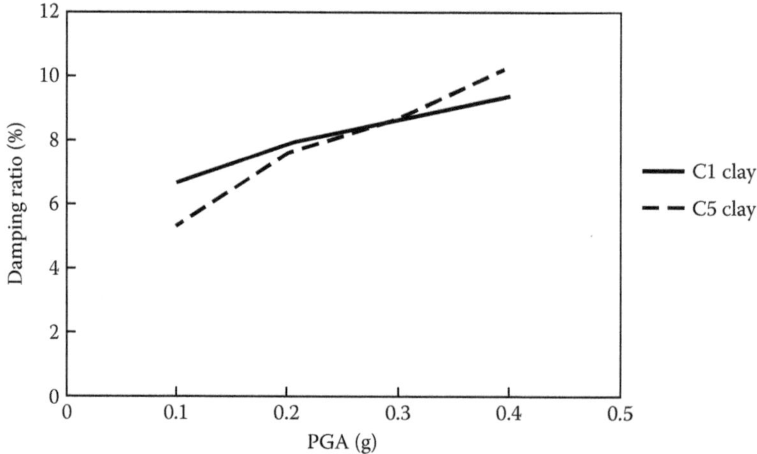


*Figure 11.7* Damping ratio related to PGA. (Adapted from McKenzie, N.P. and Pender, M.J. 1996. Representative shear modulus for shallow foundation seismic soil-structure interaction. *Proceedings of the 11th World Conference on Earthquake Engineering*, Acapulco, Mexico.)

For a layered soil deposit, Ts can be approximated as follows (Klimis et al. 2004):

$$Ts = \sum \left( \frac{4H_i}{v_{si}} \right) \tag{11.8}$$

where $H_i$ and $v_{si}$ are, respectively, the thickness and shear wave velocity of a layer i within the profile, and the summation is carried out for all the layers above bedrock.

Some more refined approaches have been suggested to obtain Ts which can also take account of the shear wave velocity of the bedrock (e.g., Sawada, 2004).

## 11.4.7 Seismic site response analyses

### *11.4.7.1 One-dimensional analyses*

For cases in which site-specific response spectra are required, or where detailed information on the propagation of the earthquake from bedrock through the soil is important, the customary method for analysing the response of a site to seismic excitation at bedrock level is via a one-dimensional (1D) analysis. In such an analysis, only the vertical propagation of shear waves from bedrock to the ground surface is considered. There are a number of computer programs available, including SHAKE and DEEPSOIL. The former considers only a single-phase soil model whereas the latter is capable of incorporating the generation and dissipation of excess pore pressures during and after the earthquake.

A simple direct time-domain solution, using a program called Earthquake Response of Layered Soils (ERLS) has been described by Poulos (1991b). The model employed by ERLS is shown in Figure 11.8. This program solves the equations of motion by a forward marching

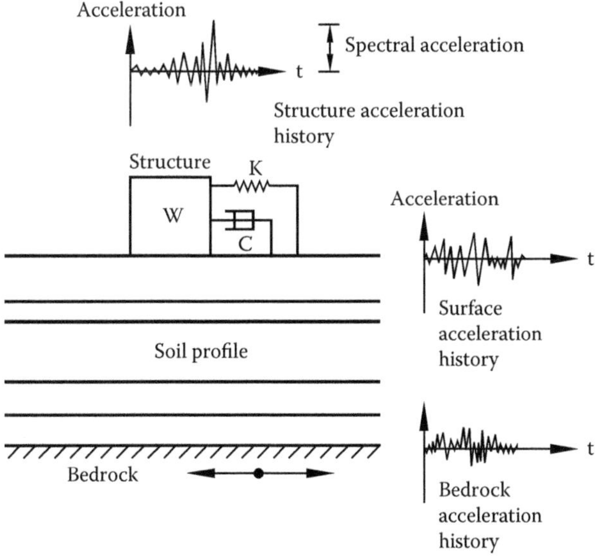


*Figure 11.8* Computational model for 1D site response analysis.

finite difference process, using a lumped parameter representation of the ground profile. The analysis incorporates the following features:

1. Values of shear modulus and damping ratio for each layer within the ground profile that vary with shear strain within the layer, as per the relationships derived by Hardin and Drnevich (1972). Allowance for the radiation damping provided by the bedrock.
2. Output of the maximum ground displacements and shear stresses during the earthquake at various depths within the ground profile.
3. Response spectra (for acceleration, velocity and displacement) are computed for both the input bedrock excitation and for the computed surface excitation. These can be used to verify or modify the response spectra specified by the relevant code.

#### 11.4.7.2 Definition of bedrock

In cases where there is no distinct boundary between soil and rock strata, it is necessary to define 'bedrock' so that a site response analysis can be carried out. A straightforward definition is provided by the U.S. Bureau of Reclamation, who define rock as 'any sedimentary, igneous or metamorphic material represented as a unit in geology; being a sound and solid mass, layer or ledge of mineral matter; and with shear wave velocities greater than 2500 feet per second'. Thus, on this basis, it seems reasonable to take the bedrock level as that having a shear wave velocity of about 760 m/s (approximately equivalent to 2500 ft/s).

#### 11.4.7.3 Simplified methods to estimate ground motion amplification

When only an approximate estimate is required of the amplification factor for ground motions travelling from bedrock to the surface, a number of approximate approaches can be considered. For example, Ansal and Pinto (1999) quote the following approximate expression for the amplification factor A:

$$A = 68\, v_{sav}^{-0.6} \tag{11.9}$$

where

A is the ratio of peak ground motion at the surface to that at bedrock and
$v_{sav}$ is the average shear wave velocity of surface layers (in m/s).

Kokusho and Matsumoto (1997) suggest the following relationship for A:

$$A = 2.0 \exp\left(\frac{-1.7\ a_{max}}{98}\right) \tag{11.10}$$

where $a_{max}$ is the peak bedrock acceleration, in m/s.

Neither Equation 11.9 nor 11.10 explicitly considers the depth of the soil profile or the amount of internal damping within the soil, and in order to incorporate this important factor, another possible approach is to adopt an amplification factor that is derived from the simple theory of harmonic motion. In this case, the amplification factor is given by the following expression (Ohsaki, 1969):

$$A = \left[ \frac{\left(1 + 4D^2(f/f_n)^2\right)}{\left\{[1 - (f/f_n)^2]^2 + (2\zeta f/f_n)^2\right\}} \right]^{-0.5} \tag{11.11}$$

where

f is the predominant frequency of the earthquake at bedrock level,
$f_n$ the natural frequency of the soil profile = $4H/v_{sav}$,
H the total thickness of soil profile above bedrock,
$v_{sav}$ the average shear wave velocity in soil profile and
$\zeta$ is the soil damping ratio.

In applying Equation 11.11, it is possible to consider the effect of cyclic shear strain on the shear wave velocity (via the degradation of the shear modulus) and the increase in soil damping with increasing shear strain, for example, via the expression given by Hardin and Drnevich (1972) or alternative expressions such as those by Okur and Ansal (2001) for shear modulus degradation, and Cavallaro et al. (2001) for damping ratio. This approach can also then reproduce the effect, noted by Idriss, Seed and others, that, as the maximum acceleration at bedrock level increases, the amplification factor tends to reduce because the increasing strain generates an increasing degree of damping which serves to reduce the motion amplification.

#### *11.4.7.4 2D and 3D effects*

1D analyses are unable to capture some important 2D and 3D effects that may arise from the geological origin of the ground profile, for example, the so-called 'basin effect'. In this case, a sediment-filled valley can significantly amplify Rayleigh waves that have a wave length shorter than twice the width of the valley. As the period, and hence the wave length, of the surface wave increases, the sediment-filled valley starts to 'ride' on the wave and the site amplification effects become less significant. In such cases, it is possible to carry out 2D site response analyses in order to obtain more realistic assessments of ground motion modifications through the ground profile.

A study of 2D effects has been made by Ciliz et al. (2007), who have carried out finite element analyses of the model shown in Figure 11.9. Figure 11.10 shows an example of their results, in which the ratio of the 2D and 1D PGAs is plotted against a normalised distance,

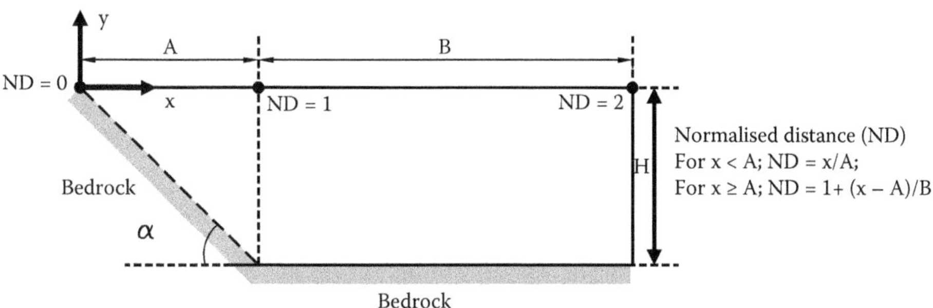


*Figure 11.9* Geometry of basin model analysed. (Adapted from Ciliz, S., Ozkan, M.Y. and Cetin, K.O. 2007. Effect of basin edge effects on the dynamic response of soil deposits. *Proceedings of the 4th International Conference on Earthquake Geotechnical Engineering*, Paper No. 1309, Thessaloniki, Greece.)

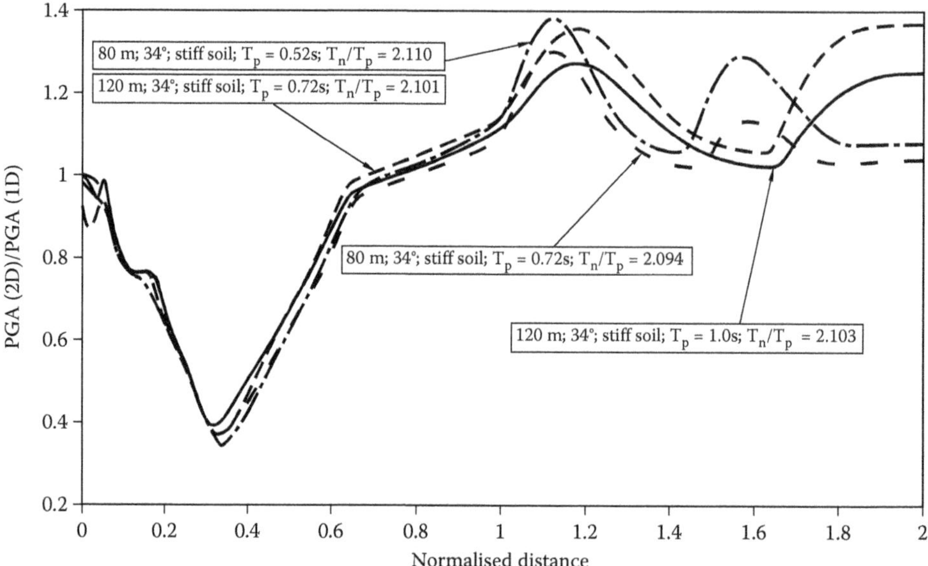


*Figure 11.10* Ratio of 2D and 1D PGA values for basin problem. (Adapted from Ciliz, S., Ozkan, M.Y. and Cetin, K.O. 2007. Effect of basin edge effects on the dynamic response of soil deposits. *Proceedings of the 4th International Conference on Earthquake Geotechnical Engineering*, Paper No. 1309, Thessaloniki, Greece.)

as defined in Figure 11.9. A 34° slope and soil layer depths of 80 and 120 m have been considered, and the effects of the ratio of natural period of the site, $T_n$, to predominant period of ground motion, $T_p$, have been examined. Figure 11.10 shows that there is an amplification of motion (as compared with a 1D analysis) within the basin, but a de-amplification along the slope.

## 11.5 STRUCTURAL RESPONSE SPECTRA

### 11.5.1 Introduction

A response spectrum describes the *maximum* response (displacement, velocity, acceleration) of an SDOF system to a particular input motion, as a function of the natural period (or natural frequency) and damping ratio of the system. The amplitude, frequency content and the duration of the input motion all affect the spectral values. In addition, the ground conditions above the bedrock level can have a very significant influence on the response spectra.

Response spectra play an important role in providing a means of computing earthquake-induced forces on structures, as discussed later in this section. Velocity and displacement spectra can be derived from the acceleration spectrum, as follows:

$$S_v = (T/2\pi) \cdot S_a \tag{11.12}$$

$$S_d = (T/2\pi) \cdot S_v \tag{11.13}$$

where $S_v$ is the spectral velocity, T the natural period of structure, $S_d$ the spectral displacement and $S_a$ is the spectral acceleration.

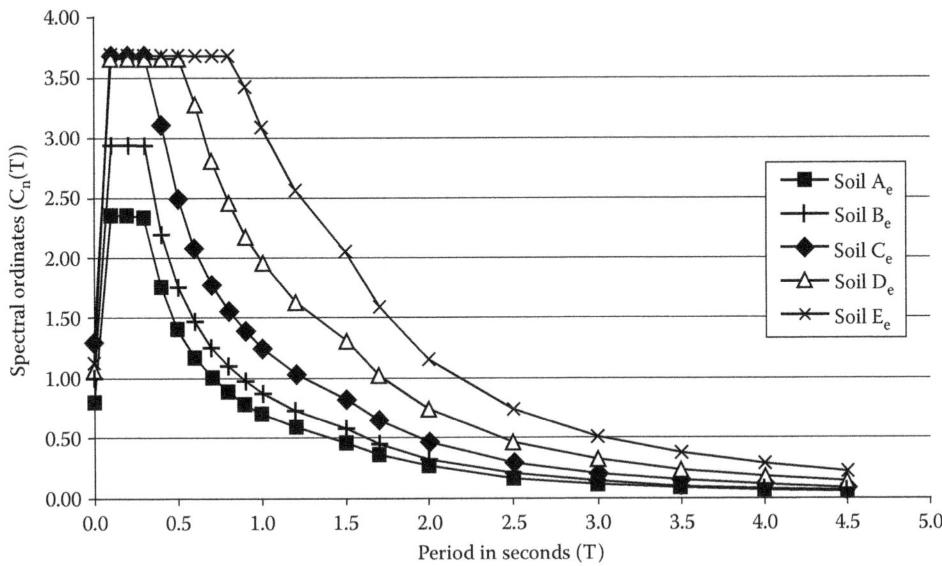


*Figure 11.11* Normalised response spectra. (From AS1170.4. 2007. *Structural Design Actions—Earthquake Actions*. Standards Australia.)

## 11.5.2 Example: Provisions of the Australian Standard AS1170.4-2007

As an example of a national standard, the Australian Standard AS1170.4-2007 provides a set of normalised response spectra for various ground conditions, characterised by a subsoil class. Figure 11.11 shows these spectra, in terms of 'spectral ordinates' or 'spectral shape factors', $C_h(T)$, which (in effect) represent the ratio of the maximum acceleration of the SDOF system to the peak bedrock acceleration from the earthquake. The ground conditions indicated in Figure 11.11 are summarised in Table 11.7.

As an alternative to the use of the spectra in Figure 11.11, site-specific design response spectra can be obtained using an analysis that considers the soil profile in detail and in

*Table 11.7* Site subsoil classes

| *Class* | *Ground type* | *Remarks* |
|---|---|---|
| $A_e$ | Strong rock | UCS > 50 MPa, $V_s$ > 1500 m/s for top 30 m. Not underlain by material with UCS < 18 MPa or $V_s$ < 600 m/s |
| $B_e$ | Rock | UCS = 1–50 MPa, $V_s$ > 300 m/s. Not underlain by material with UCS < 0.8 MPa or $V_s$ < 300 m/s |
| $C_e$ | Shallow soil | Site period <0.6s; depths not exceeding 20 m soft soil, 25 m firm soil, 40 m stiff soil, 60 m very stiff-hard soil. No very soft soil |
| $D_e$ | Deep or soft soil | Underlain by < 10 m very soft soil ($s_u$ < 12.5 kPa or SPT < 6), site period > 0.6s, soil depths do not exceed those for Class $C_e$ |
| $E_e$ | Very soft soil | >10 m very soft soil ($s_u$ < 12.5 kPa or SPT < 6), > 10 m with $V_s$ < 150 m/s, > 10 m combined depth of these soils |

Source: AS1170.4. 2007. *Structural Design Actions—Earthquake Actions*. Standards Australia.

Note: UCS = unconfined compressive strength, $V_s$ = average shear wave velocity, $s_u$ = undrained shear strength, N = SPT value.

which a bedrock ground motion is applied. Such an analysis is described briefly in Section 11.4.5. This bedrock motion should be such that it is compatible with the bedrock spectra (i.e., for soil type $A_e$) shown in Figure 11.11.

### 11.5.3 Estimation of earthquake-induced forces

The Australian Standard AS1170.4-2007 sets out the following expression to calculate the horizontal equivalent static shear force (V) acting at the base of a structure:

$$V = \left[\frac{k_p Z C_h(T_1) S_p}{\mu}\right] W_t \tag{11.14}$$

where

$k_p$ is the probability factor appropriate to limit state under consideration (see Table 11.4),
Z the earthquake hazard factor (acceleration coefficient for 10% probability of exceedance in 50 years,
$C_h(T_1)$ the spectral shape factor for fundamental natural period ($T_1$) of the structure (see Figure 11.11),
$S_p$ the structural performance factor,
$\mu$ the structural ductility factor and
$W_t$ is the total seismic weight of structure (kN).

For some typical structures, the ratio $S_p/\mu$, which appears in Equation 11.14, is given in Table 11.8.

### 11.5.4 Natural period of structure

For detailed design purposes, the natural period of a structure is generally estimated from a dynamic structural analysis. However, for preliminary design, approximations can be made

*Table 11.8* Values of $S_p/\mu$ for typical structures

| *Structural system* | *Description* | $S_p/\mu$ |
|---|---|---|
| Steel structures | Fully ductile moment-resisting or eccentrically braced frames | 0.17 |
| | Moderately ductile moment-resisting frames | 0.22 |
| | Limited ductility frames | 0.38 |
| Concrete structures | Fully ductile moment-resisting frames, or fully ductile walls | 0.17 |
| | Moderately ductile moment-resisting frames or ductile shear walls | 0.22 |
| | Ordinary moment-resisting frames, or limited ductile walls | 0.38 |
| Timber structures | Shear walls | 0.22 |
| | Braced frames, moment-resisting frames | 0.38 |
| Masonry structures | Close-spaced reinforced masonry | 0.38 |
| | Wide-spaced reinforced masonry | 0.50 |
| | Unreinforced masonry | 0.62 |
| | Other, not complying with AS 3700 | 0.77 |
| Specific structures | Cast in place silos and chimneys with continuous walls | 0.33 |
| | Trussed towers | 0.33 |
| | Storage racking | 0.33 |
| | Cooling towers | 0.33 |
| | Tanks, vessels, pressurised spheres, inverted pendulum structures, amusement structures and monuments | 0.50 |

on the basis of the number of stories or the height of the building. Some of these approximations are given in Section 2.8.1.

### 11.5.5 Effect of foundation stiffness on natural period

The natural period of a structure will increase with decreasing stiffness of the foundation system. For horizontal translation, Dowrick (2009) quotes the following expression for the effective natural period, T′, of a structure–foundation system:

$$T' = T\left[1+\left(\frac{k_{str}}{k_x}\right)\left(1+\frac{k_x {h_{eff}}^2}{k_\phi}\right)\right]^{0.5} \tag{11.15}$$

where

T is the fundamental period of fixed-base structure,
$k_x$ the horizontal stiffness of foundation in the direction being considered,
$k_\phi$ the rocking stiffness of foundation in direction being considered,
$k_{str}$ the stiffness of structure when fixed at the base $= 4\pi^2 W/gT^2$,
W the effective weight of structure vibrating in the fundamental natural mode $\approx 0.7$ times gravity load,
g the acceleration due to gravity and
$h_{eff}$ is the effective height of structure $\approx 0.7$ times total height.

For an example of a 80 m tall building, Dowrick shows that on a soft soil, the natural period may be increased by 73%, while on a stiff soil site, a 22% increase in natural period occurs.

The estimation of foundation stiffness and damping has been discussed in Section 10.8 for shallow foundations, and Sections 10.9 and 10.10 for deep foundations.

### 11.5.6 Effective damping

The effective damping of a soil–structure system incorporates the combined material and radiation damping in the soil, which can sometimes lead to significant reductions in response (Dowrick, 2009). The effective damping, β′, can be estimated as follows:

$$\beta' = \zeta_0 + \frac{\beta}{(T'/T)^3} \tag{11.16}$$

where

β is the damping ratio for fixed-base structure,
$\zeta_0$ the foundation damping factor,
T the fundamental period of fixed-base structure and
T′ is the fundamental period of structure-foundation system (Equation 11.15).

The foundation damping factor $\zeta_0$ is related to the period ratio, the aspect ratio (ratio of height to width of the structure) and the acceleration level. It varies from zero to 0.25 or more, with the larger values being associated with smaller aspect ratios, larger period ratios and larger acceleration levels.

## 11.6 ASSESSMENT OF LIQUEFACTION POTENTIAL

### 11.6.1 Introduction

Liquefaction can create a number of problems for foundations, including the following:

1. The lateral support for deep foundations from the surrounding soil will decrease and the possibility of pile buckling will thus be increased.
2. Deep foundations will be subjected to lateral ground movements from the liquefied soil during and after the earthquake.
3. For piles in liquefiable soils, there will be a loss of axial capacity during and after the earthquake, although this loss should be temporary and the capacity should be largely restored once the liquefaction-induced excess pore pressures have dissipated.
4. The natural period of the ground will tend to increase as the soil becomes 'softer' and more flexible during liquefaction.
5. The damping of pile-supported structures will also increase during and after liquefaction.

If is therefore very important to be able to make an assessment of the potential for a site to liquefy and then to assess the potential effects of liquefaction on the foundation system. Liquefaction assessment is a critical part of contemporary foundation design, and following a great deal of research over the past several decades, well-established empirical methods have been developed to enable such assessments to be made. The most common methods involve the use of one or more of the following in situ testing methods:

- Standard Penetration Tests
- Static Cone Penetration Tests
- In situ shear wave velocity measurements, either via a seismic cone or via geophysical methods such as cross-hole or down-hole testing
- Dilatometer test (DMT)

In all cases, when a deterministic assessment is being made, the factor of safety against liquefaction, $F_L$, is expressed as follows:

$$F_L = \frac{CRR}{CSR} \quad (11.17)$$

where

CRR is the cyclic resistance ratio = normalised cyclic shear resistance and
CSR is the cyclic shear stress ratio = normalised shear stress imposed by the earthquake.

Estimation of CSR is common to all methods and will be outlined below, and then the key aspects of some available methods for estimating CRR will be summarised.

### 11.6.2 Cyclic shear stress ratio

In earlier approaches, the following simple expression for the cyclic shear stress ratio, CSR, caused by an earthquake was employed:

$$CSR = 0.65\left(\frac{a_{max}}{g}\right)\left(\frac{\sigma_{vo}}{\sigma'_{vo}}\right)r_d \quad (11.18)$$

where

$a_{max}$ is the PGA,
g the acceleration due to gravity,
$\sigma_{vo}$ the initial total vertical stress,
$\sigma_{vo}'$ the initial effective vertical stress and
$r_d$ is the depth reduction factor $\approx 1 - 0.015$ z, where z is the depth (m).

In more recent applications of the simplified method commonly employed for liquefaction assessment, the cyclic shear stress ratio, CSR, is expressed as follows:

$$CSR = 0.65\left(\frac{a_{max}}{g}\right)\left(\frac{\sigma_{vo}}{\sigma_{vo}'}\right)\frac{r_d}{(MSF \cdot K_s)} \tag{11.19}$$

where

MSF is the magnitude scaling factor and
$K_\sigma$ is the overburden correction factor.

Vessia and Venisti (2011), summarising the work of Idriss and Boulanger (2006), give the following approximations:

$$MSF = 6.9\exp\left(\frac{-M}{4}\right) - 0.058 \le 1.8 \tag{11.20}$$

$$r_d = \exp\{\alpha(z) + \beta(z)M\} \tag{11.21}$$

where

M is the moment magnitude of earthquake,
$\alpha(z) = -1.012 - 1.126 \sin(z/11.73 + 5.133)$,
$\beta(z) = 0.106 + 0.118 \sin(z/11.28 + 5.142)$,
$K_\sigma = 1 - C_\sigma \ln(\sigma_{vo}'/p_a) \le 1$,
$C_\sigma = 1/(18.9 - 2.55\sqrt{[(N_1)_{60}]}) \le 0.3$ and
z is the depth (m).

The above expression tends to give smaller values of $r_d$ than the original expression.

### 11.6.3 Methods based on SPT data

The cyclic resistance ratio, CRR, is most frequently related to the standard penetration resistance SPT-N value. One of the earlier relationships is shown in Figure 11.12 for earthquakes with a magnitude of 7.5 (Seed and Idriss, 1982). In this figure, $(N_1)_{60}$ is the SPT-N value corrected for an energy ratio of 60% and an overburden pressure of 1 atm (about 100 kPa), and can be approximated as follows:

$$(N_1)_{60} = \frac{N C_N Me}{(0.60\ E_{ff})} \tag{11.22}$$

where

N is the measured SPT,
$C_N$ the stress correction factor $= \sqrt{(p_a/\sigma_{vo}')}$,

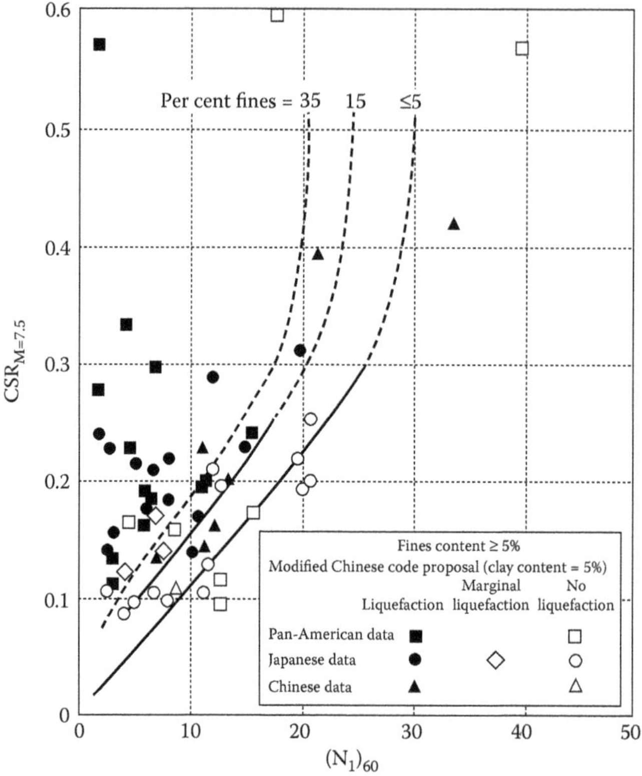


*Figure 11.12* CRR for M = 7.5 earthquakes versus SPT. (After Seed, H.B. and Idriss, I.M. 1982. Ground Motions and Liquefaction during Earthquakes. *Earthquake Engineering Research Institute*, ISBN 0-943198-24-0. Courtesy of ASCE.)

$p_a$ the atmospheric pressure,
$\sigma_{vo}'$ the vertical effective stress,
Me the actual hammer energy and
$E_{ff}$ is the theoretical free-fall hammer energy.

More recently, for the case of a horizontal ground surface, Vessia and Venisti (2011) give the following approximate expression for CRR:

$$\text{CRR} = \exp\left(\left(\frac{(N_1)_{60\text{-cs}}}{14.1}\right) + \left(\frac{(N_1)_{60\text{-cs}}}{126}\right)^2 + \left(\frac{(N_1)_{60\text{-cs}}}{23.6}\right)^3 + \left(\frac{(N_1)_{60\text{-cs}}}{25.4}\right)^4 - 2.8\right) \tag{11.23}$$

where

$$\begin{aligned}(N_1)_{60\text{-cs}} &= \text{equivalent clean sand value of SPT} - \text{N value} \\ &= (N_1)_{60} + \Delta(N_1)_{60}\end{aligned} \tag{11.24}$$

$(N_1)_{60}$ is the SPT-N value for 60% hammer efficiency and overburden pressure,
$C_N$ the overburden correction factor = $(p_a/\sigma_{vo}')^{\alpha} \leq 1.7$,

$p_a$ the atmospheric pressure and
$\alpha = 0.784 - 0.0768((N_1)_{60})^{0.5}$ and $(N_1)_{60} \leq 46$.
$\Delta(N_1)_{60}$ = correction for fines content FC (in per cent)

$$= \exp\left(1.63 + \frac{9.7}{(FC+0.1)} - \left(\frac{15.7}{(FC+0.1)^2}\right)\right) \tag{11.25}$$

A simpler approximate expression for CRR has been suggested by Fellenius (2016) as follows:

$$CRR = \alpha_{1.}\left(e^{\alpha_2 (N_1)_{60}}\right) \tag{11.26}$$

where $\alpha_1$, $\alpha_2$ are factors depending on fines content, and given in Table 11.9, and $(N_1)_{60}$ = corrected SPT value, as in Equation 11.22.

### 11.6.4 Methods based on CPT data

Because of its greater sensitivity and repeatability in softer or looser soils, the static CPT is often a more satisfactory means of assessing liquefaction potential than SPT. Among the many available methods is the one developed by Robertson (2004). The relationship between the CRR and the corrected CPT value, $q_{c1N}$ is shown in Figure 11.13.

For clean sand, the following approximate correlations can be used:

1. For $50 \leq (q_{c1n})_{cs} \leq 160$ and magnitude M = 7.5,

$$CRR = 93\left[\frac{(q_{c1n})_{cs}}{1000}\right]^3 + 0.08 \tag{11.27}$$

2. For $(q_{ciN})_{cs} < 50$ and magnitude M = 7.5,

$$CRR = 0.833\left[\frac{(q_{ciN})_{cs}}{1000}\right] + 0.05 \tag{11.28}$$

where
$(q_{c1n})_{cs}$ is the equivalent clean sand normalised cone penetration resistance = $K_c$ Q,
$K_c$ the correction factor depending on grain characteristics of soil,
Q the normalised CPT penetration resistance = $\{(q_c - \sigma_{vo})/p_a\}(p_a/\sigma_{vo}')^n$,
$q_c$ the measured cone resistance and

*Table 11.9* Factors $\alpha_1$ and $\alpha_2$

| *Fines content FC (%)* | $\alpha_1$ | $\alpha_2$ |
|---|---|---|
| <5 | 0.050 | 0.072 |
| 10 | 0.060 | 0.084 |
| 30 | 0.070 | 0.092 |

Source: Fellenius, B.H. 2016. *Basics of Foundation Design.* Electronic Ed. www.Fellenius.net, 451p.

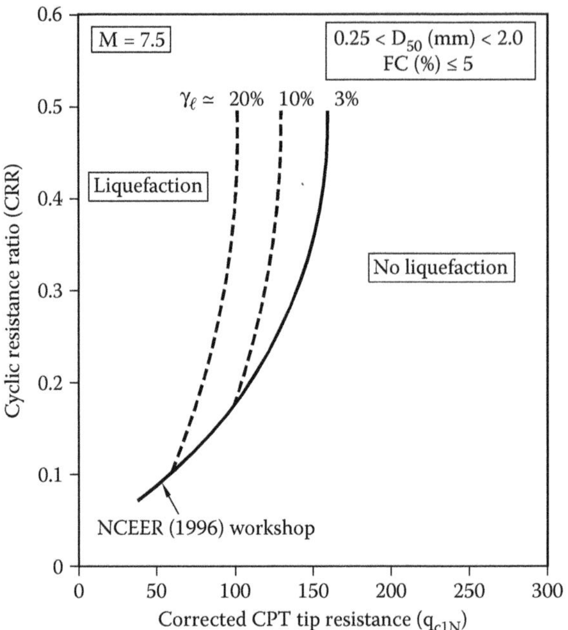


*Figure 11.13* CRR versus corrected CPT resistance. (Adapted from Robertson, P.K. 2004. Evaluating soil liquefaction and post-earthquake deformations using the CPT. *International Conference on Site Characterization 2*, Porto, Portugal. Courtesy of Dr P.K. Robertson.)

$\sigma_{vo}$ and $\sigma_{vo}'$ are the total and effective overburden stresses,
$p_a$ is a reference pressure, in the same units as the other stresses (e.g., 0.1 MPa) and
n = stress exponent.

The grain characteristic correction factor $K_c$ can be related to another factor, the soil behaviour-type index, $I_c$, where

$$I_c = [(3.47 - \log Q)^2 + (\log F + 1.22)^2]^{0.5} \tag{11.29}$$

where
$F = f_s/[(q_c - \sigma_{vo})]/100\%$ is the normalised friction ratio and
$f_s$ = CPT sleeve resistance.

For $I_c \leq 1.64$: $K_c = 1.0$

$$\text{For } I_c > 1.64: \quad K_c = -0.403\, I_c^4 + 5.581\, I_c^3 - 21.63\, I_c^2 + 33.75\, I_c - 17.88 \tag{11.30}$$

The stress exponent n is also related to $I_c$ as follows:

For $I_c < 1.64$ $\quad n = 0.5$
For $I_c > 3.30$ $\quad n = 1.0$
For $1.64 < I_c < 3.30$ $\quad n = (I_c - 1.64)\, 0.3 + 0.5$

Finally, to correct CRR for earthquake magnitudes, M, other than 7.5, the following correction factor, MSF, is applied to CRR:

$$MSF = 174/M^{2.56} \tag{11.31}$$

### 11.6.5 Methods based on shear wave velocity

The use of measurements of shear wave velocity, $V_s$, to assess liquefaction potential has been explored over the past three decades. Such an approach is attractive because $V_s$ can be measured in gravel and other soils in which SPT or CPT tests cannot be conducted. Andrus and Stokoe (2000) have developed the following relationship between the CRR and the shear wave velocity corrected for overburden, $v_{s1}$.

$$CRR = \left\{ a\left(\frac{v_{s1}}{100}\right)^2 + b\left(\frac{1}{\left(v_{s1}^* - v_{s1}\right)} - \frac{1}{v_{s1}^*}\right)\right\} MSF \tag{11.32}$$

where
$v_{s1}^*$ is the limiting upper value of $v_{s1}$ for cyclic liquefaction occurrence,
$v_{s1}$ the corrected shear wave velocity,
a, b are curve fitting parameters, where $a = 0.022$, $b = 2.8$ and
MSF is the earthquake magnitude scaling factor.

$v_{s1}^*$ is related to the fines content (FC, in per cent) as follows:

1. For $FC \leq 5\%$ $\quad v_{s1}^* = 215$ m/s
2. For $5\% < FC < 35\%$ $\quad v_{s1}^* = 215 - 0.5(FC-5)$ m/s
3. For $FC \geq 35\%$ $\quad v_{s1}^* = 200$ m/s.

$v_{s1}$ is obtained from the following expression:

$$v_{s1} = v_s \left(\frac{p_a}{\sigma_v'}\right)^{0.25} \tag{11.33}$$

where
$v_s$ is the measured shear wave velocity,
$p_a$ the atmospheric pressure and
$\sigma_v'$ is the effective overburden pressure.

The magnitude scaling factor is

$$MSF = \left(\frac{Mw}{7.5}\right)^{-2.56} \tag{11.34}$$

where
Mw is the moment magnitude.

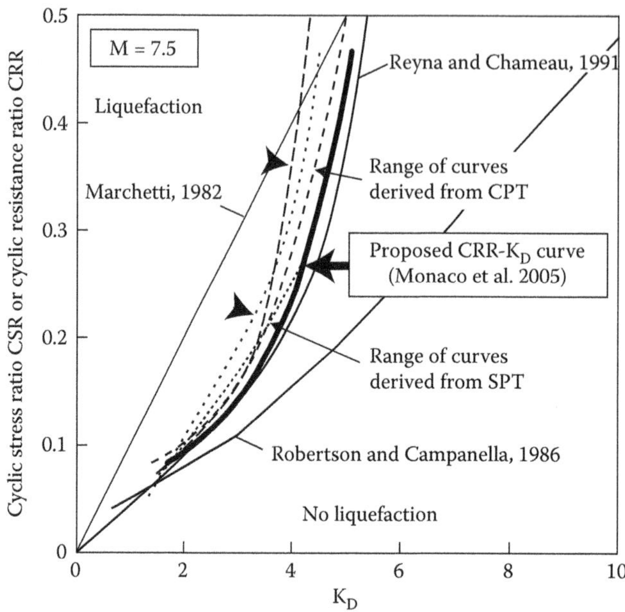


*Figure 11.14* CRR versus dilatometer modulus $K_D$. (Adapted from Monaco, P. and Marchetti, S. 2007. Evaluating liquefaction potential by seismic dilatometer (SDMT) accounting for aging/stress history. In: *Paper presented to 4th International Conference on Earthquake Geotechnical Engineering*, Thessaloniki, Greece.)

## 11.6.6 Method based on dilatometer testing

Monaco and Marchetti (2007) have correlated the CRR to the dilatometer modulus $K_D$, as shown in Figure 11.14.

## 11.6.7 Method based on relative density

Okamoto (1984) quotes the following relationship between relative density $D_R$ and CRR:

$$CRR = \frac{0.21D_R}{(50\beta)} \tag{11.35}$$

where

$D_R$ is the relative density (%) and
$\beta$ the factor depending on the form of earthquake wave
= 0.70 for a 'vibration' type wave
= 0.55 for an 'impulse' type wave.

This expression should be considered as very approximate as it takes no account of the presence of fines or other factors that may influence liquefaction resistance.

## 11.6.8 Probability of liquefaction

Vessia and Venisti (2011) report a simplified method for estimating the probability of liquefaction based on the value of $F_L$ obtained from a deterministic analysis. This approach, first developed by Juang et al. (2002), is as follows:

$$P(L) = \frac{1}{[1 + (F_L/1.05)^{3.8}]} \tag{11.36}$$

where

P(L) is the probability of liquefaction and
$F_L$ is the calculated factor of safety against liquefaction.

Fellenius (2016) has presented simplified expressions for the CRR in terms of the probability of liquefaction, based on CPT data. The following general expression has been derived:

$$CRR = A\left(e^{0.14q_{c1}}\right) \tag{11.37}$$

where A is the factor depending on the probability of liquefaction (PL), and given in Table 11.10, and $q_{c1}$ is the stress-corrected cone resistance, obtained as follows:

$$q_{c1} = q_c\left(\frac{\sigma_r}{\sigma_v'}\right) \tag{11.38}$$

where $q_c$ is the measured cone resistance, in kPa, $\sigma_v'$ the effective vertical stress and $\sigma_r$ is the reference stress =100 kPa (atmospheric pressure).

Fellenius notes that the relationship for a probability of 0.5 is very similar to that derived from Equations 11.24 and 11.25.

Shen et al. (2016) have developed a probabilistic approach in which the probability of liquefaction, $P_L$, is related to the shear wave velocity $V_{s1}^*$, corrected for overburden pressure and fines content, and the cyclic stress ratio $CSR_{7.5}$ for a magnitude 7.5 event. From the various models developed, the following is convenient to use:

$$PL = [1 + \exp\left(-14.3931 + 0.0552V_{s1}^* - 2.8628\ln(CSR_{7.5})\right]^{-1} \tag{11.39}$$

### 11.6.9 Liquefaction potential index

The concept of the liquefaction potential index (LPI) was developed by Iwasaki et al. (1984) and is meant to provide a more meaningful measure of the overall risk of liquefaction of a site by considering the liquefaction potential within the upper 20 m of a ground profile. Adopting modifications set out by Vessia and Venisti (2011), LPI is expressed as follows:

$$LPI = \int F(z)w(z)\,dz \tag{11.40}$$

*Table 11.10* Values of probability factor A

| *Probability of liquefaction PL* | *Probability factor A* |
|---|---|
| 0.1 | 0.025 |
| 0.2 | 0.033 |
| 0.3 | 0.038 |
| 0.5 | 0.046 |
| 0.7 | 0.057 |
| 0.9 | 0.085 |

*Table 11.11* Liquefaction risk based on liquefaction potential index, LPI

| *LPI* | *Liquefaction risk* |
|---|---|
| 0 | Non-liquefiable |
| 0–2 | Low |
| 2–5 | Moderate |
| 5–15 | High |
| >15 | Very high |

where

$F(z) = 1 - F_L$ when $F_L < 0.95$, and $F_L$ is the factor of safety against liquefaction,
$F(z) = 2 \times 10^{-6} \exp(-18.427F_L)$ when $0.95 < F_L < 1.2$,
$F(z) = 0$ when $F_L \geq 1.2$,
$W(z) = 10-0.5z$ and
z is the depth below surface, in metres.

The integration in Equation 11.40 is carried out from the surface to a depth of 20 m. Table 11.11 summarises the general levels of risk based on LPI.

### 11.6.10 Fine-grained soils

It is generally recognised that fine-grained soils will be more resistant to liquefaction than coarse-grained soils. Figures 11.15 and 11.16 indicate the ranges of plasticity characteristics for which soils may be liquefiable. In general, a soil with a plasticity index (PI) greater than about 20 will tend to be non-liquefiable, because it will tend to become dilatant as the effective stresses reduce. However, it may be susceptible to 'softening' and a consequent reduction of strength and stiffness because of the development of excess pore water pressures.

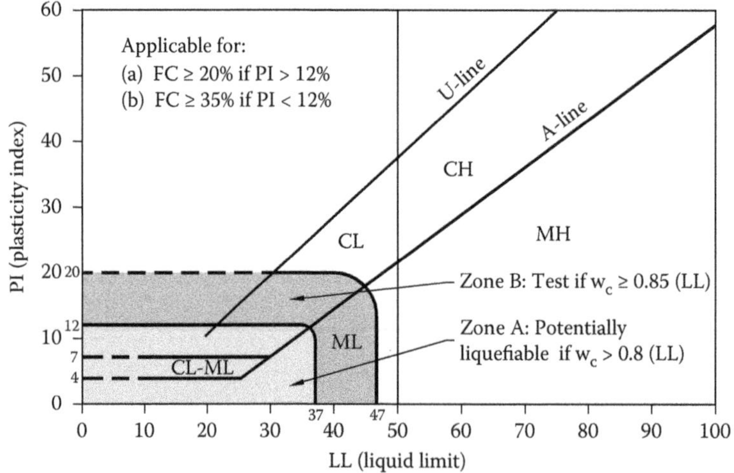


*Figure 11.15* Susceptibility of soils to liquefaction. (Adapted from Seed, R.B. et al. 2003. Recent advances in soil liquefaction engineering: A unified and consistent framework. *Keynote Presentation, 26th Annual ASCE Los Angeles Geotechnical Spring Seminar*, Long Beach, CA.)

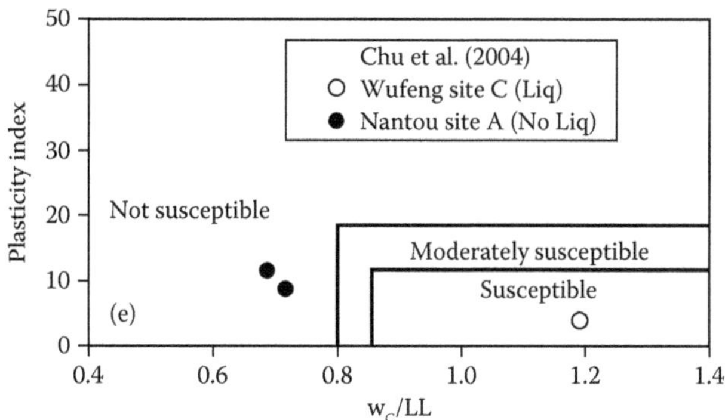


*Figure 11.16* Susceptibility of soils to liquefaction. (Adapted from Bray, J.D. and Sancio, R.B. 2006. *Journal of Geotechnical and Geoenvironmental Engineering*, 132(9): 1165–1177. Courtesy of ASCE.)

### 11.6.11 Pore pressure generation

The pore pressures generated in a soil during seismic shaking have been related to the number of loading cycles and the number of cycles to initiate liquefaction in the soil. However, a useful approximation suggested by Tokimatsu and Seed (1987), and quoted by Towhata (2008), relates the pore pressure ratio $r_u$ to factor of safety against liquefaction, $F_L$, as

$$r_u = F_L^{-n} \leq 1.0 \tag{11.41}$$

where
$r_u$ is the pore pressure ratio $= \Delta u/\sigma_{vo}'$,
$\Delta u$ the increase in pore pressure due to cyclic loading,
$\sigma_{vo}'$ the initial vertical effective stress and
n is a factor between 4 and 10, typically 7.

Dobry and Ladd (1980) have related the pore pressure ratio to the cyclic shear strain, and their data, for two soil types subjected to 10 cycles of loading in a strain-controlled triaxial test, are shown in Figure 11.17. It has been found that pore pressures will not be generated unless the cyclic strain exceeds a threshold value, which appears to be in the order of 0.003%. It is also found that the generated pore pressure is insensitive to a number of factors, including soil type and method of sample preparation.

A convenient means of estimating generated pore pressures has been given by Marcuson et al. (1990), who have related the pore pressure ratio $r_u$ to the factor of safety against liquefaction, $FS_L$, as shown in Figure 11.18.

In fine-grained soils, the above approaches may not be applicable, and an alternative expression suggested by Matsui et al. (1980) may be employed:

$$r_u = \beta\left[\log\left\{\frac{\gamma_{cmax}}{(A_1(OCR-1)+B_1)}\right\}\right] \tag{11.42}$$

where
$\gamma_{cmax}$ is the single-amplitude maximum cyclic shear strain,
OCR the over-consolidation ratio and
$\beta$, $A_1$ and $A_2$ are experimentally obtained parameters.

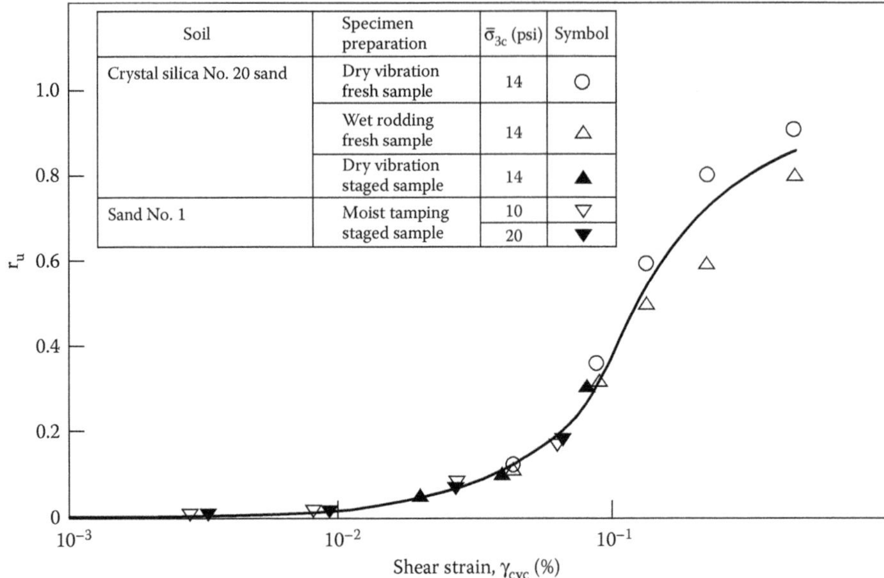


*Figure 11.17* Pore pressure ratio as a function of cyclic shear strain. (Adapted from Dobry, R. and Ladd, R.S. 1980. *Journal of Geotechnical Engineering Division*, 106(GT6): 720–724. Courtesy of ASCE.)

$\beta$ is found to be 0.45 for a wide range of clays, while $A_1$ and $B_1$ are dependent on PI, as given in Table 11.12.

### 11.6.12 Post-liquefaction strength

For practical applications, it has become customary to relate the post-liquefaction or residual strength of a sand to the SPT value, corrected for fines, $(N_1)_{60\text{-}cs}$. This correlation has been derived from back-calculation of observed flow slides. The most frequently employed

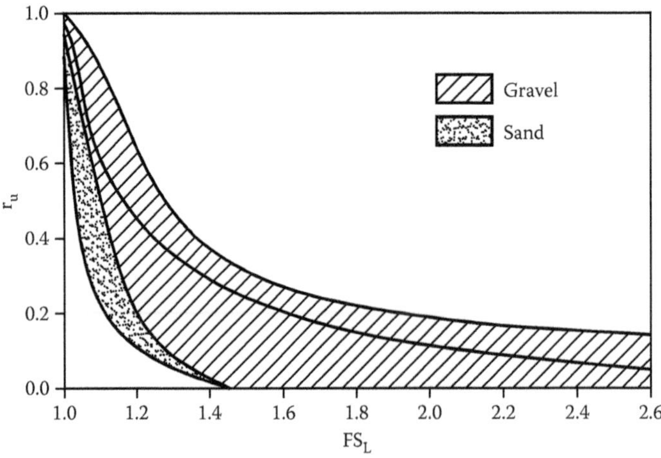


*Figure 11.18* Relationship between pore pressure ratio and factor of safety against liquefaction. (Adapted from Marcuson, W.F., Hynes, M.E. and Franklin, A.G. 1990. *Earthquake Spectra*, 6(3), 529–572. Courtesy of ASCE.)

*Table 11.12* Parameters $A_I$ and $B_I$ for Matsui et al. (1980) expression

| *Plasticity Index PI* | $A_I$ | $B_I$ |
|---|---|---|
| 20 | 0.0004 | 0.0006 |
| 40 | 0.0011 | 0.0012 |
| 55 | 0.0025 | 0.0012 |

relationship is that initially developed by Seed and Harder (1990), and which is shown in Figure 11.19.

Stark and Mesri (1992) suggested a further correction to the SPT value $(N_1)_{60}$ (the SPT value corrected for 60% energy ratio and 100 kPa overburden pressure), to allow for fines within the soil, and their recommended fines correction, $N_{corr}$, is given in Table 11.13. The equivalent clean sand SPT, $(N_1)_{60\text{-cs}}$ is then obtained as

$$(N_1)_{60\text{-cs}} = (N_1)_{60} + N_{corr} \tag{11.43}$$

Lumbantoruan (2005) has derived the following expression for the strength of the liquefied soil, $s_{u\text{-Liq}}$ is shown in Equation 11.44:

$$s_{u\text{-Liq}} = p((N_1)_{60\text{-cs}})^3 + q((N_1)_{60\text{-cs}})^2 + r(N_1)_{60\text{-cs}} + s(\text{kPa}) \tag{11.44}$$

where the coefficients (with an $R^2$ value of 0.652) are as follows: $p = 0.0050$, $q = 0.02777$, $r = 0.1697$, $s = 3.3376$.

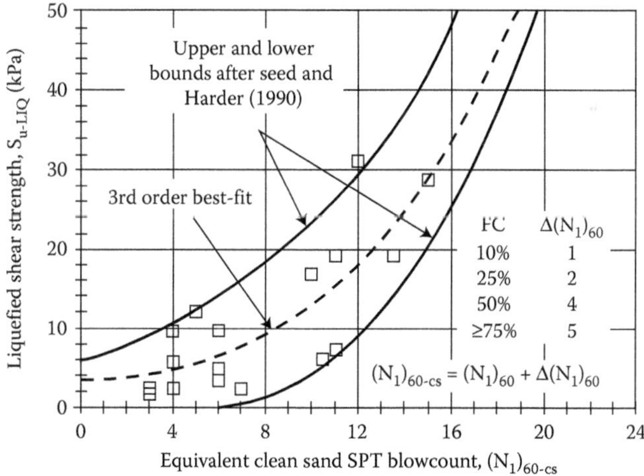


*Figure 11.19* Strength of liquefied soil versus corrected SPT resistance. (After Seed, H.B. and Harder, L.F. 1990. *Proceedings of the H. Bolton Seed Memorial Symposium*, University of California, Berkeley, Vol. 2, pp. 2: 351–376.)

*Table 11.13* Fines correction $N_{corr}$

| % Fines | 0 | 10 | 15 | 20 | 25 | 30 | 35 | 50 | 75 |
|---|---|---|---|---|---|---|---|---|---|
| $N_{corr}$ | 0 | 2.5 | 4 | 5 | 6 | 6.5 | 7 | 7 | 7 |

Source: Stark, T.D. and Mesri, G. 1992. *Journal of Geotechnical Engineering*, 118(11): 1727–1747. Courtesy of ASCE.

### 11.6.13 Repeated liquefaction

After liquefaction and the subsequent dissipation of excess pore pressures, sand will tend to become denser and would be expected to be more resistant to further liquefaction. However, as noted by Towhata (2008), examples of repeated liquefaction have been experienced in Japan, especially in young sandy deposits. However, after many earthquakes, the sand should become sufficiently dense to be more resistant to liquefaction.

## 11.7 ESTIMATION OF LATERAL SPREADING

A liquefaction-induced lateral spread generally involves large areas of ground translating laterally for distances ranging from a few centimetres to several metres along or through a layer of liquefied soil. These ground failures are a major source of damage to bridges, buildings, and other structures during earthquakes. An empirical method for estimating the amount of lateral spread has been proposed by Youd et al. (2002), who have developed an empirical expression based on available measurements at the time. They give the following equations for ground profiles with a free-face condition, and also those with a gently sloping surface.

For free-face conditions:

$$\begin{aligned}\log D_N &= -16.713 + 1.532\,M - 1.406\log R^* - 0.012R^* + 0.592\log W \\ &\quad + 0.540\log T_{15} + 3.413\log(100 - F_{15}) - 0.795\log(D50_{15} + 0.1\,mm)\end{aligned} \tag{11.45}$$

For gently sloping ground conditions:

$$\begin{aligned}\log D_N &= -16.213 + 1.532\,M - 1.406\log R^* - 0.012R^* + 0.338\log S \\ &\quad + 0.540\log T_{15} + 3.413\log(100 - F15) - 0.795\log(D50_{15} + 0.1\,mm)\end{aligned} \tag{11.46}$$

where

$D_N$ is the estimated lateral ground displacement (m),
M the moment magnitude of earthquake,
$R^*$ the nearest map distance from seismic energy source (km),
$T_{15}$ the cumulative thickness of saturated granular layers with corrected blow counts $(N_1)_{60} < 15$,
$F_{15}$ the average fines content (fraction passing No. 200 sieve) for granular materials included in $T_{15}$ (per cent),
$D50_{15}$ the average mean grain size for granular materials within $T_{15}$ (mm),
S the ground slope (per cent) and
W the free-face ratio = height of free face, divided by the distance from the base of the free face to the point in question, in per cent.

The above expressions are for displacements up to about 6 m.

Equations 11.45 and 11.46 have been derived for stiff soil sites in the western United States and Japan, where attenuation of strong ground motion from the causative fault is relatively high. For other regions, including those for liquefiable sites underlain by soft soils that may amplify ground motions, the equivalent distance ($R^* = R_{eq}$) can be obtained from Figure 11.20. This figure plots the mean acceleration expected at the site for the design earthquake versus earthquake magnitude.

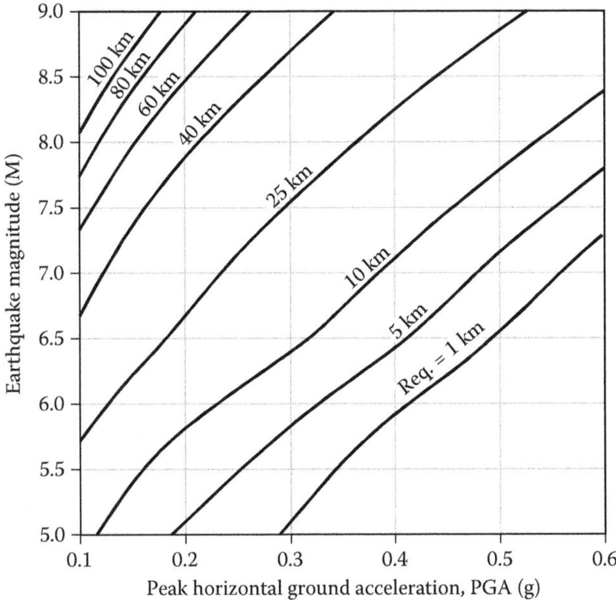


*Figure 11.20* Equivalent source distance ($R_{eq}$) versus PGA and M.

## 11.8 EFFECTS OF EARTHQUAKES ON SHALLOW FOUNDATION DESIGN

### 11.8.1 Introduction

The following effects of an earthquake should be addressed when designing shallow foundations for seismic loadings:

1. The ultimate bearing capacity of the foundation when seismic forces are acting
2. The possible reduction in soil strength and bearing capacity due to the build-up of pore pressures during seismic action
3. Foundation stiffness and damping, and the movements (vertical, horizontal and rotational) of the foundation
4. The effects of liquefaction on foundation capacity
5. The settlements that may be developed if liquefaction occurs

### 11.8.2 Seismic bearing capacity

When seismic forces act on a foundation, there will generally be a tendency for the bearing capacity to be reduced because of the presence of a component of horizontal force, resulting in an inclination of the applied load. Analyses to estimate the effects of seismic action on bearing capacity have been presented by Kumar and Mohan Rao (2002), who have presented their results in a convenient graphical form in which the bearing capacity factors $N_c$, $N_q$ and $N_\gamma$ are plotted as functions of the horizontal earthquake acceleration coefficient $a_h$ and the angle of internal friction of the soil, $\phi$. These factors are plotted in Figures 11.21 through 11.23. It can be seen that all three factors decrease as $a_h$ increases.

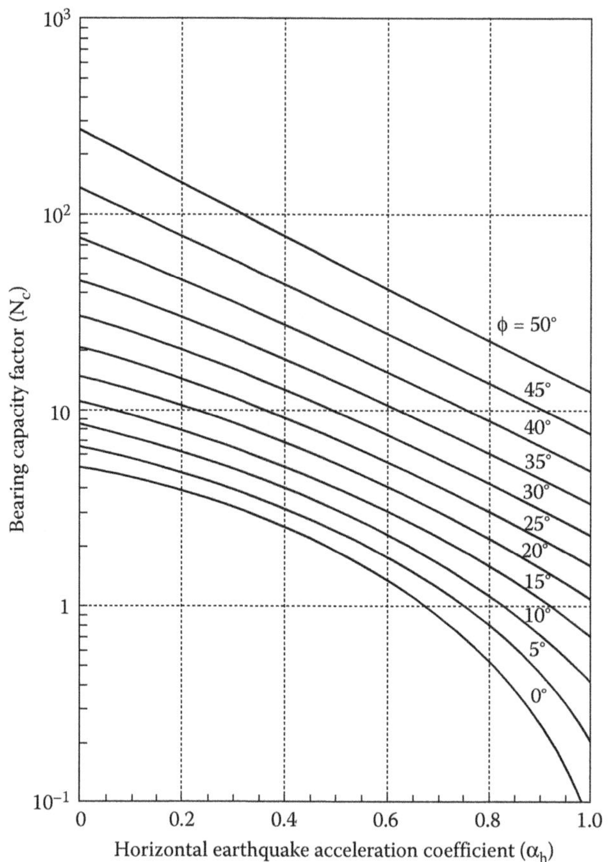


*Figure 11.21* Bearing capacity factor $N_c$ as a function of $a_h$. (Adapted from Kumar, J. and Mohan Rao, V.B.K. 2002. *Geotechnique*, 52(2): 79–88. Courtesy of ICE Publishing.)

The foundation bearing capacity, $p_u$, is expressed via the traditional equation first presented by Terzaghi (1943):

$$p_u = c\,N_c + q\,N_q + 0.5\gamma N_\gamma \tag{11.47}$$

where

c is the cohesion,
q the overburden pressure at level of foundation base and
$\gamma$ is the soil unit weight and
$N_c$, $N_q$ and $N_\gamma$ are bearing capacity factors which depend on the angle of internal friction $\phi$ and footing shape.

As is the case in conventional bearing capacity assessments, the values of c, $\phi$, $\gamma$ and q are total stress values if an undrained analysis is being carried out, and effective stress values if an effective stress analysis is being undertaken. In fine-grained soils, a total stress analysis would normally be employed, and only in very permeable soils would an effective stress analysis be appropriate in most cases of seismic loading.

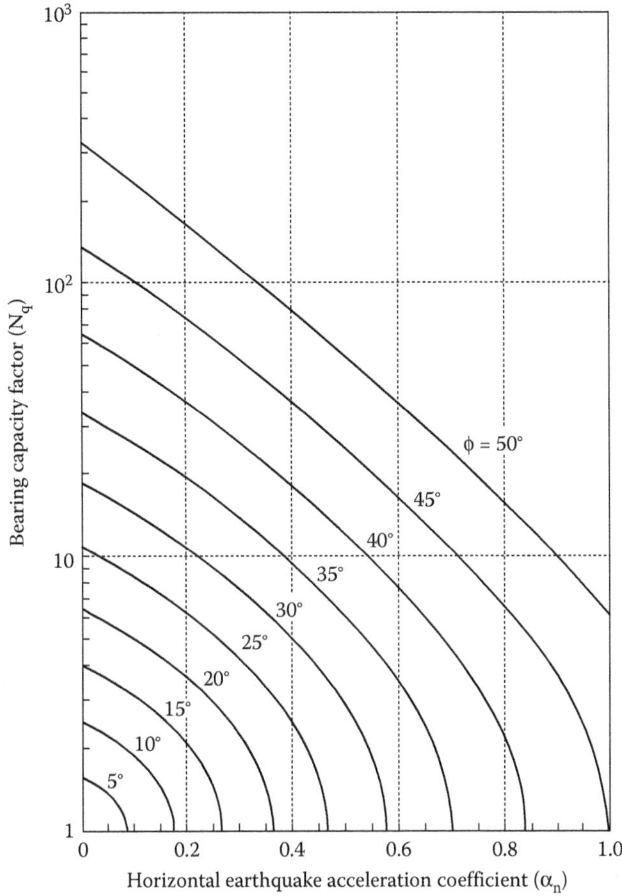


*Figure 11.22* Bearing capacity factor $N_q$ as a function of $a_h$. (Adapted from Kumar, J. and Mohan Rao, V.B.K. 2002. *Geotechnique*, 52(2): 79–88. Courtesy of ICE Publishing.)

Budhu and Al-Karni (1993) have considered the effect of both horizontal and vertical ground accelerations and have expressed the bearing capacity factors for earthquake loading, $N_{cE}$, $N_{\gamma E}$ and $N_{qE}$, in terms of the corresponding values for static loading, $N_{cS}$, $N_{\gamma S}$ and $N_{qS}$, as follows:

$$N_{cE} = N_{cS}\exp(-\beta_c) \tag{11.48}$$

where $\beta_c = 4.3a_h^{1.4}$

$$N_{\gamma E} = (1 - 2k_v/3)N_{\gamma s}\exp(-\beta_\gamma) \tag{11.49}$$

where $\beta_\gamma = 9a_h^{1.1}/(1 - a_v)$

$$N_{qE} = (1 - a_v)N_{qS}\exp(-\beta_q) \tag{11.50}$$

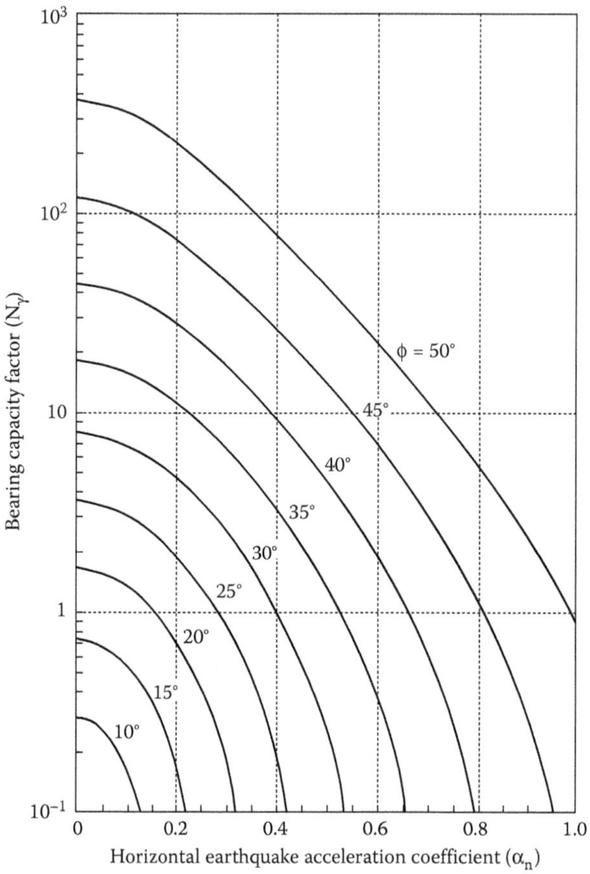


*Figure 11.23* Bearing capacity factor $N_\gamma$ as a function of $a_h$. (Adapter from Kumar, J. and Mohan Rao, V.B.K. 2002. *Geotechnique*, 52(2): 79–88. Courtesy of ICE Publishing.)

where

$\beta_q = 5.3a_h^{1.2}/(1 - a_v)$ and
$a_h$ is the horizontal acceleration coefficient and
$a_v$ the vertical acceleration coefficient.

## 11.8.3 Soil strength reduction due to pore pressure build-up

Seismic shaking will generate excess pore pressures within saturated soil, and these pore pressures will tend to reduce the shear strength of the soil. In coarse-grained soils that are saturated, if the excess pore pressure ratio $r_u$ can estimated, the influence of the induced pore pressure on the bearing capacity can be estimated from Equation 11.47 by reducing both q and $\gamma$ by a factor $(1 - r_u)$.

In fine-grained soils, if the strength is expressed in terms of total stress and the SHANSEP approach described by Ladd and Foott (1974) is used, it may be shown that the reduction factor $RF_{su}$ for undrained shear strength can be expressed as follows:

$$RF_{su} = \frac{s_{uE}}{s_{us}} = (1 - r_u)^{1-m} \tag{11.51}$$

where

$s_{uE}$ is the undrained shear strength after seismic shaking,
$s_{us}$ the undrained shear strength for static loading, before earthquake shaking,
$r_u$ the pore pressure ratio, estimated, for example, from Equation 11.42 and
m is an exponent, generally taken as 0.8.

In addition, in comparing the undrained shear strength during the earthquake with the value for static loading, it may be reasonable to apply a rate factor to the former value. Typically, the undrained shear strength increases approximately linearly with the logarithm of loading rate, at a rate of between 5% and 20% per log cycle of time. As a first approximation, it would seem reasonable to apply a rate factor of about 1.2 to the value under static loading representing a rate effect equivalent to a loading rate of 100 times the static rate with a 10% increase per log cycle of time. In this way, it should be possible to obtain a more realistic estimate of the undrained shear strength during the earthquake. Thus, for example, if $r_u$ is 0.7, then the value of $RF_{su}$ from Equation 11.50 is 0.786 (implying a loss of 21.4% in undrained strength), but applying a rate factor of 1.2, the overall reduction factor for undrained shear strength is 0.943, that is, there is only a reduction of 4.7% in undrained shear strength during the earthquake.

### 11.8.4 Reduction in bearing capacity due to reduced strength layer

In many cases, there may be soil above the water table that is not affected significantly by seismic shaking, with a liquefaction-susceptible layer below the water table in which strength reduction may occur. Figure 11.24 shows the case of a soil profile in which the strength in the lower layer is less than that in the upper layer, perhaps due to the effects of pore pressure build-up during seismic shaking. Also shown is the bearing capacity factor $N_c$ as a function of the ratio of shear strengths of the lower and upper layers, and the relative thickness of the upper layer. This figure clearly shows that the greater the reduction in shear strength of the lower layer, and/or the thinner is the upper layer, the smaller will be the bearing capacity factor $N_c$, that is, the greater will be the reduction in bearing capacity as compared with the case where the strengths of the upper and lower layers are the same.

If there is no crust and the saturated liquefiable sand extends to the ground surface or to the base of the foundation, then the bearing capacity will be very low and can be estimated approximately, for a circular or square footing, as about 5.8 times the undrained shear strength of the liquefied soil.

### 11.8.5 Foundation stiffness and damping

Estimation of the foundation stiffness and damping is required for estimating foundation movements due to seismically induced forces, and also for considering the effects of structure–foundation interaction on the natural period and damping of a structure (see Chapter 10). Foundation stiffness and damping can be calculated by numerical methods, including the finite element method. However, for preliminary estimates, if an equivalent value of Young's modulus of a soil profile can be estimated, the vertical, horizontal and rotational movements can be estimated via the values of foundation stiffness for a homogeneous semi-infinite layer given in Chapter 10.

In assessing the equivalent Young's modulus of the soil profile, consideration needs to be given to the strain level within the soil profile. Dowrick (2009) suggests that, as a first approximation, the modulus value for vertical response can be taken as about 50% of

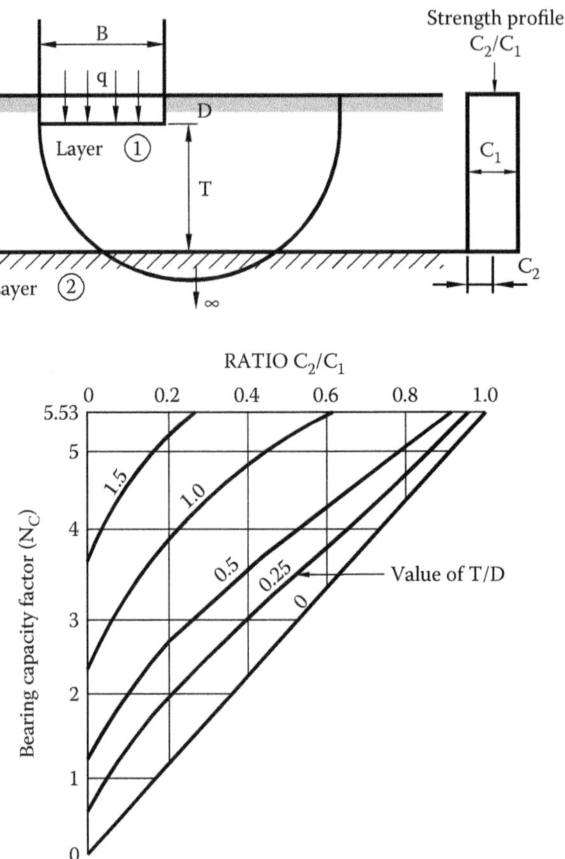


*Figure 11.24* Bearing capacity factor $N_c$ for weaker lower layer. (Adapted from NAVFAC 1982. DM-7.03. Soil dynamics and special design aspects. In: *Handbook*, USA Dept. Defense, Washington.)

the small-strain value, while the value for lateral response can be taken as 3%–40% of the small-strain value. Alternatively, Figure 11.6 can be used as a guide, depending on the PGA.

Values of radiation damping for the various modes of vibration are given in Chapter 10. As mentioned therein, for a first approximation, the internal damping ratio, for example from Figure 11.7, may be added to the radiation damping to obtain the overall damping ratio.

## 11.8.6 Effects of liquefaction on site settlement

If liquefaction of a saturated sand layer occurs, then there may be a significant settlement of the site after the excess pore pressures have dissipated. Estimates of this settlement can be made using the approach developed by Tokimatsu and Seed (1987). Figure 11.25 reproduces their chart for the estimation of volumetric strain of a saturated sand, as a function of the cyclic stress ratio (CSR) (for a magnitude 7.5 earthquake) and the corrected SPT value, $(N_1)_{60.}$ The CSR can be estimated from Equation 11.18, for various layers or sublayers at the site. By summing the product of the volumetric strain and the layer or sublayer thickness, an estimate can be made of the site settlement.

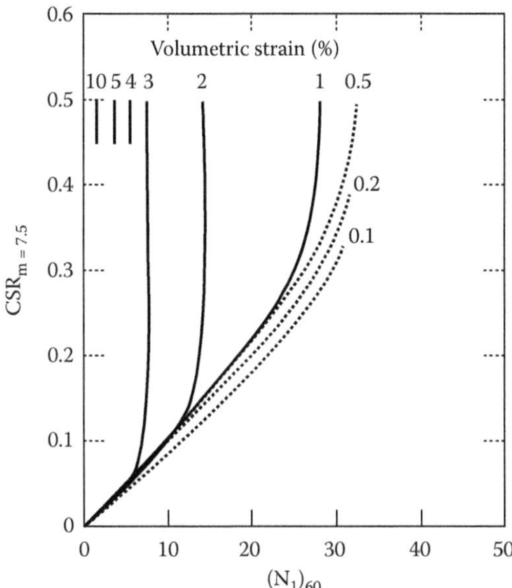


*Figure 11.25* Volumetric strain in saturated sands. (Adapted from Tokimatsu, K. and Seed, H.B. 1987. *Journal of Geotechnical Engineering*, 113(8): 861–878. Courtesy of ASCE.)

An alternative approach for estimating the volumetric strain of clean sands has been suggested by Ishihara and Yoshimine (1992), who have related the volumetric strain of clean sand to the factor of safety against liquefaction ($FS_L$) and the relative density of the sand (or the SPT to initiate liquefaction, $N_l$, or the CPT value). Their chart is shown in Figure 11.26.

Tsukamoto and Ishihara (2010) have also discussed an approach for estimating the settlement of soil deposits following an earthquake, using the results of Swedish weight sounding tests. This approach has been used for three case histories involving the post-liquefaction settlement of saturated soil deposits observed during earthquakes in Japan.

## 11.9 EFFECTS OF EARTHQUAKES ON PILE FOUNDATIONS: NO LIQUEFACTION

### 11.9.1 Introduction

There are two sources of earthquake-induced forces that need to be considered in the design of piles subjected to seismic activity:

1. Inertial loadings – these are forces that are induced in the piles because of the accelerations generated within the structure by the earthquake. The lateral inertial forces and moments are assumed to be applied at the pile heads.
2. Kinematic loadings – these are forces and bending moments that are induced in the piles because of the ground movements that results from the earthquake. Such movements will interact with the piles, and because of the difference in stiffness of the piles and the moving soil, there will be lateral stresses developed between the pile and the soil, resulting in the development of shear forces and bending moments in the piles. These actions will be time-dependent and also need to be considered in the structural design of the piles.

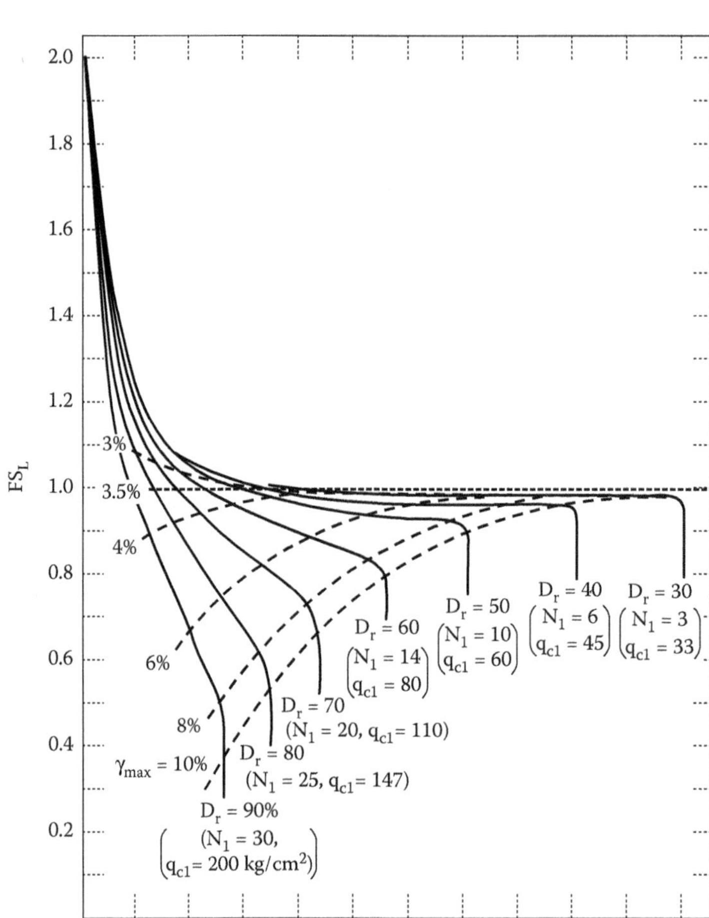


*Figure 11.26* Volumetric strain in saturated sands. (Adapted from Ishihara, K. and Yoshimine, M. 1992. *Soils and Foundations*, 32(1): 173–188.)

In addition to the inertial and kinematic forces induced in the piles, consideration must also be given to possible reductions in soil strength arising from the build-up of excess pore pressures during and after the earthquake. In extreme cases, the generation of pore pressures may lead to liquefaction in relatively loose sandy and silty soils. The case of piles in liquefiable soils will be dealt with in Section 11.11.

Consideration will be given first to issues related to lateral pile response, and then to axial pile response in Section 11.10.

### 11.9.2 Assessment of inertial loadings

For geotechnical analysis, the inertial forces imposed on the foundation system by the structure are usually provided to the geotechnical designers by the structural designers. The maximum lateral force on the structure is generally estimated by multiplying the mass of the structure by the peak spectral acceleration, $a_{smax}$. $a_{smax}$ can be obtained either via using code-specified values for bedrock acceleration and then applying site factors to obtain surface acceleration (see Section 11.4). For a piled foundation, the lateral inertial force can be estimated as $P \cdot a_{smax}/g$ where P is the axial force on pile and g is the gravitational acceleration.

Rather than adopting a code-specified site factor, a site response analysis using representative earthquake records at 'bedrock' level (e.g., where $v_s > 760$ m/s) may be carried out to assess the ground response. From this analysis, response spectra can be obtained for the acceleration versus time history at an appropriate depth in the layer (e.g., at about 2/3 of the pile length) to reflect the effect of the structure being founded on a pile foundation system.

If an elastic analysis is applied to the pile and a linearly varying Young's modulus with depth is assumed to apply within the soil, then the maximum bending moment due to inertial loading, $M_{imax}$, can be estimated approximately from the following expression given by Randolph (1981):

1. For a free-head pile:

$$M_{imax} = 0.1 H_i L_c / \rho_c \tag{11.52}$$

2. For a fixed-head pile: (fixing moment at the pile head):

$$M_{imax} = -0.1875\ H_i L_c / (\rho_c) 0.5 \tag{11.53}$$

where $H_i$ is the inertial force on pile

$$L_c = \text{effective pile length} = d(E_p/G_c)^{2/7} \tag{11.54}$$

$G_c$ the average shear modulus of soil over a depth equal to the effective length of the pile, and

$\rho_c$ the ratio of soil modulus at a depth of 1/4 of effective length to that at a depth of 1/2 of the effective pile length.

### 11.9.3 Assessment of kinematic loadings

To estimate the possible additional bending moments and shears in the piles arising from the kinematic ground movements during and after a seismic event, there are at least two design approaches that may be adopted:

1. A simple approach employing the results of analyses reported by Nikolaou et al. (2001), among others
2. A more detailed approach involving the use of a pseudostatic analysis in which the results of a site response analysis are combined with a pile–soil interaction analysis (e.g., Tabesh and Poulos, 2001).

### 11.9.4 Simplified analysis methods for kinematic moments in piles

A number of simplified approaches have been developed to allow estimation of the kinematically induced bending moments in a pile. Four of these approaches are outlined below.

#### *11.9.4.1 Method of Nikolaou et al. (2001)*

A convenient design approach for estimating the maximum moment induced in a pile by kinematic bending has been provided by Nikolaou et al. (2001). They found that the induced moments were a maximum at interfaces between layers of different stiffness and

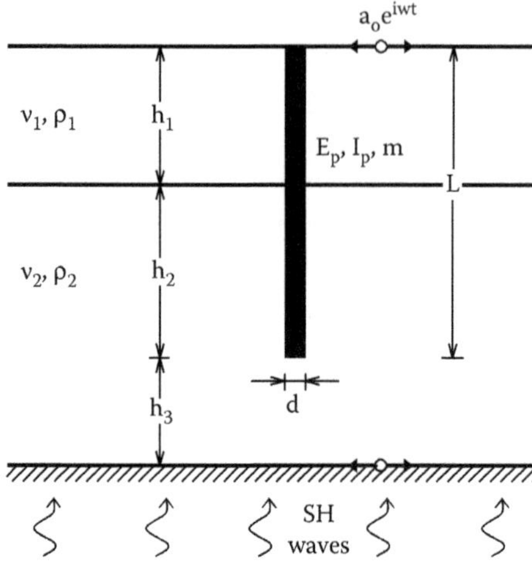


*Figure 11.27* Model adopted by Nikolau et al. (2001) for kinematic moments in pile.

then undertook a series of analyses to compute the bending moment at the interface between two layers (see Figure 11.27). They recognised that a distinction must be made between the maximum bending moment under steady-state harmonic motion and the bending moment that would be developed under transient excitation, such as during an earthquake. The latter would generally be smaller than the steady-state value, which would only be developed after a very large number of cycles. This distinction was also emphasised by Sica et al. (2011) and Figure 11.28 shows, diagrammatically, the relationship between the bending moment and the frequency ratio (ratio of the predominant frequency of the earthquake to the natural frequency of the subsoil.)

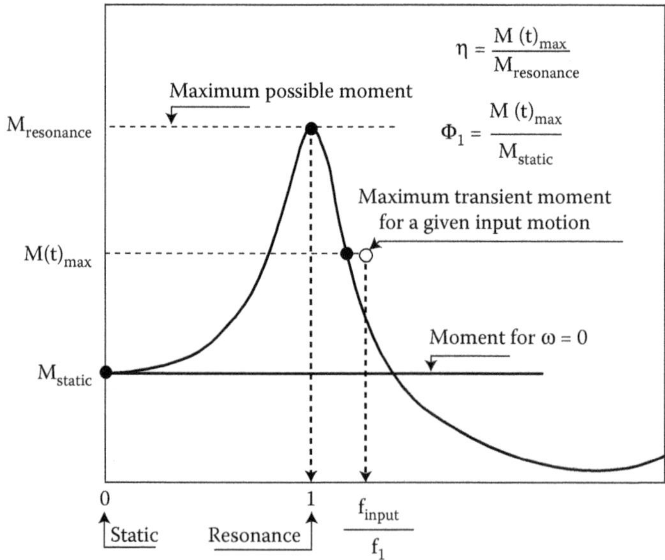


*Figure 11.28* Relationship between pile bending moment and frequency ratio. (Adapted from Sica, S., Mylonakis, G. and Simonelli, A.L. 2011. *Soil Dynamics and Earthquake Engineering*, 31: 891–905.)

The following approximate relationship was developed for the peak bending moment, $M_{pk}$, during the transient phase of seismic excitation:

$$M_{pk} = \eta \cdot M_{res} \tag{11.55}$$

where

$M_{res}$ is the bending moment developed under resonant conditions and
$\eta$ is the reduction factor to allow for non-resonant conditions.

From the results of a frequency domain analysis, Nikolaou et al. developed the following fitted formula for $M_{res}$:

$$M_{res} = 0.042\tau_c d^3 (L/d)^{0.30} (E_p/E_1)^{0.65} (V_{s2}/V_{s1})^{0.50} \tag{11.56}$$

with

$$\tau_c = a_s \rho_1 h_1 \tag{11.57}$$

where

d is the pile diameter,
L the pile length,
$E_p$ the Young's modulus of pile,
$E_1$ the Young's modulus of upper layer,
$V_{s1}$ the average shear wave velocity in upper layer,
$V_{s2}$ the average shear wave velocity in lower layer,
$a_s$ the peak ground surface acceleration,
$\rho_1$ the mass density of upper layer and
$h_1$ is the thickness of upper layer.

In the original paper, Nikolaou et al. give the following expressions for the reduction factor $\eta$:

*Case 1:* For resonant conditions in which the fundamental period of the deposit lies within the range of predominant periods of the excitation:

$$\eta = 0.04 N_c + 0.23 \tag{11.58}$$

*Case 2:* For non-resonant conditions where the fundamental period of the deposit lies outside the range of predominant periods of the excitation:

$$\eta = 0.015 N_c + 0.17 \approx 0.2 \tag{11.59}$$

where $N_c$ is the effective number of cycles within the earthquake record.

Subsequently, Sica et al. (2011) have suggested the following alternative expression for the average value of $\eta$ based on a series of parametric analyses:

$$\eta = 0.68 (f_{input}/f_1)^{-1.5} \tag{11.60}$$

where

$f_{input}$ is the predominant frequency of earthquake and
$f_1$ is the natural frequency of soil profile.

The above expression holds for $(f_{input}/f_1) \geq 1.5$, but for frequency ratios less than 1.5, $\eta = 0.37$, with a standard deviation of 0.17.

The value of $f_{input}$ can be obtained from the input earthquake record, but can also be estimated (very approximately) as the inverse of the predominant period from Figure 11.5. The value of the natural period of the soil deposit, $f_1$, can be obtained from the equations in Section 10.6 or the inverse of Equation 11.8.

Maiorano et al. (2009) have found that some of the simplified approaches, such as that of Nikolaou et al., tend to be conservative and can predict bending moments at the soil-layer interface with adequate accuracy only within certain depths of the soil-layer interface. On the basis of a parametric study, a modified criterion to evaluate the transient peak bending moments at interfaces between layers was proposed, but this required a preliminary assessment of the peak soil shear strain via a free-field site response analysis.

#### *11.9.4.2 Method of Di Laora et al. (2012)*

Di Laora et al. (2012) found that the Nikolaou et al. (2001) approach can underestimate kinematically induced bending moments, a characteristic also noted subsequently by Martinelli et al. (2016). As an alternative procedure, Di Laora et al. developed an approach which followed that previously proposed by Dobry and O'Rourke (1983) and Mylonakis (2001) in which a 'strain transmissibility function' is employed. The following is the most convenient of various expressions provided for the maximum bending moment, $M_{max}$, at the interface between two layers:

$$M_{max} = (2E_pI_p/d) \cdot (\varepsilon_p/\gamma) \cdot a_s.\rho_1.h_1.\Phi_{1,s}/G_1 \tag{11.61}$$

where $E_pI_p$ is the pile bending stiffness, d the pile diameter, $a_s$ the maximum surface acceleration, $\rho_1$ the mass density of upper layer, $h_1$ the thickness of upper layer, $G_1$ the shear modulus of upper layer, $\Phi_{1,s}$ a factor usually less than 1, but which can be taken as 1 for a conservative estimate and $(\varepsilon_p/\gamma)$ is the strain transmissibility function, given as

$$(\varepsilon_p/\gamma) = -0.5(h_1/d)^{-1} + (E_p/E_1)^{-0.25}(c-1)^{0.5} \tag{11.62}$$

where $E_1$ is the Young's modulus of upper layer and c is the ratio $G_2/G_1$ of shear modulus of lower layer ($G_2$) and upper layer ($G_1$).

De Laora et al. also made the following observations in relation to kinematically induced moments:

1. Kinematic bending near a layer interface can be viewed as a superposition of two components: a negative contribution imposed by the softer layer, and a positive contribution provided by the restraint of the stiffer layer.
2. Small interface depths and stiffness contrasts tend to reduce the absolute values of bending moment, as compared to bending in a homogeneous layer having a stiffness equal to that of the softer layer.
3. Conversely, deep interfaces and sharp stiffness contrasts lead to higher bending moments than in homogeneous soil.
4. The bending moments tend to increase with increasing pile-soil stiffness ratio.

### 11.9.4.3 Dezi et al. (2010) method

Dezi et al. (2010) have used an extensive parametric study to develop simplified expressions for the kinematic bending moments within a pile, for two locations:

- At the pile head
- At the interface between the lower-most soil layer and the underlying bedrock

The bending moment in each case is given by the following expression:

$$M = (PGA/0.25g) \cdot M_{400}(d,h) \cdot e^{f(d,h)\cdot(V_s - 400)} \quad (11.63)$$

where PGA is the peak ground acceleration, d the pile diameter, h the soil layer thickness, $V_s$ the shear wave velocity in the soil layer, $M_{400}$(d,h) the moment for a soil shear wave velocity of 400 m/s, and is given in Equations 11.64 and 11.65 for the pile head, and Equations 11.66 and 11.67 for the soil–bedrock interface.

For the pile head:

$$M_{400}(d,h) = (85d^3 - 85.75d^2 + 30.93d - 3.37)\cdot(0.000133h^2 - 0.00042h + 1.091) \quad (11.64)$$

$$f(d,h) = (0.000067h - 0.0113)\cdot(-0.07d + 1.002) \quad (11.65)$$

For the soil–bedrock interface:

$$M_{400}(d,h) = (77.7d^3 + 409d^2 - 192d + 24.5)\cdot(-0.0009h^2 + 0.068h - 0.2) \quad (11.66)$$

$$f(d,h) = (0.000124h - 0.01106)\cdot(-0.05d + 0.864) \quad (11.67)$$

In deriving the above approximate expressions, the following parameters have been assumed: Poisson's ratio = 0.4 for both soil and bedrock, internal damping ratio – 5% for soil and 2% for bedrock.

### 11.9.4.4 Dezi and Poulos (2016): Pile groups

On the basis of a parametric study of a series of square pile groups, and the previous results for a single pile, Dezi and Poulos (2016) have developed an empirical expression for the bending moments, both at the head and at the deposit–bedrock interface, with the following form:

$$M^G_{max} = M^S\alpha\left(n, \frac{s}{d}\right) \quad (11.68)$$

where $M^G_{max}$ is the maximum bending moment arising in the piles of the group at the head or at the deposit–bedrock interface, $M^S$ the relevant single pile bending moment, α a group factor depending on the number of piles and the pile spacing and n is the number of piles constituting the square group. $M^S$ has to be determined from a dynamic analysis or by means of simplified methods set out above. The following expressions are proposed for the group factor α:

$$\alpha\left(n, \frac{s}{d}\right) = a\left(\frac{s}{d}\right)\log(n) + b\left(\frac{s}{d}\right) \tag{11.69}$$

in which coefficients a and b assume different expressions depending on the considered pile cross section.

1. For the pile head:

$$a\left(\frac{s}{d}\right) = 0.16\left(\frac{s}{d}\right)^{-0.28} \tag{11.70}$$

$$b\left(\frac{s}{d}\right) = 0.58\left(\frac{s}{d}\right)^{0.23} \tag{11.71}$$

2. For the deposit–bedrock interface:

$$a\left(\frac{s}{d}\right) = -0.12\left(\frac{s}{d}\right)^{-0.3} \tag{11.72}$$

$$b\left(\frac{s}{d}\right) = 0.88\left(\frac{s}{d}\right)^{0.04} \tag{11.73}$$

The following conclusions may be drawn from the parametric analysis:

- With reference to a single pile, at the deposit–bedrock interface the kinematic bending moment in the most stressed pile of the group reduces as the number of piles in the group increases, while at the pile head, the bending moment generally increases, as a consequence of the group effect
- s/d has a minor effect on bending moments even if, as expected, the group effect increases by reducing the s/d ratio
- External piles experience greater stresses than inner piles; by considering groups with a high number of piles (e.g., 4 × 4 and 5 × 5 groups), stresses in piles diminish moving from edge to inner piles; corner piles are generally the most stressed piles in the group, while those within the group core are protected

The application of the above approximate expressions is straightforward due to the low number of parameters involved and only require the prior knowledge of the kinematic bending moments, which may be obtained from Equations 11.63 through 11.67.

### 11.9.5 Pseudostatic analysis for pile response

Tabesh and Poulos (2001) have proposed a pseudostatic approach for estimating the maximum response of a pile during an earthquake. The approach involves the following steps:

1. A free-field site response analysis is carried out to obtain the time history of surface motion and the maximum horizontal displacement of the soil along the length of the pile. In general, a 1D analysis can be employed, using commercially available codes such as SHAKE or DEEPSOIL, or else custom codes such as ERLS (Poulos, 1991b).
2. The surface motion obtained in the above step is used in a spectral analysis of a single degree of freedom system whose natural period is equal to that of the supported structure. The spectral acceleration $a_s$ is thus obtained.
3. A static analysis of the pile is carried out in which the pile is subjected simultaneously to the application of the following loadings:
   a. A lateral force at the pile head equal to $a_sP$, where P is the vertical load acting on pile head;
   b. The maximum ground movements along the pile length, as obtained from Step 1.

The analysis will give the maximum moment and shear force developed in the pile by the simultaneous application of the inertial and kinematic loadings.

### 11.9.6 Combined inertial and kinematic effects

The approach employed by Tabesh and Poulos (2001) may be conservative as the analysis implicitly assumes that both loadings are in phase. However, they have found that this approach gives reasonable agreement with the results of a more complete dynamic analysis, although it does tend to be conservative, and also good agreement is found when applied to a case history in Japan.

Tokimatsu et al. (2005) have suggested the following approach to deal with combined inertial and kinematic loadings:

- If the natural period of the superstructure is less than that of the ground, the kinematic force tends to be in phase with the inertial force, increasing the stress in the piles. The maximum pile stress occurs when both the inertial force and the ground displacements take the peaks and act in the same direction. In this case, the maximum moment is the sum of the values for inertial and kinematic effects.
- If the natural period of the superstructure is greater than that of the ground, the kinematic force tends to be out of phase with the inertial force. This tends to restrain the pile force, rather than increasing it. The maximum pile stress tends to occur when both inertial force and ground displacement do not become maxima at the same time.In this case, the maximum moments are taken to be the square root of the sum of the squares of the moments due to each effect.

The moments via this approach have been found to be in good agreement with model tests, both with and without the effects of liquefaction.

### 11.9.7 Design charts

Using the pseudostatic approach, Tabesh and Poulos (2007) produced some simple design charts for piles within a uniform soil profile. Figure 11.29 shows the problem addressed and Figure 11.30 gives examples of these charts for the case of a 20 m long pile with a vertical load corresponding to a factor of safety of 2.5 against geotechnical failure. The main value of such simplified charts is to obtain a preliminary idea of whether a more detailed analysis may be required.

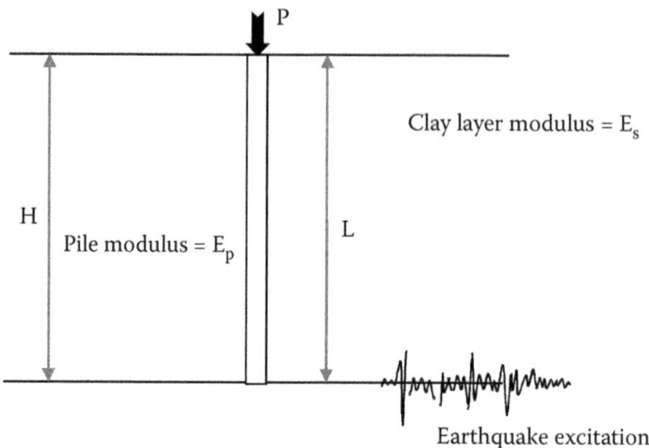


*Figure 11.29* Problem considered for design charts. (Adapted from Tabesh, A. and Poulos, H.G. 2007. *Geotechnical Engineering, ICE*, 160(GE2): 85–96. Courtesy of ICE Publishing.)

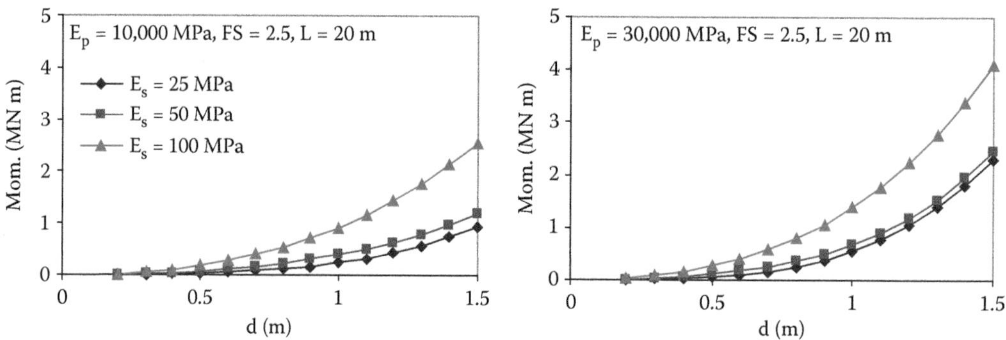


*Figure 11.30* Typical design charts for maximum moment in 20 m long pile. (Adapted from Tabesh, A. and Poulos, H.G. 2007. *Geotechnical Engineering, ICE*, 160(GE2): 85–96. Courtesy of ICE Publishing.)

## 11.10 OTHER ASPECTS

### 11.10.1 Axial pile response

Relatively little attention has been paid to the axial response of piles during and after an earthquake. Poulos (1993b) has provided an example of a pile in clay subjected to earthquake action. The generation of excess pore pressures, and the consequent loss of strength of the clay, has been incorporated into the analysis. The main conclusions drawn from this study are as follows:

1. Earthquakes with a Richter Magnitude in excess of about 6 have the potential to generate significant excess pore pressures in the clay, causing subsequent consolidation settlement.
2. The rate of development of these settlements is similar to that obtained from Terzaghi's consolidation theory, using the coefficient of consolidation for the soil in an over-consolidated state.

3. Piles in clay may be subjected to a short-term loss of axial capacity due to the 'softening' of the surrounding soil arising from pore pressure build-up.
4. Piles may also experience a long-term increase in settlement and axial force, due to the effects of negative friction arising from the ground settlements induced by the earthquake.

### 11.10.2 Foundation stiffness and damping

Piles will experience vertical and horizontal movements during an earthquake due to the inertial loadings imposed by the structure which is supported by the piles. These movements will generally be dynamic in nature and therefore it is necessary to consider the pile head stiffnesses under dynamic loading, and the radiation damping that will be generated by the dissipation of energy away from the piles into the surrounding soil. Knowledge of foundation stiffness and damping is also required to assess the effects of structure–foundation interaction on the natural period and damping of a structure (Sections 11.5.5 and 11.5.6).

Various solutions for the stiffness and damping of deep foundations have been presented in Chapter 10.

### 11.10.3 Measured foundation performance

Yamashita et al. (2012) have described a case history of a building supported by a piled raft foundation on medium to dense sand, underlain by over-consolidated silty soil. Field measurements were made of the foundation settlements and the load sharing between the piles and raft. During the monitoring period, the site was subjected to the 2011 Tohoku Pacific earthquake. A PGA of 3.24 m/s$^2$ (0.33 g) was observed at a site about 0.9 km from the building site. The building settlement increased by 4–25 mm, but no significant changes were found in the load sharing between the raft and the piles. There did not appear to have been liquefaction at this site.

Hamada et al. (2014) reported the outcomes of measurements of a 7-storey building on a piled raft foundation subjected to the above earthquake. Based on the seismic records, it was found that the lateral inertial force imposed on the building was supported by frictional resistance beneath the raft as well as by the piles. It was also found that the ratio of the lateral load carried by the piles to the lateral inertial force of the building was estimated to be about 10%–20%.

These cases have demonstrated that piled raft foundations appear to be resilient when subjected to seismic events, provided that appropriate allowances for such events have been made in the foundation design process.

## 11.11 EFFECTS OF EARTHQUAKES ON PILE FOUNDATIONS: INCLUDING LIQUEFACTION

### 11.11.1 Introduction

Soil liquefaction during a seismic event results in almost a complete loss of strength and stiffness in the liquefied soil, and consequent large lateral ground movements. The consequences of liquefaction on pile foundations can be very significant and can result in failure of the piles, as evidenced by many case histories during the 1995 Kobe earthquake (e.g., Ishihara and Cubrinovski, 1998). Madabhushi et al. (2010) provide a comprehensive review and exposition of pile design for liquefiable soils, and have discussed modes of failure of piles in liquefiable soils, both for single piles and pile groups. For single piles, failure can occur by excessive bending moments, by loss of vertical bearing capacity or by buckling instability caused by loss of lateral support. Similar mechanisms can cause the failure of pile groups, and additionally, pile group instability can be caused by the formation of plastic hinges at the pile-cap connection or at depth along the piles. These mechanisms can be exacerbated by the occurrence of lateral spreading.

A rational analysis of the behaviour of a pile in a liquefiable soil during and after an earthquake should take account of the effects of liquefaction on the soil properties, the natural period of the ground and the ground movements.

### 11.11.2 Simplified analyses

A procedure that has been employed in Japan to allow for the effects of liquefaction is to reduce the soil parameters by a factor which depends on the following factors:

1. The factor of safety against liquefaction, $F_L$
2. The depth below the ground surface
3. The original SPT (N) value of the soil

An example of values of a reduction factor ($r_k$) for the modulus of subgrade reaction (k) for building foundation design in Japan is given in Table 11.14 (JGS, 1998). This approach, while simple, considers only the effects of inertial loading and does not incorporate the effects of kinematic loading.

### 11.11.3 Pseudostatic analyses

#### *11.11.3.1 Cubrinovski and Ishihara (2004)*

The model developed by Cubrinovski and Ishihara (2004) and Cubrinovski et al. (2009) is shown in Figure 11.31. The soil profile contains three components:

1. A surface layer or crust which does not liquefy
2. A liquefiable layer
3. An underlying base layer that does not liquefy

Each layer is characterised by a stiffness (expressed in terms of a modulus of subgrade reaction) and a limiting pile-soil pressure. The pile is analysed as a beam-spring model, and non-linear behaviour of the pile, as well as the soil, can be incorporated into the analysis by use of a finite element approach. Account is taken of both inertial and kinematic loadings.

*Table 11.14* Reduction factor for modulus of subgrade reaction – building foundations

| *Factor of safety against liquefaction, $F_L$* | *Depth z below ground surface (m)* | *Reduction factor $r_k$, applied to modulus of subgrade reaction k* | | | |
|---|---|---|---|---|---|
| | | $N \leq 8$ | $8 < N \leq 14$ | $14 < N \leq 20$ | $N > 20$ |
| $\leq 0.5$ | $0 \leq z \leq 10$ | 0 | 0 | 0.05 | 0.1 |
| | $10 < z \leq 20$ | 0 | 0.05 | 0.1 | 0.2 |
| $0.5 < F_L \leq 0.75$ | $0 \leq z \leq 10$ | 0 | 0.05 | 0.1 | 0.2 |
| | $10 < z \leq 20$ | 0.05 | 0.1 | 0.2 | 0.5 |
| $0.75 < F_L \leq 1.0$ | $0 \leq z \leq 10$ | 0.05 | 0.1 | 0.2 | 0.5 |
| | $10 < z \leq 20$ | 0.1 | 0.2 | 0.5 | 1.0 |

Source: JGS. 1998. *Remedial Measures Against Liquefaction*. Japanese Geotechnical Society (JGS), Balkema, Rotterdam.

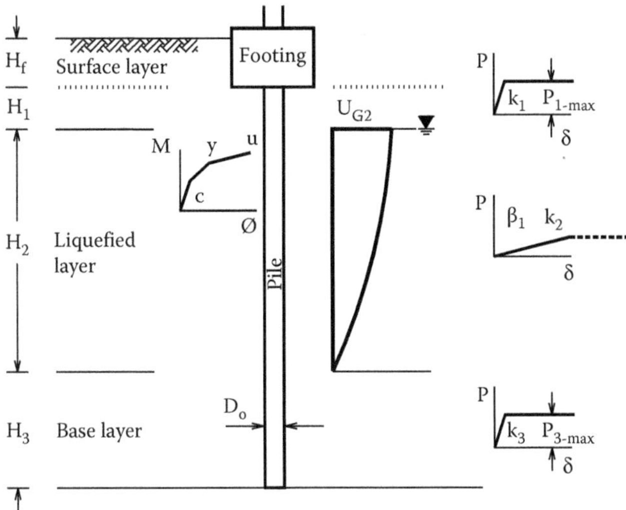


*Figure 11.31* Model for pile in liquefied soil. (Adapted from Ishihara, K. and Cubrinovski, M. 2004. Case studies of pile foundations undergoing lateral spreading in liquefied deposits. Paper SOAP 5, *Proceedings of the 5th International Conference on Case Histories in Geotechnical Engineering*, Paper SOAP 5. New York. CD volume.)

### *11.11.3.2 Liyanapathirana and Poulos (2005)*

For piles in soils subject to liquefaction, Liyanapathirana and Poulos (2005) have developed an extension of the approach used by Tabesh and Poulos (2001) for piles in non-liquefiable soils. Account has been taken of the degradation of shear modulus of the soil that occurs with the generation of pore water pressure in the soil. The shear modulus of the soil is assumed to vary with the effective stress level of the soil as shown below:

$$G_s = G_0((1+2Ko)\sigma_{v'}/p_a)^{0.5} \text{MPa} \tag{11.74}$$

where $\sigma'_v$ is the effective stress level of the soil, $K_0$ the coefficient of earth pressure at rest, $p_a$ the atmospheric pressure and $G_0$ is a constant which varies with the relative density, $D_r$, of the soil.

It has been found that the inertial force at the pile head calculated using the spectral acceleration, as used by Tabesh and Poulos (2001), may overestimate the pile response when the surrounding soil starts to liquefy. Hence, the maximum acceleration at the ground surface, rather than the spectral acceleration, has been used to calculate the inertial force at the pile head.

The calculation steps involved in this approach can be summarised as below:

1. First, a free-field site response analysis is performed by taking into account the pore pressure generation and dissipation in the soil deposit due to the earthquake loading (Liyanapathirana and Poulos, 2002). From this analysis, the maximum ground surface acceleration, the maximum ground displacement along the length of the pile, the minimum shear modulus and effective stress level attained during the seismic activity can be obtained.
2. The superstructure is modelled as a concentrated mass at the pile head. Generally superstructures supported by pile foundations are multi-degree-of-freedom systems, but in the design of pile foundations, the superstructure is reduced to a single mass at the pile head to simplify the analysis.

3. The lateral force to be applied at the pile head is the cap-mass (vertical load divided by g), multiplied by the maximum ground surface acceleration obtained from the ground response analysis.
4. The pile–soil interaction is modelled using the spring coefficients calculated from the minimum shear modulus of the soil deposit at each depth, at any time, given by the free-field site response analysis (Step 1).
5. A non-linear static analysis is carried out to obtain the profile of maximum pile displacement, bending moment and shear force along the length of the pile by applying the lateral forces calculated in Steps 3 and 4, and the profile of maximum soil movement calculated in Step 1, simultaneously to the pile.

This approach has been verified by comparison with the results of centrifuge tests undertaken by Wilson et al. (1999) and Abdoun et al. (1997).

Comparisons have also been made with the field measurements made in the piles at the Pier 211 in Uozakihama Island after the Hyogoken-Nambu earthquake occurred on 17 January 1995, and reported by Ishihara and Cubrinovski (1998). This case has been simulated using the pseudostatic approach presented above. Figure 11.33 shows the crack distributions observed in piles after the earthquake. The reinforced concrete piles at bridge Pier 211 were 46 m long and the diameter was 1.5 m. The water table was 2.0 m below the ground surface and the upper 20 m of this site consisted of Masado sand with an initial shear modulus of 57.8 $MN/m^2$ and density of 2 $t/m^3$. Soil liquefaction was observed in the Masado sand layer below the water table only. Therefore only the top 20 m layer was analysed using the effective stress method incorporating pore pressure generation and dissipation. After liquefaction, the effective stress level in the soil was reduced to a minimum of 2% of the initial effective overburden pressure. For this analysis the cyclic shear strength curve for the Masado sand given by Ishihara (1997) was used. It was assumed that the base rock had a density of 2200 $kg/m^3$ and a shear modulus of 75 $GN/m^2$.

Figure 11.32 shows the maximum bending moment profile along the pile obtained from the pseudostatic approach and also that calculated by Ishihara and Cubrinovski (1998). The lower end of the pile was assumed to be fixed while the pile head was assumed to be fixed to the footing but free to move in the horizontal direction. The predictions made by the pseudostatic approach agree reasonably well with the results given by Ishihara and Cubrinovski (1998). The yield moment for these piles was about 5 MN m. The computed maximum bending moment profile exceeded the yield moment near the pile head and in the vicinity of the boundary between the liquefied and non-liquefied layers. This is consistent with the location of cracks observed after the earthquake shown in Figure 11.32.

### 11.11.4 Simplified approach

Despite the simplifications involved in the pseudostatic approaches, they still require a considerable amount of computational effort, and it is therefore worthwhile to consider a simpler approach in which the amount of computational effort is relatively limited. This approach involves the following steps:

1. Estimate the reduction in shear modulus via the approach suggested by Ishihara and Cubrinovski (1998), in which the post-liquefaction shear modulus, $G_{pl}$, is given by

   $$G_{pl} = \beta G \tag{11.75}$$

   where $\beta$ is the reduction factor and G is the shear modulus prior to liquefaction.

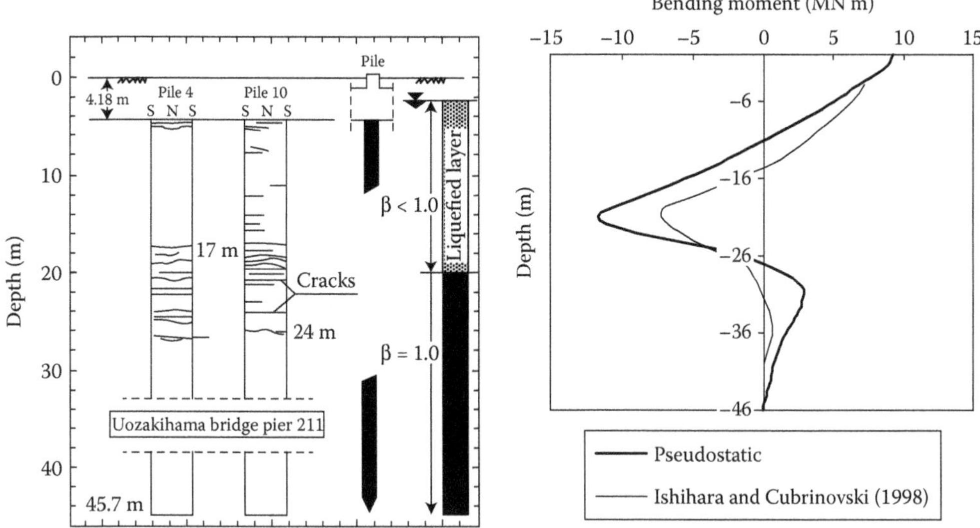


*Figure 11.32* (a) Cracks observed in piles at the bridge pier 211 in Uozakihama Island. (Adapted from Ishihara, K. and Cubrinovski, M. 1998. Performance of large-diameter piles subjected to lateral spreading of liquefied soils. *Proceedings of the 13th Southeast Asian Geotechnical Conference*, Keynote Lecture, Taipei, Vol. 2, pp. 13–26.) and (b) bending moment of the pile at Bridge Pier 211 in Uozakihama Island calculated from the pseudostatic approach and Ishihara and Cubrinovski (1998) results ($\beta = 0.01$).

2. Estimate the bending moment for resonant conditions, if using the Nikolaou et al. (2001) method, via Equations 11.56 and 11.57. The shear wave velocity in the (upper) liquefied layer is now $V_{s1\text{-liq}} = V_{s1}\sqrt{\beta}$ where $V_{s1}$ is the original shear wave velocity of the layer.
3. Estimate the modified natural frequency of the liquefied layer from Equation 11.8, using the reduced shear wave velocity $V_{s1\text{-liq}}$.
4. Compute the correction factor $\eta$ from either Equations 11.58 and 11.59 or Equation 11.60.
5. Compute the bending moment from Equations 11.55 to 11.57.

The other methods outlined in Section 11.9.4 may also be used in place of the Nikolaou et al. approach.

#### *11.11.4.1 Modulus reduction factor* β

A key parameter choice in the above method is the value of the modulus reduction factor $\beta$. Cubrinovski (2006) suggests that $\beta$ values range between 0.02 and 0.10 for cyclic liquefaction, and 0.001–0.02 for lateral spreading.

An alternative approach to estimating $\beta$ is to relate it to the factor of safety against liquefaction $F_L$, or perhaps more logically, to the LPI, as defined in Equation 11.39. To do this, use can be made of the recommendations given in Table 11.14, from which ranges of values of LPI can be obtained from the range of values of $F_L$ given in that table for the upper 10 m of the profile and profile from 10 to 20 m depth. Figure 11.33 shows the values of the modulus reduction factor $\beta$ can be plotted against the derived average values of LPI, and from this plot, an empirical relationship can be drawn as follows:

$$\beta = \beta_{lim} + e^{b.LPI}(1 - \beta_{lim}) \tag{11.76}$$

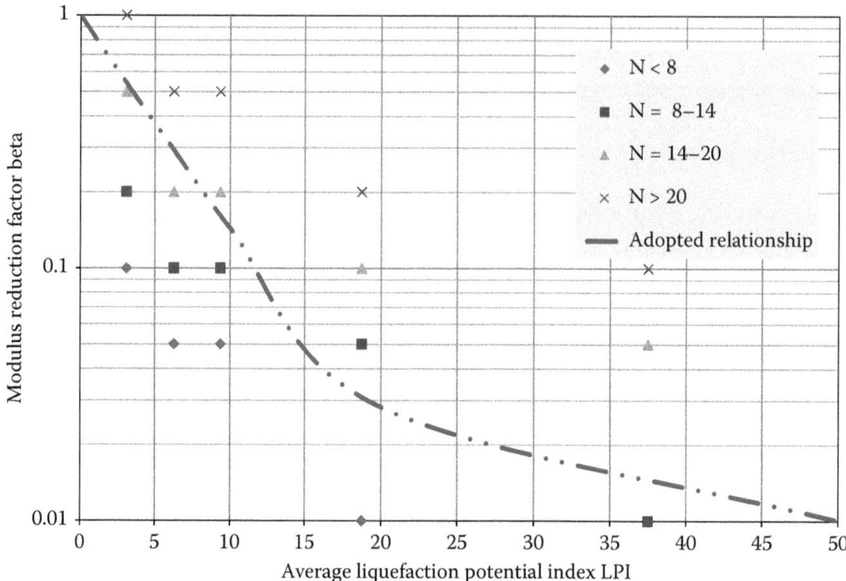


*Figure 11.33* Relationship between modulus reduction factor β and LPI.

where

$\beta_{lim}$ is the lower limit value of β,
b the index derived from fitting through the points in Figure 11.33, and found to be approximately −0.20 and
LPI is the liquefaction potential index.

This relationship is also shown in Figure 11.33 for b = −0.2 and $\beta_{lim} = 0.01$, and shows reasonable agreement with the points derived from Table 11.14, despite the considerable scatter in the points.

### *11.11.4.2 Upper limit to kinematic bending moment*

There will be an upper limit to the kinematic bending moment, which will occur if the layer has completely liquefied and flows past the pile. In this case (assuming that the applied pressure from the liquefied layer acts in the same direction along the whole layer), the limiting kinematic bending moment, $M_{klim}$, will be given by the following approximate expression:

$$M_{klim} = n \cdot s_{u-Liq.} d\, h_1(h_1 + dh) \tag{11.77}$$

where

n is the undrained shear strength multiplier for limiting lateral pile–soil pressure,
$s_{u\text{-}Liq.}$ the shear strength of liquefied soil (see Section 11.6.12),
d the pile diameter,
$h_1$ the thickness of liquefied layer and
dh is the additional distance within underlying layer at which the maximum moment occurs (typically expected to be 0.5d–1d).

Data presented by Cubrinovski (2008) have indicated that the parameter n in Equation 11.77 may be about 9 (i.e., 4.5 times the Rankine passive pressure).

#### *11.11.4.3 Estimation of inertial bending moment in pile*

In cases where complete liquefaction of the upper layer occurs, with resultant lateral spreading of the liquefied soil, it has been suggested by Klimis et al. (2004) that inertial effects can be neglected. However, the analyses and comparisons carried out by Liyanapathirana and Poulos (2005) indicate that inertial effects can be present and should be considered. Accordingly, it appears prudent to include inertial effects in the analysis.

Following the recommendations of Liyanapathirana and Poulos (2005), the inertial force $H_i$ can be estimated as follows:

$$H_i = a_s \cdot P \tag{11.78}$$

where

$a_s$ is the peak ground acceleration and
P is the vertical load acting on pile.

Assuming (albeit boldly) that an elastic analysis can be applied to a pile in a liquefied layer, and that a constant Young's modulus applies to the liquefied layer, the maximum bending moment due to inertial loading, $M_{imax}$, can be estimated from the following expression given by Randolph (1981):

1. For a free-head pile:

$$M_{imax} = 0.1H_i(L_c + h_1) \tag{11.79}$$

2. For a fixed-head pile (fixing moment at the pile head):

$$M_{imax} = -0.1875\ H_i(L_c + h_1) \tag{11.80}$$

where $H_i$ is the inertial force on pile, $h_1$ the depth of liquefied soil, $G_{red}$ the shear modulus of the non-liquefied layer, $L_c$ the critical pile length in the non-liquefied soil, and approximated as

$$L_c = d\left(\frac{E_2}{G_{red}}\right)^{2/7} \tag{11.81}$$

#### *11.11.4.4 Effect of near-surface crust above liquefiable layer*

In cases where a stiffer (and non-liquefiable) crust may exist above the liquefiable layer, account needs to be taken of the effect of the stiffness of this crust on the behaviour of the pile under inertial loading, and its effect on the limiting bending moment that can be imposed on the pile.

An approximate, but conservative, allowance for the stiffness of the crust may be made by using a weighted average Young's modulus, $E_{cl}$, of the crust and the liquefiable layer, as follows:

$$E_{cl} = \frac{(E_c \cdot\ h_c + E_l \cdot h_l)}{(h_c + h_l)} \tag{11.82}$$

where
$E_c$ is the Young's modulus of crust,
$h_l$ the thickness of crust,
$E_l$ the Young's modulus of liquefiable layer and
$h_l$ is the thickness of liquefiable layer.

The maximum inertial bending moment can then be obtained from Equation 11.78 for a free-head pile, or from Equation 11.79 for a fixed head pile.

To obtain the maximum kinematic bending moment at the interface between the liquefiable and lower non-liquefiable layers, the crust should not influence this value, and the simplified approach may again be used, using as before, the modulus of the liquefiable layer $G_{pl}$.

When estimating the limiting bending moment that can be imposed on the pile, an additional shear force, $H_{cr}$ and moment $M_{cr}$ will be applied to the pile by the crust. These can be estimated as follows:

$$H_{cr} = h_{cr} \cdot p_{ucr} \cdot d \tag{11.84}$$

$$M_{cr} = H_{cr}(h_1 + 0.5\,h_{cr} + L_c) \tag{11.84}$$

where
$h_{cr}$ is the thickness of non-liquefied crust,
$p_{ucr}$ the ultimate lateral crust–pile pressure, and which can be estimated as approximately 4.5 times the Rankine passive pressure exerted by the crust,
d the pile diameter,
$h_1$ the thickness of liquefied layer (below crust) and
$L_c$ is the critical pile length in the non-liquefied soil layer (Equation 11.81).

The combination of inertial and kinematic effects can then be considered as per the recommendations of Tokimatsu et al. (2005) in Section 11.9.6.

#### *11.11.4.5 Example*

To illustrate the application of the simplified approach, the case shown in Figure 11.32 of the bridge pile BP211 that was damaged in the 1995 Kobe earthquake (Ishihara and Cubrinovski, 1998) will be considered. This case has also been analysed by Liyanapathirana and Poulos (2005) using a pseudostatic analysis.

The pile is one of a group of 22 bored piles, 1.5 m in diameter and is 41.5 m long below the pile cap. The liquefiable layer is a reclaimed deposit 16 m thick (below the pile cap) and has an average SPT-N value of about 10. The underlying layers consist of strata of sandy silt, with gravel, with SPT-N values ranging between about 10 and 50, with an average value of about 25. The average vertical load on a pile is 2.94 MN. A Young's modulus of the pile of 30,000 MPa will be assumed.

Using a relationship between shear wave velocity and SPT-N (see Chapter 6), the shear wave velocities of the upper and lower layers are estimated to be 173 and 234 m/s, respectively, and the consequent small-strain shear moduli are 53.7 and 103.9 MPa. Minimum and maximum damping ratios of 0.025 and 0.26 have been assumed for the strata at the site.

The earthquake characteristics at the site have not been documented, but based on other available information, it is assumed that the peak bedrock acceleration is 0.45 g and that the predominant period of the earthquake is 1s.

Assuming that the piles have a fixed head, the inertial and kinematic maximum bending moments can be estimated for two cases:

- During the cyclic loading stage and prior to liquefaction
- After liquefaction of the upper layer has developed

1. *Cyclic loading phase*: For the first case, the first step is to assess the amplification of ground motion, and based on the above assumptions, the natural period of the site is about 0.79s. Using the approximate expression in Equation 11.11, the amplification factor is found to be 2.64. Thus, the ground surface acceleration is estimated to be $0.45 \times 2.64 = 1.188$ g.

   Using the small-strain modulus values, the following maximum moments are computed:

   Inertial: −7.65 MN m (using Equation 11.80)

   Kinematic: 3.10 MN m (using the method of Nikolaou et al., 2001, Equations 11.55 through 11.59)
2. *Post-liquefaction phase:* The small-strain modulus is used for the lower non-liquefiable layer, but a degraded modulus is used for the liquefied layer. To obtain this degraded modulus, given that liquefaction did indeed occur, an LPI, of 50 was used. From Equation 11.75, the reduction factor $\beta$ was 0.01 (the limiting value assumed). The corresponding value of shear modulus was 0.54 MPa. The site period now becomes 1.10 s, while the damping ratio has increased to its assumed maximum value of 0.26. Accordingly, the amplification factor from Equation 11.13 then decreases from 2.64 (for no shear modulus degradation) to 1.64 (for LPI = 50), and the PGA is then $0.45 \times 1.64 = 0.738$ g.

   From the expressions for inertial and kinematic bending moments (Equations 11.79 and 11.55 through 11.60), the following values are obtained:
   a. Inertial moment (at pile head): −11.96 MN m
   b. Kinematic moment: 121.1 MN m

However, a check needs to be made with the limiting kinematic bending moment that can occur when the liquefied layer flows past the pile. Adopting an undrained shear strength of 3 kPa for the liquefied soil, the limiting bending moment is found, from Equation 11.84, to be 12.3 MN m. It is clear that the assumption of elastic behaviour in the Nikolau et al. method is not valid in this case.

The values obtained from this simple analysis compare reasonably well those reported by Liyanapathirana and Poulos (2005) (approximately −9 MN m at the pile head and about 12 MN m at the base of the liquefied layer). Ishihara and Cubrinovski (1998) incorporated pile cracking into their analysis, and their computed maximum moments of about 7 MN m reflect the occurrence of pile cracking, which limited the moment that could be induced in the piles. Figure 11.32 shows details of two piles that were excavated and shows the cracks in the piles at or near these two locations.

The simplified analysis therefore clearly indicates that large moments would have been generated at both the pile head and at the interface between the liquefied and non-liquefied layer. Both of these locations are 'hotspots' for kinematically induced bending moments in the piles.

## 11.12 OTHER DESIGN ISSUES

### 11.12.1 Axial response of piles

Most design attention appears to focus on the lateral response of piles to seismic ground movements, but the axial response of piles can also be of concern. Some issues that require consideration are as follows:

1. Seismic excitation will cause settlements as well as lateral ground movements and thus there will be a tendency for the development of negative skin friction on those parts of the pile which tend to settle less than the soil. Methods of estimating soil settlement have been discussed in Section 11.8.6.
2. There will tend to be a substantial reduction in effective stress in the soil due to the generation of excess pore pressures, and this will lead to a reduction in the lateral effective stress between the pile and the soil, and a consequent reduction in the ultimate shaft friction. This will reduce the axial capacity of the pile and the factor of safety against geotechnical pile failure. This reduction of capacity is however temporary and the initial shaft resistance should be largely re-instated after the excess pore pressures have dissipated.

### 11.12.2 Pile buckling

Bhattacharya and Bolton (2004) have identified the mechanism of failure of piles in liquefiable soils by buckling under axial loading due to loss of lateral support. This mechanism is generally overlooked in design, yet can be an important contributor to foundation failure. They emphasise that buckling of a pile is an unstable and destructive failure mechanism, whereas pile bending is a more stable mechanism. They recommend that the pile design process should incorporate the following considerations:

- The ratio of the axial load to the critical buckling load should be limited to about 1/3 to provide a safety margin on buckling.
- The slenderness ratio of the piles, $SR = L/(I/A)^{0.5}$ (where L is the effective pile length within the liquefiable layer, I the minimum moment of inertia, A the pile cross-sectional area), in the buckling zone should be no greater than 50 to avoid buckling instability.

Estimates of the pile buckling load can be made using the results summarised by Poulos and Davis (1980).

### 11.12.3 Group effects

Under inertial loading, it is now well recognised that group effects are detrimental in that they tend to reduce the stiffness of piles within the group and decrease the overall group capacity. Methods of dealing with these effects are given by Poulos and Davis (1980) and Fleming et al. (2009).

In contrast, under kinematic loading, group effects however tend to be beneficial, due to the 'shielding' action of the piles. In particular, inner piles within a group tend to be subjected to smaller forces and moments developed by ground movements than outer piles, and all piles in the group tend to experience less effect than a single isolated pile. As a consequence, the consideration of a single isolated pile will be conservative when considering kinematic effects.

Towhata (2008) has presented an approximate approach for estimating the reduction in lateral earth pressure on a pile due to soil movement or flow past a group of piles. This reduction can be expressed in terms of two components:

1. A 'shadow factor' which reduces the 'downstream' pressure by a factor of 0.8 for each successive row of piles
2. A spacing factor, $\alpha_p$, which can be expressed as follows:
   a. *For a non-liquefied layer*:

$$\alpha_p = \frac{(0.877(1-\eta)f)}{(0.877\eta + 0.123)} + 1.0 \tag{11.85}$$

where
$f = 1.0$ when $h_l/d \leq 3.8$,
$f = 14(d/h_l)_2 + 0.03$ when $h_l/d > 3.8$,
$h_l$ is the depth of flowing soil,
d the pile diameter and
$\eta$ is the ratio of pile diameter to pile spacing (d/s).

   b. *For a liquefied layer*:

$$\alpha_p = \frac{1}{(0.599\eta + 0.401)} \tag{11.86}$$

For practical use, it is convenient to derive a group reduction factor for the pressure developed by moving soil. If, from a practical viewpoint, s/d = 20 is considered as being sufficiently widely spaced to represent a single pile, then the group reduction factor for pressure, RFp, can be defined as

$$RFp = \frac{\alpha_p}{\alpha_p(s/d = 20)} \tag{11.87}$$

Figures 11.34 and 11.35 plot the resulting values of RFp for the non-liquefied and liquefied soil cases. The beneficial effects of group action can clearly be seen from these figures. For the non-liquefied case, the effects of grouping become increasingly beneficial as the ratio $h_l/d$ decreases, that is, as the thickness of the moving soil decreases relative to the pile diameter.

When both inertial and kinematic effects are present, the group effects may tend to counteract each other.

## 11.13 CATEGORY 3 ANALYSIS METHODS

Category 3 methods of analysing the seismic response of foundations can be subdivided into three groups:

- Site response analyses
- Pile-based analyses
- Complete analyses

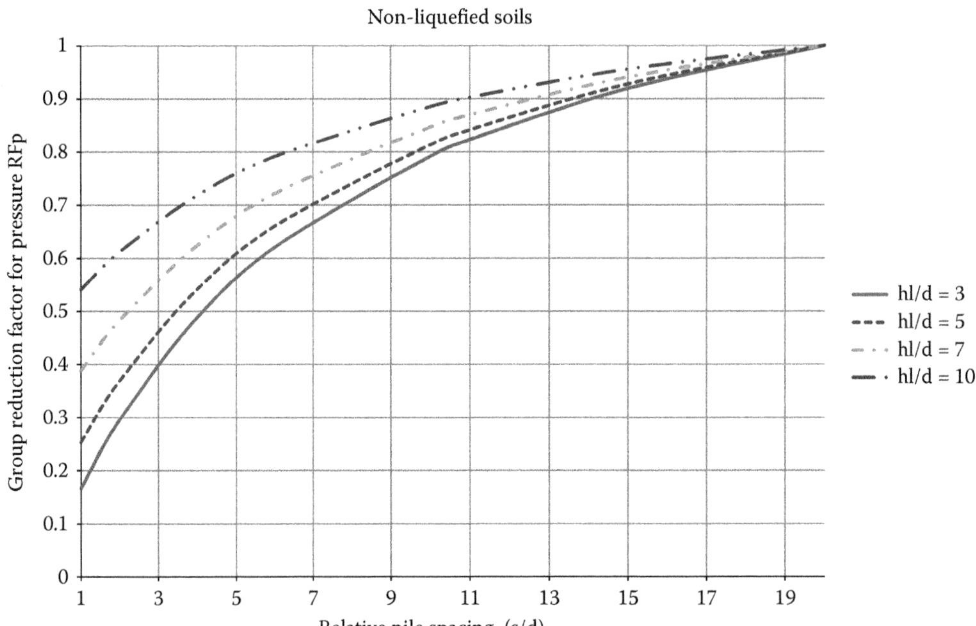


*Figure 11.34* Group reduction factor for lateral pile–soil pressure – non-liquefied soil.

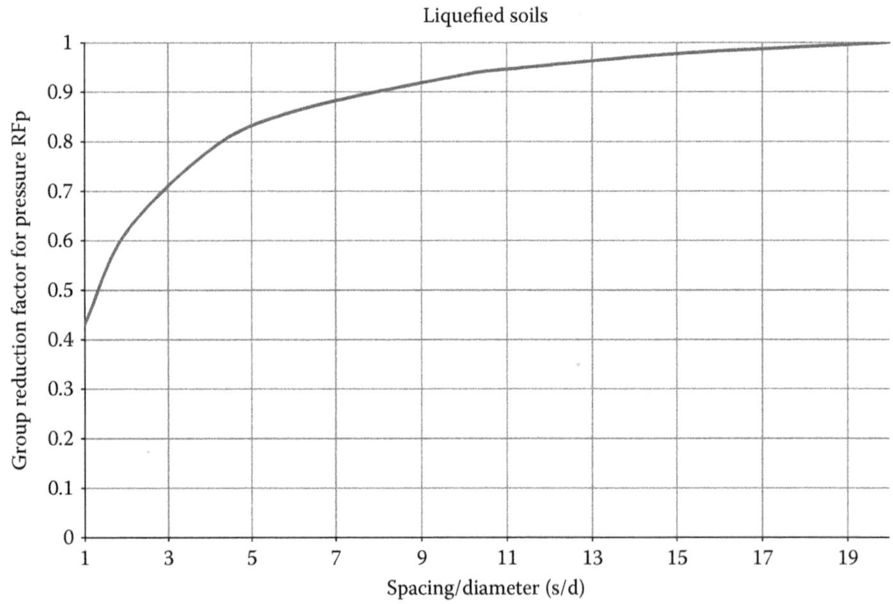


*Figure 11.35* Group reduction factor for lateral pile–soil pressure – liquefied soil.

The characteristics of these analyses, and examples of available software, are set out briefly below.

### 11.13.1 Site response analyses

Site response analyses have been discussed in Section 11.4.5. Such analyses compute the response of a ground profile that is subjected to a specified acceleration–time history at the base of the stratum. 1D programs that are commercially available include SHAKE and DEEPSOIL, and implicitly assume a horizontally layered ground profile of infinite lateral extent. SHAKE assumes a single-phase soil layers in which there is no degradation of soil strength, although both stiffness and damping are dependent on the shear strain with the various layers. DEEPSOIL assumes a two-phase soil model below the water table, and thus allows for the generation and dissipation of pore pressures during and after the passage of the earthquake. It can therefore be used to simulate the development of liquefaction.

An example of a 2D program that can be used to compute site response is VERSAT-2D (Wu, 2008). This also allows consideration of two-phase soils with pore pressure generation and dissipation, with various models available for calculating the generated pore pressures.

### 11.13.2 Pile-based analyses

In this class of analysis, a sub-structuring approach is adopted. The ground movements developed by an earthquake are computed via a site response analysis, and then these movements are imposed on the pile or pile group to examine the pile response. Examples of this class of analyses are boundary element analyses such as those described by Poulos and Davis (1980) and Hull (1987). Such programs have been used to analyse problems of static ground movements on piles, as discussed in Chapter 9, and the pseudostatic analyses described in this chapter. In addition, analyses based on the p–y and t–z concepts have been employed and implemented in programs such as LPile and GROUP8 (Ensoft, 2015).

### 11.13.3 Complete analyses

The most important aspect of boundary conditions for numerical modelling of pile foundations is the simulation of semi-infinite boundary conditions. When the proper boundary conditions are not applied, the input motion generates a reflected wave, resulting in an inaccurate simulation of the actual motion. Programs such as FLAC (Itasca, 2015) can couple the soil model with structural elements representing the piles, and thus soil–structure interaction due to ground shaking can be simulated. It is also possible to incorporate groundwater flow and thus analyse time-dependent pore pressure changes during the development of liquefaction.

Soil modelling is often simplified by use of the equivalent-linear method, but non-linear methods can also be employed. In the equivalent-linear method (Seed and Idriss 1969), a linear analysis is performed, with some initial values assumed for damping ratio and shear modulus in the various regions of the model. The maximum cyclic shear strain is recorded for each element and used to determine new values for damping and modulus, by reference to laboratory-derived curves that relate damping ratio and secant modulus to amplitude of cycling shear strain. An empirical scaling factor is usually used when relating laboratory strains to model strains. The new values of damping ratio and shear modulus are then used in a new numerical analysis of the model. The whole process is repeated several times, until there is no further change in properties. At this point, it is said that 'strain-compatible' values of damping and modulus have been found, and the simulation using these values is representative of the response of the real site.

In contrast, only one run is done with a fully non-linear method. Provided that an appropriate non-linear law is used, the effects of strain level on damping and apparent modulus are incorporated.

Both methods have their strengths and weaknesses. The equivalent-linear method takes some liberties with physics but is user friendly and accepts laboratory results from cyclic tests directly. The fully non-linear method correctly represents the physics but demands more user involvement and needs a comprehensive stress–strain model in order to reproduce some of the more subtle dynamic phenomena. Further discussion of the characteristics of both types of soil modelling is given in Itasca (2015).

Celebi et al. (2012) have found that the nature of the soil constitutive model used for the soil can play an important role in the computed seismic response, and that there is a significant difference between analysis results using a linear elastic model and a Mohr–Coulomb model. The analysis sensitivity decreases as the soil becomes stiffer or as the slenderness ratio of the pile increases.

Examples of the use of Category 3 methods for analysing seismic pile–soil interaction are given by Wu and Finn (1997); Klar et al. (2004); Maheshwari et al. (2004); Luan et al. (2015).

## 11.14 MITIGATION OF LIQUEFACTION EFFECTS

### 11.14.1 Categories of mitigation measures

Conventional remedial measures to mitigate the effects of liquefaction can be divided into three broad categories (JGS, 1998):

- Treatment of the liquefiable soil to strengthen it
- Treatment of the soil to accelerate the dissipation of seismically induced excess pore pressures
- Measures to reduce liquefaction-induced damage to the structure or facility

Figure 11.36 summarises the above measures and some of the means of implementing them.

### 11.14.2 Conventional measures to strengthen the liquefiable soil

An extensive discussion of measures that can be used to strengthen liquefaction-susceptible soils is given in JGS (1998). Some of the more common measures are as follows:

- Soil densification, by a variety of means, for example, vibroflotation, blasting
- Insertion of stiffer columns
- Provision of drainage via stone columns

A major limitation of all of these methods is that they cannot be used to remediate sites on which structures or facilities already exist. A number of the methods for soil densification may also not be feasible if the site is within a commercial or residential area, because of the noise and vibrations involved.

Methods involving the insertion of stiffer columns can be effective and generally involve less noise and vibration than conventional methods of densification.

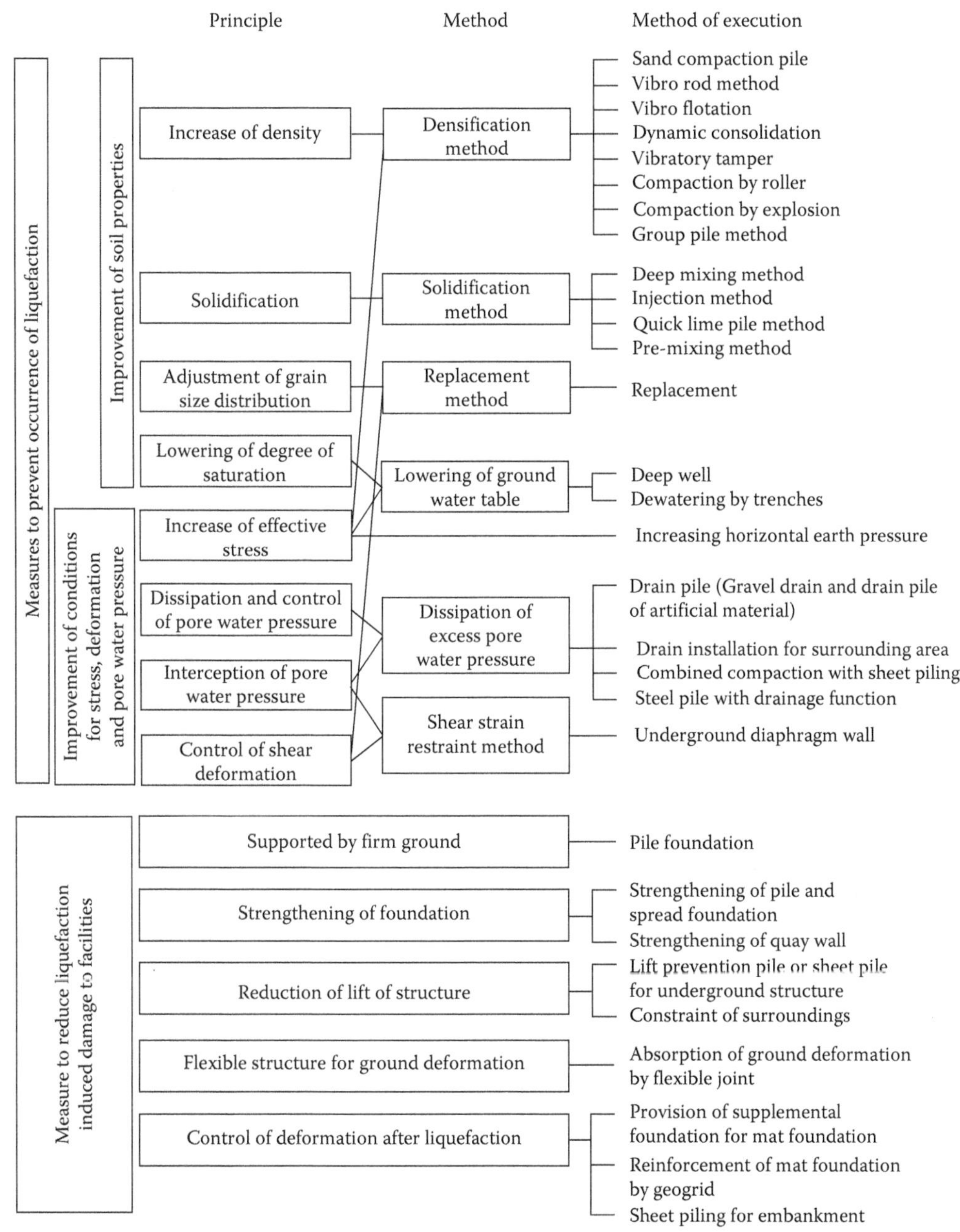


*Figure 11.36* Techniques for remedial measures against liquefaction. (Adapted from JGS. 1998. *Remedial Measures Against Liquefaction*. Japanese Geotechnical Society (JGS), Balkema, Rotterdam.)

## 11.14.3 Conventional measures to accelerate excess pore pressure dissipation

The use of stone columns or gravel drains to accelerate pore pressure drainage was developed by Seed and Booker (1977) and has been used successfully in a number of cases. In the design of stone columns for liquefaction mitigation, it is common to specify a limiting maximum pore pressure ratio (excess pore pressure divided by vertical effective stress), and this is often taken as 50%. The required length of the columns will depend on the depth and thickness of the liquefiable layer, while the required spacing of the columns will depend on the following factors:

1. The specified maximum pore pressure ratio
2. The permeability and compressibility of the liquefiable layer
3. The earthquake duration
4. The equivalent number of cycles of loading from the earthquake
5. The number of cycles to cause liquefaction
6. The column diameter
7. The finite permeability of the column and the effects of well resistance

### 11.14.4 Mitigation measures for pile foundations

Sato et al. (2004) have suggested various countermeasures for pile foundations subjected to lateral flow of liquefiable soils. Three of these measures are illustrated in Figure 11.37.

The 'drain piles' method aims to prevent the liquefaction of the lower layers while reducing the stiffness of the stiffer crust above the liquefiable layers. The drain piles should extend only about half-way into the upper crust, and the permeability of the drain piles should be greater than that of the liquefiable layers so that the excess pore pressures that are generated are transmitted to the upper layer.

In the 'earth retaining wall' method, an earth retaining wall is constructed in front of the pile foundation, to block the lateral flow.

The 'streamlined shield block' method involves the casting of an angled face on the upstream side of the pile cap. This angled face is meant to disperse the flowing soil and reduce its effect on the foundation.

Centrifuge tests were carried out by Sato et al. (2004) to examine the effectiveness of the above countermeasures. Figure 11.38 summarises the test results and indicates that the 'streamlined

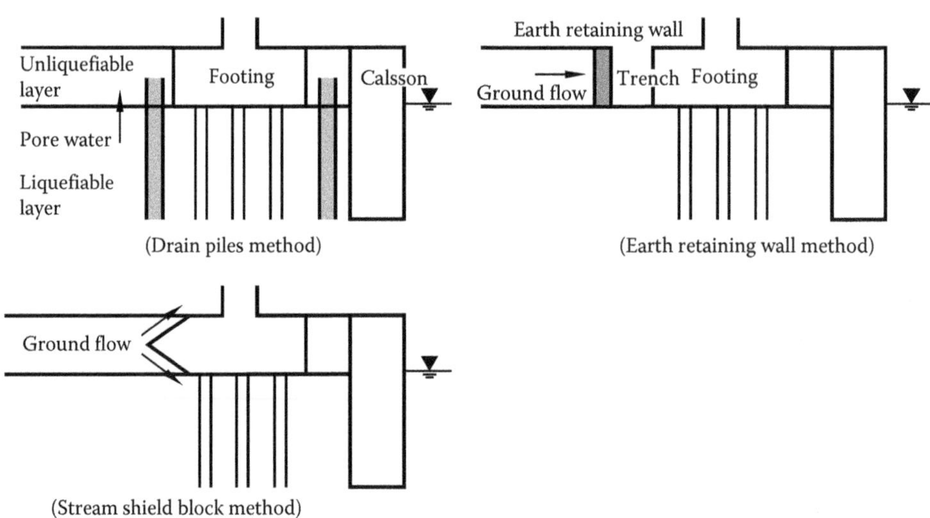


*Figure 11.37* Countermeasures for lateral ground flow. (Adapted from Sato, K., Higuchi, S. and Matsuda, T. 2004. A study of the effect of countermeasures for pile foundation under lateral flow caused by ground liquefaction. *Proceedings of the 13th World Conference on Earthquake Engineering,* Paper No. 452, Vancouver. Courtesy of WCEE.)

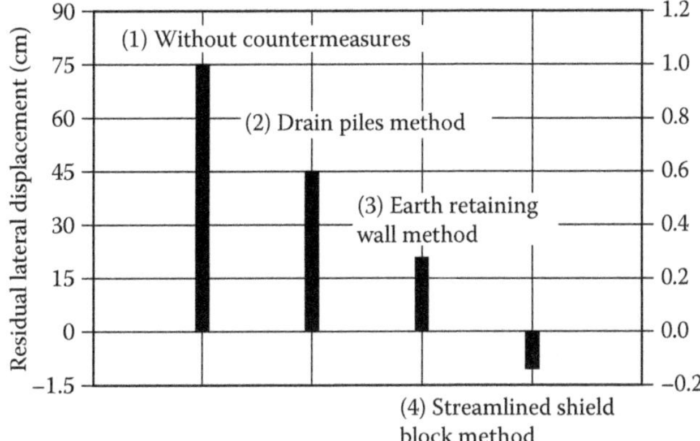


*Figure 11.38* Measured residual lateral movements for various countermeasures. (Adapted from Sato, K., Higuchi, S. and Matsuda, T. 2004. A study of the effect of countermeasures for pile foundation under lateral flow caused by ground liquefaction. *Proceedings of the 13th World Conference on Earthquake Engineering*, Paper No. 452, Vancouver. Courtesy of WCEE.)

shield block' method is very effective in reducing the residual lateral displacement, and also the bending strain in the piles. It has the advantage of not requiring any ground improvement.

## 11.14.5 Some innovative methods for liquefaction risk mitigation

There are a number of recent methods that have been explored for reducing the risk of liquefaction, several of which have the potential to be used for sites on which structures or facilities exist. A brief review of some of these innovative approaches is given below.

### *11.14.5.1 Passive remediation via infusion of colloidal silica*

Colloidal silica is a dispersion of silica particles in water. With about 5% weight of silica, a colloidal silica aqueous dispersion has density and viscosity values that are similar to water, and this dispersion becomes a permanent gel abruptly after a period of time, generally a few months at most. Colloidal silica is non-toxic, biologically and chemically inert, and relatively durable. The influence of the gel is to reduce the strains in the treated soil developed by cyclic loading and to reduce the soil permeability.

Figure 11.39 shows results of cyclic triaxial tests on Monterey sand, both untreated, and treated with 10% colloidal silica. The increased resistance to liquefaction of the latter is very clearly demonstrated in this figure.

Gallagher et al. (2007) have described field tests to assess the performance of a dilute colloidal silica stabiliser in reducing the settlement of liquefiable soils. Slow injection methods were used to treat a 2 m thick layer of liquefiable sand, using eight injection wells around the perimeter of a 9 m diameter test area. The gel times ranged between 10 and 30 days. A subsequent blasting test revealed that the settlement of the treated soil was only about 60% that of an adjacent untreated area. Interestingly, there appeared to be no significant increase in the CPT resistance or the shear wave velocity, and the mechanism of improvement was considered to be due to the development of cohesion within the treated soil due to the formation of interparticle siloxane bonds.

Gallagher and Lin (2009) demonstrated that colloidal silica could be successfully delivered through 0.9 m diameter columns packed with loose sand. The main factors influencing

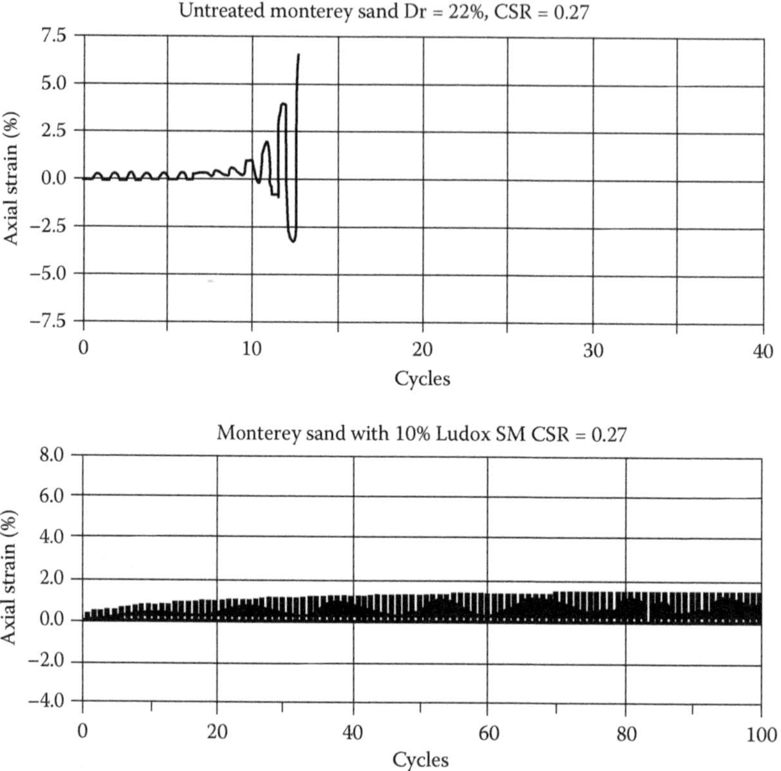


*Figure 11.39* Comparison between cyclic behaviour of untreated sand and sand treated with 10% colloidal silica. (Adapted from Gallagher, P.M. 2000. Passive site remediation for mitigation of liquefaction risk. PhD thesis, Virginia Polytechnic institute and State University, Blacksburg, VA.)

the transport of the stabiliser were the viscosity of the colloidal silica stabiliser, the hydraulic gradient, and the hydraulic conductivity of the liquefiable soil.

Rasouli et al. (2016) have described the application of controlled permeation of colloidal silica to ground improvement under a runway at Fukuoka International Airport in Japan. While increasing liquefaction resistance was not the main motive of the ground improvement, the increase in unconfined compressive strength resulting from the treatment indicated the effectiveness of the process.

#### *11.14.5.2 Biogeochemical remediation*

Over the past decade, new biogeochemical techniques have been explored for the improvement of ground properties. DeJong et al. (2011) summarise some of these techniques, and at least two of these have the potential to improve the liquefaction resistance of soils.

The first of these involves the use of microbes to induce calcite precipitation between soil particles. The second involves the use of microbes to generate small gas bubbles within the soil and thus increase the resistance to liquefaction.

Microbially induced calcite precipitation (MICP) occurs through a variety of microbiological processes, such as urea hydrolysis and denitrification. The process of calcite precipitation appears to have been discussed initially by van Meurs et al. (2006). Specific bacteria create a reaction which results in the precipitation of calcium carbonate in the form of crystalline calcite on the surface of sandy particles. When this reaction occurs under the right conditions, crystals are formed between two adjacent particles, producing a connection between them. This connection increases the strength and stiffness of the soil. The longer the nutrients, bacteria and reactants are present, the thicker the layer of mineralisation. An important characteristic of this process of cementation is that the permeability of the porous material only reduces slightly. The biomineralisation process progresses slowly in natural circumstances, but the rate of mineralisation can be enhanced by stimulating the conditions.

Test data presented by Meurs et al. indicated that the strength of treated sand could be increased by a factor of almost 5, while the stiffness was increased by a factor of 3.

DeJong et al. (2010) provide a more detailed description of the process of calcite precipitation and its consequences on the geotechnical properties of the treated soil. Significant increases can be obtained in both strength and stiffness, and the volumetric behaviour can be altered from contractive to dilative, thus improving the resistance to liquefaction.

DeJong et al. (2011) point out that there are a number of major challenges in implementing MICP technology. One is related to upscaling to a field scale, with issues related to cost, the stimulation of native biota, the uniformity of treatment and the management of potentially harmful by-products. Another issue relates to the alternatives of employing bio-stimulation of native bacteria species, or the augmentation of a specific species of bacteria for a treatment zone. They suggest that bio-stimulation may be the preferred approach. A further issue may be the longevity of the treated soils, which is related to the pH of the groundwater. Tests suggest that MICP products will remain stable provided that the pH remains above 6.3.

With respect to the second technique involving microbial gas generation, Chu et al. (2011) discuss some types of microorganisms that may contribute to biogas effects. This latter approach, sometimes referred to as the 'Induced Partial Saturation' technique, involves the injection of non-hazardous chemicals into the ground to create gas bubbles and hence to reduce saturation. This in turn reduces the potential for liquefaction. This again has the potential to be used to improve liquefaction resistance below existing structures and buildings.

A further approach has been discussed by Kavazanjian and De Jong (2016), enzyme-induced carbonate precipitation (EICP). This is a variant of MICP which uses urease enzyme derived from agricultural sources instead of microbial urease to catalyse the hydrolysis of urea and so induce carbonate precipitation. This process has an advantage over MICP in that the free enzyme is much smaller than ureolytic microbes and so can penetrate into the pores of finer grained soils. However, it has the disadvantage that the rate of reaction can be too fast and can result in precipitation of less stable forms of calcium carbonate than calcite.

Chapter 12

# Design of basement walls and excavations

## 12.1 INTRODUCTION

The vast majority of modern tall buildings contain basements, which are generally underground enclosures, extending up to eight levels or more below ground level. Among the factors influencing the extent and nature of the basement are

1. The anticipated use of the below ground space and the relationship between the estimated expenditure and the expected revenue from the space
2. The plan area and the shape of the basement
3. The influence of the subsoil and groundwater conditions on the choice of construction method
4. The time required for completion

The basement walls serve at least three purposes: temporary ground support, a permanent wall in the final structure and a means of transferring some of the structural load from the superstructure to the ground.

The design of the walls forming the sides of the basement requires a number of issues to be addressed, the key ones being:

1. The choice of construction method
2. The choice of type of basement wall
3. Support of the wall during and after construction
4. The possible use of the wall to support building loads
5. Control of the wall and ground movements arising from the construction
6. Control of groundwater during construction
7. Groundwater conditions after construction, and during the life of the structure

This chapter will discuss these aspects and provide some guidance on both relatively simple design methods and also on more complex methods of analysis and design. A comprehensive treatise on retaining wall design has been provided in CIRIA (2003).

## 12.2 WALL CONSTRUCTION METHODS

Two methods of basement construction are commonly used:

1. 'Bottom-up' construction, in which an excavation is made to the bottom level of the basement, and then the basement floors are constructed in sequence to ground level.

A cantilever retaining wall is constructed to support the ground during excavation, after which the permanent works are constructed within the excavation. It is usually necessary to provide additional temporary support for the excavation, either via temporary bracing or via ground anchors. Once the permanent works are in place, it may be possible to remove the temporary support system.
2. 'Top-down' construction, in which the basement walls are constructed first, and then the ground floor slab is cast, which acts as an initial lateral support for the wall. Succeeding levels of floors are then constructed via mining techniques, with the floors again acting as lateral supports, until the lowermost basement floor is constructed. This method therefore removes the need for temporary support during construction.

   The top-down approach can have the following advantages:

   a. The structures above ground can be carried out simultaneously with the structures below ground. This greatly reduces the time for construction.
   b. The settlements and lateral deflections can be reduced. Wang et al. (2015) indicate that, from experience in Shanghai, maximum average wall displacements are reduced by about 30% as compared with the conventional bottom-up method.
   c. Since the permanent columns and slabs can be utilised to support loadings during construction, the cost of formwork is saved.

If the foundation system includes piles, it is often most convenient to construct the piles before the excavation commences, with the piles being concreted only to the base of the raft. Construction of the basement slab is carried out soon after the excavation process is completed, to reduce the amount of heave that the piles will have to withstand.

In top-down construction, the empty pile hole from the pile head to below the lowermost slab is usually supported by temporary casing. Accuracy in drilling is essential to avoid errors in the position of the pile head and consequent difficulties in locating the columns on the pile heads.

## 12.3 WALL TYPES

There are four common types of wall construction for buildings:

1. Contiguous pile walls
2. Secant pile walls
3. Soil mix walls
4. Diaphragm walls

Steel sheet piles may under some limited circumstances be an option, but are unlikely to be viable for deep excavation in urban areas.

### 12.3.1 Contiguous piles

A contiguous pile wall is formed by the installation of bored piles in either a single or double row. The piles are positioned such that they either touch or are in close proximity to each other. Alternate piles are drilled first, and then the intermediate piles are installed. Gaps between the piles can be grouted to try and obtain a water seal, although this process is not always fully effective.

Contiguous pile walls are economical relative to diaphragm walls, but are generally more suited to medium scale excavations above the water table. Below the water table, they tend not to be suitable, as their ability to retain the ground and contain water inflow is limited.

### 12.3.2 Secant piles

Secant piles are interlocking bored piles which are constructed as for contiguous piles, but at a closer spacing than the contiguous pile method. The intermediate piles are then installed by drilling out the soil between each installed pair and then chiselling a groove down the sides of their shafts. Concrete is then placed to fill the drilled hole and the grooves, thus forming an interlocking and nominally watertight wall.

Secant pile wall construction involves relatively little ground disturbance to adjacent property and relatively low levels of noise and vibration. The main disadvantages of secant pile walls are

- Verticality tolerances may be hard to achieve for deep piles.
- Total waterproofing is very difficult to obtain in joints.
- Increased cost compared to sheet pile walls.

### 12.3.3 Soil mix walls

Soil mix walls are constructed by mixing and partly replacing the in situ soils with a stronger cement material. Various methods of soil mixing have been used, including mechanical, hydraulic, with and without air and combinations of both types. Such methods include Jet Grouting, Soil Mixing, Cement Deep Mixing (CDM), Soil Mixed Wall (SMW), Geo-Jet, Deep Soil Mixing (DSM), Hydra-Mech, Dry Jet Mixing (DJM) and Lime Columns. Each of these methods aims at finding the most efficient and economical method to mix cement (or in some cases fly ash or lime) with soil to transform it toward becoming more like a soft rock.

Some of the advantages of a sheet pile wall system can be gained by constructing an SMW reinforced with closely spaced steel soldier piles (Pearlman and Himick, 1993). However, the issues associated with the design and the construction of a permanent wall system remain. Installation of an SMW temporary excavation support system generates substantial spoil that must be managed and removed for disposal. While soil mixing has been used for many temporary and permanent deep excavation projects, including the Central Artery project in Boston, it has found little application for tall building projects.

### 12.3.4 Diaphragm Walls

A diaphragm wall is constructed by excavation of a series of panels in a trench which is temporarily supported by slurry or polymer fluid. Figure 12.1 (Xanthakos, 1994) shows the construction process, which consists of the following main steps:

1. Excavation of a rectangular trench which is supported by a suitable slurry, generally bentonite or polymer.
2. Insertion of a tube to form the panel joint to which the adjacent panel is to be attached.
3. Insertion of a reinforcement cage.
4. Placement of concrete via tremie pipes. The slurry is displaced and is pumped into a storage area for reconditioning and reuse.
5. The round tube is withdrawn.

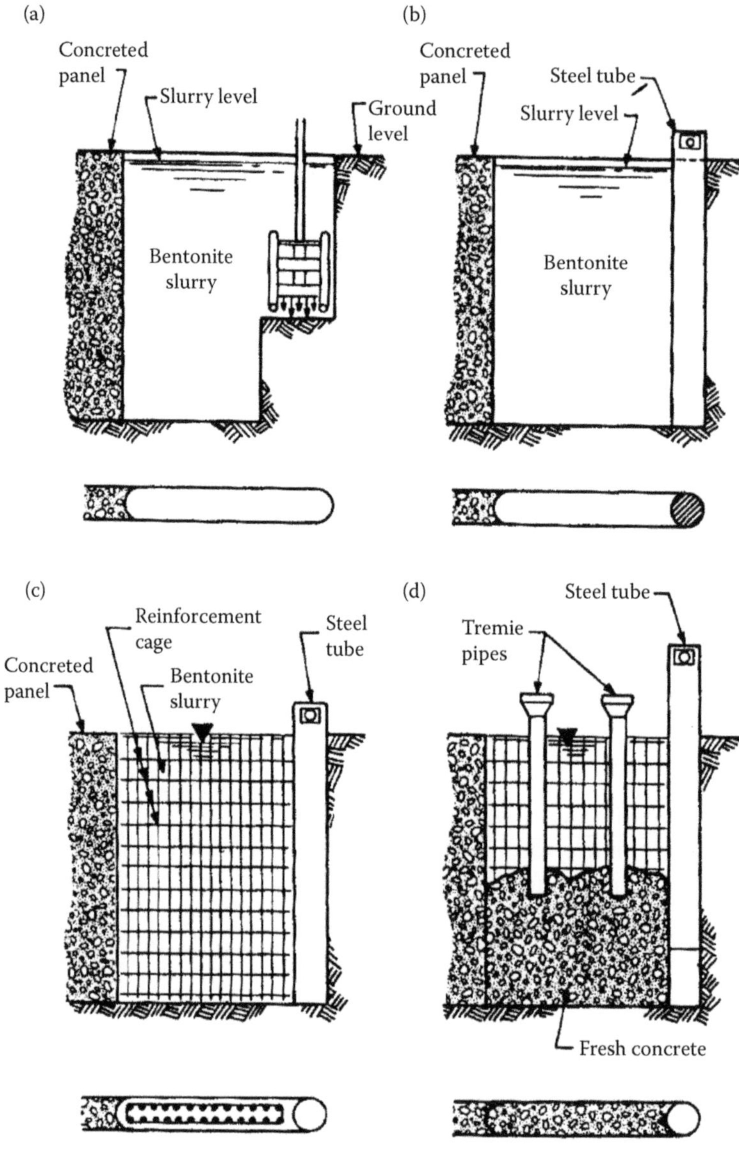


*Figure 12.1* Typical construction sequence for a diaphragm wall (a) Excavation of panel; (b) Completion of panel excavation, and insertion of tube; (c) Placement of reinforcement; (d) Concrete placement. (Adapted from Xanthakos, P.A. 1994. *Slurry Walls as Structural Systems*. McGraw-Hill, New York. Courtesy of McGraw-Hill.)

As pointed out by Pearlman et al. (2004), diaphragm walls are an excellent solution for structures requiring deep basements, particularly where a high water table is present. They provide the following advantages:

1. Temporary and permanent groundwater cut-off
2. Zero lot line construction
3. Considerable stiffness to reduce movements
4. Easily adaptable to both anchors and internal bracing systems
5. Expedited construction, because only interior columns and slabs need to be constructed

## 12.3.5 Wall support systems

### *12.3.5.1 Tieback anchors*

Anchors or tiebacks provide a very convenient means of excavation support and eliminate obstructions in the excavation that are inherent to rakers or struts. They are generally effective in reducing movements of the excavation walls. If left in place, tiebacks can be cut to relieve tension when the permanent structure can safely support the load. Unfortunately, it is often not possible to use this system because of intrusion onto adjacent properties. In such cases, it is necessary to employ some form of internal bracing using struts.

### *12.3.5.2 Bracing using struts*

Cross-lot or internal bracing transfers the lateral earth and water pressures between opposing walls through compressive struts. Such systems are suitable for relatively narrow excavations. For wider excavations, it may be necessary to use rakers resting on a foundation mat or rock as an alternative to internal bracing. Typically the struts are either pipe or I beam sections and are usually preloaded to provide a very stiff system. Installation of the bracing struts is done by excavating soil locally around the strut and only continuing the excavation once preloading is complete The struts rest on a series of wale beams that distribute the strut load to the diaphragm wall.

Figure 12.2 illustrates three different types of internal bracing that can be used with diaphragm walls. Figure 12.2a shows cross-bracing with struts, while a raker system suitable for relatively large building excavations is shown in Figure 12.2b. It requires a support system to transfer the vertical and lateral load components to the ground. Figure 12.2c shows

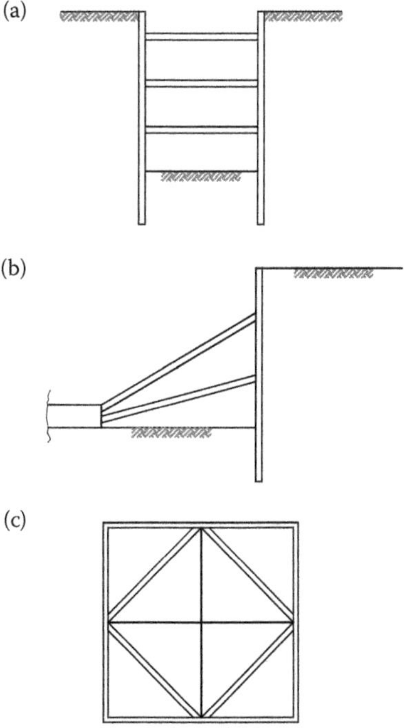


*Figure 12.2* Interior bracing systems: (a) cross-lot bracing, (b) raker supports and (c) diagonal bracing (plan).

a diagonal bracing system that can be used in rectangular cuts, and which allows a large unobstructed area within the excavation.

Preloading ensures a rigid contact between interacting members and is accomplished by inserting a hydraulic jack at each side of an individual pipe strut between the wale beam and a special jacking plate welded to the strut. The strut load can either be measured with strain gauges or can be estimated using equations of elasticity by measuring the increased separation between the wale and the strut. In some earlier projects the struts were not preloaded, and as a result when the excavation progressed deeper the soil and the wall movements were large. Thus, it has become standard practice to preload struts in order to minimise wall movements.

Cross-lot bracing can be a good option in relatively narrow excavations (e.g. 20–40 m wide) when tieback installation is not feasible. The struts can bend excessively under their own weight if the excavation spacing is too large. In addition, special provisions have to be taken into account for thermal expansion and contraction of the struts.

The typical strut spacing is in the order of 4 m, both in the vertical and horizontal direction. This is larger than the typical spacing when tiebacks are used, because the preloading levels are much higher. A clear benefit of using struts is that there are no tieback openings in the slurry wall, thus eliminating one potential source of leakage.

## 12.4 DESIGN ISSUES

The key issues that must be addressed during design of the basement structure include the following:

- Geotechnical stability and bearing capacity, which require assessment of
    - The required depth of the retaining wall below the basement slab to maintain an adequate safety margin against instability.
    - The required depth of the retaining wall to resist the vertical and lateral loads during and after construction.
- Structural design requirements for the walls and the basement slab, including
    - The extent and nature of the support system required for stability during and after construction
    - The forces acting on this system
    - The bending moments and shear forces in the walls
    - The bending moments and shear forces in the basement slab during and after construction, including those arising from upward water pressures if the basement is founded below the water table
- Control of induced vertical and horizontal ground movements during and after construction, inside and outside the excavation, including the effects of such movements on adjacent structures and facilities.
- Control of groundwater during construction and during operation, including
    - Prevention of piping and hydraulic failure at the base of the excavation
    - The rate of water inflow into the excavation during construction
    - The control of groundwater drawdown outside the excavation

Burland et al. (2012) also discuss the following important matters that need to be considered:

1. Accidental loadings, such as the accidental loss of a temporary prop, the loss of a permanent prop and flooding of the excavation.

2. Out-of-balance forces which may arise from several sources, for example, asymmetric ground or groundwater conditions, surcharges adjacent to the basement, varying superstructure characteristics over the area of the basement, adjacent tunnels, loading from adjacent construction plant, construction storage or stockpiles and asymmetrical construction sequences.

## 12.5 DESIGN CRITERIA

### 12.5.1 Stability and ultimate limit state

In the conventional factor of safety approach, a minimum factor of safety is specified, and typical values are shown in Table 12.1.

A somewhat different of criteria is provided by TMR (2015), as shown in Table 12.2. This is more closely aligned to the usual calculations carried out for checking the stability of walls.

When a limit state design approach is adopted, a reduction factor is applied to the soil strength parameters, while the applied forces are increased by a load factor. Typical of this approach is that set out in CIRIA (2003), in which three design approaches are considered, each involving the selection of soil parameters, groundwater conditions, loads and geometry, as follows:

- A – Moderately conservative: A cautious estimate of the relevant values
- B – Worst credible: The worst values that the designer reasonably believes might occur, but are very unlikely. Note that this approach is not appropriate for serviceability limit state (SLS) calculations.
- C – Most probable: Values that have a 50% probability of being exceeded. This approach should only be used in conjunction with an 'Observational Method' where the design can be re-assessed based on observed deflections of the wall.

*Table 12.1* Typical factors of safety

| *Failure type* | *Item* | *Factor of safety* |
|---|---|---|
| Shearing | Earthworks | 1.3–1.5 |
| | Retaining structures, excavations | 1.5–2.0 |
| | Foundations | 2.0–3.0 |
| Seepage | Uplift, heave | 1.5–2.0 |
| | Exit gradient, piping | 2.0–3.0 |

Source: Xanthakos, P.A. 1994. *Slurry Walls as Structural Systems*. McGraw-Hill, New York. Courtesy of McGraw-Hill.

*Table 12.2* Factors of safety for walls

| *Mode of failure* | *Required minimum factor of safety* |
|---|---|
| Sliding | 2.0 |
| Overturning | 2.0 |
| Bearing | 2.5 |
| Global | 1.5 |

Source: TMR. 2015. *Geotechnical design standard – minimum requirements*. Department of Transport and Main Roads, Queensland, Australia.

*Table 12.3* Partial safety factors

| | *Ultimate limit states* | | |
|---|---|---|---|
| | *Effective stress* | | *Total stress* |
| *Design approach* | $F_{c'}$ | $F_{\phi}$ | $F_{cu}$ |
| A: Moderately conservative | 1.25 | 1.25 | 1.5 |
| B: Worst credible | 1.0 | 1.0 | 1.0 |
| C: Most probable | 1.25 | 1.25 | 1.5 |

Source: After CIRIA. 2003. *C580, Embedded Retaining Walls – Guidelines for Economic Design*. RP29, CIRIA, London.

Soil parameters are factored as follows:

$$\tan\phi'_d = \frac{\tan\phi'}{F_{\phi'}} \tag{12.1}$$

$$c'_d = \frac{c'}{F_{c'}} \tag{12.2}$$

$$s_{ud} = \frac{s_u}{F_{cu}} \tag{12.3}$$

where $\phi'$, $c'$, $s_u$ are soil parameters based on Design Approach A, B or C, $F_{\phi'}$, $F_{c'}$, $F_{cu}$ are partial factors of safety on friction angle, cohesion intercept and undrained shear strength, $\phi'_d$, $c'_d$, $s_{ud}$ are design soil parameters.

The relevant partial factors of safety are detailed in Table 12.3. In relation to any imposed surcharge loadings, the load factors in Table 12.4 are applied for ULS calculations.

Bending moments, shear forces and prop forces are calculated from both the ULS calculations, and also from SLS calculations, for which factors of unity are employed in the analysis. The ultimate forces for which the wall are structurally designed are then:

- Bending moment and shear forces: The greater of the ULS or 1.35*SLS values
- Prop forces: The greater of 1.35*SLS or 1.85*ULS. This is because ULS calculations can underestimate the actual prop force generated to keep deflections to acceptable limits.

An additional prop load due to potential temperature effects should also be included.

*Table 12.4* Load factors for surcharges

| *Permanent/variable* | *Condition* | *Value* |
|---|---|---|
| Permanent | Unfavourable | 1.5 |
| Permanent | Favourable | 1.0 |
| Variable | Unfavourable | 1.5 |
| Variable | Favourable | 0 |

*Table 12.5* Example of settlement criteria used for Taipei Metro

| *Building type* | *Settlement (mm)* | *Absolute rotation (rad)* | *Angular distortion (rad)* | *Hogging ratio D/L (rad)* | *Sagging ratio D/L (rad)* |
|---|---|---|---|---|---|
| Multistorey frame on raft | 45 | $2 \times 10^{-3}$ | $2 \times 10^{-3}$ | $0.8 \times 10^{-3}$ | $1.2 \times 10^{-3}$ |
| Concrete frame on footing | 40 | $2 \times 10^{-3}$ | $2 \times 10^{-3}$ | $0.6 \times 10^{-3}$ | $0.8 \times 10^{-3}$ |
| Brick building on footing | 25 | $2 \times 10^{-3}$ | $0.4 \times 10^{-3}$ | $0.2 \times 10^{-3}$ | $0.4 \times 10^{-3}$ |
| Temporary structures | 40 | $2 \times 10^{-3}$ | $2 \times 10^{-3}$ | $0.8 \times 10^{-3}$ | $1.2 \times 10^{-3}$ |

Source: After Negro, A. et al. 2009. *Proceedings of the 17th International Conference on Soil Mechanics and Geotechnical Engineering*, Alexandria, Egypt, Vol. 4, pp. 2930–3005.

### 12.5.2 Allowable serviceability movements

Negro et al. (2009) reproduce a set of criteria used for excavations for the Taipei Metro, and these are shown in Table 12.5.

## 12.6 ANALYSIS AND DESIGN METHODS

The necessary calculations may be undertaken by empirical (Category 1) methods, simplified (Category 2) methods or more complete (Category 3) methods. Most modern designers employ Category 3 methods, with Category 1 and/or Category 2 methods used to check the design outcomes derived from the more complex analyses. Examples of approaches within each category are given below.

### 12.6.1 Category 1 methods

#### *12.6.1.1 Structural design*

'Apparent pressure' diagrams, such as those in Terzaghi and Peck (1967) have been widely used for structural design of walls. Such diagrams were originally developed on the basis of measurements on flexible walls, and so their use on the much stiffer piled or diaphragm walls may not be appropriate.

More recent apparent pressure diagrams have been developed by Twine and Roscoe (1999), for various types of soil. Most are for flexible walls, but the case of a stiff wall in very stiff clay is also considered. The general form of these apparent pressure diagrams is illustrated in Figure 12.3, and the pressures and depths for various cases are shown in Table 12.6.

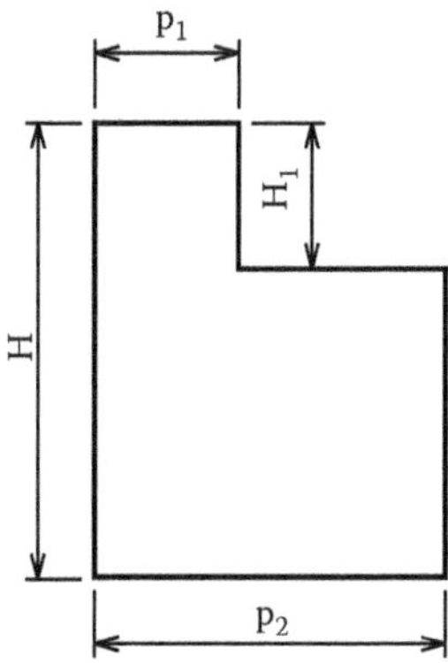


*Figure 12.3* General form of apparent pressure diagrams for walls.

*Table 12.6* Pressures and depths for apparent pressure diagrams in Figure 12.3

| *Wall type* | *Soil type* | $p_1/\gamma H$ | $p_2/\gamma H$ | $H_1/H$ | *Notes* |
|---|---|---|---|---|---|
| Flexible | Firm clay | 0.2 | 0.3 | 0.2 | – |
| Flexible | Soft clay | 0.5 | 0.65 | 0.2 | Stable base |
| Flexible | Soft clay | 0.65 | 1.15 | 0.2 | Enhanced base stability |
| Flexible | Stiff–very stiff clay | 0.3 | 0.3 | – | – |
| Stiff | Stiff–very stiff clay | 0.5 | 0.5 | – | – |
| – | Coarse soils, dry | 0.2 | 0.2 | – | – |
| – | Coarse soils, submerged | 0.2 | 0.2 | Above water table (WT) | – |
| | | $0.2\gamma'/\gamma$ | $0.2\gamma'/\gamma + u_w/\gamma H$ | Below WT | $u_w$ = water pressure at base of wall. Water pressure increases linearly with depth below WT |

Source: After Twine, D. and Roscoe, H.C. 1999. *Temporary Propping of Deep Excavations-Guidance on Design*. CIRIA, C517.

#### *12.6.1.2 Settlements*

Peck (1969) summarised data from measurements of ground settlements adjacent to unsupported excavations, and his diagram has been reproduced in Figure 9.13. The movements shown would tend to be conservative in the case of properly supported excavations, but the figure nevertheless gives a useful indication of both the extent and order of magnitude of settlement that might be expected. In this figure, $N_b$ represents the stability factor at the base of an excavation in clay, defined as

$$N_b = \frac{\gamma \cdot H_e}{s_{ub}} \tag{12.4}$$

where $s_{ub}$ is the undrained shear strength at base of excavation, $\gamma$ the unit weight of clay and $H_e$ is the depth of excavation.

It is clear that, the greater the stability number, the less stable is the excavation and the greater the expected movements. The figure also indicates that, for relatively stable excavations, the settlements extend horizontally to about twice the depth of excavation, whereas for less stable excavations, the lateral extent of settlement may be significantly greater.

### 12.6.2 Category 2 methods

#### *12.6.2.1 Geotechnical stability*

The following three aspects of the geotechnical stability of basement walls require consideration in design:

1. Lateral and overturning stability
2. Axial capacity to carry the vertical loads
3. Base heave of the excavation

##### *12.6.2.1.1 Lateral and overturning stability*

This aspect is generally considered via the use of theoretical earth pressure diagrams based on conventional earth pressure theory. On the landward side, active earth pressure and

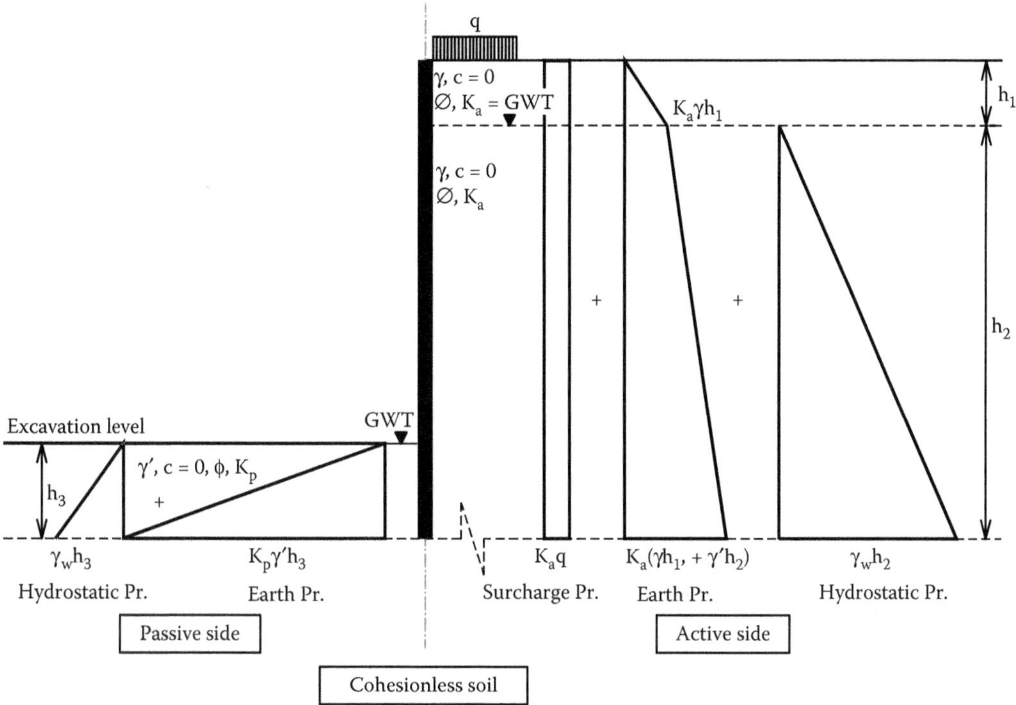


*Figure 12.4* Pressure distributions on wall in uniform cohesionless soils.

water pressure are assumed to act, together with any surface surcharge loading that may be present. On the passive side (i.e. within the excavation), passive earth pressure and water pressure (if present) are assumed to act. Figure 12.4 shows typical earth pressure diagrams for a uniform cohesionless soil. The active and passive earth pressures, $p_a$ and $p_p$ respectively, at a particular depth below the surface are usually estimated as follows:

$$p_a = (q + \gamma z_a - u)K_a - 2cK_a^{0.5} \quad (12.5)$$

$$p_p = (q + \gamma z_p - u)K_p + 2cK_p^{0.5} \quad (12.6)$$

where q is the surcharge pressure on surface, $\gamma$ the soil unit weight, u the water pressure, $z_a$ the depth below surface on active side, $z_p$ the depth below surface on passive (excavation) side, $K_a$ the active earth pressure coefficient, $K_p$ the passive earth pressure coefficient and c the soil cohesion.

$K_a$ and $K_p$ have usually been estimated from the classical Rankine theory, but there are more accurate solutions available, especially for $K_p$, that take into account wall roughness, and soil surface inclination, for example, Clayton et al. (1993).

The traditional analysis is based on the inherent assumption that the wall is very flexible, and this is not usually a good assumption for basement walls, especially diaphragm walls. Moreover, in addition to the earth and water pressures, consideration needs to be given to the supports for the excavation. The case of a single tie can be handled by the free earth support method, which considers the horizontal load and moment equilibrium of the system. When there is more than one tie, anchor or support, Xanthakos (1994) refers to a method whereby the system is converted to a single-anchored wall. However, such cases ideally require a more complete Category 3 analysis (see Section 12.6.3).

#### *12.6.2.1.2 Earthquake effects*

Allowance should be made for the effects of earthquakes in increasing the active earth pressures on embedded walls. Mikola et al. (2014) have found that the approximation suggested by Seed and Whitman (1970) using the peak ground acceleration (PGA) is a reasonable upper bound for the value of the seismic earth pressure increment for both fixed base cantilever structures (U-shaped walls) and cross-braced, basement-type walls. The traditional Mononabe–Okabe (M–O) solution and the Mylonakis et al. (2007) solutions have been found to give considerably higher design values than measured, at accelerations above about 0.4 g. The data also shows that the seismic earth pressure increments increase with depth, consistent with the static earth pressure distribution and consistent with that implicit in the M–O solution which forms the upper bound for the experimental results. The use of 0.85 PGA in the Seed and Whitman analysis produces values very close to the mean of the experimental data. In contrast, the dynamic earth pressure increment on a free standing cantilever wall is significantly smaller and corresponds to using 0.35 PGA in the Seed and Whitman approximation.

The experimental and analytical results also show that applying the moment at 0.33H (H = wall height), as recommended in the M–O method, gives amply conservative results over the full range of accelerations, and that applying the seismic earth pressure increment at 0.6H, as recommended by Seed and Whitman (1970) and many others, leads to a significant overestimate. Moreover, at PGA values less than 0.3 g, the dynamic earth pressure increment does not exceed the static design capacity for a design with a static factor of safety of 1.5 for both non-displacing basement walls and for non-displacing U-shaped cantilever structures. This effect is even more pronounced for free standing cantilever structures. Similar conclusions were reached by Seed and Whitman (1970) who observed that a wall designed to a reasonable static factor of safety should be able to resist seismic loads for ground accelerations up of 0.3 g.

#### *12.6.2.1.3 Axial capacity*

The axial capacity of a wall can be calculated in the same manner as that for a large diameter pile, by summation of the side frictional resistances on each side, plus the end bearing capacity of the base of the wall.

#### *12.6.2.1.4 Base heave*

This mode of failure is analogous to the problem of bearing capacity of a foundation. The soil outside the excavation acts as an applied load at the level of the base of the excavation. Thus, it is usual to apply conventional bearing capacity theory to assess this aspect of stability. Figure 12.5 shows the well-known diagrams initially developed by Bjerrum and Eide (1956) for undrained failure of a clay soil.

For the case of shallow soil below the base of the excavation (D/B < 0.7), the factor of safety $F_b$ against base heave is approximated as

$$F_b = \frac{5.7s_u}{[H(\gamma - s_u/D)]} \qquad (12.7)$$

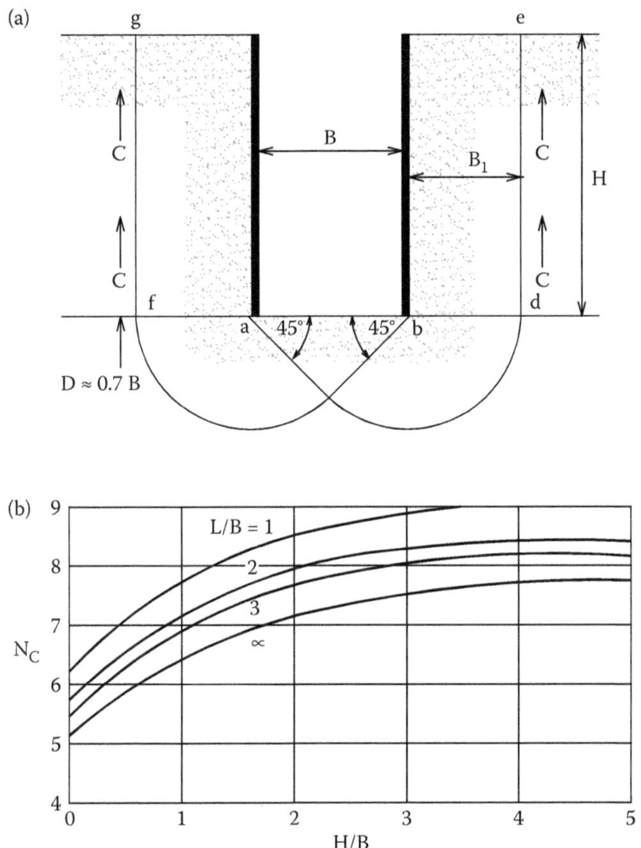


*Figure 12.5* Assessment of base heave for excavations: (a) D/B < 0.7 and (b) D/B > 0.7. (After Bjerrum, L. and Eide, O. 1956. *Geotechnique*, 6(1): 32–47.)

where $s_u$ is the undrained shear strength of clay near base, H the depth of excavation, $\gamma$ the unit weight of soil and D is the depth to hard layer.

For the case of a relatively long and narrow (small B) excavation, the factor of safety $F_b$ is

$$F_b = \frac{s_u \cdot N_c}{\gamma H} \tag{12.8}$$

where $N_c$ is the bearing capacity factor from Figure 12.5b.

If the factor of safety $F_b$ is found to be inadequate (e.g. less than 1.5), then the wall will either need to be made deeper or else piles may be installed within the excavation in order to enhance the soil capacity within the excavation, as described below.

#### *12.6.2.1.5 Effect of piles on base stability*

If the sides of an excavation are adequately supported and if no piles are present, the maximum depth of excavation is dictated by bottom heave considerations. In the case of a deep layer of soft clay whose undrained shear strength increases linearly with depth, the factor of safety decreases as the depth of the excavation increases, as expected. However, the factor of safety also increases with increasing excavation width, in contrast to the more commonly described case of a constant undrained shear strength with depth.

When piles are installed prior to the excavation being carried out, an expression can be derived for the equivalent cohesion for pile-reinforced soil, and this equivalent cohesion can be used in the conventional equations for base heave stability, in place of the undrained cohesion. Assuming that the clay layer is deep, the Davis and Booker (1973) expression may be used for the foundation bearing capacity of the footing on the equivalent soil mass with a 'crust'. The following expression can be derived for the equivalent cohesion for vertical loading of the reinforced soil:

$$c_{ev} = \frac{1}{N_c}\left[\frac{P_{vu}}{F \cdot Ar} - \frac{pB}{t}\right] \tag{12.9}$$

where $N_c$ is the bearing capacity factor (5.14 for strip foundations, 6 for circular foundations), $P_{vu}$ the ultimate vertical bearing capacity of the piled foundation, F the correction factor, which can be approximated as 1.0 for many cases, Ar the plan area of footing or raft, p the rate of increase of soil shear strength with depth, B the footing or raft width and t is the shape factor (4 for strip, 6 for circular foundation).

$P_{vu}$ is computed from normal pile capacity theory (Chapter 7), and can be taken as the lesser of the sum of the individual pile capacities and the capacity of a block containing the piles and the soil between them.

As a simple example of the beneficial effect of piles on the stability of an excavation, Figure 12.6 shows the computed factor of safety against bottom heave for a 20 m square excavation, 6 m deep, when 300 mm square precast concrete piles, at spacings of 1.0 and 1.5 m, are present. As would be expected, the factor of safety tends to increase with increasing pile length, but only for lengths in excess of 5–10 m, depending on spacing. As might be expected, the factor of safety increases as the pile spacing is reduced.

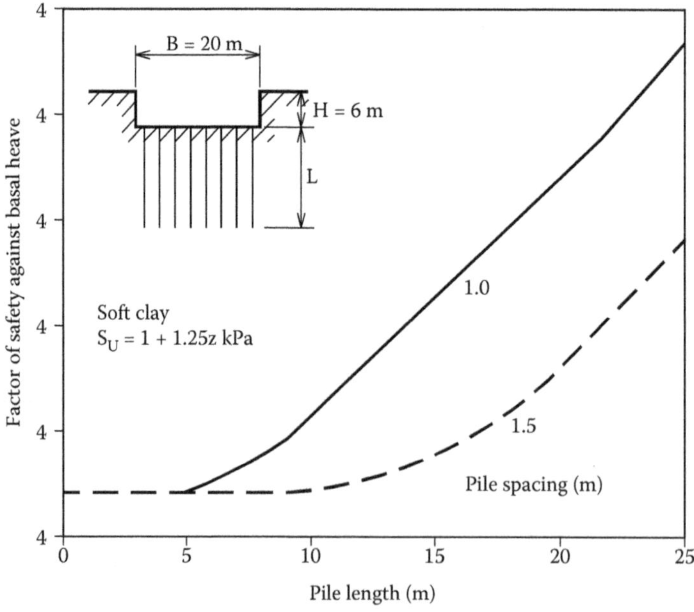


*Figure 12.6* Example of influence of pile length on factor of safety against basal heave.

#### 12.6.2.2 Structural design

The structural design of the wall and anchor or support system requires assessment of the maximum bending moments and shears in the walls, the anchor or support forces and the bending moments and shears in the basement raft. Category 2 methods for assessing geotechnical stability will provide a means of estimating wall moments and shears, and also the axial forces in the supports. However, the basement raft requires a more detailed analysis which is usually carried out in conjunction with the foundation design. A detailed analysis of raft or piled raft behaviour is generally needed for this aspect of design.

#### 12.6.2.3 Ground movements

Some valuable insights into ground movements arising from excavations have been obtained from parametric studies via finite element analyses. Among these are the results produced by Goldberg et al. (1976); Mana and Clough (1981); Clough and Schmidt (1981); Clough et al. (1989) and Clough and O'Rourke (1990).

The general pattern of deformation behind a wall is shown in Figure 12.7 (Clough and Schmidt, 1981). The patterns differ, depending on the factor of safety and the consequent magnitude of movements.

Figure 12.8 shows how the relative magnitude of horizontal movement varies with the factor of safety against base heave. From both field measurements and finite element analyses, it is clear that the lateral movements accelerate rapidly once this factor of safety becomes less than about 1.5.

The stiffness of the wall support system also plays an important role in controlling lateral movements. Figure 9.14 in Chapter 9 shows how increasing the support stiffness reduces the lateral movements, especially if the factor of safety against base heave is relatively low. Idealised profiles of surface settlement adjacent to excavations in various soil types are shown in Figure 9.15 (Clough and O'Rourke, 1990). These profiles are valuable in indicating the lateral extent of the settlement adjacent to the excavation.

The magnitude of the maximum settlement, $\rho_{vmax}$, is often related to the maximum horizontal movement, $\rho_{hmax}$, via the following expression:

$$\rho_{vmax} = R \cdot \rho_{hmax} \tag{12.10}$$

where R is the deformation ratio.

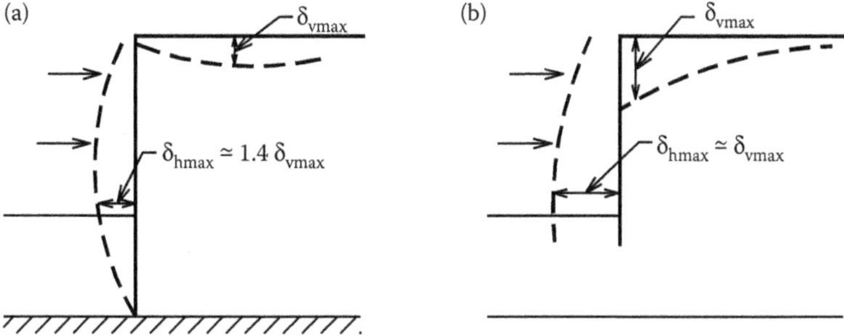


*Figure 12.7* Pattern of excavation-induced movements. (a) Pattern of behaviour when movements are 'small'. (b) Pattern of behaviour when movements are 'large'. (Adapted from Clough, G.W. and Schmidt, B. 1981. *Soft Clay Engineering*. Elsevier, Amsterdam, pp. 569–634. Courtesy of Elsevier.)

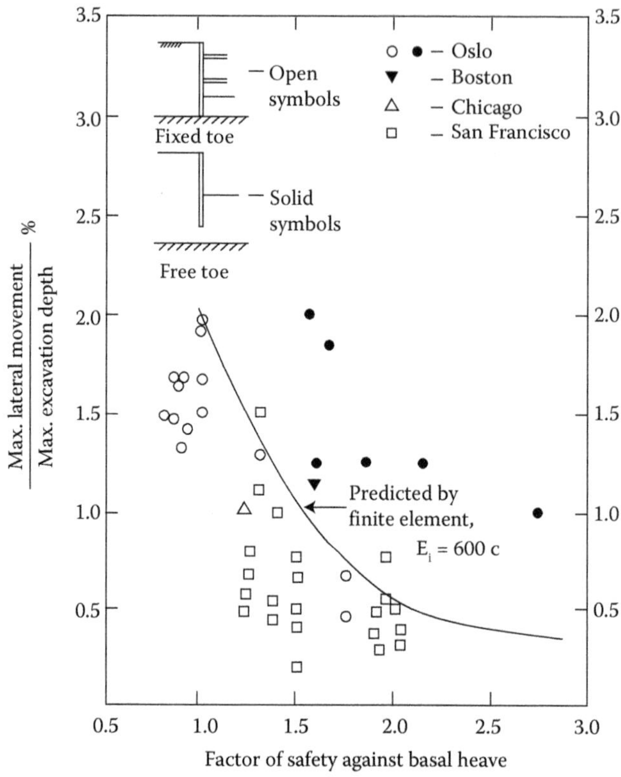


*Figure 12.8* Effect of factor of safety against base heave on lateral movements. (Adapted from Clough, G.W. and Schmidt, B. 1981. *Soft Clay Engineering*. Elsevier, Amsterdam, pp. 569–634. Courtesy of Elsevier.)

Kung et al. (2007) indicate that R generally lies between 0.5 and 1.0 for excavations in clay, and that R is strongly influenced by three parameters: the soil shear strength, Young's modulus and the clay layer thickness relative to the wall length. They develop an empirical expression for R in terms of dimensionless values of these three parameters. They also suggest the modified surface settlement profile which is shown in Figure 12.9.

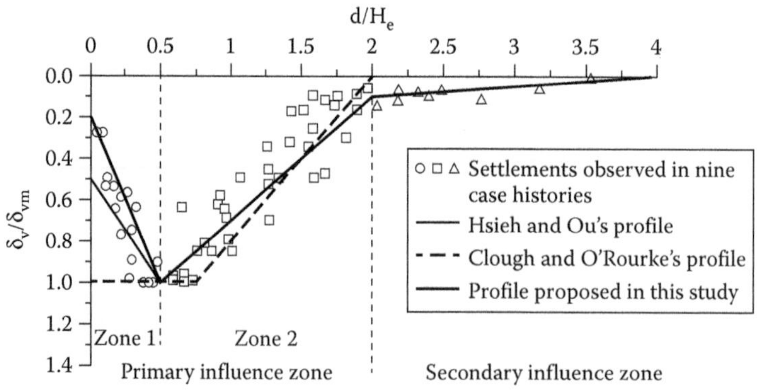


*Figure 12.9* Surface settlement profile proposed by Kung et al. (2007). (Courtesy of ASCE.)

#### *12.6.2.3.1 3D effects*

The foregoing analyses have all been based on the assumption of 2D plane strain conditions. While such conditions will be relevant for the central parts of relatively long walls, for short walls, or near the corners of longer walls, or in the vicinity of cross-walls, 3D effects will come into play and will tend to reduce the wall deflections. Such effects can be examined via Category 3 analyses (see Section 12.6.3), but some useful approximations for correcting 2D analyses are provided by Finno et al. (2007). They conducted a parametric study of basement excavations in clay via a 3D finite element analysis, and examined the effects of a number of parameters on the computed lateral wall movement, including excavation size and depth, wall stiffness and factor of safety against basal heave. They derived the following approximate expression for the plane strain ratio (PSR), defined as the ratio of the maximum movement at the centre of an excavation wall from 3D analysis to the maximum movement from a plane strain analysis:

$$PSR = (1 - e^{kC(L/He)}) + 0.05\left(\frac{L}{B-1}\right) \qquad (12.11)$$

where C is a factor depending on the factor of safety against basal heave, k a factor depending on system stiffness, L the length of wall, B the width of excavation and $H_e$ is the excavation depth.

The deformation is computed on the side of the L dimension, and the following expressions give the factors k and C:

$$k = 1 - 0.0001(S) \qquad (12.12)$$

$$S = \text{system stiffness} = \frac{EI}{\gamma_w h^4} \qquad (12.13)$$

where EI is the bending stiffness of wall, $\gamma_w$ the unit weight of water and h is the average vertical spacing of lateral support elements.

$$C = 1 - \{0.5(1.8 - FS_{BH})\} \qquad (12.14)$$

where $FS_{BH}$ is the factor of safety against basal heave.

Finno et al. (2007) also provide an approximate expression for the shape of the distribution of lateral deflection along the wall.

#### ***12.6.2.4 Hydraulic stability and groundwater inflow***

To prevent piping and maintain hydraulic stability at the base of the excavation, the hydraulic gradient, i, must be less than the critical hydraulic gradient, $i_{cr}$, where

$$i_{cr} = \frac{(G_s - 1)}{(1 + e)} \qquad (12.15)$$

where $G_s$ is the specific gravity of soil particles and e is the void ratio.

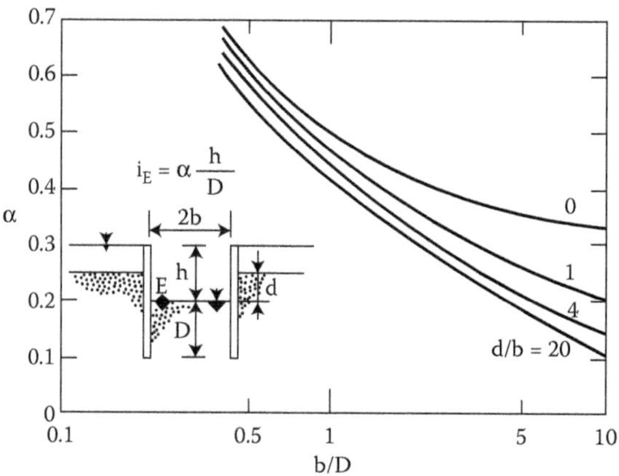


*Figure 12.10* Hydraulic gradient at base of excavation.

The factor of safety against piping, $F_p$, is then

$$F_p = \frac{i_{cr}}{i} \tag{12.16}$$

and it is desirable to have $F_p \geq 2$ (Xanthakos, 1994).

In a uniform soil, the hydraulic gradient i can be conveniently estimated from Figure 12.10. Alternatively, it may be obtained from a traditional flow net if suitable software is not available.

The required depth of penetration of a wall to prevent piping in sand can be estimated from Figure 12.11.

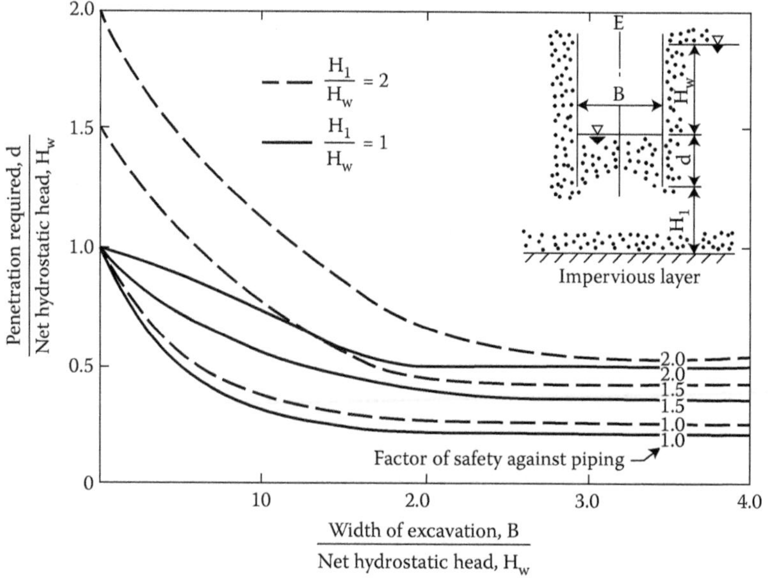


*Figure 12.11* Required wall penetration to avoid piping.

### 12.6.3 Category 3 methods: Simulation of excavation process

Staged excavation analyses use numerical approaches to model the actual sequence of excavation and support installation by considering each stage as it is constructed, and the excavation support is installed and, if appropriate, removed. As set out by Richards (2005), the soil and water pressures applied to the wall should be representative of the actual pressures acting (*not* the apparent pressure envelope) at each stage, and calculated loads are representative of the actual loads (*not* upper bound loads). The models can consider soil–structure interaction, with the earth pressures varying with displacement. As with all geotechnical analyses, the quality of the result depends both on the quality and relevance of the geotechnical parameters, and the nature of the problem idealisation. Two methods will be discussed briefly below:

1. The beam on elastic foundation method.
2. The finite element method.

#### *12.6.3.1 'Beam on elastic foundation' method*

In this approach, the wall is modelled as a beam, while the soil is modelled as a series of independent springs, in effect via a subgrade reaction model. In this type of analysis, the soil mass is not modelled, and so the analysis is effectively a one-dimensional analysis. Thus, it can only consider the wall and its supports, and cannot be used to estimate ground movements away from the wall.

Initially, the springs are compressed to create an initial load representing the at-rest earth pressures. At each stage of excavation or support installation, the loads in the soil springs change as the soil, water and support system loads are applied or removed, and lateral wall displacement occurs. The modulus of subgrade reaction is generally derived from the input values of soil stiffness and govern the spring displacement until the limiting value of active or passive pressure is reached.

The required soil input parameters include the unit weight, the at-rest, active and passive pressures along the wall, and the values of modulus of subgrade reaction for the various layers within the soil profile. The latter values are not fundamental soil parameters, but are dependent on the model dimensions and the geometry of the excavation.

Typically, the computed wall displacements are more sensitive to the modulus of subgrade reaction used in the analysis than are the computed support loads and wall moments. Thus, the use of conservative values of modulus of subgrade reaction will lead to conservative displacements, without greatly increasing the wall moments and the support forces.

A number of commercially available computer programs are available for carrying out this type of analysis. An example of such a program is WALLAP, which uses input values of Young's modulus to obtain values of the modulus of subgrade reaction, via closed-form elastic solutions. Another program, DeepXcav (www.deepexcavation.com/en/Plaxis-wallap-DeepXcav), analyses a wall with limit-equilibrium or elastoplastic methods. In the elastoplastic analysis, the lateral soil pressures depend on the soil properties and the construction stage history.

Beam on elastic foundation analyses can be useful for providing insights into the wall behaviour, and are very convenient for carrying out multiple analyses for optimising the wall and excavation design and undertaking sensitivity analyses. They can also be used as a check on more refined Category 3 analyses.

#### 12.6.3.2 Category 3: finite element method

Most finite element analyses involve the use of 2D models that include the wall, the support system and the surrounding soil. Various soil models can be employed in commercially available programs, including a variety of non-linear models, defined either in terms of effective stress or total stress. Non-linear soil models for the soil are essential in order to track local yielding and failure within the soil mass. In some cases, the ability to model volumetric changes in the soil, either consolidation or dilation) may be useful.

In contrast to the beam on elastic foundation analyses, a finite element analysis can provide direct information on the ground movements inside and outside the excavation. It can also consider the response of nearby structures to the excavation-induced ground movements.

With such an analysis, it is therefore not necessary to carry out separate analyses of stability and deformation, and the potential also exists to simulate piping and hydraulic failure. If an estimate of the conventional factor of safety against failure is required, it is usual to factor down the strength parameters and run the analyses until failure occurs. The factor of safety is then approximated as the reciprocal of the strength reduction factor at which failure occurred.

PLAXIS is an example of a 2D finite element analysis that is widely used in geotechnical practice. It is relatively user-friendly, but as with all such programs, care must be exercised to develop an appropriate mesh for the problem in hand. Furthermore, default values of some of the parameters need to be scrutinised to make sure that they are relevant to the problem being analysed.

In recent years, 3D analyses have become increasingly used. Such analyses may be useful for confined excavations where corner effects may need to be considered. In general, 3D analyses will lead to smaller wall and soil movements and wall bending moments than a 2D analysis.

#### 12.6.3.3 Comparison between WALLAP and PLAXIS analyses

A comparison has been made between WALLAP and PLAXIS and also the program DeepXcav. The problem studied involved a 10 m excavation with an 800 mm thick diaphragm wall. Table 12.7 shows comparisons between the computed support reactions, wall moments and displacements. There is generally fair agreement between the results from the three programs, although there are considerable differences in the lower support reactions and in the computed deflections. There is a tendency for WALLAP to predict larger moments and smaller deflections than the other programs.

#### 12.6.3.4 Some potential problems with Category 3 numerical analyses

An interesting case was documented by Schweiger (1998) in which 15 experienced geotechnical specialists were asked to undertake an analysis of a tieback wall in Berlin sand, as shown in Figure 12.14. Limited field measurements were available for this example, providing information on the order of magnitude of the deformations to be expected. The choice of

*Table 12.7* Comparison between alternative analyses of wall behaviour

| | *Support reactions (kN/m)* | | | *Wall bending moment (kN m/m)* | | |
|---|---|---|---|---|---|---|
| *Program* | *El 2.5 m* | *El −1 m* | *El −4 m* | *External* | *Internal* | *Latl. defln. (mm)* |
| WALLAP | 242 | 499 | 299 | 1277 | 465 | 49 |
| PLAXIS | 255 | 583 | 476 | 1020 | 279 | 77 |
| DeepXcav | 252 | 598 | 551 | 1165 | 264 | 52 |

constitutive model was left to the user and the parameter values had to be selected either from the literature, on the basis of personal experience or estimated from laboratory tests which were made available to the analysts. Some additional results from one-dimensional compression tests on loose and dense samples were given to the participants, together with the results of triaxial tests on dense samples. Thus, the exercise represented closely the situation that is often faced in practice. Inclinometer measurements made during construction provided information of the actual behaviour in situ, although due to the simplifications involved in the analysis sequence, a one-to-one comparison was not possible.

Additional specifications for this example were as follows:

- Plane strain conditions could be assumed.
- Any influence of the diaphragm wall construction could be neglected, that is, the initial stresses were established without the wall, and then the wall was 'wished-in-place' and its different unit weight incorporated appropriately.
- The diaphragm wall could be modelled using either beam or continuum elements.
- Interface elements existed between the wall and the soil, the domain to be analysed was as suggested in Figure 12.12, the horizontal hydraulic cut-off that existed at a depth of −30.00 m was not to be considered as structural support, and the prestressing anchor forces were given as design loads.

The following computational steps had to be performed by the various analysts:

- The initial stress state was given
- The wall was 'wished-in-place' and the deformations reset to zero

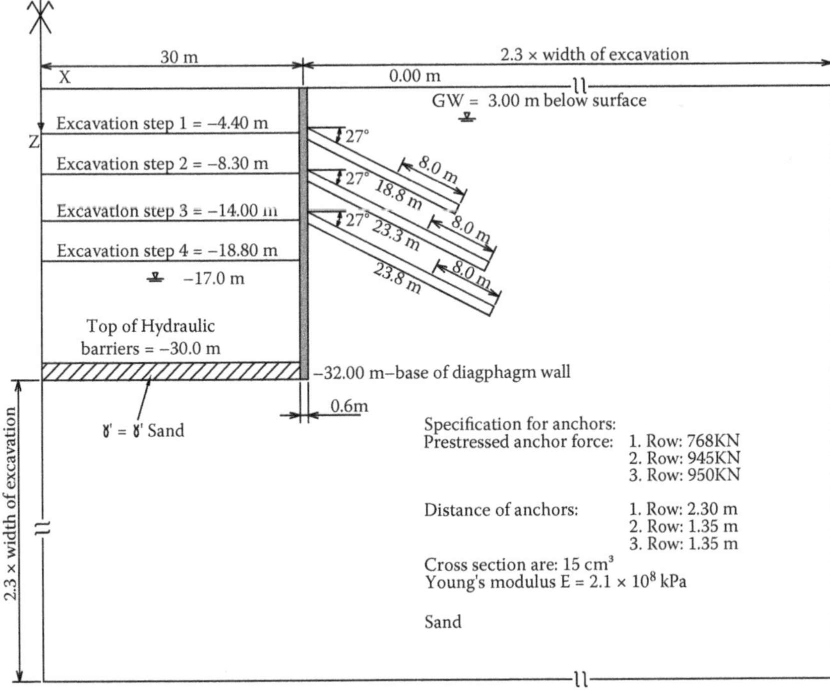


*Figure 12.12* Geometry and excavation stages for tieback wall exercise. (Adapted from Schweiger, H.F. 1998. *Proceedings of the 4th European Conference on Numerical Methods in Geotechnical Engineering*. pp. 645–654.)

The construction stages specified were as follows:

- Stage 1: groundwater-lowering to −17.90 m
- Stage 2: excavation step 1 (to level −4.30 m)
- Stage 3: activation of anchor 1 at level −4.30 m and prestressing
- Stage 4: excavation step 2 (to level −9.30 m)
- Stage 5: activation of anchor 2 at level −5.50 m and prestressing
- Stage 6: excavation step 3 (to level −14.35 m)
- Stage 7: activation of anchor 3 at level −13.55 m and prestressing
- Stage 8: excavation step 4 (to level −16.50 m)

A wide variety of computer programs and constitutive models were employed to solve this problem. Details may be found in Schweiger (1998) and Carter et al. (2000). Only a limited number of analysts utilised the laboratory test results provided in the specification to calibrate their models. Most of the analysts used data from the literature for Berlin sand, or their own experience to arrive at input parameters for their analysis. Only marginal differences existed in the assumptions made about the strength parameters for the sand (everybody believed the laboratory experiments in this respect), and the angle of internal friction $\phi'$ was taken as 36° or 37° and a small cohesion was assumed by many authors to increase numerical stability. A significant variation was observed however in the assumption of the dilatancy angle, with values ranging from 0° to 15°. An even more significant scatter was observed in the assumption of the soil stiffness parameters.

Figure 12.13 shows the computed deflection curves of the diaphragm wall for all analysts. It is obvious from the figure that the results scatter over a very wide range, which is unsatisfactory and probably unacceptable to most critical observers. For example, the predicted horizontal displacement of the top of the wall varied between −229 mm and +33 mm (−ve means displacement toward the excavation). Looking into more detail in Figure 12.13, it can be observed that entries B2, B3, B9a and B7 are well out of the 'mainstream' of results. These are the ones that derived their input parameters mainly from the oedometer tests provided to all analysts, but it should be remembered that these tests showed very low stiffnesses as compared to the values given in the literature.

A similar scatter of predicted results was found for the vertical displacement profile where the predictions varied from settlements of up to approximately 50 mm to surface heaves of about 15 mm.

Maximum anchor forces for the final excavation stage ranged from 106 to 634 kN/m, while predicted bending moments, important from a design perspective, also differed significantly from 500 to 1350 kN m/m. The exercise clearly indicated the need for guidelines and training for numerical analysis in geotechnical engineering in order to achieve reliable solutions for practical problems, and the importance of selection of relevant geotechnical parameters.

Teo and Wong (2012) examined the influence of the soil model on the computed behaviour of deep supported excavations for three case histories in Singapore. They first identified a number of limitations of the widely used Mohr–Coulomb (MC) model, including the following:

1. The MC model cannot model stress-dependent stiffness
2. The MC model cannot properly model unloading-reloading behaviour
3. The MC model cannot generate the correct one-dimensional compression behaviour
4. The MC model may produce an incorrect response under certain stress paths

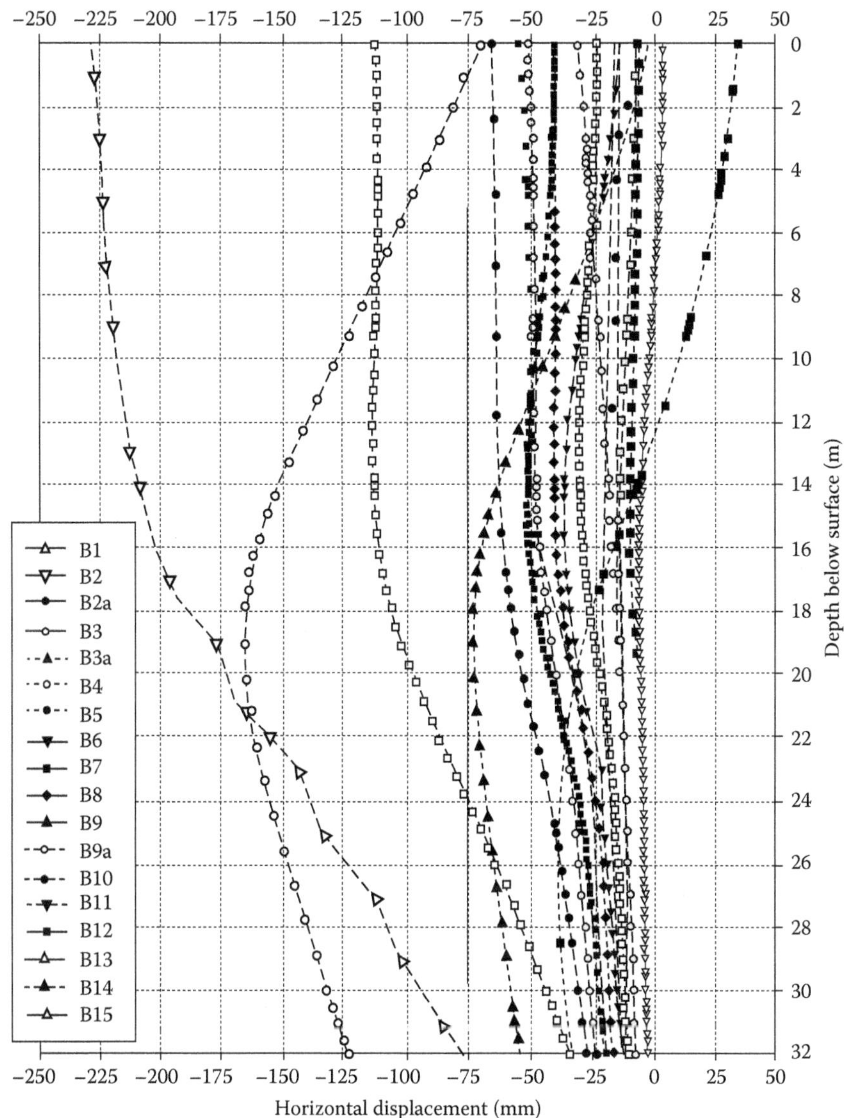


*Figure 12.13* Predicted lateral movements for tieback wall exercise. (Adapted from Schweiger, H.F. 1998. *Proceedings of the 4th European Conference on Numerical Methods in Geotechnical Engineering.* pp. 645–654.)

5. The MC model underestimates the horizontal stress in certain stress paths
6. The MC model results may be sensitive to the chosen Poisson's ratio in a drained analysis

They then compared the performance of the MC model and the Hardening Soil (HS) model and came to the following conclusions:

1. The HS model overcomes some of the shortcomings of the MC model, allowing a reasonable consistent ratio of $E_u/s_u$ of 250 to be applied successfully to all three cases.

2. In contrast, to achieve a similar agreement with the measured deflection using the MC model, the ratio $E_u/s_u$ had to be varied between 300 and 400.
3. The HS model produced a more realistic ground settlement profile than the MC model.
4. The HS model produced smaller toe movements and bottom heave than the MC model.
5. The HS model generated less plastic points within the finite element mesh because of its ability to take account of the softer soil behaviour as failure was approached. The MC model gave a false impression of the extent of plastic yielding.

This and other studies clearly demonstrate that a successful Category 3 analysis requires not only selection of appropriate parameters but also an appropriate model of soil behaviour. Problems involving excavation are more prone to be sensitive to the soil model than problems involving only compressive loading of the soil.

## 12.7 CONTROL MEASURES

### 12.7.1 Control of ground movement effects

If the ground movements due to excavation are assessed to be excessive, and may impact on adjacent structures, it will be necessary to take measures to reduce the movements or their impacts on the adjacent structure. Some such measures have been discussed by Wang et al. (2015), and are illustrated in Figure 12.14. Three broad strategies are shown:

- Controlling the excavation-induced movement via ground improvement, increasing the stiffness of the excavation support system, or modifying the excavation strategy to a top-down approach.
- Installing a barrier, generally via piles, between the excavation and the adjacent structures to reduce the ground movements.
- Providing additional foundation support for the adjacent structures via underpinning, such that they are better able to cope with the ground movements.

The effectiveness of each of these measures can be accessed via suitable analyses, which may often fall into Category 3 because of the complexity of the problem.

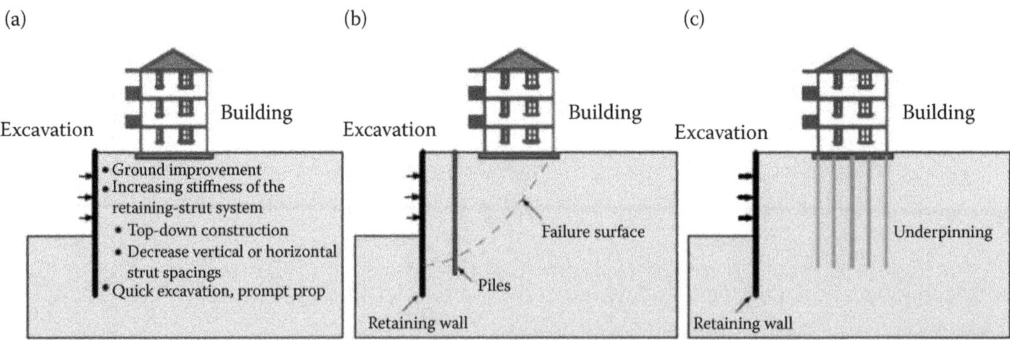


*Figure 12.14* Strategies for mitigation of adverse effects of excavations: (a) controlling ground deformations; (b) installation of barrier; (c) increasing foundation support. (Adapted from Wang, W.D., Xu, Z.H. and Li, Q. 2015. *ISSMGE Bulletin*, 9(6): 18–39. Courtesy of ISSMGE.)

### 12.7.2 Groundwater control

Key issues related to groundwater control near excavations are

- Creating a dry area for construction work, and preventing excess inflow of water into the excavation that could impede construction
- Reducing water pressure on retaining structures
- Prevention of 'blowout' of excavation if water pressure exceeds overburden stress
- Avoiding piping or sand boils within the excavation
- Control of water table drawdown so that areas adjacent to excavation do not suffer additional settlement
- Reducing uplift pressures on the building, especially on lower-rise podium areas where the dead loads are relatively small

The groundwater level can be controlled by dewatering, which involves some form of pumping to locally lower groundwater levels in the vicinity of the excavation. If it is possible, hydraulically isolating the internal part of the excavation from the outside is to be preferred, as it can reduce or even avoid unwanted drawdown, and possible consequent settlement, outside the excavation. Such isolation may be possible by either extending the wall down to a relatively impermeable layer, or else grouting the ground beneath, and in the vicinity of, the excavation to reduce its permeability. Grouting may impede or stop the penetration of water in subsoil with high permeability, such as in fissured and jointed rock strata. Rows of holes are bored on the soil, and grout, usually cementitious, is injected under high pressure. The cement grout will penetrate into the voids of the subsoil and form a relatively impermeable curtain vertically separating the ground water. Sometimes chemical grout (often silica-based) can be used; this creates a gel which can increase strength and reduce the permeability of the soil.

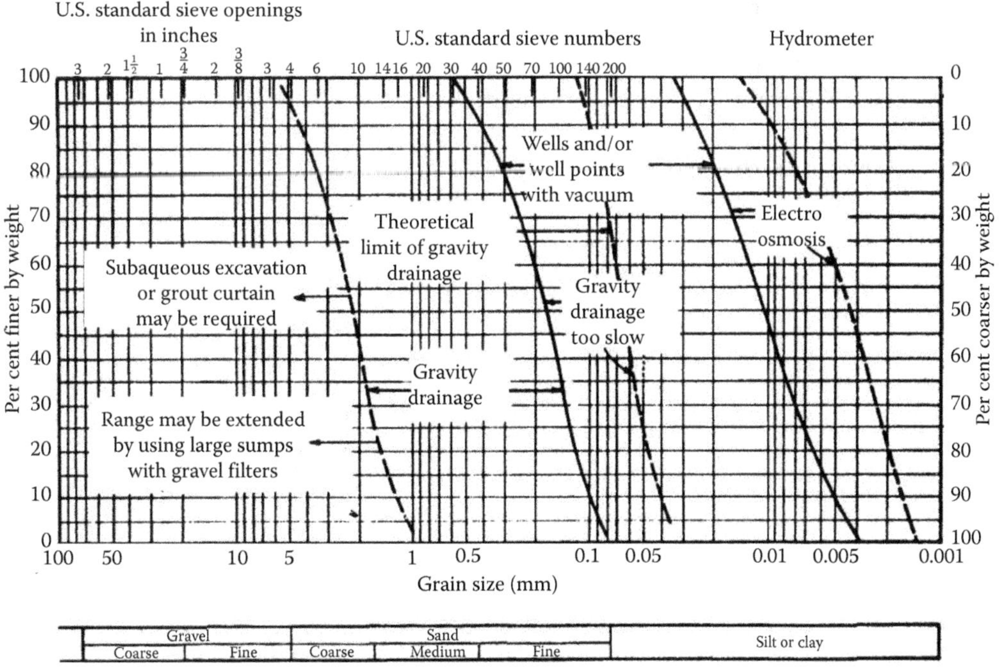


*Figure 12.15* Dewatering systems applicable to various ranges of soil permeability.

If there is a need to lower the water table significantly within the excavation, and grouting or other methods are not feasible, it may be necessary to use recharge wells outside the excavation to reduce the risk of causing additional settlements and distortions of adjacent structures. Further details related to groundwater control are provided in several sources, including CIRIA (2000); UFC (2004); Puller (2003) and Powers et al. (2007).

The simplest form of dewatering is sump pumping, where groundwater is allowed to enter the excavation where it is then collected in a sump and pumped away by robust solids handling pumps. Sump pumping can be effective in many circumstances, but seepage into the excavation can create the risk of instability and other construction problems.

To prevent significant groundwater seepage into the excavation and to ensure stability of excavation side slopes and base it may be necessary to lower groundwater levels in advance of excavation. Available methods include

- Deep wells
- Wellpoints
- Eductors
- Vacuum wells
- Horizontal wells

The method adopted will depend to a large extent on the permeability of the surrounding ground. Figure 12.15 provides some guidance on ranges of permeability for which various methods may be suitable.

Chapter 13

# Pile load testing

## 13.1 INTRODUCTION

Pile testing is a fundamental part of deep foundation design, and one of the more effective means of dealing with uncertainties that inevitably arise during the design and construction of piles. Pile testing is usually undertaken to provide relevant information on one or more of the following issues:

- The ultimate load capacity of a single pile
- The load–settlement behaviour of a pile
- The acceptability of the performance of a pile, as-constructed, according to specified acceptance criteria
- The structural integrity of a pile, as constructed

Such information may be used in a number of ways, including

1. Construction and quality control
2. As a means of verification of design assumptions
3. As a means of obtaining design data on pile performance which may allow for a more effective and confident design of the piles

The last two uses are of particular interest to the foundation designer, and will be the main focus of this chapter. In addition, consideration will be given to methods of estimating the characteristics of existing piles in the ground, and the possible reasons for pile imperfections which may lead to unsatisfactory behaviour.

## 13.2 THE DESIGNER'S VIEWPOINT

From the designer's viewpoint, pile load testing should ideally be able to satisfy the following requirements:

- Provide information on the various design issues
- Be able to be undertaken on pre-production piles
- Be able to be undertaken on any of the production piles without special preparation
- Be relatively inexpensive
- Provide reliable and unequivocal information which can be applied directly to the design process

In relation to the last point, there are some additional criteria which the designer may require the test to satisfy. These include

1. The test should load the pile in the same way as the structure will load the prototype piles
2. The test set-up should not induce inappropriate stress changes in the ground, that is, it should not have significant 'side effects'
3. The test set-up should not cause inaccuracies in the measurement of settlement or deflection
4. The test set-up should allow accurate measurement of the applied load
5. The duration of loading should be similar to that which will be experienced by the prototype piles

In reality, it is highly unlikely that any one test procedure can simultaneously satisfy all of the above requirements of the designer. The common types of test which may be employed, and the extent to which these tests can satisfy the above requirements, are discussed in the following sections.

## 13.3 TYPES OF TEST

A number of types of pile load test have been used in practice, and these are reviewed briefly in this section. Particular attention is paid to the inevitable problems which each test setup causes for accurate interpretation of the test data.

### 13.3.1 Static vertical load test

This is the most fundamental type of test and involves the application of vertical load directly to the pile head, usually via a series of increments. The ideal load test would be one which simulates the way in which a structural load is applied to the pile such that it is subjected to 'pure' vertical loading, and does not require any reaction system. Unfortunately, the ideal test cannot usually be achieved in practice, and the reaction system inevitably interacts with the test pile, thus creating some potential problems with the interpretation of the test data.

Test procedures have been developed and specified by various codes, for example, ASTM D1143. The static load test is generally regarded as the definitive test and the one against which other types of tests are compared. The test may take a variety of forms, depending on the means by which the reaction for the applied loading on the pile is supplied. Figure 13.1 illustrates some of the types of set-up commonly used, with the reaction being supplied by reaction piles, kentledge or ground anchors (either vertical or inclined).

Figures 13.2 shows details of the reaction pile option, Figures 13.3 and 13.4 show examples of the kentledge option and Figure 13.5 shows an example of the option employing inclined anchor cables.

The usual basic information from such a test is the load–settlement relationship, from which the load capacity and pile head stiffness can be interpreted. However, such interpretation should be carried out with caution, as the measured pile settlement may be influenced by interaction between the test pile and the reaction system. In the case of the test with kentledge, the stresses arising from the weight of the kentledge will initially cause an increase in the vertical and lateral stresses along the pile shaft and also at the pile base. These stresses will tend to cause an increase of the shaft friction and end bearing, compared to the case of 'pure' pile loading. As the load on the pile is increased, via jacking against the

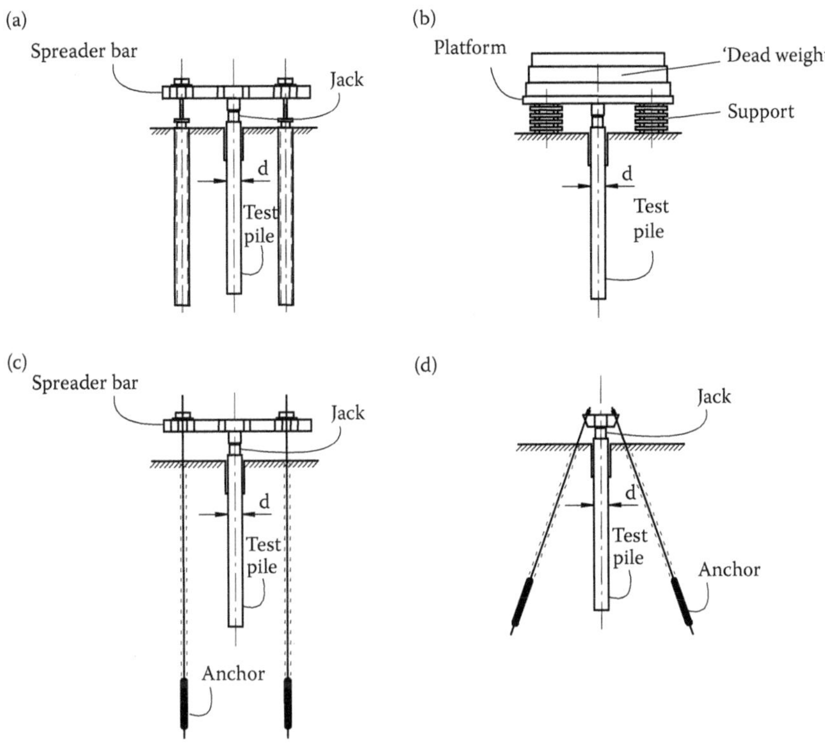


*Figure 13.1* Typical static load testing reaction systems: (a) reaction piles, (b) kentledge, (c) vertical anchors and (d) inclined anchors.

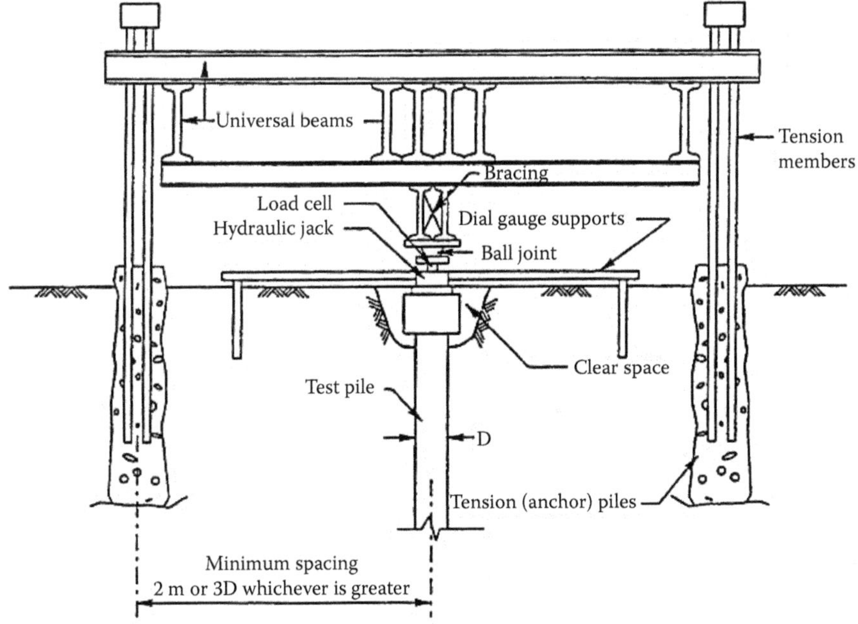


*Figure 13.2* Reaction provided by anchor piles.

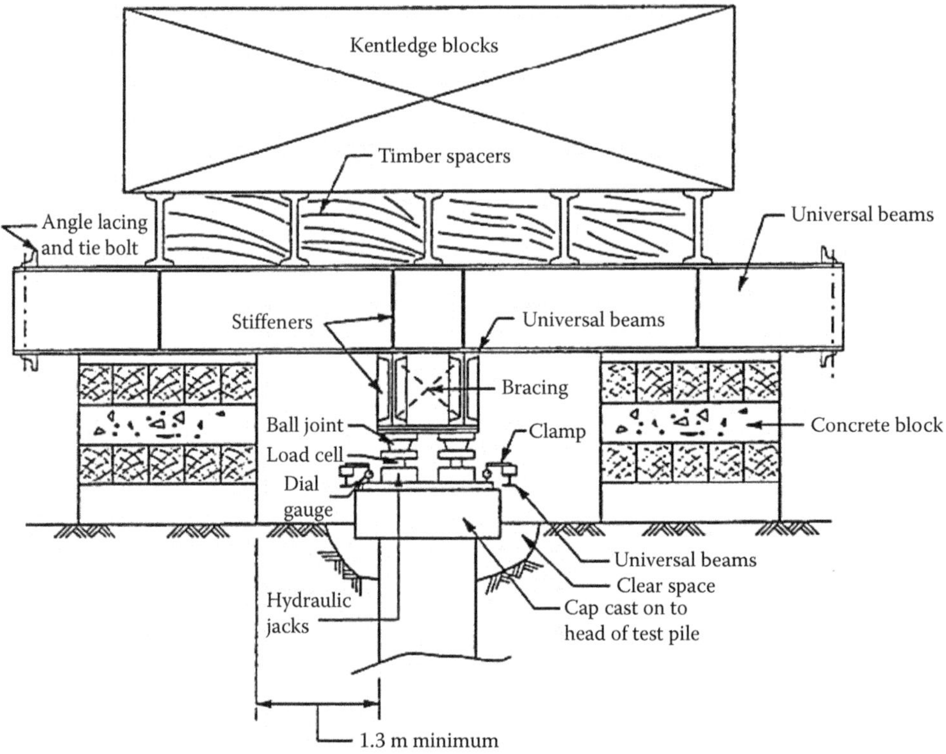


*Figure 13.3* Reaction via kentledge.

kentledge, these stresses will reduce, and there will be a tendency for upward 'free-field' soil movements to occur around the pile, while the pile itself undergoes settlement. As a result, the measured settlement of the pile will be less than the true settlement which would occur solely under the action of the load. The pile head stiffness will thus be overestimated if no allowance is made for the interaction effects. As the load on the pile is increased, the stress caused by the kentledge on the soil surface will decrease, and at failure (depending on the

*Figure 13.4* Example of use of kentledge for pile load test reaction.

*Figure 13.5* Example of use of anchor cables for pile load test reaction.

difference between the weight of the kentledge and the load on the pile), the pile capacity may be relatively close to that of the ideal pile test.

As an example of the possible consequences of kentledge on the pile performance, Figure 13.6 shows the computed load–settlement curves for an 'ideal' pile test subjected to pure vertical loading, and a test in which kentledge is used. These curves have been obtained from non-linear finite element analysis, assuming the soil to be a uniform sand exhibiting an ideal elasto—plastic behaviour. Figure 13.6 shows that the effect of the kentledge is to increase both the apparent load capacity and stiffness of the pile because of the stresses induced along the pile shaft by the kentledge. Yi (2004) refers to a series of field pull-out tests on tension piles carried out to investigate the effects of ground reaction stresses on the pile performance. It was found that the interaction between the kentledge support and the test pile resulted in an over-prediction of the ultimate uplift capacity of the pile (typically by about 10%–20%) and an underestimate of the pile head displacement (i.e. an overestimate

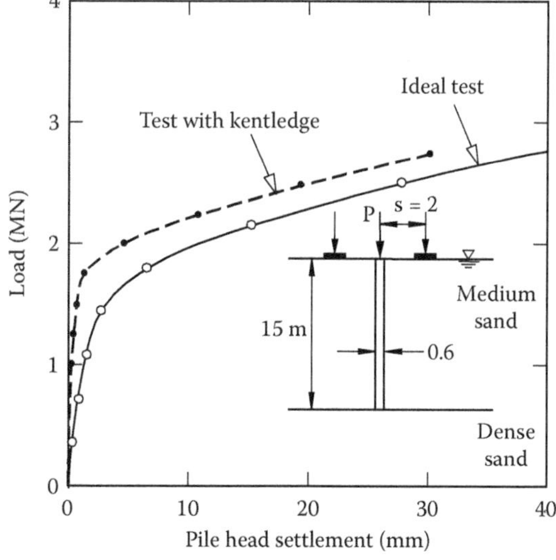


*Figure 13.6* Example of the influence of kentledge on a pile test in sand. (Adapted from Poulos, H.G. 2000b. *Proceedings of the 2nd International Statnamic Seminar*, Tokyo. A.A. Balkema, Rotterdam, pp. 3–21.)

of the pile head stiffness). These field tests were therefore consistent with the theoretical results in Figure 13.6.

For the case of a test employing two reaction piles, effects of interaction between the reaction piles and the test pile have been examined theoretically by Poulos and Davis (1980). Because the reaction piles are subjected to uplift, they will tend to cause a reduction in settlement of the test pile, and hence will result in an overestimation of the pile head stiffness. Typical theoretical results for an end bearing test pile to rock, jacked against identical reaction piles, is shown in Figure 13.7, where the ratio $F_c$ of the apparent to the true pile head stiffness is plotted against spacing between the test pile and the reaction piles. The pile head stiffness may be overestimated considerably if the test pile is close to the reaction piles, and/or the piles are relatively slender (i.e. have a large length-to-diameter ratio).

The conclusions from the theoretical analyses have been verified by centrifuge tests on piles in dense sand carried out by Latotszke et al. (1997). These tests indicate that the test with the reaction piles gives a significantly higher stiffness, and also a higher pile capacity, than the 'ideal' pile test. The effect of the reaction piles is most noticeable for the pile base.

Theoretical results for the case of a test pile jacked against ground anchors is shown in Figure 13.8. In this case, the overestimation of the pile head stiffness is significantly less than when reaction piles are used, especially if the anchors are located well below the pile base.

Another potential source of inaccuracy can occur if the settlement is measured with respect to a beam placed near the pile. Because the beam supports on the soil surface will tend to move downward as the pile settles, there will tend to be an under-registration of pile head settlement, and a consequent overestimation of the pile head stiffness. The extent of this overestimation will depend on the distance of the beam supports from the test pile; the greater this distance, the less will be the overestimation. Some theoretical results are given by Poulos and Davis (1980).

In summary, while the static load test may be considered as the definitive test, it is subject to several potential sources of inaccuracy which may affect the interpreted pile capacity and

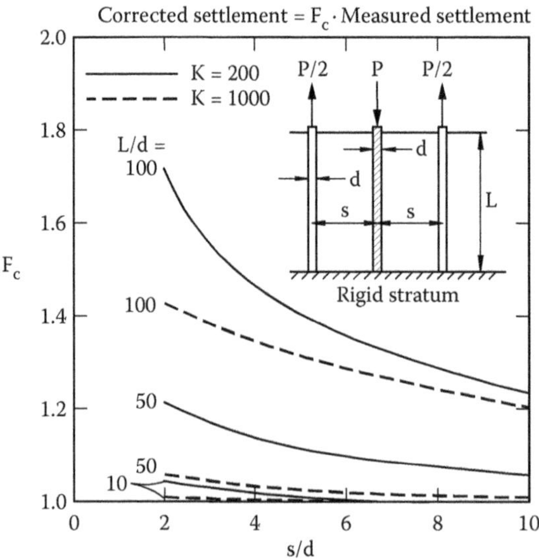


*Figure 13.7* Correction factor $F_c$ for end bearing test pile on rigid stratum. (After Poulos, H.G. and Davis, E.H. 1980. *Pile Foundation Analysis and Design*. John Wiley, New York.)

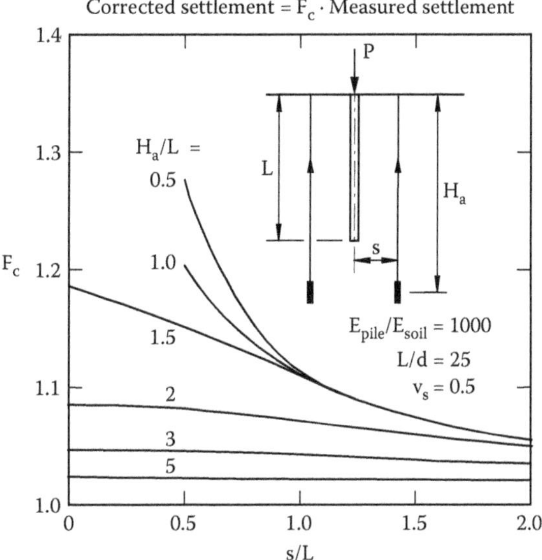


*Figure 13.8* Correction factor $F_c$ for floating test pile in deep layer with ground anchor reaction system. (After Poulos, H.G. and Davis, E.H. 1980. *Pile Foundation Analysis and Design*. John Wiley, New York.)

stiffness. Of concern is the fact that such inaccuracies tend to lead to overestimates of both capacity and stiffness, and are therefore unconservative, unless appropriate allowances are made for the effects of the interaction between the test pile and the reaction and/or settlement measuring system.

### 13.3.2 Static lateral load test

There are several forms of the lateral load test, but the most common and convenient is that which involves the jacking of one pile against one or more other piles; for example, ASTM Standard D3966 outlines a procedure for lateral load testing and for test interpretation.

As with the static vertical load test, there are 'side effects' if two piles are jacked against other piles. In particular, because the direction of loading of each pile is different, the interaction between the piles will tend to cause a reduced head deflection of each pile, and as a consequence, the measured lateral stiffness of the pile will be greater than the true value. An example is shown in Figure 13.9. Depending on the spacing between the piles, and the relative flexibility of the piles, the stiffness of the test pile may be overestimated by up to about 40% in the case considered. Clearly, it is desirable to allow for interaction effects when interpreting lateral load test data.

### 13.3.3 Dynamic load test

The principles of the dynamic load test are now very well established (Goble and Rausche, 1970; Rausche et al., 1985; Goble, 1994), and Figure 13.10 illustrates these principles. A hammer having sufficient energy to mobilise the pile resistance is necessary if a pile is to be tested dynamically. In this case, the energy of the blow applied to the pile should be large enough to mobilise the equivalent of at least 150% of the working pile load, or in terms of limit state design, 150% of the design load.

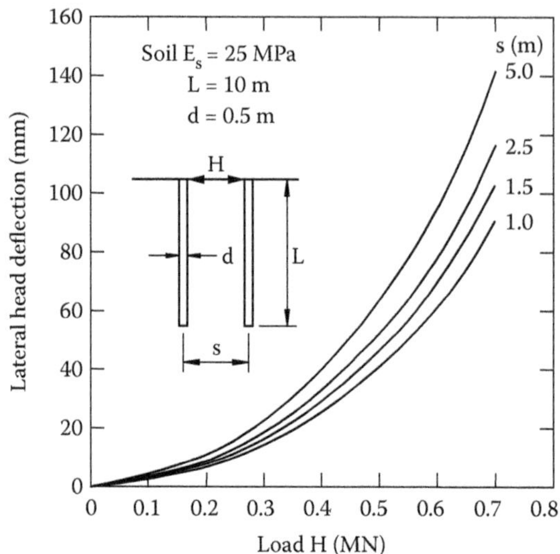


*Figure 13.9* Lateral load test – example of effects of pile interaction.

The test is illustrated in Figure 13.10, and is now accepted as a routine procedure, especially for quality control and design confirmation purposes. The pile head is instrumented with accelerometers and strain gauges, and from the recorded values, plots are made of force versus time and velocity versus time. There are a number of alternative approaches for interpreting the results, the most common being the 'CAPWAP', procedure (Rausche et al., 1985) and the TNO procedure (Middendorp and van Weele, 1986). In each case, the test data are interpreted via the use of a dynamic wave equation analysis and a curve matching

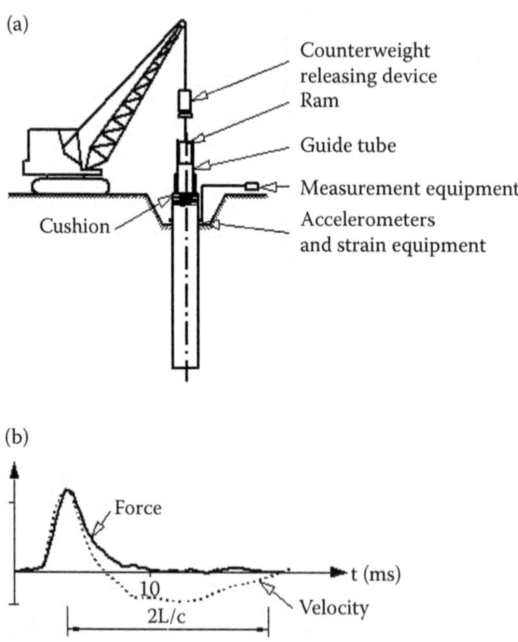


*Figure 13.10* Dynamic pile testing: (a) typical equipment, (b) typical force and velocity records.

procedure which provides an estimate of the static load–settlement behaviour. A number of comparisons with the results of static load tests, both on driven piles and bored piles, have indicated the ability of the dynamic pile load test procedure to produce meaningful estimates of the load–settlement behaviour of piles under static loading conditions, particularly if measurements are taken of the behaviour on re-striking the pile well after initial installation.

Despite its widespread use, the dynamic pile load test has a number of potential limitations, including the fact that the load–settlement behaviour estimated from the test is not unique, but is a best-fit estimate. Two measurements (strain and acceleration versus time) are taken, and from these, the complete distribution of resistance along the pile, as well as the load–settlement behaviour, are interpreted. Also, the load is applied far more rapidly than in most actual situations in practice, and hence time-dependent settlements are not developed during the test. Fortunately, under normal design load levels, the amount of time-dependency (from both consolidation and creep) is relatively small as most of the settlement arises from shear deformation at or near the pile–soil interface. Hence, the dynamic test may give a reasonable (if overestimated) assessment of the pile head stiffness at the design load. However, it may be expected to be increasingly inaccurate as the load level approaches the ultimate value.

For tall buildings, the foundation piles are likely to be large in diameter, and hence it is unlikely that, if full prototype-size piles are tested, the resistance could be fully mobilised using even the heaviest hammer. Accordingly, this type of test may be limited to testing for integrity.

### 13.3.4 The bi-directional or Osterberg cell test

This test was developed by Osterberg (1989) while a similar test was developed in Japan (Fujioka and Yamada, 1994). It has been used increasingly over the past decade or so, and is illustrated schematically in Figure 13.11. A special cell is cast at or near the pile base, and pressure is applied. The base is jacked downward while the shaft provides reaction and is jacked upward. The test can continue until the element with the smaller capacity reaches its ultimate resistance. Using the Osterberg cell (O-cell), test loads in excess 150 MN have been applied. It is common for two levels of cell to be installed, as this provides the ability to better define both the ultimate shaft and base capacities of the test pile. A major advantage of this test is its ability to fully mobilise the end bearing capacity of the pile, something that can rarely be achieved with tests involving loading at the pile head.

Despite its ability to provide 'self-reaction', the O-cell test (like all tests) has its limitations and shortcomings, including the following:

- It is applicable primarily to bored piles.
- The cell must be pre-installed prior to the test.
- There is interaction between the base and the shaft, and each will tend to move less than the 'real' movement so that the apparent shaft and base stiffnesses will tend to be larger than the real values.

The results of a numerical analysis with the commercial program FLAC are shown in Figure 13.12. The hypothetical case of a pile in medium sand bearing on a denser sand layer is considered. The results of an 'ideal' static compression test are shown together with the results of the O-cell test. The results overall are comparable, with the ideal test appearing to give slightly larger ultimate and base capacities. However, both the shaft and base responses from the O-cell are stiffer than in the ideal test in the early part of the test, and such differences could have unconservative consequences on the predicted settlement of the pile.

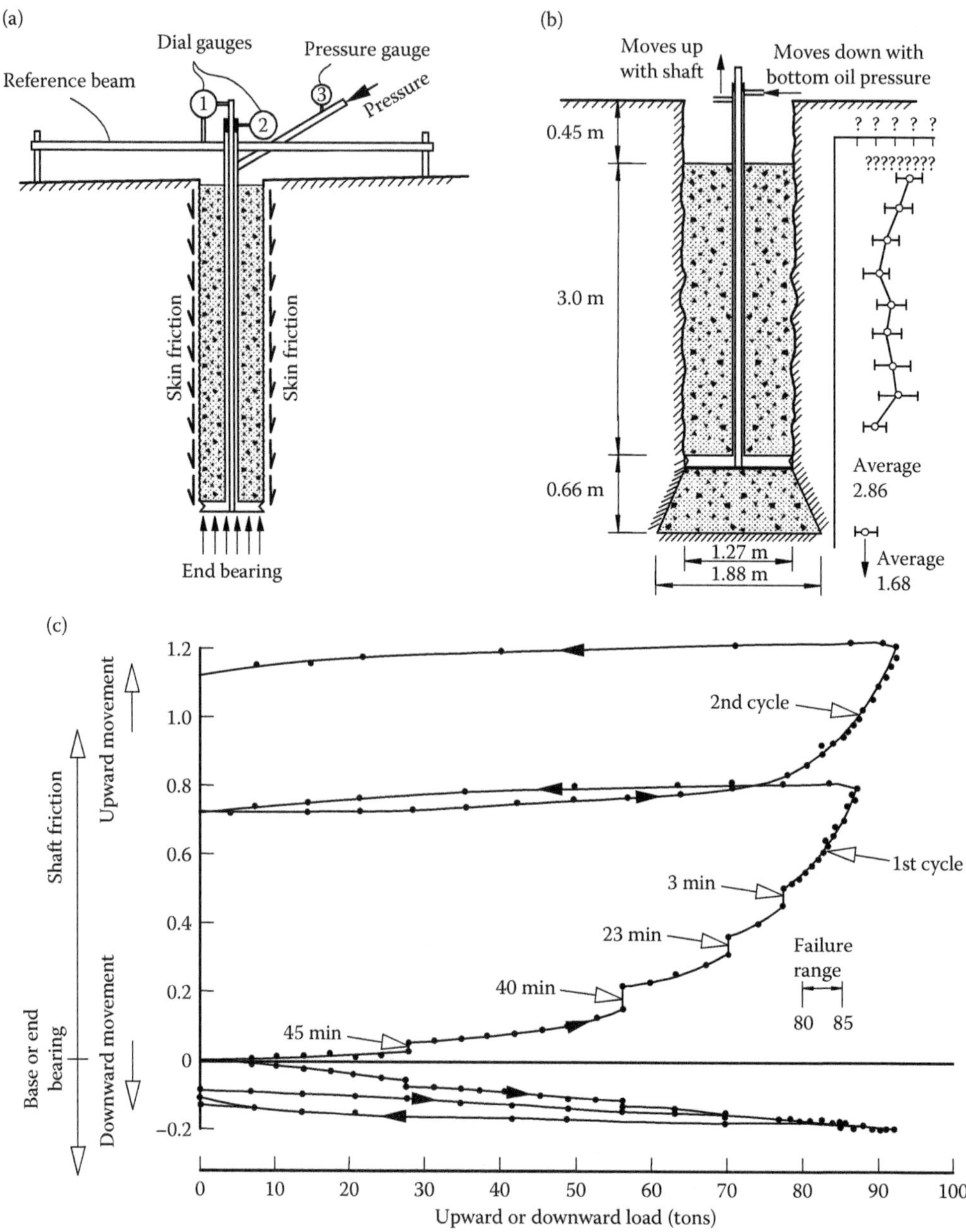


*Figure 13.11* Typical O-cell test: (a) test set-up, (b) test pile and (c) test result.

### 13.3.5 Statnamic test

Statnamic testing was jointly developed in Canada and the Netherlands (Middendorp et al., 1992; Bermingham et al., 1994). It has also found considerable use and development in Japan (Matsumoto and Tsuzuki, 1994). The principle of the test is illustrated in Figure 13.13 and involves the application of a downward force on the pile head via the burning of fast-expanding solid fuel in a combustion chamber, resulting in a large pressure acting upward on a reaction mass. The mass is accelerated to about 20 g, in turn producing an equal and

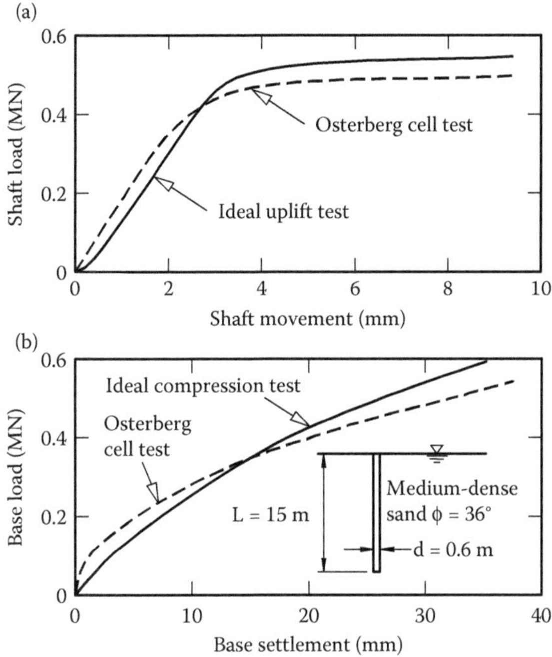


*Figure 13.12* Theoretical comparisons between ideal test and O-cell test for pile in sand: (a) shaft behaviour, (b) base behaviour. (Adapted from Poulos, H.G. 2000b. *Proceedings of the 2nd International Statnamic Seminar*, Tokyo. A.A. Balkema, Rotterdam, pp. 3–21.)

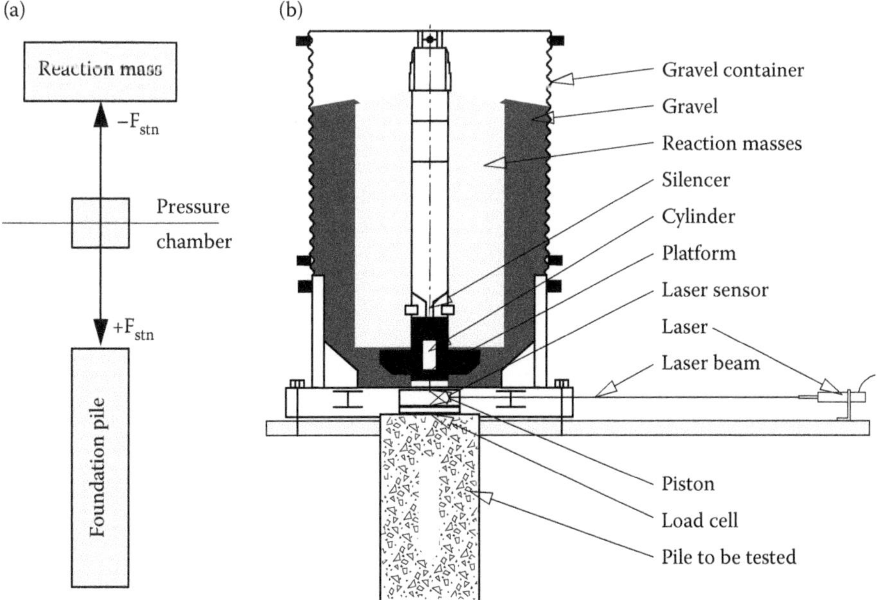


*Figure 13.13* Statnamic test set-up: (a) principe of test, (b) test set-up.

opposite force acting downward on the pile head. The load is applied in a linearly increasing manner, followed by a gradual unloading which is achieved by controlled venting of the pressure. The reaction mass, usually rings of concrete or steel, provides the resistance and needs to be only 5% of the total load to be applied to the pile. During the test, a load cell and laser sensor act in concert with a high-speed laptop computer to measure load and pile head movement directly, taking up to 4000 readings per second.

Comparative tests on piles subjected to conventional static testing and Statnamic testing have shown good agreement in load–settlement performance. Statnamic testing appears to offer a number of advantages over other test types, including

1. The test is quick and easily mobilised
2. High loading capacity is available
3. The loading is accurately centred and can be applied to both single piles and pile groups
4. The test does not require any pre-installation of the loading equipment
5. It can be adapted to apply lateral loading
6. The test is quasi-static, and does not involve the development of potentially damaging compressive and tensile stresses in the test pile
7. The test can be carried out on both uninstrumented and instrumented piles
8. The load is measured via a calibrated load cell and does not rely on pile material and cross-section properties

Inevitably, there are also some potential shortcomings, including

1. Certain assumptions need to be made in the interpretation of the test, especially in relation to the unloading of the pile
2. If used for cyclic or repetitive loading, it must be used in conjunction with a special catching device
3. It cannot provide information on time-dependent settlements or movements. While this may not be of great importance for single piles, it can be a major limitation when testing pile groups, especially if compressible layers underlie the pile tips

## 13.4 SUMMARY OF CAPABILITIES OF PILE TEST PROCEDURES

Based on the comments made above in relation to the various types of test, Tables 13.1 and 13.2 summarise the perceived capabilities of the various tests to satisfy the needs of the designer. It

*Table 13.1* Summary of capabilities of various pile load tests with respect to the results obtained

| *Test procedure* | *Ult. axial geot. capacity* | *Ult. lateral geot. capacity* | *Load–settlement* | *Lateral defln.* | *Group effects* | *Struct. capacity and integrity* | *Special loadings* | *Ground movements* |
|---|---|---|---|---|---|---|---|---|
| Static—uninstrumented | 3 | 0 | 3 | 0 | 1 | 1 | 1 | 0 |
| Static—instrumented | 3 | 0 | 3 | 0 | 2 | 2 | 2 | 2 |
| Static lateral | 0 | 3 | 0 | 3 | 1 | 2 | 2 | 0 |
| Dynamic (PDA) | 3 | 0 | 2 | 0 | 0 | 3 | 1 | 0 |
| Osterberg cell | 3 | 0 | 2 | 0 | 0 | 1 | 1 | 0 |
| Statnamic (instrumented) | 3 | 2 | 2 | 2 | 2 | 2-3 | 2 | 1 |

Note: 3 = very suitable; 2 = may be suitable under some circumstances; 1 = possible but unlikely to be suitable; 0 = not suitable or not applicable.

*Table 13.2* Summary of capabilities of various pile load tests with respect to the accuracy and relevance of the results

| *Test procedure* | *Pile loaded in same way?* | *Additional stress changes (side effects)* | *Accuracy of movement measurement* | *Accuracy of load measurement* | *Similar duration of loading to prototype?* |
|---|---|---|---|---|---|
| Static—uninstrumented | 3 | 2 | 2 | 3 | 3 |
| Static—instrumented | 3 | 2 | 2 | 3 | 3 |
| Static lateral | 3 | 2 | 2 | 3 | 3 |
| Dynamic (PDA) | 3 | 2 | 1 | 1 | 1 |
| Osterberg cell | 3 | 2 | 2 | 3 | 3 |
| Statnamic | 3 | 3 | 3 | 3 | 2 |

Note: 3 = good; 2 = may be adequate; 1 = generally not good.

will be seen that no single test can satisfactorily supply all the information which the designer may require, and that the static load test, which is usually considered to be the 'benchmark' test, usually provides only single pile capacity and stiffness. In addition, no test can provide the 'perfect' load test without 'side effects', and as discussed below, the interpretation of the static test should allow for the interaction between the test pile and the reaction system.

The testing system chosen will depend on the information that is required from the test, the cost of the test and the availability of equipment to perform the test.

## 13.5 PILE LOAD TEST INSTRUMENTATION

### 13.5.1 Conventional pile tests: Top loading

For piles loaded at the head, the deflection is measured either by dial gauges, or by electronic transducers attached to a reference beam. Movements can also be measured with precise levels or laser beams placed at some distance from the pile head. A traditional dial gauge used for measurement of deflection with reference to a support beam is shown in Figure 13.14. Generally four gauges are placed at equal intervals around the pile head. They should be accurate enough to measure pile head deflection to about 0.25 mm and have 50 mm of travel.

Measurements are affected by temperature and the effects can be quite pronounced in regions where the early morning and midday temperatures vary widely. Care should be taken to minimise temperature differences during the test by shielding the measuring equipment and the pile head from the sun in such circumstances.

*Figure 13.14* Dial gauge measurement of pile head settlement.

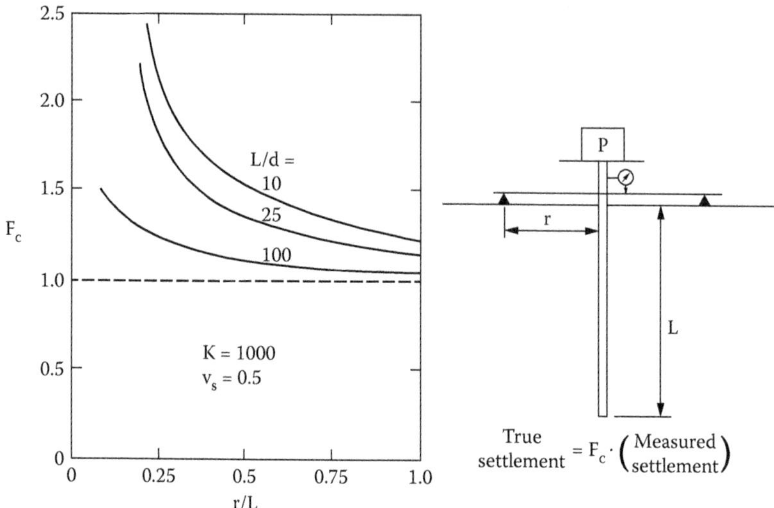


*Figure 13.15* Correction factor for the effect of movement of reference beam supports—deep uniform soil layer. (Adapted from Poulos, H.G. and Davis, E.H. 1980. *Pile Foundation Analysis and Design*. John Wiley, New York.)

When a pile is loaded, it causes the ground around it to deflect as well as the pile, and so the location of the supports for the reference beam will also be affected. For this reason, the supports for the reaction beam should be as far as possible from the loaded pile. Standards from different countries specify different distances that the reference beam supports should be from the pile head. ASTM D1143-81 says that the supports should be at least 2.5 m from the pile. This may however be inadequate if large diameter piles are being tested.

If the soil properties or the pile head stiffness are to be back figured from the pile load test, the relative movement of the pile head and the reference beam supports can become critical. Erroneous values of soil modulus can be calculated if this is not taken into account.

Methods of correcting for interaction effects have been presented by Poulos and Davis (1980) for various pile reaction systems, and one such system is shown in Figure 13.15. An estimate of the true settlement of the pile can be calculated by multiplying the measured settlement by a correction factor $F_c$. In Figure 13.15 the correction factor is plotted against the r/L factor where r is the distance of the support from the pile, and L is the pile length. It should be noted that these factors are based on the assumption of an infinitely deep uniform soil layer, and so the interaction effects in this case are likely to be larger than in most practical cases. For more accurate assessments of the testing side effects, numerical analyses can be carried out for the specific pile and soil profile conditions.

### 13.5.2 O-cell test

The O-cell test is now commonly used for testing piles for tall buildings as it does not need kentledge for the reaction and is capable of providing additional information about the pile behaviour.

A typical instrumented O-cell test set-up is shown in Figure 13.16 where two levels of O-cells are used. In this case the base of the lower O-cell assembly is 4 m from the toe of the pile and the upper assembly is 21 m from the toe.

Strain gauges are provided at 12 different levels within the pile to provide information on pile compression throughout its length. From this information, pile skin friction with depth

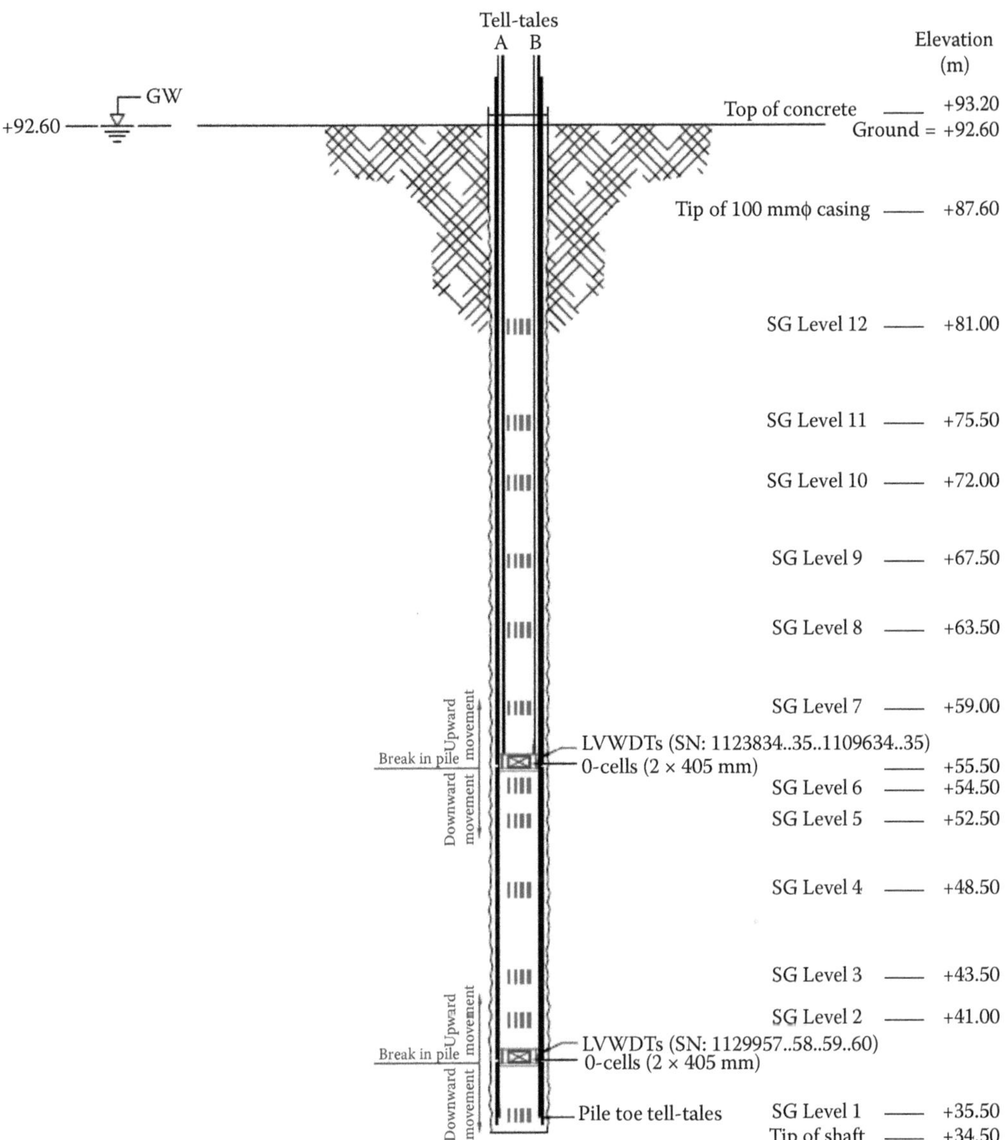


*Figure 13.16* O-Cell test using two levels of hydraulic cells.

can be back figured. Two tell-tales are installed to monitor pile toe movement and upper pile compression (with the end of tell-tale being just above the upper O-cell).

Strain gauges can be attached to the reinforcing cage of a pile, or concrete type gauges can be cast into the concrete. The strain gauges allow strains, and therefore stresses, to be calculated at various depths within the pile shaft and therefore the load in the pile shaft at those locations.

Extensometers may also be used to measure vertical movement within a pile shaft. Vibrating wire or DCDT displacement transducers are installed inside 51 mm steel or PVC sonic testing pipe. In one type of gauge, anchors at the top and bottom of the gauge can be expanded using compressed air to attach the gauge to the sides of the pipe. The relative movement between the top and bottom anchor of each gauge can be read to assess strains. The gauges can be retrieved after testing by releasing the air pressure in the anchors.

Measurement of movement can be performed using tell-tale rods that are placed inside casings within the pile. The end of the tell-tale rod can be placed at any level, and the movement recorded at that level. This allows measurement of the movement of the base and top of the pile as well as at intermediate points, and therefore more data can be collected on pile behaviour. The movements are still measured relative to a reference beam and so the supports of the beam are subjected to movement as they are in conventional tests.

### 13.5.3 Fibre optic measurements

Fibre optic cables provide a relatively new means of measuring strains in test piles. The principles of this system are set out in Section 14.6.1, and involve installation of a loop of protected fibre optic cable attached to the bottom of the reinforcement cage to provide a zero strain reference point. Sensing cable is then fastened along the entire length of the cages and pre-strained to a predetermined strain level. The system provides a continuous strain profile as the test pile is loaded, thus allowing the interpretation of the skin friction and end bearing resistances. The full strain profile provides useful insights into pile behaviour that are not easily measured with conventional instrumentation systems, and the measurements tend to be more reliable as they are insensitive to local structural defects such as crack openings or air pockets that can affect conventional methods of measurement.

## 13.6 TEST INTERPRETATION

### 13.6.1 Ultimate load capacity

From the load–settlement curve derived from a load test, it may be difficult to estimate where the pile reaches its ultimate load, as the deflection curve may continue to climb with increasing loading and not show any clear-cut failure. In this case, it is more usual to define the failure load as the load for a specific displacement.

One of the enduring problems in pile load testing is to define the failure load of a pile. There have been a plethora of suggestions, many involving constructions of uncertain origin and dubious validity. As pointed out by Abdrabbo and El-Hansy (1994) and England and Fleming (1994), the lack of a standardised definition of failure can cause disputes between the designer and the contractor.

There are two common situations from which a failure load may need to be derived:

1. Cases in which the loading is carried to a relatively large displacement.
2. Cases in which the pile is loaded to a relatively small displacement and the applied load is clearly less than the failure load, for example, a proof load test.

Ideally, in the first case, the failure load should be taken as that load at which there is no further increase in load with increasing displacement. In reality, such a well-defined load–displacement behaviour is not common, and it is more usual to define the failure load as the load for a specific displacement. For example, for conventional compression load tests, Eurocode 7 defines the failure (or 'limit') load as that causing a gross settlement of 10% of the equivalent base diameter. Such a definition has the attraction of being consistent with the earlier suggestion of Terzaghi, and is simple to interpret, in contrast to some of the alternative approaches.

For the second case, a frequently used approach is that of Chin (1970) which, in effect, assumes that the load–settlement curve is hyperbolic, and extrapolates the load–settlement data on this basis. It has been found that Chin's method commonly tends to overestimate the failure load, and it is occasionally modified so that the failure load is taken as a proportion (typically 90%) of the value derived from Chin's construction. It is also possible to adopt a consistent approach and extrapolate the load–settlement curve via Chin's approach, but to define the failure load as the value at a head settlement of 10% of the diameter.

Hwang et al. (2003) have reviewed a number of suggested methods of interpreting the ultimate axial pile load capacity, and have concluded that the above approach attributed to Terzaghi is reasonable in terms of consistency and physical meaning.

### 13.6.2 Axial load distribution along shaft

Other information can be obtained from pile load tests as well as the usual ultimate pile load and the load–deflection behaviour (or pile stiffness). With instrumented piles, the load in the pile at various locations along the shaft, and hence the skin friction distribution, may be obtained.

In interpreting the data from such instrumentation, the effects of residual stresses within the pile must be given careful consideration. It has been recognised for some time that the installation of most pile types results in residual stresses being developed in the pile, in contrast to the usual assumption that a pile is stress-free prior to loading. The presence of residual stresses generally does not affect the ultimate load capacity of the pile, but may influence the stiffness of the pile and the apparent sharing of the load between the base and the shaft. A simplified indicative analysis of these effects has been presented by Poulos (1987a), which shows that the following effects may arise from the presence of residual stresses:

1. The stiffness of the pile in compression may be increased
2. The stiffness of the pile in uplift may be reduced, considerably in the case of piles in sand
3. The proportion of load carried by the base may appear to be smaller than it is in reality

The latter point is illustrated in Figure 13.17. The installation of the pile results in a residual load distribution and a compressive load at the pile base. If readings of load distribution are commenced only after the pile is installed, then there will be a tendency for the shaft resistance to appear to reach a limiting value with depth, while the base resistance will appear to be smaller than the real value.

The effects of residual stresses tend to be more severe for piles with a relatively large end bearing resistance, such as piles in sand or piles bearing on a layer that is much stiffer than the overlying soils.

While empirical procedures have been developed for making allowances for residual stress effects (e.g. Briaud and Tucker, 1984), the most satisfactory approach is to use instrumented piles, and to take readings of the load distribution along the pile from the commencement of installation. In this way, the residual stresses can be measured, and more correct distributions of shaft and base resistance can be obtained.

### 13.6.3 Pile stiffness

The deflection of a pile under load may be found from a pile load test and used to refine predictions of pile group or piled raft behaviour. If the deflection at working load is required,

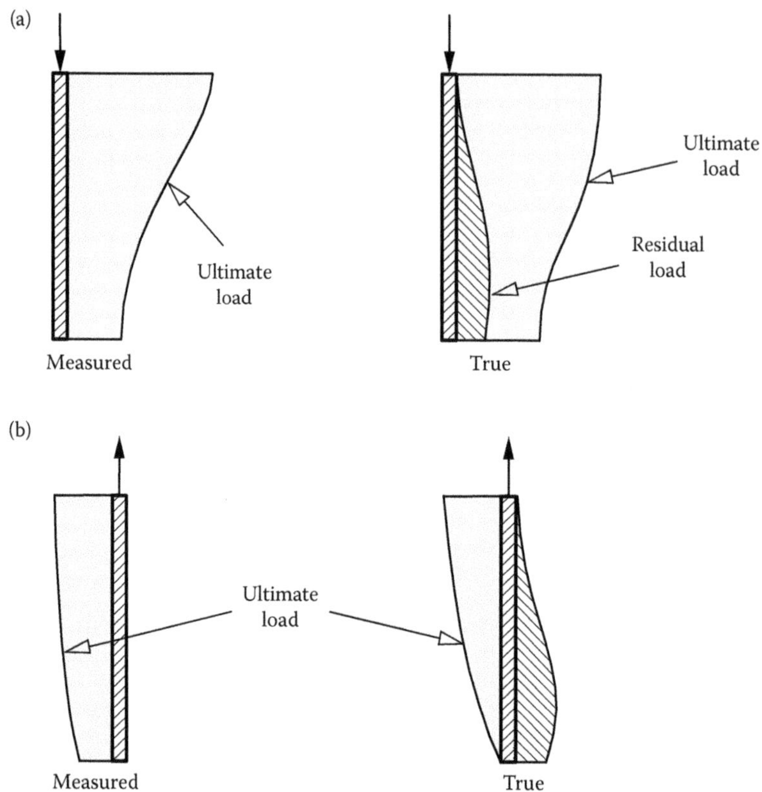


*Figure 13.17* Effect of residual stresses in pile tests: (a) compression, (b) tension.

it may be adequate to backfigure a secant pile stiffness, and to assume the pile has a linear load–deflection behaviour over the range of loads anticipated.

At higher loads the load–deflection behaviour measured for the pile will be non-linear. Often, a hyperbolic relationship is used to model the pile stiffness in this case. Parameters for the hyperbolic relationship can be changed until a good fit to the measured load–deflection behaviour of the pile is found.

For the model of ground behaviour assumed in the pile analysis, the relevant ground parameters need first to be interpreted from the measured load–settlement behaviour. For example, if a load transfer (t–z) approach is adopted, the initial slope and subsequent shape of the load transfer curves must be assumed and then the parameters for the curves derived via a process of trial and error. If an elasto-plastic soil model is assumed, then a distribution of Young's modulus and ultimate shaft friction with depth must be assumed and again, a trial and error process will generally be required to obtain a fit between the load–settlement behaviour from the theoretical model and the measured load–settlement behaviour. If there is no instrumentation along the pile, and hence no detailed load transfer information along the pile shaft, an assumption has to be made regarding the distribution of soil stiffness and strength with depth. This needs to be done in relation to the geotechnical profile in order to obtain reliable results.

If instrumentation has been installed in the pile, and if proper account is taken of residual stresses in the interpretation of the results, then the value of Young's modulus of the ground,

$E_s$, between each adjacent set of instrumentation can be interpreted by use of the following relationship developed by Randolph and Wroth (1978):

$$E_s = \left(\frac{\tau}{w_s}\right) d(1+\nu)\ln\left(\frac{2r_m}{d}\right) \tag{13.1}$$

where $\tau$ is the local shear stress, $w_s$ the local settlement, d the pile diameter, $\nu$ the ground Poisson's ratio, $r_m$ the radius at which displacements become very small and $\tau/w_s$ the slope of the derived load transfer (t–z) curve.

Randolph and Wroth (1978) give an expression for $r_m$ (see Section 8.5) and indicate that it is in the order of the length of the pile.

In obtaining the pile stiffness, whether linear or non-linear, it is necessary to allow for the interaction of the pile with the datum for the measuring system. Misleading values of pile head stiffness may be obtained if this is not done. Correction is also needed for interaction with the reaction system. As mentioned previously, for conventional top-loading pile tests, where there are reaction piles adjacent to the test pile, interaction between the upward moving reaction piles and the downward moving test pile will lead to an under-registration of the test pile settlement. Kitiyodom et al. (2004) have presented charts to allow correction of pile head stiffness found from pile load tests. The charts are for the case where anchor piles are used. Poulos and Davis (1980) also present charts to correct for the effects of the reaction system (see Figure 13.8).

Failure to take the test set-up into account in interpreting the load test results can lead to a significant overestimation of the real stiffness of the pile and the stiffness of the surrounding ground. This is illustrated in the example below.

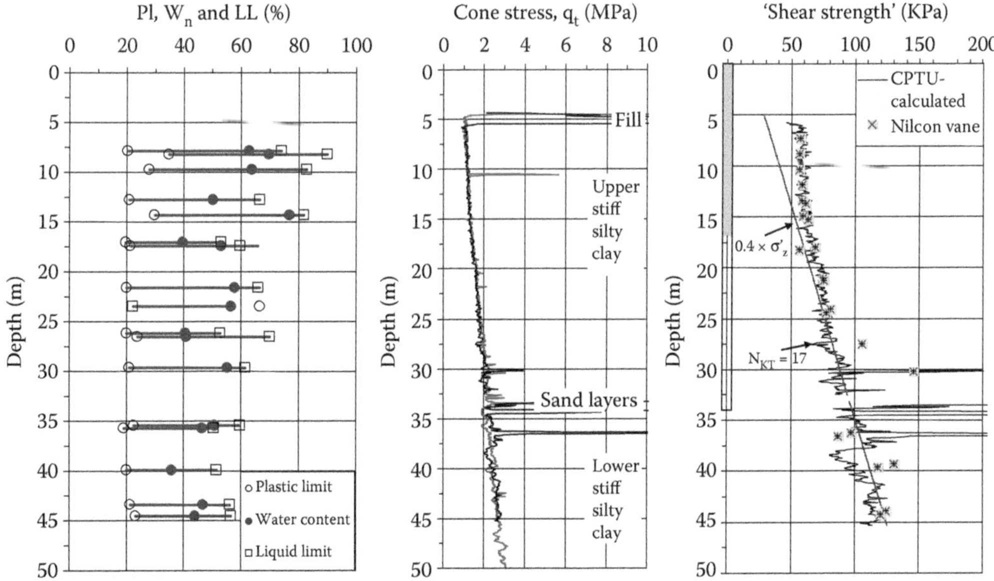


*Figure 13.18* Soil conditions for pile test. (Adapted from Amini, A. et al., 2008. Pile loading tests at Golden Ears Bridge. *Proceeding of the Canadian Geotechnical Conference*, Edmonton. Courtesy of Dr. A. Amini.)

#### 13.6.3.1 Example of load test interpretation

Figure 13.18 shows an example of the ground profile in which a load test was carried out on a large diameter bored pile, 2.5 m in diameter and 32 m long (Amini et al., 2008). In this case, based on the shear strength data interpreted from cone penetration testing, the ground profile has been characterised as one in which the Young's modulus, $E_s$, of the soil along the shaft increased linearly with depth and was related to the undrained shear strength, $s_u$, via the relationship $E_s = As_u$. The value of $s_u$ at the pile tip was about 100 kPa, so that the Young's modulus at the level of the pile tip is 0.1A MPa. The measured load–settlement curve is shown in Figure 13.19. At a load of 8 MN, the measured pile head settlement was about 1.8 mm.

In the test set-up, there were two reaction piles, each 2.5 m diameter, and 50 m long, located about 3 diameters from the test pile.

In interpreting the load test data, two sets of calculations were made: one in which no account was taken of the effects of the reaction piles, and the other in which the interaction between the test pile and the reaction piles was allowed for. The interpretation analyses were carried out using the computer program PIES (Poulos, 1989), assuming that the soil behaviour was linear up to the 8 MN load. Figure 13.20 shows the computed relationship between the assumed Young's modulus at the level of the pile tip and the pile head settlement at a load of 8 MN, for the two sets of calculations. By fitting the computed settlements to the measured settlement of 1.8 mm, the following backfigured values of Young's modulus at the pile tip, $E_{sb}$, are obtained:

- Ignoring the effects of the reaction piles: $E_{sb} = 400$ MPa (i.e. the factor A = 4000).
- Accounting for the effects of the reaction piles: $E_{sb} = 270$ MPa (A = 2700).

The latter value is considered to be more appropriate, and it can be seen that ignoring the effect of the reaction piles results in an overestimate of the soil modulus by almost 50%. Consequently, foundation settlements based on this erroneous value of Young's modulus would tend to be underestimated.

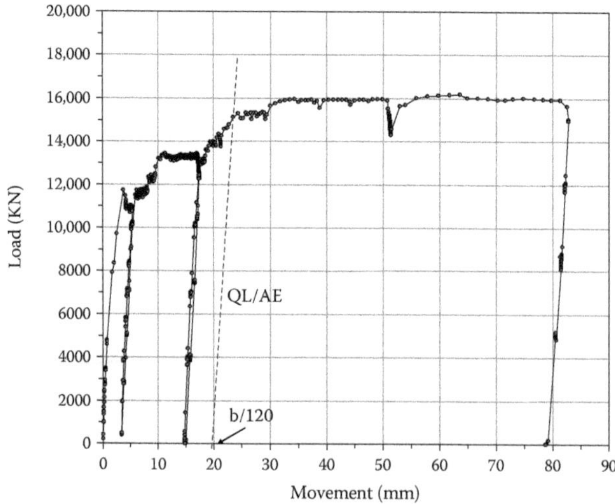


*Figure 13.19* Measured load–settlement curve. (Adapted from Amini, A. et al., 2008. Pile loading tests at Golden Ears Bridge. *Proceeding of the Canadian Geotechnical Conference*, Edmonton. Courtesy of Dr. A. Amini.)

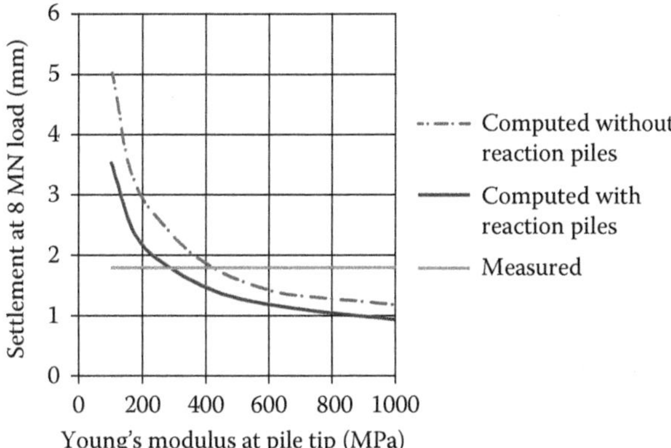


*Figure 13.20* Backfigured Young's modulus values.

### 13.6.4 Acceptance criteria

The definition of the acceptance criterion for a pile which is load tested can be contentious. Ideally, acceptance criteria should be specified by the designer, taking into account the requirements of the design and the need to verify that the design performance is being met by the piles as-installed. For example, in modern piled raft design, piles may be designed to develop a particular stiffness as well as a minimum load capacity. Piles that are too stiff may therefore be as unacceptable as those that are too 'soft'. In such cases, the acceptance criterion should be framed in terms of the target pile stiffness.

In many cases, acceptance criteria are specified for quality control purposes, and are taken from a code, without necessarily being related directly to the design. Typical criteria as specified in the Australian Piling Code AS2159-2009, for example, are shown in Table 13.3. The designer may need to react to the load testing in that, if the piles are deemed to be unacceptable, a decision then needs to be made on the future course of action, for example:

- Redesign the pile foundation using more appropriate assumptions
- Replacement of the piles which have shown inadequate performance
- Addition of extra piles to compensate for the piles which have performed inadequately
- Re-analyse the proposed foundation with the inadequate piles carefully to assess whether the performance of the foundation system as a whole will perform adequately

*Table 13.3* Acceptance criteria for vertical pile load tests

| *Load* | *Maximum settlement (mm)* *Static load test* |
|---|---|
| Serviceability Load, $P_s$ | $P_sL/AE + 0.01d$ |
| After removing serviceability load | Maximum $(0.01d_t, 5)$ |
| Maximum test load, $P_g$ | $P_gL/AE + 0.05d + 10$ |
| After removing maximum test load | $0.05d + 10$ |

Source: AS 2159. 2009. *Piling – Design and Installation.* Standards Australia.

Notes: d = shaft diameter (mm); $d_t$ = base diameter (mm); L = pile length; E = pile Young's modulus; A = pile cross-sectional area. The above settlements are default values that may be overwritten by an alternative specification.

While there may be circumstances in which one of the first three options is inevitable, there may also be instances where the group action may allow redistribution of some of the loads from the inadequate piles to the other piles, without causing unacceptable consequences to the group performance.

### 13.6.5 Allowing for negative friction effects

All conventional tests involve direct loading of the test pile, and thus cannot be used directly to provide information on piles subjected to ground movements (e.g. negative friction). There are suggestions made on how to allow for negative friction in conventional load tests, for example, Eurocode 7 recommends that the maximum load applied to a working pile should be greater than the sum of the design external load plus twice the design downdrag force. However, as shown by Wong and Teh (1996), such approaches are not valid as loading at the pile head cannot properly simulate loading of the pile via ground movements.

In reality, as discussed in Section 9.2.4, the behaviour of a pile subjected to negative friction will depend on both the magnitude and distribution of soil settlement. The most satisfactory approach is to test an instrumented pile and measure the distribution of shaft friction with depth. Assuming that the positive and negative skin friction values are similar in the settling layers, it is then possible to calculate the behaviour of the pile at the design load, with the effects of the soil settlements taken into account (Fellenius, 1989; Poulos, 1997a). In such cases, careful consideration should be given to the possibility of the water content and effective stresses in the soil being different in service from the values existing at the time of the pile test. Also, negative friction involves long-term loading, whereas pile testing is generally short term. Fortunately, there appears to be relatively little difference between short-term and long-term skin friction values in a number of soils.

Wong and Teh (1996) have suggested a modified testing procedure which involves isolating the pile shaft from the settling soil so that only the portion of the pile in the non-settling soil carries the load. An applied load equal to the sum of the working load and the estimated downdrag force at the bottom of the settling soil is applied to the pile to simulate working load conditions. A correction to the measured settlement is necessary to allow for the fact that, in the prototype pile, the load along the upper part of the shaft is not constant but increases with depth.

## 13.7 PILE INTEGRITY TESTS

There are a number of different tests that can be carried out on bored concrete piles in order to assess if the piles have been constructed correctly without defects. Common defects can be cavities in the pile shaft or inclusions caused by material falling from the sides of the drilled shaft. Pile integrity tests include high strain and low strain integrity tests.

High strain tests are primarily undertaken via dynamic pile testing, as described in Section 13.3.3. Anomalies in the measured traces of force and acceleration with time can be interpreted in terms of defects in the pile shaft. The load–settlement behaviour inferred from the test will indicate the performance of the pile which contains these defects.

Low strain integrity testing is a non-destructive form of testing in which the main objective is to detect the presence of any defects in a pile (e.g. cracks, waists, voids or soil inclusions). A comprehensive review of integrity tests is given by Turner (1997). Such tests fall into two classes:

- Tests that may be applied to an existing pile, and that therefore do not require pre-planning.

- Those that require pre-construction planning and that require the insertion of equipment into the pile during construction, to enable a test to be carried out on completion.

Some of the available types of small-strain integrity test will be described briefly below.

### 13.7.1 Sonic integrity testing

The most common form of test that does not necessarily require pre-planning is the sonic integrity test. The test involves the application of a blow to the pile head (usually with a plastic mallet) and the measurement of the time of arrival of reflected waves, via a transducer connected to the pile head. Details of the test are given in ASTM D5582-07.

If the pile is sound, the reflected wave should return at a time which is dependent on the wave velocity of the pile material and the length of the pile. If the pile contains defects, premature reflections of the stress waves will occur. Typical records of sound and unsound piles are shown in Figure 13.21 (Tchepak, 1998). Interpretation of the wave traces (or 'reflectograms') requires both experience and caution, as reflections can occur not only because of defects, but also because of changes in soil stratigraphy and changes in pile geometry.

Sonic integrity testing has a number of attractive features; the tests can be performed quickly and economically, an immediate indication of pile integrity can be obtained, and no special treatment (other than a sound pile head surface for the hammer blow) is required prior to the test. Analytical studies to evaluate the capabilities of sonic integrity tests in detecting details of defects have been reported by Liao and Roesset (1997). However, it must also be borne in mind that the test has several limitations, in that it cannot detect gradual changes in cross sections, curved forms, small defects or inclusions and local loss of concrete

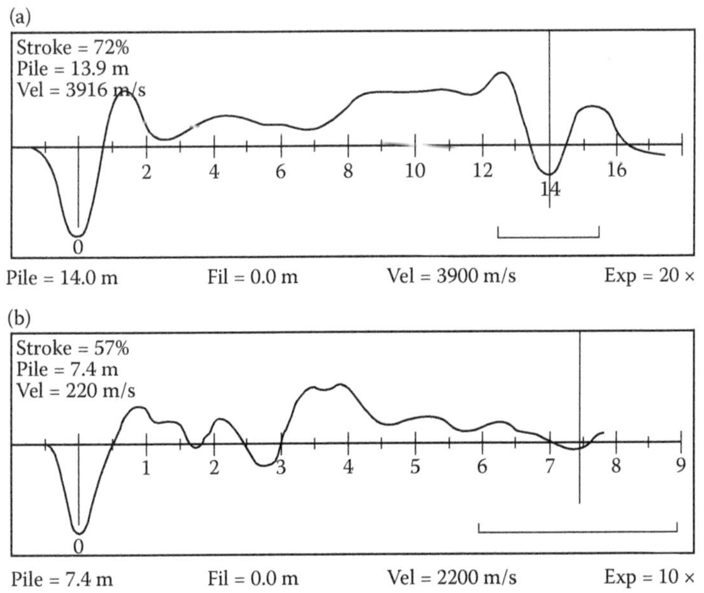


*Figure 13.21* Typical results for sonic integrity tests on bored piles: (a) reflectogram of a sound pile, (b) reflectogram of an unsound pile. (Adapted from Tchepak, S. 1998. Pile testing. *Mini-symposium on Recent Developments in Piling Practice in Sydney*, Australian Geomechanics Society, Sydney, Chapter.)

cover. In addition, it is generally limited to piles having a length of no more than about 20–25 m. In addition, for large diameter piles, the wave transmission will involve some 3D effects, which may complicate the test interpretation. These effects have been examined theoretically by Zheng et al. (2016).

### 13.7.2 Cross-hole sonic logging

One of the most effective methods in the second category is the cross-hole sonic logging method (Levy, 1970; Stain and Williams, 1991). The test involves the lowering of two piezo-electric probes, one a sonic emitter and the other a receiver, down two parallel access tubes embedded within the pile. Figure 13.22 illustrates a typical test set-up. The tubes are filled with water prior to the test to ensure good acoustic coupling (ASTM D6760-02). The system is restricted to bored piles, and tests the integrity of the concrete between the tubes by measuring its effect on the propagation of the sonic wave between the emitter and receiver. Sound concrete shows consistently good transmission characteristics, but the presence of soil, voids or other foreign material can affect the transmission signal. In general, two or more pairs of tubes are installed within the pile. It is possible to utilise the data to construct 3D tomographic images of the pile and hence identify any areas of defective concrete (Poulos et al., 2013a).

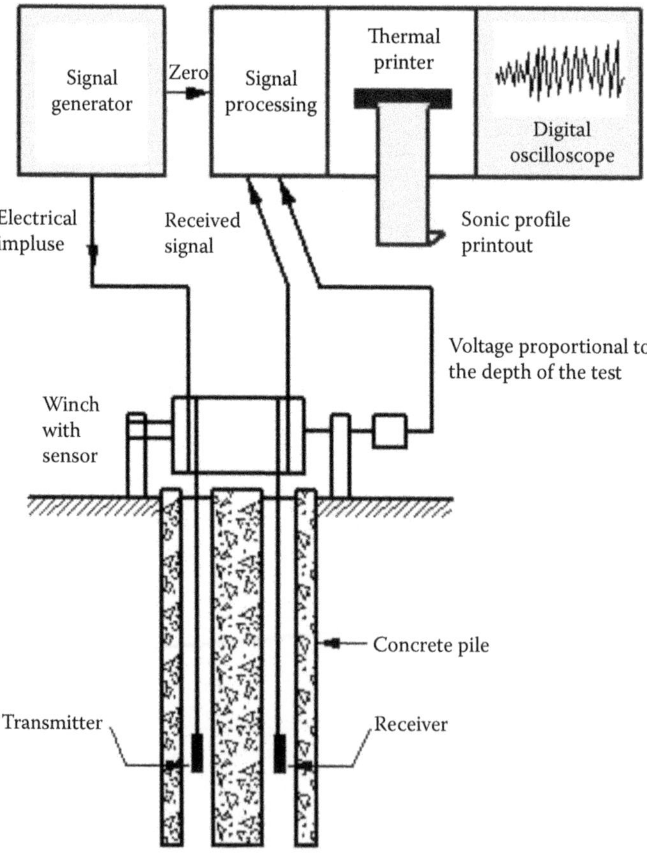


*Figure 13.22* Elements of a cross-hole sonic logging system.

### 13.7.3 Gamma logging

Another technique that can be used for pile integrity testing is gamma logging, which uses a radioactive source and a detector that can be used to measure variations in the density of concrete in a drilled pile.

The radioactive source is generally Caesium 137 that emits gamma radiation. The detector is a Geiger–Mueller probe, and the source and detector are placed into different PVC pipes (50 mm diameter) that are cast into the pile in the same way as is done for cross-hole sonic tests. The PVC pipes must be free of any water, and so are sealed to prevent water ingress. The pipes should be inspected to make sure that they are free from water and other obstructions before testing.

### 13.7.4 Thermal integrity profiling

Thermal integrity profiling (TIP) is a recent non-destructive test method that has gained popularity in post-construction evaluation of bored piles. It uses the heat generated by curing cement to indicate the quality of concrete in bored piles and other types of non-displacement pile. It is able to detect anomalies across the entire cross section of a shaft as well as provide a measure of lateral cage alignment. The expected temperature at any location depends on the shaft diameter, the mix design, the time of measurement, and the distance from the centre of the shaft. Early developments showed that the shape of a temperature profile with depth matched closely with the shape of the shaft, thus allowing for a fairly straightforward interpretation of data. The relationship between shape and temperature however had two exceptions, first near the ends of the shaft where heat can escape both radially and longitudinally, and second, where drastic changes in the surroundings are encountered (e.g. soil to water, soil to air). Methods of analysing these data in these regions are available, although they may often involve considerable parameter iterations and trial-and-error thermal modelling. Johnson (2016) presents a comparison of model and field results that provide further insight into the temperature distributions and also, the difficulties associated with the analysis. He also presents case studies that demonstrate the application of this approach.

### 13.7.5 Probe for pile base assessment

Another recent development is a device known as the shaft quantitative inspection device (SQUID) (PDI, 2016). It can be attached to the Kelly bar of most drill rigs. The rig lowers the body of the device to the base of the pile where three cone penetrometers move through the underlying debris layer and into the bearing material under the weight of the Kelly bar. The system measures the cone tip resistance and displacement, and the measurements are sent to a recording tablet in real time. The device can measure both the thickness of the debris above the bearing layer and the penetration resistance of the bearing layer. It thus provides a basis for deciding if further cleaning of the pile base is required and also the extent to which the base bearing layer may have been softened during construction.

## 13.8 ASSESSMENT OF NUMBER OF PILES TO BE TESTED

### 13.8.1 Load testing

For tall buildings that may be supported by large numbers of piles, the question arises as to the number of piles that should be tested so that the test results are representative of the whole pile group.

*Table 13.4* Pile testing requirements for serviceability

| *Average risk rating (ARR)* | *2.50–2.99* | *3.00–3.49* | *3.50–3.99* | *4.00–4.49* | *≥4.5* |
|---|---|---|---|---|---|
| Percentage of piles to be tested for serviceability | 1 | 2 | 3 | 5 | 10 |

Source: AS 2159. 2009. *Piling – Design and Installation*. Standards Australia.

The Australian Piling Code AS 2159-2009 specifies the percentage of piles to be tested for serviceability conditions. This depends on the Risk Rating which is a number calculated from Risk Factors such as the variability of the geology and the extent of the site investigation, experience with design in similar conditions, the extent of soil testing and the quality of construction supervision. From Tables in the code, an average risk rating (ARR) can be calculated (higher risk has a higher ARR) and from this the number of piles to be tested estimated as shown in Table 13.4. Testing of piles is only specified if the strength reduction factor applied in design (for the ultimate limit state geotechnical design) is greater than 0.4 (i.e. the soil strengths are factored down by a value >0.4).

The Federation of Piling Specialists (UK) *Handbook on Pile Testing* (2006) gives guidelines for the number of piles tested, and these are shown in Table 13.5. As set out in the Australian code, the amount of testing depends on the amount of risk associated with the project.

The number of tests performed can also be estimated on a cost basis as described by Kay (1976). More tests mean that the pile sizes can be refined to save money, but too many tests will raise the cost due to the cost of testing. Equation 13.2 may be used to calculate the cost C:

$$C = \frac{XF_m}{F_0} + mY \tag{13.2}$$

where $F_m$ is the factor of safety for m load tests, $F_0$ the original factor of safety for no pile load tests, X the cost of the total number of piles and Y the cost of a single load test.

For tall buildings where the geological conditions are uniform and construction control is good, typically up to 5 vertical pile load tests are performed, with perhaps 1–2 lateral load tests, and some cyclic load testing as well for both the axial and lateral load tests. A tension

*Table 13.5* Pile testing requirements according to risk

| *Characteristics of the piling works* | *Risk level* | *Pile testing strategy* |
|---|---|---|
| Complex or unknown ground conditions.<br>No previous pile test data.<br>New piling technique or very limited relevant experience. | High | Both preliminary and working pile tests essential.<br>1 preliminary pile test per 250 piles.<br>1 working pile test per 100 piles. |
| Consistent ground conditions.<br>No previous pile test data.<br>Limited experience of piling in similar ground. | Medium | Pile tests essential.<br>Either preliminary and/or working pile tests can be used.<br>1 preliminary pile test per 500 piles.<br>1 working pile test per 100 piles. |
| Consistent ground conditions.<br>Previous pile test data available.<br>Extensive experience of piling in similar ground. | Low | Pile tests not essential.<br>If using pile tests either preliminary and/or working tests can be used.<br>1 preliminary pile test per 500 piles.<br>1 working pile test per 100 piles. |

Source: FPS. 2006. *Handbook on Pile Load Testing*. Federation of Piling Specialists, UK.

test may also be required if some of the piles are subjected to uplift forces; this may often occur for piles under a podium area adjacent to the tower.

### 13.8.2 Number of integrity tests

Guidance for the number of integrity tests is given in the Australian Piling Code AS21599-2009. The amount of testing depends on the pile type (e.g. precast or cast in place). Lower percentages of piles are specified for testing if the pile design load is governed by soil strength rather than pile structural capacity. For bored piles, the percentage of pile integrity testing depends on how carefully drilling fluid, base cleaning and concrete tremie pouring is monitored.

As an example, for a bored pile constructed using a casing or drilling fluid with good construction monitoring and the design load governed by geotechnical capacity, 5%–15% of piles should be tested. If the design load is governed by pile shaft structural capacity and there is minimal construction monitoring, 15%–25% of piles need to be tested.

## 13.9 ESTIMATING THE LENGTH OF EXISTING PILES

The above methods of integrity testing can provide estimates of the length of a pile when there is direct access to the pile head. However, in cases where a foundation system has to incorporate existing piles, and there may be uncertainties as to the length and condition of these piles, it is highly unlikely that there will be access to the pile head. For such cases, various non-destructive methods have been developed to evaluate the pile geometry, and some of the more effective methods rely on subsurface measurements from boreholes alongside the piles. Such methods include parallel seismic, cross-hole sonic logging, borehole magnetometer, induction field and borehole radar, and a summary of these methods and the necessary equipment is provided by Wightman et al. (2003).

The parallel seismic method has been used in a number of cases, and is claimed to be more accurate and more versatile than other non-destructive surface techniques for the estimation of unknown foundation lengths, with a 5% accuracy often being achieved. The accuracy of the method depends on the variability of the velocity of the surrounding soil and the spacing between the borehole and the foundation element. The principles of this method are illustrated in Figure 13.23. The method involves hitting any part of the structure that is connected to the pile or foundation (or hitting the foundation itself, if accessible) and receiving compressional and/or shear waves travelling down the foundation by a hydrophone or a geophone receiver. Direct arrival times of compressional and shear waves at the receiver locations are recorded, as well as the wave amplitudes. The investigation is performed at 30–60 cm vertical receiver intervals in the borehole. Some portion of the structure that is connected to the foundation must be exposed for the hammer impacts. A borehole is required, and typically a 50–100 mm diameter hole is drilled as close as possible to the foundation, typically within 1.5 m. The borehole should extend at least 3–4.5 m below the expected bottom of the foundation. If a hydrophone is used, the hole must be cased, capped at the bottom and the casing and hole filled with water. For geophone use, the hole must usually be cased and grouted to prevent the soil from caving in during testing.

Coe and Kermani (2016) provide the following comments on some of these methods:

1. Cross-hole sonic logging is effective, but is complex, time consuming and costly, since it relies on measurements from multiple boreholes on either side of the pile.

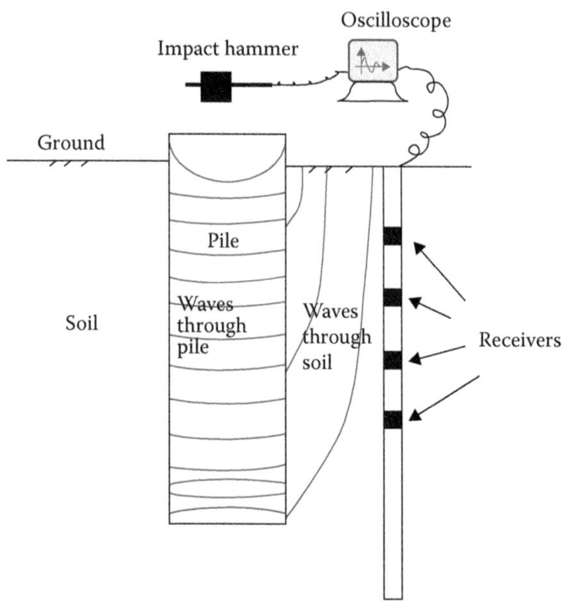


*Figure 13.23* Parallel seismic method.

2. The parallel seismic method, which is illustrated in Figure 13.23, has been used reasonably widely, but suffers when there is no direct access to impact the pile, particularly at sites with high background seismic noise such as traffic.
3. Borehole magnetometer and induction field methods are only useful when the foundation contains a significant amount of continuously connected steel.
4. Borehole radar is ineffective in clay soils, as clays have high electrical conductivity, which increases the attenuation of radar waves.
5. The borehole ultrasonic P-wave imaging system holds promise but shows limited potential for imaging a foundation, due to transducer directivity and the limitations of ultrasonic wave propagation distance.

Clearly, scope exists for the development of improved or new methods of estimating the length of existing and in-service piles.

Chapter 14

# Performance monitoring and control

## 14.1 INTRODUCTION

Monitoring of the foundation performance of tall buildings should be an essential part of the design and construction process. It can be used to assess the accuracy of predictions of performance (e.g. settlements, pile loads), and if the predictions are found to be over-optimistic, the monitoring can provide warning of impending problems. In such cases, if the deviations from expectation are identified sufficiently early, corrective action can be taken. On the other hand, if the monitored performance exceeds expectations, the data can be used to refine the geotechnical models, and perhaps the analysis techniques, for use in the design of the current project, if possible, or for future projects in the same area or in similar soil profiles.

Marr (2000) quotes Dunnicliff (1993) who sets out a number of reasons for implementing an instrumentation and monitoring program, namely:

- Evaluate critical design assumptions
- Assess contractor's means and methods
- Minimise damage to adjacent structures
- Control construction
- Control operations
- Provide data to help select remedial methods to fix problems
- Document performance for assessing damages
- Inform stakeholders
- Satisfy regulators
- Reduce litigation
- Reveal unknowns
- Indicate impending failure
- Provide a warning
- Advance the state-of-knowledge

All of these reasons have some relevance to tall building projects. Marr (2000) discusses each of these reasons and then suggests a method of quantifying, in monetary terms, the benefits of setting up a monitoring scheme, based on concepts of decision theory and risk analysis.

In this chapter, the main objectives of monitoring for tall building foundations will be outlined, and then the use of the Observational Method will be summarised. A brief description will be given of some available techniques for monitoring the key aspects of tall building foundation behaviour, and then some examples of the results of monitoring programs will be presented. Methods of dealing with unsatisfactory performance will then be discussed.

## 14.2 OBJECTIVES OF MONITORING

Before developing a monitoring scheme for a high-rise project, the objectives of the monitoring process should be clearly defined and communicated to the parties involved. A useful list of items to be considered is given by BTS (2011) in relation to tunnelling, but the principles are more generally applicable. The questions to be considered include the following:

- What is the reason for the monitoring?
- What needs to be monitored?
- Who is monitoring to inform?
- When is the monitoring required to be active?
- What monitoring techniques are anticipated?
- What is the frequency of monitoring?
- How is the monitoring data to be used, including if it is required to trigger any form of contingency response?
- What are the requirements for system reliability, data accuracy and data processing and usage?

The Observational Method provides a useful framework for implementing and utilising a monitoring program, and is described very briefly below.

## 14.3 THE OBSERVATIONAL METHOD

Peck (1969) is usually considered to have first advocated and developed the use of the Observational Method in ground engineering. This is a rational approach to dealing with geological and geotechnical uncertainty, and involves a continuous, managed, integrated, process of design, construction control, monitoring and review. It enables previously defined modifications to be incorporated during or after construction, if and as appropriate.

In brief, the key components of this approach are as follows:

1. Carry out sufficient ground investigation to establish general nature and properties of the strata.
2. Assess the most probable and most unfavourable conditions.
3. Establish a design based on the most probable properties.
4. Select the most significant monitoring parameters and calculate their values.
5. Calculate their values for the most unfavourable conditions.
6. Select design modification options that can be adopted if the monitoring outcomes are not favourable.
7. Monitor and evaluate the actual behaviour and conditions.
8. Modify the design, if necessary, to suit the actual monitoring results.

The Observational Method thus involves the following processes:

1. Establishing the limits of behaviour.
2. Developing a design which has an acceptable probability of the actual behaviour being within these limits.
3. Developing a monitoring plan, a response strategy and a contingency plan.
4. Utilising the contingency plan if the monitored behaviour is outside the assessed limits.

## 14.4 QUANTITIES MEASURED

From a foundation design viewpoint, the most relevant measurements are those related to the settlement and deflection of the foundation system, together with the distribution of load (usually vertical load) within the foundation system, and attention below will be focussed on such measurements. However, it should also be recognised that it is also possible to make measurements of the ground behaviour around and below the foundation system, and the response of the structure itself to lateral loads arising from wind, and possibly seismic, action.

### 14.4.1 Foundation system

The key foundation performance parameters are as follows, in descending order of importance:

1. The settlement of various points around the foundation system, preferably covering the areas in which the maximum and minimum settlements are anticipated.
2. The angular rotation at various points around the foundation system. Although approximate values of rotation may be derived from the differences in settlement, it is more accurate to measure rotations directly at key locations.
3. The lateral deflection of the basement walls.
4. The load being carried at the head of a number of piles.
5. The contact pressures between the raft/mat and the underlying ground at a number of locations in the vicinity of the piles at which the head load is being measured.
6. The distribution of axial load along the pile shaft, for a limited number of piles.

The above quantities should be measured from the commencement of construction.

### 14.4.2 In-ground

Monitoring of the ground in which the foundation is located may also be required for tall building projects, and some circumstances in which it may be desirable include

- Measuring excess pore pressures to check on the process of consolidation around and below the foundation.
- Measuring ground movements outside the foundation area to check on the possible 'side effects' of foundation construction.

### 14.4.3 Structure

The primary objective of structural monitoring is to improve safety and reliability of building systems by providing data to improve computer modelling and enable damage detection for post-event condition assessment. Such data can enable engineers to accurately estimate input ground motions, spectral accelerations, effects of soil–structure interaction, overturning, inter-storey drifts, torsional effects, modal properties (periods, damping ratios, mode shapes) and peak floor accelerations, velocities and displacements. A typical instrumentation scheme is shown in Figure 14.1 (Kinemetrics, 2013).

The minimum number of channels required for floors above ground is proportional to building height as per Table 14.1. The first two columns of this table are taken from 'An alternative procedure for seismic analysis and design of tall buildings located in the Los Angeles Region', Los Angeles Tall Buildings Structural Design.

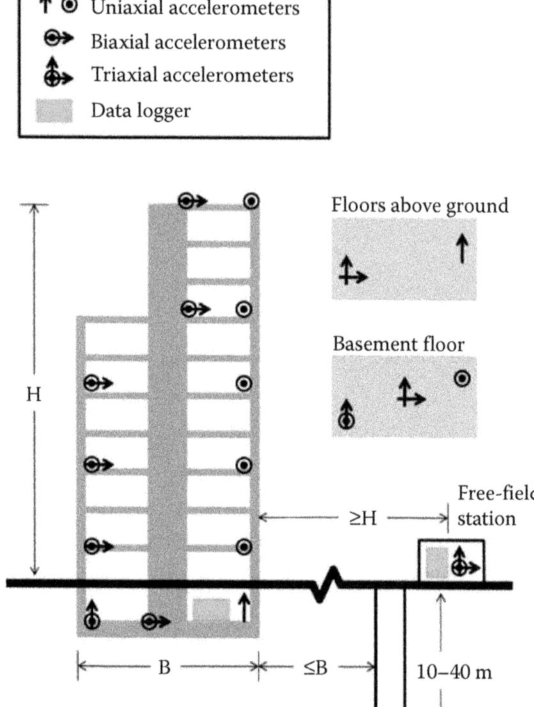


*Figure 14.1* Example of structural instrumentation system. (Courtesy of Kinemetrics. 2013.)

## 14.5 CONVENTIONAL MEASUREMENT TECHNIQUES FOR FOUNDATIONS

### 14.5.1 Settlements and lateral movements

Probably the most common and important measurements taken are of the vertical and horizontal movements of the foundation. As extra storeys are added to the structure, the foundation will settle, and there may be immediate, consolidation, and creep settlements that take place. As well, the foundation settlement may not be uniform and the structure may rotate. Rotation can be serious for tall buildings since a small rotation at foundation level

*Table 14.1* Minimum number of instrumentation channels

| *Number of stories above ground* | *Minimum number of channels above ground* | *Total number of channels* |
|---|---|---|
| 10–20 | 15 | 26 |
| 20–30 | 21 | 32 |
| 30–50 | 24 | 35 |
| >50 | 30 | 41 |

Source: Courtesy of Kinemetrics. 2013.

may mean large lateral movements at the top of the structure. Often corrections to the verticality of a tall structure can be made if the building is diverging from the vertical as more storeys are added.

Settlement measurements may be taken with accurate levels onto a measurement marker placed on the foundation. A benchmark that is not affected by the settlement of the structure needs to be used as the datum for the measurements. Total station theodolites may be used to obtain both vertical and lateral movements of markers.

Other equipment such as lasers and electronic inclinometers can be used to record lateral movements and tilt of buildings. Dynamic behaviour of structures can be measured using GPS techniques that are capable of measuring distances to sub-centimetre accuracy and collecting data at 10 Hz (Luo et al. 2000).

### 14.5.2 Pressure cells

Pressure cells may be used to monitor the pressure beneath raft foundations or the load in piles. This may be of interest if the load sharing between the raft and the piles is to be measured and compared with design estimates.

For measurement of pressure beneath a raft, Glötzl-type cells may be used as reported in Hemsley (2000). The Glötzl cell has a thin sealed chamber containing oil or fluid that causes a membrane to deflect when the fluid is pressurised. A fluid pressure is then applied to the membrane to return it to its null position, and that pressure is taken as the pressure applied to the cell.

Load cells for measuring pressure at the base of piles are available and consist of a fluid filled cell between two plates. The pressure of the fluid is measured by pressure transducers.

### 14.5.3 Strain gauges

Strain gauges have been mentioned previously in the section on pile testing (see Section 13.5). The gauges allow loads in the pile shaft to be calculated. These types of gauges are generally used with bored piles where the gauge can be attached to the steel reinforcing cage, and cast into the concrete. Alternatively, extensometers can be placed in tubes within the pile shaft. For steel piles, strain gauges can be welded to the pile shaft.

### 14.5.4 Piezometers

Piezometers may be installed beneath rafts or piled rafts to monitor excess pore water pressures generated during loading. In the case where ground water has been lowered to allow excavation of a basement, the total ground water pressures will fall and then rise again as the pumping is ceased. The water pressures need to be suppressed by pumping until the weight of the structure can counter the water uplift pressure, and so water pressure monitoring is important.

Various types of piezometer may be used (see Dunnicliff, 1993) including standpipe, hydraulic and vibrating wire devices. The vibrating wire piezometers have the advantage that they are connected to the readout location and have a short lag time.

### 14.5.5 Extensometers and inclinometers

Extensometers are sometimes placed beneath foundations so as to obtain the settlement of the foundation soils with depth. There are many different kinds of extensometer available commercially, but most involve a hollow tube that can telescope and move with the ground. Either magnets or steel rings are placed around the tube, and the position of these rings is detected with a probe lowered into the tube. The probe can accurately locate the position of the rings, and so the soil movement at the locations of the rings can be found.

Inclinometers may be used to measure lateral soil movement. A plastic casing is placed into a borehole and grouted in place. The casing has grooves in the sides (generally two sets at right angles) in which the wheels of a probe can run. As the probe is lowered down the tube, an accelerometer takes readings of the inclination of the probe, and from these readings the lateral movements of the casing may be found. Some inclinometers can have a dual role as an extensometer and an inclinometer.

Different types of extensometers and inclinometers are discussed in the book on instrumentation by Dunnicliff (1993).

## 14.6 MODERN MEASUREMENT TECHNIQUES

Finno (2014) has described a number of developments in performance monitoring, including the following:

1. Remotely operated robotic total survey stations to monitor the displacements of optical prisms
2. In-place inclinometers to remotely measure lateral movements with depth, or vertical displacements within the ground
3. Tiltmeters placed on structural elements to monitor the distortion of a structure
4. Fibre optic instrumentation using Brillouin optical time domain reflectometry (BOTDR) to measure deformations and strains
5. In-ground time-domain reflectometry (TDR) to detect local shearing
6. Internet-accessible weather resistant video cameras to allow remote visualisation of the construction process in real time, and to provide a dated record of construction
7. 3D laser scans to capture an accurate image of the geometry of an excavation and provide a digital record of construction progress

Finno (2014) has pointed out that wireless communications have enabled real-time data transmission to host computers where data are uploaded on websites accessible to a number of interested parties.

### 14.6.1 Fibre optic technology

Fibre optic technology is of increasing importance for engineering instrumentation, and has been used to make direct measurements of strain for geotechnical and structural applications. Mair (2008) summarises the technology involved. Optical fibre sensing relies on the interaction between a laser light and the glass material in an optical fibre. Strains and deformations alter the refractive index and geometry of the fibre optical material, and these changes perturb the intensity, phase and polarisation of the light wave propagating along the probing fibre. When a pulse of light is launched through the fibre, the majority travels through, but a small fraction is scattered back. Different components of light power, each with distinctive peaks at certain wavelengths, are identified, as shown in Figure 14.2. In the case of Brillouin scattering, the frequency of the backscattered light is shifted by an amount linearly proportional to the strain applied at the scattering location. By resolving the backscattered signal in time and frequency, a complete strain profile

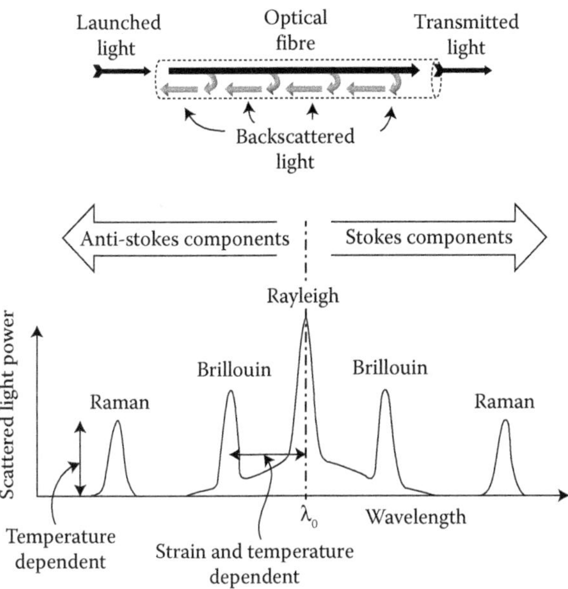


*Figure 14.2* Principle of distributed optical fibre sensing. (Adapted from Mair, R.J. 2008. Tunnelling and geotechnics: New horizons. 46th Rankine Lecture. *Geotechnique*, 58(9): 695–736. Courtesy of ICE Publishing.)

along the full length of the fibre can be obtained. Strain can be measured along the full length of the fibre of a suitably installed optical fibre by attaching a BOTDR analyser at one end of the fibre.

Mair (2008) states that the optical fibre system offers the following features:

- The average strain over 1 m is measured every 200 mm.
- The range over which the system can work is 5–10 km.
- The resolution is 0.003%.
- The sensors are very low cost, since the optical fibre is very cheap, although the analyser itself is expensive.
- The system is almost 'real time', typically taking 25 min per measurement.
- It is possible to link or switch between fibres.

Fibre optic technology has been used widely in monitoring tunnels (Mair, 2008; Mohamad, 2008) while other geotechnical applications have been mentioned by Negro et al. (2009) and Li et al. (2004). In relation to tall buildings, Mikami et al. (2015) have described the use of fibre optics to undertake structural health monitoring (SHM) of a 33-storey building in Tokyo, while Glisic et al. (2013) have described a 10-year program of structural monitoring for a 19-storey building in Singapore. Applications to foundation monitoring appear to be less common, although fibre optic monitoring of test piles has been described by Inaudi and Glisic (2007).

## 14.7 FREQUENCY OF MEASUREMENTS

Measurements of displacements, pile loads, etc. need to be taken as the structure increases in height, since the increased loads cause changes in the measurements of all instruments.

Once the construction is complete, the structure may continue to settle due to consolidation and creep of the ground supporting the foundation system. Measurements may need to be taken for several years to assess if the rate of settlement is slowing down. During this period, there may also be changes in pile loads and raft moments.

It is therefore desirable to take several measurements for each storey that is constructed of a tall building. For example, for the One Shell Plaza building constructed in Houston, Texas (Focht et al. 1978), regular readings of instruments were taken during construction. After the structural frame was completed, observations were made every 4–6 months. Two years after completion of the structure, readings were taken at yearly intervals, up to 10 years post-construction.

## 14.8 PORTRAYAL OF MEASUREMENTS

Measurements are often portrayed as a function of time. For instance, the settlement of a structure can be plotted against time (or log time), thus showing how it increases with building height and how it continues to increase after construction is complete due to consolidation and creep.

Loads in piles and raft contact pressures may also be plotted against time to monitor changes as the structural loads increase.

## 14.9 MONITORING SUPPORTED EXCAVATIONS AND BASEMENT WALLS

Negro et al. (2009) have provided assessments of the relative benefits of various forms of monitoring of supported excavations and basement walls. The measurements considered include horizontal movements of the wall, horizontal and vertical movements of the ground behind the wall, the settlement and tilt of surrounding structures, loads in struts and anchors, strut temperatures, strains in the wall, pore pressures inside and outside the excavation and earth and pore pressures against the wall. Each type of measurement was rated on a scale of 5–1, with 5 being of very high value and 1 being of low value, and the rating was made in relation to each of the following end points of the monitoring process:

- Verification of the basis of design
- Warning against failure
- Observational design approach
- Influence on surroundings
- Verification of quality of construction
- Improvement of design approach
- Enhancement of knowledge

The assessments are shown in Table 14.2, from which it can be seen that the most productive measurements are horizontal movement of the wall, ground movements behind the wall and the load in the struts or anchors.

Schwamb et al. (2016) describe the instrumentation and data interpretation for a deep circular excavation, and have found that the measured wall and ground movements were much smaller than those based on empirical methods of prediction measurements can be judged are discussed in Chapter 12.

*Table 14.2* Relative value of monitoring of supported excavations

| *Category* | *Hor. disp. at wall* | *Hor. disp. behind wall* | *Ground surf. settl.* | *Vert. distribn. of ground movs.* | *Settl. of surr. structs.* | *Tilt and strain in surr. structs.* |
|---|---|---|---|---|---|---|
| **Movements (5 = most important, a 'must'; 1 = low importance)** | | | | | | |
| Verify basis for design | 5 | 3 | 4 | 1 | 3 | 1 |
| Warning against failure | 5 | 3 | 2 | 1 | 2 | 1 |
| Observational design approach | 5 | 3 | 4 | 1 | 4 | 2 |
| Influence on surroundings | 4 | 3 | 5 | 2 | 5 | 4 |
| Verify construction quality | 4 | 2 | 4 | 1 | 3 | 2 |
| Improve design rules | 4 | 4 | 4 | 2 | 4 | 3 |
| Enhance knowledge | 5 | 5 | 5 | 5 | 4 | 3 |
| | *Loads in struts. or anchors* | *Temp. in struts.* | *Strain in wall* | *Pore press. within excavation* | *Pore press. outside excavation* | *Earth and pore press. against wall* |
| **Other measurements** | | | | | | |
| Verify basis for design | 3 | 3 | 3 | 1–5 | 3–5 | 1 |
| Warning against failure | 5 | 4 | 3 | 1–5 | 1–3 | 1 |
| Observational design approach | 5 | 3 | 4 | 1–5 | 1–3 | 1 |
| Influence on surroundings | 1 | 1 | 1 | 1–2 | 3–5 | 1 |
| Verify construction quality | 4 | 2 | 1 | 1–4 | 1–4 | 1 |
| Improve design rules | 5 | 3 | 3 | 1–4 | 1–4 | 4 |
| Enhance knowledge | 5 | 3 | 5 | 2–4 | 3–5 | 5 |

Source: Negro, A. et al., 2009. Prediction, monitoring and evaluation of performance of geotechnical structures. *Proceedings of the 17th International Conference on Soil Mechanics and Geotechnical Engineering*, Alexandria, Egypt, Vol. 4, pp. 2930–3005.

## 14.10 EXAMPLES OF MONITORING

### 14.10.1 Westendstrasse tower 1, Frankfurt, Germany

The Westendstrasse 1 tower is a 51 storey, 208 m high building in Frankfurt, Germany, and has been described by Franke et al. (1994, 2000). A cross section and foundation plan of the building is shown in Figure 14.3. The foundation for the tower consists of a piled raft with 40 piles. The central part of the raft is 4.5 m thick, decreasing to 3 m at the edges. The raft below the main tower covers an area of about 47 × 62 m$^2$. The piles are bored piles, about 30 m long and 1.3 m in diameter. Full details of the geotechnical profile at the site were not available in the published literature, but it appears that a thick deposit of Frankfurt Clay is present.

#### *14.10.1.1 Instrumentation*

Franke et al. (2000) give details of the instrumentation employed for this building, and this information is reproduced in Figure 14.4. The instrumentation installed included the following:

- Three extensometers, measuring settlements with depth; two of these were combined with inclinometers to measure lateral soil movements.
- Six piles along which the distribution of axial force with depth could be measured.
- Raft–soil contact pressures at eight locations.
- Raft–soil contact pressures at five further points, combined with pore pressure measurements.

#### *14.10.1.2 Typical measurements*

Figure 14.5 shows measured time-dependent loads and load distributions on the piles, and reveals that the piles carry substantial loads ranging from about 10 to 16 MN.

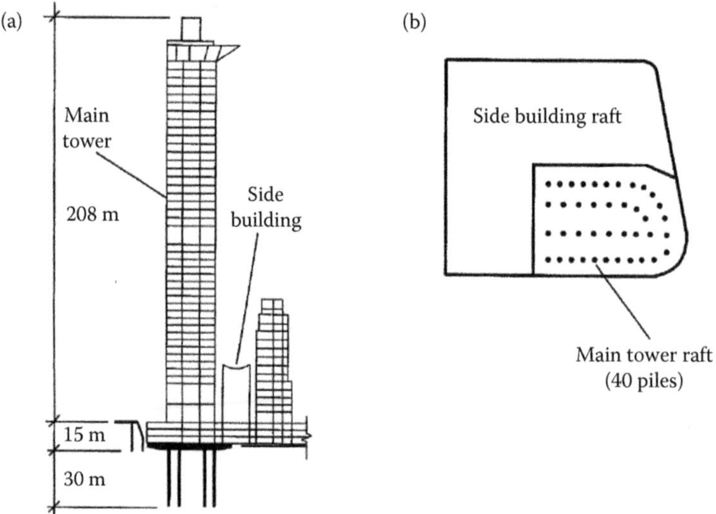


*Figure 14.3* Cross section and foundation plan for Westendstrasse 1 building. (Adapted from Franke, E., Lutz, B. and El-Mossallamy, Y. 1994. *Vertical and Horizontal Deformation of Foundations and Embankments, ASCE Geotechnical Special Publication No. 40,* 2: 1325–1336. Franke, E., El-Mossallamy, Y. and Wittmann, P. 2000. *Design Applications of Raft Foundations*, Thomas Telford, London.)

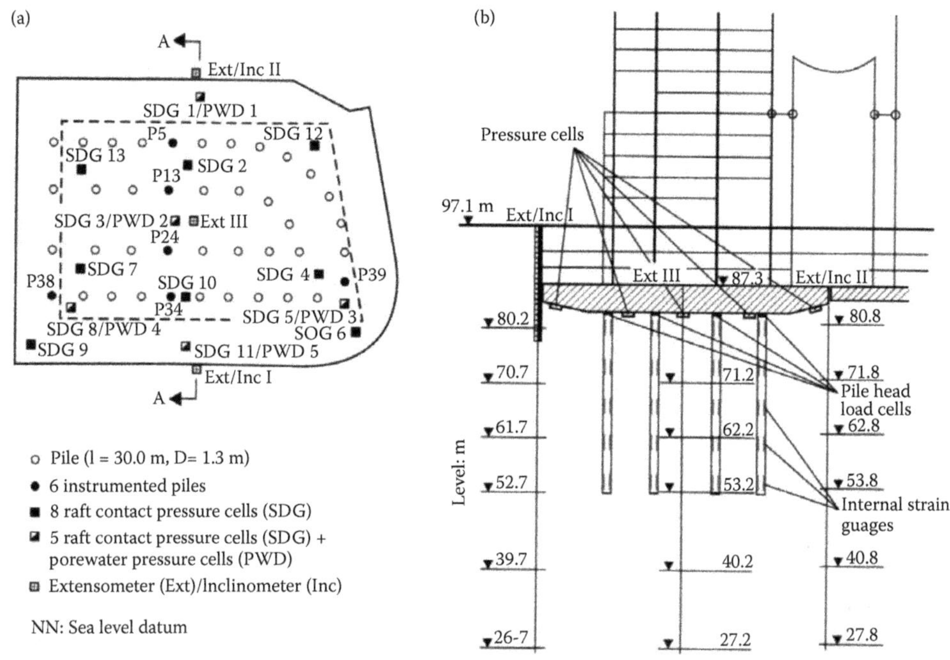


*Figure 14.4* Foundation instrumentation for Westendstrasse I building. (a) Plan of foundation; (b) Section A-A. (Adapted from Franke, E., El-Mossallamy, Y. and Wittmann, P. 2000. *Design Applications of Raft Foundations*, Thomas Telford, London. Courtesy of ICE Publishing.)

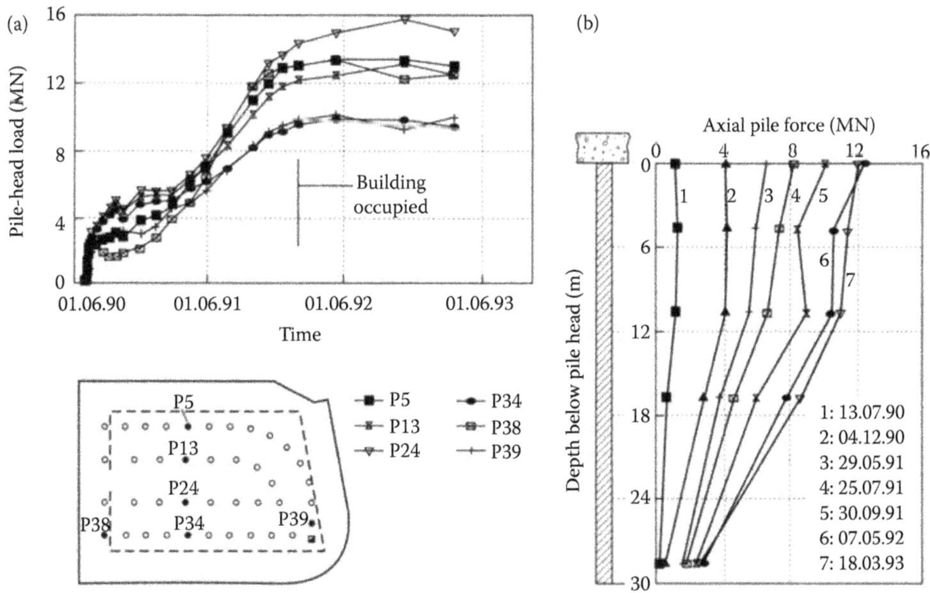


*Figure 14.5* Measured time-dependent behaviour of piles: (a) pile head loads; (b) average axial force along pile length. (Adapted from Franke, E., El-Mossallamy, Y. and Wittmann, P. 2000. *Design Applications of Raft Foundations*, Thomas Telford, London. Courtesy of ICE Publishing.)

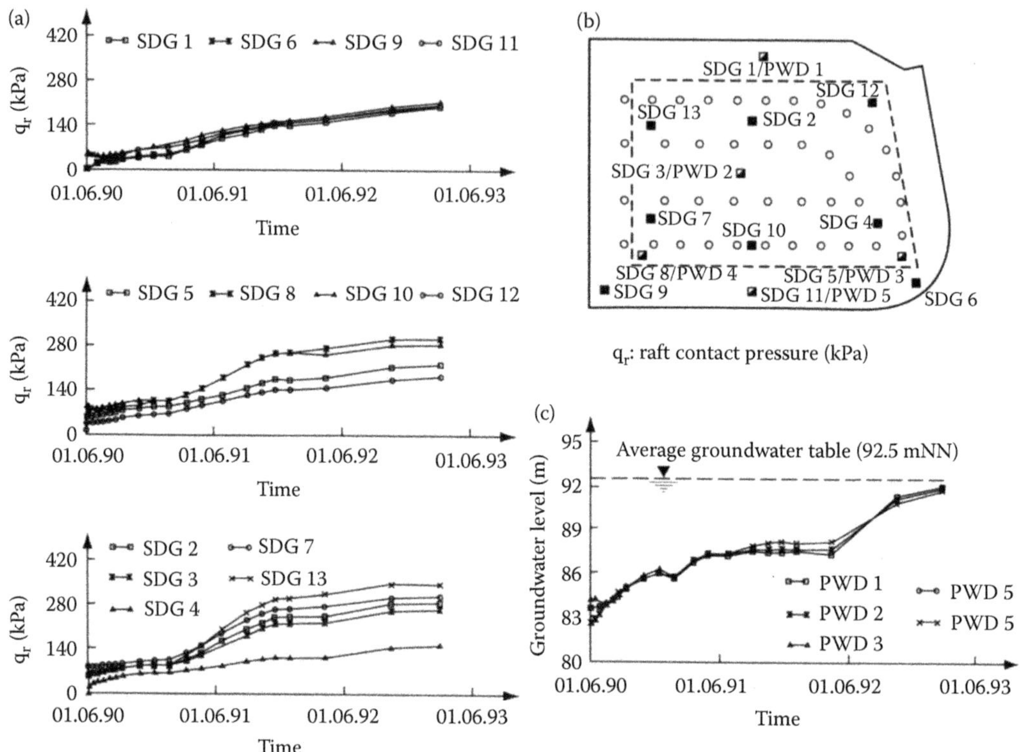


*Figure 14.6* Measured time-dependent foundation pressures: (a) contact pressures; (b) Instrument locations; (c) pore water pressures at raft base. (Adapted from Franke, E., El-Mossallamy, Y. and Wittmann, P. 2000. *Design Applications of Raft Foundations*, Thomas Telford, London. Courtesy of ICE Publishing.)

Figure 14.6 shows the measured contact pressures and their variation with time, while Figure 14.7 shows the evolution of load sharing between piles and raft with time. It is interesting to note that the raft and the piles carry about 50% of the applied load at the end of construction, which is close to the predictions made by Franke et al. (2000).

#### *14.10.1.2.1 Comparison between measured and computed behaviour*

The excellent data obtained from this case history enabled an assessment to be made of the capabilities of various methods of calculation to compute the behaviour of the piled raft foundation. Calculations were carried out by Poulos et al. (1997) to predict the behaviour of the piled raft foundation using the following methods:

- The simplified approach of Poulos and Davis (1980)
- The simplified method of Randolph (1983)
- An analysis based on the idealisation of the raft as a series of strips, and the piles as non-linear springs, implemented via the program, GASP (Poulos, 1991a)
- An analysis based on the idealisation of the raft as a thin plate, and the piles as non-linear springs, implemented via the program, GARP (Poulos, 1994a)
- The finite element method of Ta and Small (1996)
- The finite element method developed by Sinha (1996)
- The analysis reported by Franke et al. (1994)

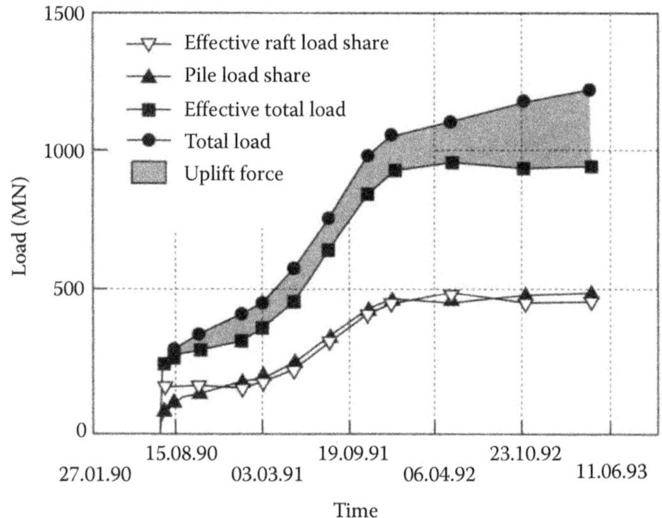


*Figure 14.7* Measured distribution of pile and raft load sharing. (Adapted from Franke, E., El-Mossallamy, Y. and Wittmann, P. 2000. *Design Applications of Raft Foundations*, Thomas Telford, London. Courtesy of ICE Publishing.)

On the basis of pressuremeter tests, an average reloading modulus of 62.4 MPa was reported by Franke et al. (1994), and this value was adopted for the purposes of calculation by Poulos et al. (1997), while a total clay thickness of 120 m was assumed. From the information available, it was estimated that the ultimate axial compressive load capacity of each pile was 16 MN. A total load of 968 MN was assumed to be applied to the foundation, as performance measurements were reported for this load (equivalent to an average applied pressure of about 323 kPa).

Figure 14.8 compares the predictions of performance for the methods mentioned above, together with those reported by Franke et al. (1994). The measured values are also shown. The following points are noted:

- There is a tendency for many of the methods to over-predict the central settlement, especially the simpler hand methods; however, most methods provide an acceptable design prediction when compared with the measured value.
- There is a tendency for most methods to over-predict the proportion of load carried by the piles, however, the extent of this over-prediction is generally acceptable from a design viewpoint.
- All methods which are capable of predicting the individual pile loads suggest that the load capacity of the most heavily loaded piles is almost fully utilised; this is in agreement with the measurements.
- There is considerable variability in the predictions of minimum pile load. The methods which over-predict the amount of load carried by the piles indicate a larger value of the minimum pile load than was actually measured.

The results of the comparisons demonstrated that at relatively low load levels, typical of working loads, when the piles are behaving more-or-less elastically, even the simpler methods are capable of providing a reasonable order-of-magnitude estimate of settlement and

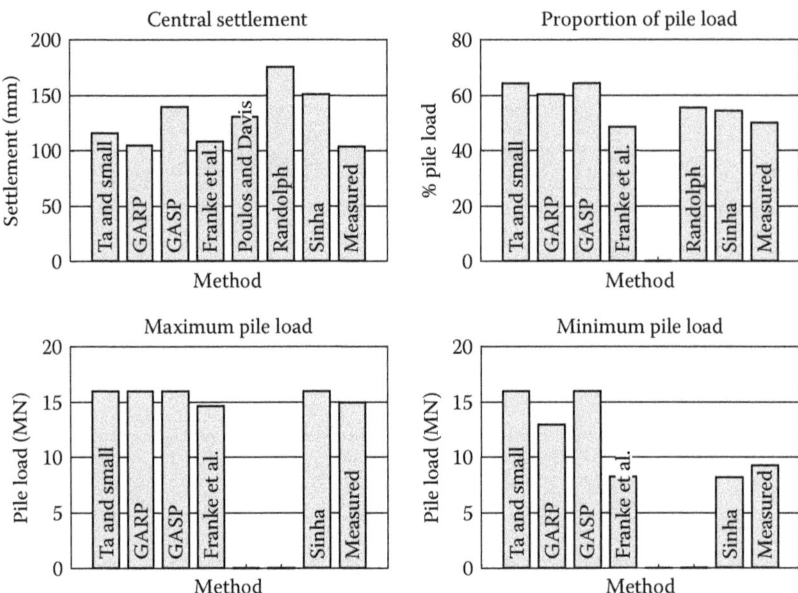


*Figure 14.8* Comparison of methods of analysis for Westendstrasse 1 building. (Adapted from Poulos, H.G. et al., 1997. Comparison of some methods for analysis of piled rafts. *Proceedings of the 14th International Conference on Soil Mechanics and Foundation Engineering*, Hamburg. Balkema, Rotterdam, Vol. 2, pp. 1119–1124.)

load sharing between the piles and the raft. This is despite the fact that a number of the piles had mobilised their full capacity.

### 14.10.2 Building in Recife, Brazil

De Seixas et al. (2006) discuss the monitoring of the settlement of the foundation for a 23-storey concrete building in Recife, Brazil. Few details are given of the ground conditions, other than that they consisted of granular soil into which compacted columns of sand and gravel were installed. They set out the development of the settlement and differential settlement criteria, and then give details of the measurement points, the load versus time and the measured settlements versus time. Figures 14.9 and 14.10 reproduce some of this information. The measured settlements and differential settlements were found to be within the acceptable values.

### 14.10.3 Burj Khalifa, Dubai

Abdelrazaq (2011, 2012) outlined the development of the very comprehensive survey and SHM program for the Burj Khalifa in Dubai. It was intended to track the structural behaviour and responses of the tower during construction and during its lifetime, and included the following:

- Monitoring the reinforced concrete bored piles and their load dissipation into the soil.
- Survey and monitoring of the tower foundation settlement, core walls and column vertical shortening, and the lateral displacements of the tower resulting from its asymmetrical geometric shape and structural system asymmetry.

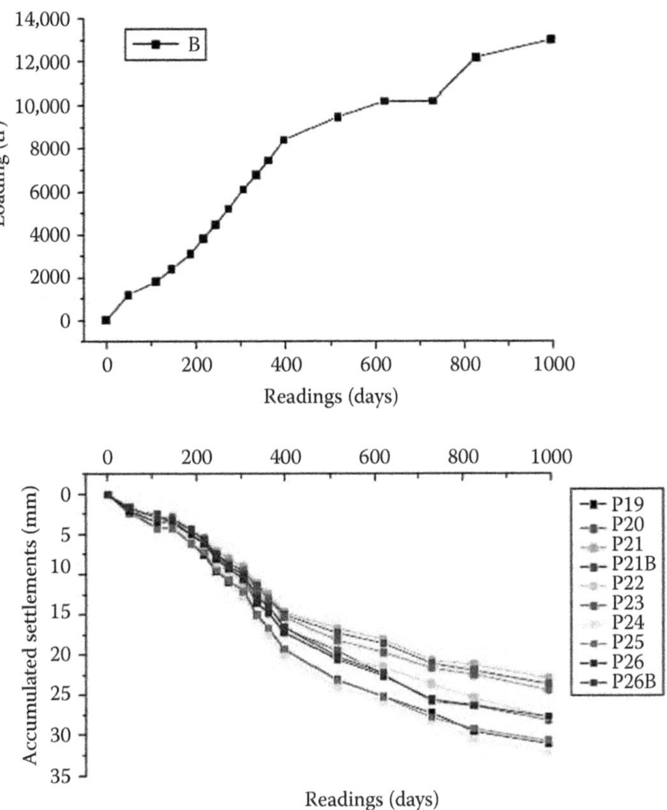


*Figure 14.9* Loading and settlements versus time—tower in Recife, Brazil. (Adapted from De Seixas, A. De Seixas, J.R. and De Seixas, J.J. 2006. Control and monitoring of buildings foundation – Applications in very high buildings structure. *Shaping the Change, XXIII FIG Congress*, Paper T78, Munich, Germany, 8–13 October, 21pp.)

- Monitoring of the tower vertical element strains and stresses due to gravity load effects.
- Installation of a temporary real-time monitoring program to monitor the building displacement and dynamic response under lateral loads (wind and seismic) during construction.
- Installation of a permanent real-time monitoring program to monitor the building displacement and dynamic response under lateral loads (wind and seismic in particular). The intent of this monitoring program was to confirm the actual dynamic characteristics and response of the building, including its natural mode of vibration, estimate of damping, measuring the building displacement and acceleration, immediate diagnosis of the change in building structural behaviour, identify potential of fatigue at structural elements that were considered fatigue sensitive and that could be subjected to severe and sustained wind induced vibration at different wind speeds and profiles, and most importantly, in providing real-time feedback on the performance of the building structure and immediate assistance in their day-to-day operations.
- Providing sufficient data to predict the fatigue behaviour of the pinnacle under low/moderate/severe wind and seismic excitations.
- Tracking the wind speed profile along the building height in an urban, but semi-open, field setting, considering the scale of the project relative to its surroundings.
- Correlating the building measured responses with the predicted behaviour of the tower.

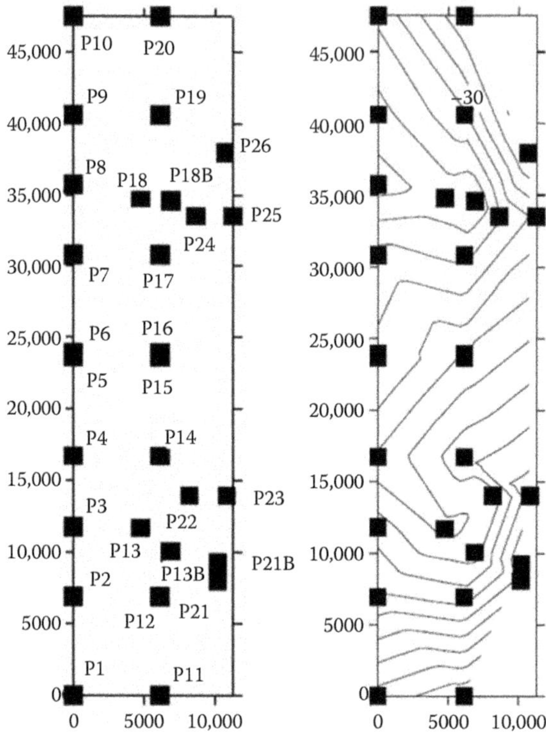


*Figure 14.10* Settlement point locations and contours of settlement (1 mm intervals). (Adapted from De Seixas, A., De Seixas, J.R. and De Seixas, J.J. 2006. Control and monitoring of buildings foundation – Applications in very high buildings structure. *Shaping the Change, XXIII FIG Congress*, Paper T78, Munich, Germany, 8–13 October, 21pp.)

Abdelrazaq (2011, 2012) provided some details of the monitoring results during construction, and stated that the extensive survey and SHM programs had, since their inception, resulted in an extensive feedback and insight into the actual in situ material properties, the tower's structural behaviour and response under wind and seismic excitations and continuous change in the building characteristics during construction. Importantly, the SHM program has provided the building owner ongoing and continuous feedback on the performance of the structure and other building systems in real time to better assist them in their day-to-day operations and facility management.

Further details of the geotechnical aspects of the Burj Khalifa are provided in Chapter 15.

## 14.11 COPING WITH THE CONSEQUENCES OF INADEQUATE FOUNDATION PERFORMANCE DURING AND AFTER CONSTRUCTION

### 14.11.1 Introduction

If monitoring of the foundation behaviour carried out during and after construction identifies irregularities or inadequacies in the performance, it is necessary to make decisions in relation to undertaking appropriate investigation works, and if necessary, remedial works. This should be done as soon as possible, as the later problems are uncovered, the more difficult and costly remedial action is likely to be.

In new foundation construction, there is usually relatively unimpeded access to the site and to the areas in which the new foundation system is to be constructed. However, the environment around or within an existing foundation system which is being investigated and/or upgraded may pose several constraints and problems, and the following characteristics can be anticipated:

1. Access to the area may be very difficult and may limit the range of remedial construction methods that can be employed.
2. The ground will often be highly stressed, and thus changes in the stress regime due to investigation or construction may result in larger ground movements than would be the case in a 'greenfield' situation.
3. The foundation system will generally be carrying loads, and it is therefore necessary to try and assess these existing loads so that a more realistic design can be carried out. One cannot assume that the foundation system is stress-free and load-free.
4. Existing piles will generally be subjected to some measure of restraint from the building which they are supporting, via attachment to the raft or pile caps and the overall foundation system.
5. Strict control of investigation and construction processes are likely to be more critical, but more difficult to achieve, than with 'greenfield' situations.
6. The consequences of uncontrolled ground movements on the existing structure and foundation system are likely to be more immediate and severe than with a 'greenfield' site (see Section 9.5).

For these reasons, it is worthwhile giving some attention to the possible reasons for unsatisfactory pile performance, and the consequent problems of investigation, design and construction within an existing foundation system, and then examining the possible consequences of inadequate control of the resulting ground movements.

### 14.11.2 Causes and consequences of pile defects

Poulos (2005a) has considered some of the causes of inadequate pile behaviour and has attributed them primarily to pile defects that arise from imperfections as a consequence of various sources, including inadequate ground investigation, construction and pile load testing. These sources of imperfection can be categorised as follows:

1. *Natural geological sources*: imperfections which arise from the geological processes at the site, and which have resulted in the present ground profile. They may include (among many others) layers which are not horizontal or continuous, boulders within a soil layer, sloping bedrock, intrusions of rock over limited areas of the site, cavities in limestone rock or the presence of softer layers below what might have been regarded as suitable founding strata for the piles.
2. *Inadequate ground investigation*: these imperfections are generally related to those that arise from natural sources, but are exacerbated because the site is not properly characterised. Inadequacies are usually related to an insufficient number or depth of boreholes or probes to identify stratigraphic variations across the site, or inadequate testing to quantify the relevant geotechnical parameters.
3. *Construction*: such imperfections arise from processes related to construction of the foundations, either from inadequate construction control, or from inevitable consequences of construction activities. They may include
   a. A soft toe on bored piles due to inadequate base cleaning (avoidable)
   b. Defects within the shaft of bored piles (avoidable)

c. Inadequate founding conditions (avoidable)
d. Ground movements developed due to drilling during the construction process (generally unavoidable)
e. Excavation and dewatering effects, especially with remedial piling projects (generally unavoidable, but controllable)

In general, construction-related imperfections in piles can be broadly classified into two main categories, structural defects and geotechnical defects. Structural defects can result in the size, strength and/or stiffness of the pile being less than assumed in design. Such defects have been discussed extensively in the literature, particularly with respect to cast-in situ concrete piles, for example, Hobbs (1957); Thorburn and Thorburn (1977); Reese (1978). Brown (2004) discusses some aspects of construction techniques and materials that can lead to defects or suboptimal pile performance. Examples of structural defects include the following: 'necking' of the shaft of bored piles, leading to a reduced cross-sectional area along part of the pile, poor quality control during the construction of bored piles, leading to some parts of the shaft having lower strength than assumed in design, tensile cracking of large diameter bored piles under the influence of thermal strains, damage during the driving of precast concrete or steel piles, leading to reduced strength and stiffness of parts of the pile, especially near the top or tip of the pile, and bending of slender driven piles.

Geotechnical defects usually arise from either a mis-assessment of the in situ conditions during design, or else from construction-related problems, and may include reduced shaft friction and end bearing resistance arising from localised softer or weaker geotechnical conditions in the vicinity of one or more of the piles in the group, reduced skin friction and end bearing resistance arising from construction operations such as the use of bentonite without due caution, and a 'soft base' arising from inadequate cleaning of the base of bored piles. The latter is one of the most common concerns in bored pile construction, and is likely to lead to a reduction in the stiffness of the soil below the base of the pile. The ultimate base bearing capacity may not be affected significantly, but may require a large movement to be fully mobilised. Several authors have investigated the influence of construction defects on the shaft friction and end bearing resistance of piles, for example, O'Neill and Hassan (1994), O'Neill (2001), Abdrabbo and Abouseeda (2002). O'Neill and Hassan suggest bounds for the effects of construction-related factors and present a framework for quantifying these parameters for design purposes.

On the basis of a series of analyses of piles containing various types of defects, Poulos (2005a) reached the following conclusions:

1. Natural imperfections, arising from the geology of the site, can create significant problems for a piled foundation, and can lead to reductions in both capacity and stiffness of a single pile or pile group. Among the more significant imperfections are
   a. Clay seams below the pile toe
   b. Compressible layers below the founding levels of the piles
   c. Compressible soil layers of uneven thickness
   d. Differences in founding conditions, which can give rise to piles of different length within a group
2. The first two sources have been shown to have the potential to reduce axial pile load capacity and decrease the stiffness. The effects on pile groups are generally more significant than on a single pile. The latter two sources of imperfection are of particular concern, as they generate uneven settlements and can induce unexpected bending moments and shears in the piles under axial loading.

3. Construction-related imperfections include both 'geotechnical' and structural defects. Geotechnical defects, such as a 'soft base' at the pile tip, may cause a reduction in pile head stiffness that depends on the soil modulus at the pile tip and the applied load level. Even a modest softening of the soil at or near the pile base can lead to significant reductions in pile head stiffness. Structural defects may lead to only modest reductions in single pile stiffness, as long as structural failure of the pile does not occur. Structural failure of a pile is more abrupt than for a pile with geotechnical defects.
4. Within a pile group or a piled raft, the ability of the stiffer undamaged piles in a group to carry additional load reduces the potential consequences of imperfections and defective piles, as compared with an isolated pile. However, careful consideration needs to be given to the additional bending moments induced in the piles and the raft by the presence of the defective piles. A critical aspect of the group response is that the presence of defective piles can result in induced lateral deflection and cap rotation of the group, and additional bending moments in the piles. This induced lateral response can become more severe as the location of the defective piles becomes more asymmetric.

It is highly desirable that methods be developed to account for such imperfections so that more realistic designs can be executed, especially if remedial design is necessary. Remedial design poses interesting and important challenges to the foundation designer in that the imperfections inherent in real problems need to be properly taken into account. Some techniques for identifying problems and developing remedial measures for improving foundation performance are discussed below.

### 14.11.3 Investigation issues

Investigations to identify possible causes of foundation performance irregularity may involve some form of drilling adjacent to, or through, the existing pile foundations. Both forms of drilling may have deleterious effects on the piles being investigated. As previously mentioned in Chapter 9, the drilling of holes adjacent to piles will generally cause vertical and lateral ground movements and these will act upon the nearby piles, inducing additional stresses and movements. These effects may be particularly severe if the ground is highly stressed. Coring through the pile itself may also create difficulties for the existing foundation, via the following mechanisms:

- Unbalanced fluid pressures inside the core hole and outside the pile. These may cause loosening or even piping of the soil beneath the pile toe when 'breakthrough' is achieved and the underlying soil is soft or loose. In turn, the settlement of the pile may then be exacerbated.
- SPT testing of the soil below the pile base may cause further disturbance if it is not carried out carefully, and if the SPT rods are withdrawn too quickly, thus causing a suction within the soil surrounding the hole.

Poulos (2005a) points out that the investigation process itself may help to accentuate the problem being investigated. Such a mechanism was thought to be responsible for additional settlements of defective bored piles in two high-rise buildings in Hong Kong. Clearly, it is imperative that such possible 'side effects' of the investigation are anticipated and that appropriate cautionary measures are adopted to minimise the negative impact of these side effects.

### 14.11.4 Design issues

There are at least three key design issues that may need to be addressed when designing remedial works for pile foundations which have been demonstrated to be inadequate or are not performing to expectations:

- Correction of uneven settlements, if the foundation has already undergone excessive tilting or differential settlement, or is likely to do so during or after the remedial works
- Design of remedial or enhancement works, which may include repair of defective piles, the installation of additional piling or extension of the pile cap to obtain additional capacity and stiffness
- Consideration of the load sharing between the existing piles and the additional foundation elements. It is possible that excessive load may be carried by the additional elements, unless the design can incorporate a means of controlling the distribution of loads in the upgraded system.

These issues are discussed in more detail below.

### 14.11.5 Correction of uneven settlement

Methods of correcting uneven settlements of buildings can be divided broadly into two categories:

1. 'Hard' methods, which rely on the application of some form of direct force to the building
2. 'Soft' methods, which rely on processes which produce corrective foundation movements by inducing appropriate ground movements

The treatment can be carried out to lower the building on the 'high' side, or alternatively, to raise the low side of the building. In both cases, the treatment may be accompanied by some form of foundation strengthening or remediation on the 'low' side. Amirsoleymani (1991a,b) lists six different methods that have been used to reduce or eliminate differential settlement and tilting. Three methods involve lifting of the low side of the structure, while three involve lowering of the high side.

#### *14.11.5.1 'Hard' methods*

1. *Application of force by anchor stressing.* This method involves the installation of a series of strategically located anchors within the foundation system. The anchors are grouted into a suitable hard stratum at depth below the building. The anchors are then stressed (typically to 60%–75% of their ultimate capacity) to obtain a corrective tilt to the foundation. In some cases, repeated stressing and de-stressing may have a beneficial effect in developing additional corrective tilting.
2. *Application of additional loading.* This method involves the application of additional loading on the high side of the foundation by water or other means. There are often limitations on the amount of loading that can be applied because of height limitations within the structure, and the limitations that may be imposed by the limited strength of the structure itself.
3. *Cutting of piles.* This method involves the cutting of some piles supporting the high side of the building, to promote load transfer to other piles on the high side and hence

promote beneficial settlement. When the process has been completed and the settlements have ceased, the piles may be re-attached so that they may carry part of the future loadings. The cutting of the piles is in itself a process that requires considerable care so that the cut pile is not destroyed in the process. A method that has been developed in Hong Kong involves the partial cutting of both sides of the pile, the placement of jacks to support the upper and lower parts of the pile and carry the load existing in the pile, and then the cutting away of the remaining central portion of the pile which is freed of load prior to the cutting by the jacks. This process has been used in conjunction with anchor stressing in correcting a building in Hong Kong. Unfortunately, because of legal constraints, it is not possible to present details of this case.

4. *Jetting of the soil beneath pile tips.* There have been anecdotal reports of the use in China of high pressure water jetting applied below the tips of piles to reduce their stiffness and capacity and promote settlement of the high side of the building. This process is in some ways similar to the cutting of piles, but is a less controlled procedure whose effects appear to be difficult to predict.
5. *Jacking of the foundations on the 'low' side.* This method generally employs compaction grouting to push the low side up and at the same time strengthen the foundation. An example of this approach is described by Tsai et al. (1991) and was used to correct a building on a raft foundation that had tilted by about 0.74° after being subjected to hurricane loading together with seismic forces. A process of staged grouting allied with careful monitoring restored the building to a plumb position. Amirsoleymani (1991a) describes the use of mechanical jacking to correct the settlement of a storage tank, and the use of chemical grouting to restore a piled foundation that tilted and failed after a deep excavation nearby caused piping of sand near the pile tips. An expansive admixture was used in the grout to promote uplift of the columns, which were raised by 28 mm. Maffei et al. (2001) describe a case of a tall building in Sao Paulo in which the uneven settlement causing a tilt of 2.2° was corrected by constructing a new pile foundation and jacking the low side up against these piles to transfer load from the old to the new foundation system.
6. *Fracture grouting.* This method involves the use of a grout under controlled high pressure to fracture the soil and cause uplift of the foundation. Amirsoleymani (1991a) has described the case of a five-storey warehouse in which 210 mm of differential settlement was corrected by hydraulic fracturing through 24 tubes installed into rock below the foundation. Cement lenses 50–100 mm in thickness were found to have been formed by the fracture grouting.

#### 14.11.5.2 'Soft' methods

1. *Soil extraction.* In this method, soil is excavated from beneath or between the piles on the high side. This process causes the ground to settle and thus induces a settlement of the pile foundation also. Soil extraction was used to arrest the tilt of the famous Pisa Tower in Italy (which was supported by a shallow foundation) (Jamiolkowski, 2001; Burland, 2004). Amirsoleymani (1991b) has described the use of a similar process to correct the tilt of a grain storage silo. In that case, thin layers of soil were extracted via specially constructed chains which cut through the soil. Brandl (1989) has described the use of soil extraction to correct uneven settlement of piles supporting bridge piers, while the use of soil extraction to correct uneven foundation settlements has been described by Tamez et al. (1997). In this case, involving two historic churches in Mexico City, soil extraction was carried out via 32 shafts 3 m in diameter, from tubes 100 mm in diameter, inserted a maximum distance of 22 m into the soil. After

4 years of treatment, more than 2600 $m^3$ of soil had been removed, and corrective settlements of about 800 mm were achieved. It was estimated by the authors that about 65% of the settlement could be attributed to the soil extraction, while the remaining 35% was due to consolidation arising from pumping from the deep wells at the bottom of the shafts. Poulos (2003) has developed a method of analysis for pile foundations subjected to ground movements from soil extraction, in which the process of drilling sub-horizontal holes is represented by the formation of small diameter tunnels. The resulting ground movements can be used in a pile–soil interaction analysis to assess the effect of the under-excavation on the pile group response.

2. *Dewatering*. In principle, lowering of the water table can be used to promote settlement of the high side and thus correct uneven settlement. Amirsoleymani (1991a) describes the use of this approach for apartment building supported by raft foundations. However, this method is fraught with difficulty as the effects of ground water lowering are highly dependent on the local hydrogeology, are time-dependent and may extend considerably beyond the building being treated. Amirsoleymani (1991a: 354) summarises this method succinctly as follows: 'lowering water table to eliminate differential settlement is one of the most unreliable methods'. Despite these misgivings, Liu (2004) has described a successful application of dewatering in conjunction with grouting to reduce the tilt of a building adjacent to a deep excavation.
3. *Compensation grouting*. van der Stoel et al. (2003) have described the use of compensation grouting via fracturing to control the vertical movement of timber pile foundations affected by tunnelling operations. The system involved the use of 22 sub-horizontal tubes–a manchette (TAM) in two levels. About 5 mm of heave was observed on a structure after post grouting was performed following the tunnelling.
4. *Removal of soil support (RSS)*. Poulos et al. (2003) have described an approach which involves the drilling of a series of vertical or sub-vertical holes just outside the high side of the structure. The removal of soil reduces the lateral support of the ground and therefore promotes settlement of the high side of the structure. The greater the stress imposed in the ground by the building, the greater will be the settlement. Excavation of a continuous trench can, in principle, cause significant settlement to occur. The advantage of this approach is that it can be applied without intruding into the building footprint. While no cases of the use of this method for pile foundations appear to have been reported in the literature, it does appear to have been used successfully for buildings on shallow foundations (Zou, 1996). Settlements of the order of 150 mm appear to have been developed using this approach. Preliminary centrifuge tests on a model shallow footing (Ng, C.W.W. 2002, personal communication) have confirmed the potential of this approach.

### 14.11.6 Foundation enhancement works

Among the options that may be considered for foundation enhancement works are the following:

1. Repair of the existing piles which contain imperfections or defects
2. The addition of new piles to strengthen and/or stiffen the existing foundation
3. Extension of existing pile caps or rafts, to provide additional bearing capacity and stiffness

In each case, attention needs to be given to the transfer of load from the existing to the enhanced foundation system.

*Repair of existing piles:* Methods of repair of existing piles are relatively limited, because access to the pile is frequently limited, the nature of the defect may be uncertain and the repair process may itself result in deleterious side effects, such as additional ground movements, which may adversely affect the remainder of the foundation system. Structural defects, especially those found near the head of a pile, have been repaired with epoxy jackets. Geotechnical defects, especially soft bases as a result of inadequate base cleaning, have been treated successfully by base grouting techniques. Examples have been described by Teparaksa et al. (1999) and Moh (1994). Moh (1994) reports a dramatic improvement in load-settlement performance arising from base grouting a bored pile 1.5 m in diameter and 22 m long. At the working load of 6.6 MN, the settlement was reduced from 80 mm to less than 10 mm.

*Addition of new piles:* The addition of piles to an existing foundation system is a remedial measure of long standing, especially in relation to underpinning operations, and has been carried out using a variety of systems for the new piles. Because of constraints with access in most cases, the methods of installation must allow for limited headroom; as a consequence, most additional piles tend to be relatively small in diameter, although their length can be significant (e.g. Bruce, 1994). The use of jet grouted piles has been described by Popa et al. (2001), while the use of compaction grouted piles to remediate an extensive warehouse raft foundation has been described by Hayward Baker (2003).

In analysing the effectiveness of a foundation with additional remedial piles, it is necessary to take account of several factors, including the capacity and stiffness of the existing foundation system, the time at which the additional piling is installed, and the interaction between the original system and the new components, including the subsequent load sharing. Makarchian and Poulos (1996) have developed an approximate method of analysis of a remediated foundation system which uses concepts of pile–raft interaction. This method has been applied to model pile tests and reasonable agreement has been found between the measured and computed ratio of settlement reduction arising from the installation of piles beneath a shallow footing.

An issue of concern with the installation of additional piles is the disturbance of the ground caused by the installation process. Installation of displacement piles will cause both vertical and lateral ground movements which will interact with the existing foundations system. Such ground movements have the potential to cause additional vertical and lateral forces and bending moments in existing piles, which may compromise their integrity. The effects of such ground movements may be particularly severe for the case where the existing piles are restrained from moving laterally or vertically, as mentioned previously in Chapter 9. For piles that are installed by drilling, there may also be potentially damaging ground movements as a consequence of the release of lateral ground stress and changes in water pressure arising from lack of control of water levels during drilling. The effects may be of particular concern if the existing foundation is heavily loaded, as the release of stress due to drilling can then be large. These side effects of foundation remediation are thought to have been significant in a case in Hong Kong, where an existing building adjacent to a building being remediated experienced some additional settlements during the installation of over 50 remedial piles.

*Extension of existing pile caps:* A remedial option which can be useful is to extend an existing pile cap and make use of its capacity and stiffness. Clearly, this option may be limited to those cases in which the near-surface soils themselves have reasonable strength and stiffness. The performance of such a remediated system can again be assessed readily using pile–raft analysis concepts (Randolph, 1994; Poulos, 2001b).

Poulos (2005a) describes a project in Queensland Australia involving jacked piles in sand which had not developed their anticipated ultimate capacity of 4000 kN (this being the jacking force employed for installation). The conventional solution was to add an additional two piles to the group (one on either side of the original pile, to avoid asymmetric loading), thus necessitating an extension of the pile cap. Another solution investigated was to simply extend the pile cap, and make use of its capacity and stiffness to complement the existing pile. Calculations were carried out for both these options.

It was found that the three-pile option was an overdesign, while the pile plus enlarged cap appeared to perform satisfactorily. It was also noted that for the three-pile option, the pile cap would have had to be extended to accommodate the two extra piles, and its area would have been a considerable proportion of the cap area in the extended cap option. Clearly, the latter had some economic, as well as logistical, advantages in this case, in that the additional piles would have been very difficult to install because part of the structure had already been constructed.

### 14.11.7 Control of load distribution in the foundation system

An issue of concern is the transfer or sharing of load between the original and the remediated foundation system. It should be expected that, following the laws of mechanics, a greater proportion of any additional load that is imposed on a system would be carried by the stiffer components. This may in turn mean that the new and stiffer piles may become over-stressed or overloaded. It is possible to control the load distribution within a pile foundation system by means of inserts that are attached to the head of selected piles. Such inserts, which may be of neoprene or a similar semi-compressible material, serve to decrease the pile head stiffness in a controlled manner, and are discussed in Chapter 3.

Zhou et al. (2016) outline the use of a similar system which they term 'deformation adjustors' for piled raft systems. They describe a case history of a piled raft on a ground profile of sandy clay overlying completely to slightly weathered granite, and in which 'deformation adjusters' were applied on the end bearing piles to modify the distribution of load between the raft and the piles.

# Chapter 15

# Case histories

## 15.1 INTRODUCTION

This chapter will describe a number of case histories of tall building design and performance, with the objective of illustrating the application of the procedures set out in this book and also providing, in some of the cases, details of the performance of tall building foundations. The cases described are ones in which the author has been involved, either directly or indirectly, and in each case, an effort will be made to summarise some of the lessons learned.

## 15.2 TYPICAL HIGH-RISE FOUNDATION SETTLEMENTS

Prior to considering specific case histories, it is useful to review more broadly the settlement performance of some high-rise buildings in order to gain some appreciation of the order of settlements that might be expected from three foundation types founded on various deposits. Table 15.1 summarises details of the foundation settlements of some tall structures founded on raft or piled raft foundations, based on documented case histories in Hemsley (2000), Katzenbach et al. (1998), and from the author's own experiences. The average foundation width in these cases ranges from about 40 to 100 m. The results are presented in terms of the settlement per unit applied pressure, and it can be seen that this value decreases as the stiffness of the founding material increases. Typically, with the exception of the soft clay case, these foundations have settled between 25 and 300 mm/MPa.

Some of the buildings supported by piled rafts in stiff Frankfurt Clay have settled more than 100 mm, and despite this apparently excessive settlement, the performance of the structures appears to be quite satisfactory. It may therefore be concluded that the tolerable settlement for tall structures can be well in excess of the conventional design values of 50–65 mm. A more critical issue for such structures may be overall tilt, and differential settlement between the high-rise and low-rise portions of a project. A discussion of design criteria is given in Section 8.2.

## 15.3 CASE 1: LA AZTECA BUILDING MEXICO

The case of the La Azteca building was described by Zeevaert (1957). Figure 15.1 shows the original building. The building exerted a total average loading of about 118 kPa, and was located on a deep highly compressible clay deposit which was also subjected to ground surface subsidence arising from groundwater extraction. The building was founded on a compensated piled raft foundation, consisting of an excavation 6 m deep with a raft supported by 83 concrete piles, 400 mm in diameter, driven to a depth of 24 m (i.e. the piles were about 18 m long below the raft).

*Table 15.1* Examples of settlement of tall structure foundations

| *Foundation type* | *Founding condition* | *Location* | *No. of cases* | *Settlement per unit pressure (mm/MPa)* |
|---|---|---|---|---|
| Raft | Stiff clay | Houston, Amman, | 2 | 227–308 |
| | Limestone | Riyadh | 2 | 25–44 |
| Piled raft | Stiff clay | Frankfurt | 5 | 218–258 |
| Compensated piled raft | Dense sand | Berlin, Niigata | 2 | 83–130 |
| | Weak rock | Dubai | 5 | 32–66 |
| | Limestone | Frankfurt | 1 | 38 |
| | Soft clay | Mexico City | 1 | 1750+ |

The challenges for the original designers in this case were to design the foundation for a relatively tall building founded on a very deep deposit of soft clay, in a pre-computer era.

Figure 15.2 shows, reproduced from Zeevaert's paper, details of the foundation, the soil profile, the settlement computed by Zeevaert, and the measured settlements. The settlement without piles computed by Zeevaert (from a one-dimensional analysis) was substantial, but the addition of the piles was predicted to reduce the settlement to less than half of the value without piles. The measured settlements were about 20% less than the calculated settlements, but nevertheless confirmed the predictions reasonably well.

An approximate analysis developed by the author (Poulos, 2005c) was applied to this case, excluding the effects of ground settlements, which were not detailed by Zeevaert in his paper. The following approach was adopted:

1. The one-dimensional compressibility data presented by Zeevaert was used to obtain values of Young's modulus of the soil at various depths, for the case of the soft clays in a normally consolidated state. A drained Poisson's ratio of 0.4 was assumed. The

*Figure 15.1* La Azteca building. (Courtesy of Dr. G. Auvinet.)

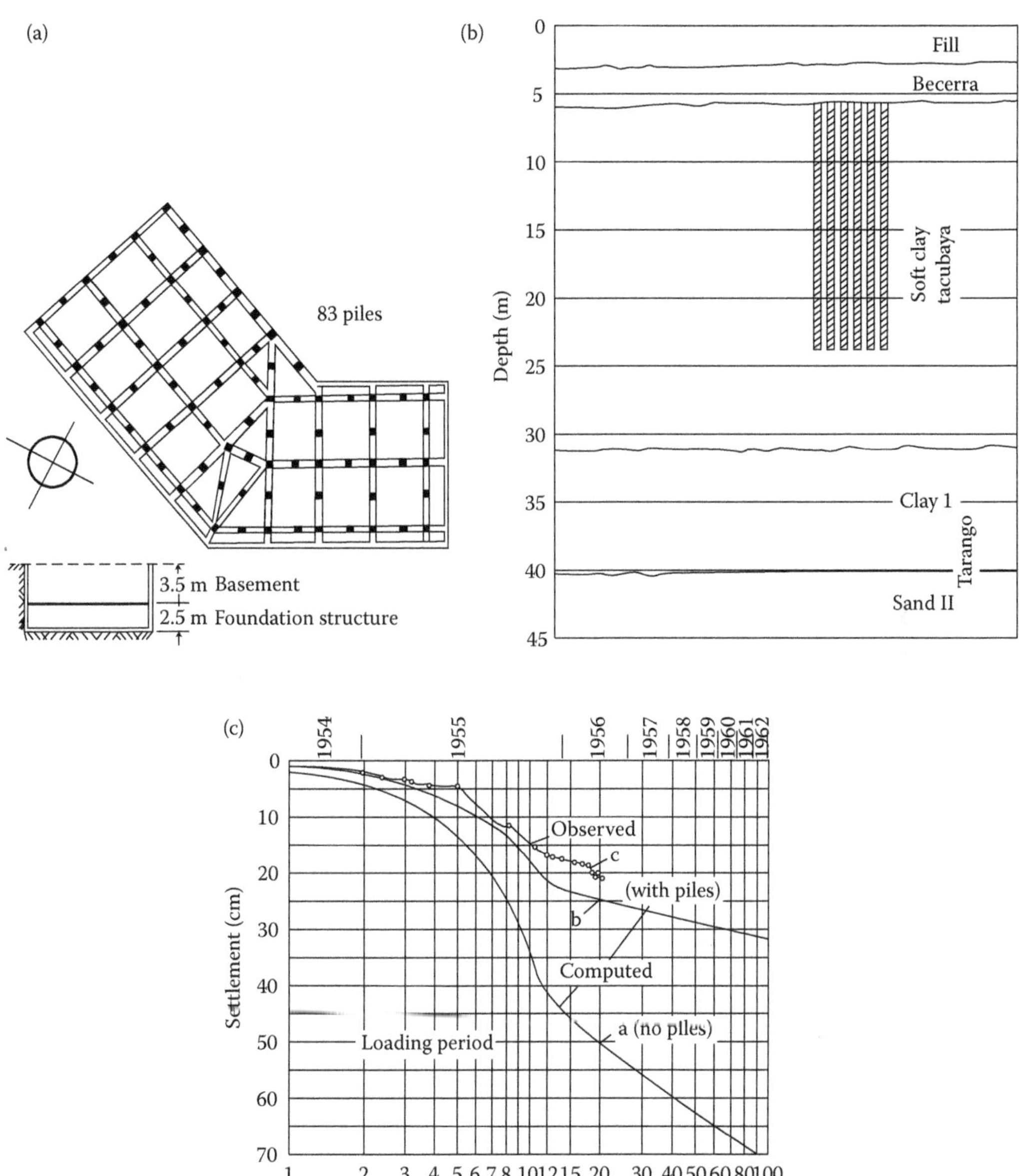


*Figure 15.2* Details of La Azteca building on compensated piled raft: (a) foundation plan and pile layout, (b) subsoil profile and (c) settlement curves. (Adapted from Zeevaert, L. 1957. Compensated friction pile foundation to reduce settlement of buildings on the highly compressible volcanic clay of Mexico City. *Proceedings of the 4th International Conference on SMFE*, London, Vol. 2, pp. 81–86.)

modulus values thus obtained were typically very low, of the order of 0.5–1.0 MPa, and lower than would have been anticipated on the basis of the measured shear strength of the clay.

2. The bearing capacity of the raft was estimated from the shear strength data provided by Zeevaert, and was found to be about 200 kPa. This represented a factor of safety of about 1.7 on the average applied loading of 118 kPa.

3. The settlement of an uncompensated raft was computed using these modulus values together with conventional elastic theory. A very large settlement, in excess of 2.3 m, was calculated for the final settlement.
4. The settlement of the compensated raft was computed, assuming a 6 m depth of excavation, and assuming that the soil modulus values for the over-consolidated state were 10 times those for the normally consolidated state (based on the oedometer data presented by Zeevaert). The additional raft pressure to recommence virgin loading conditions, $p_{ec}$, was taken to be zero. A settlement of the order of 988 mm was thus computed.
5. From the pile load tests reported by Zeevaert, values of the single pile capacity and stiffness were obtained, these being about 735 kN and 25 MN/m, respectively.
6. For the 83 piles used in the foundation, the group stiffness was computed by using the approximation of Poulos (1989) and applying a group factor of 9.1 (the square root of the number of piles i.e. $83^{0.5}$) to the single pile stiffness. A group stiffness of about 230 MN/m was calculated.
7. The average settlement of the foundation for an uncompensated piled raft was computed, using the equations developed by Randolph (1994) for the piled raft stiffness. A settlement of about 1.08 m was obtained. The analysis indicated that, in this case, the raft would carry only about 4% of the load under elastic conditions, and that the capacity of the piles would be mobilised fully under the design load of about 78 MN.
8. The effects of carrying out a 6 m deep excavation (as was actually used) was simulated by reducing the thickness of the soil profile accordingly, and again assuming that, for the raft, the soil Young's modulus for the over-consolidated state was 10 times that for the normally consolidated state. The stiffness of the raft was thus increased significantly, leading also to a significant increase in the stiffness of the piled raft foundation, to about 300 MN/m. The raft, at the design load, was found to carry about 40% of the total load, and the computed settlement under that load was reduced to about 280 mm.

The analysis results are summarised in Table 15.2. It can be seen that the settlement of the compensated piled raft is about 26% of the settlement of the piled raft without compensation, 29% of the settlement of the compensated raft alone, and only about 12% of the value for the uncompensated raft.

Zeevaert's calculations gave larger settlements than those computed above, being about 1000 mm for the compensated raft alone, and about 370 mm for the compensated piled raft. This represented a reduction in settlement of about 63% in using the compensated piled raft rather than the compensated raft alone. This compares reasonably well to the 71% reduction in settlement computed from the present approach. It is also interesting to note that the measured settlements about 2 years after the commencement of construction were about 20% less than those predicted by Zeevaert. At that stage, the measured settlement

*Table 15.2* Summary of computed average settlements

| *Case* | *Computed average final settlement (mm)* | *Ratio of settlement to settlement of compensated raft* |
|---|---|---|
| Raft alone, no compensation | 2342 | 2.37 |
| Raft alone, with compensation | 988 | 1.0 |
| Piled raft, no compensation | 1084 | 1.10 |
| Piled raft, with compensation | 283 | 0.29 |

was about 205 mm and the computed settlement from Zeevaert was 250 mm, that is, about 68% of the final predicted settlement. Assuming a similar rate of settlement, the prediction made by the current approach for the settlement after 2 years would be about 192 mm, in fair agreement with, but somewhat less than, the measured 205 mm.

### 15.3.1 Lessons learned

The La Azteca case has demonstrated that tall buildings can be constructed on deep soft clay deposits by making use of compensated piled rafts. It has also shown that reasonable estimates of the settlement of such foundations can be made without necessarily resorting to complex numerical analyses. Characterising the ground conditions accurately is more critical than the method of analysis employed to carry out the settlement calculations, provided that it is reasonably sound and properly reflects the mechanisms of behaviour.

## 15.4 CASE 2: EMIRATES TWIN TOWERS, DUBAI

### 15.4.1 Introduction

The Emirates project is a twin tower development in Dubai, one of the United Arab Emirates. The towers are triangular in plan with a face dimension of approximately 50–54 m. The taller Office Tower has 52 floors and rises 355 m above ground level, while the shorter Hotel Tower is 305 m tall (see Poulos and Davids, 2005; Poulos, 2009). The completed towers are shown in Figure 15.3.

*Figure 15.3* Emirates Towers soon after completion.

### 15.4.2 Ground characterisation

A comprehensive series of in situ tests was carried out, and in addition to standard penetration tests (SPTs) and permeability tests, pressuremeter tests, vertical seismic shear wave testing, and site uniformity borehole seismic testing was carried out.

Conventional laboratory testing was undertaken, consisting of conventional testing, including classification tests, chemical tests, unconfined compressive strength (UCS) tests, point load index tests, drained direct shear tests and oedometer consolidation tests. In addition, a considerable amount of more advanced laboratory testing was undertaken, including stress path triaxial tests for settlement analysis of the deeper layers, CNS direct shear tests for pile skin friction under both static and cyclic loading, resonant column testing for small-strain shear modulus and damping of the foundation materials, and undrained static and cyclic triaxial shear tests to assess the possible influence of cyclic loading on strength, and to investigate the variation of soil stiffness and damping with axial strain.

The geotechnical model for foundation design under static loading conditions was based on the relevant available in situ and laboratory test data, and is shown in Figure 15.4. The ultimate skin friction values were based largely on the CNS data, while the ultimate end bearing values for the piles were assessed on the basis of correlations with UCS data (Reese and O'Neill, 1988) and also previous experience with similar cemented carbonate deposits (Poulos, 1988a).

### 15.4.3 Foundation design

The number, depth, diameter and locations of the foundation piles were altered several times during the design process. There was close interaction between the geotechnical and structural designers in executing an iterative process of computing structural loads and consequent foundation response. In the final design, the piles were 1.2 m diameter, and extended 40 or 45 m below the base of the raft. In general, the piles were located directly below 4.5 m deep walls which spanned between the raft and the Level 1 floor slab. These walls acted as 'webs' which forced the raft and Level 1 slab to act as the flanges of a deep box structure. This deep box structure created a relatively stiff base to the tower superstructure, although the raft itself was only 1.5 m thick. The Office Tower foundation contained 102 piles while the shorter Hotel Tower had 92 piles.

### 15.4.4 Pile load testing

Using the geotechnical data shown in Figure 15.4, predictions were made for pile load tests, and once the tests were complete, the predicted and measured pile responses were compared. Compression tests were performed using anchor cables as the reaction system. Other tests were also performed, including tension tests, lateral load tests and cyclic load tests.

Four main types of instrumentation were used in the test piles:

- Strain gauges – (concrete embedment vibrating wire type) to allow measurement of strains along the pile shafts, and hence estimation of the axial load distribution.
- Rod extensometers – to provide additional information on axial load distribution with depth.
- Inclinometers – the piles for the lateral load tests had a pair of inclinometers, at 180°, to enable measurement of rotation with depth, and hence assessment of lateral displacement with depth.
- Displacement transducers – to measure vertical and lateral displacements.

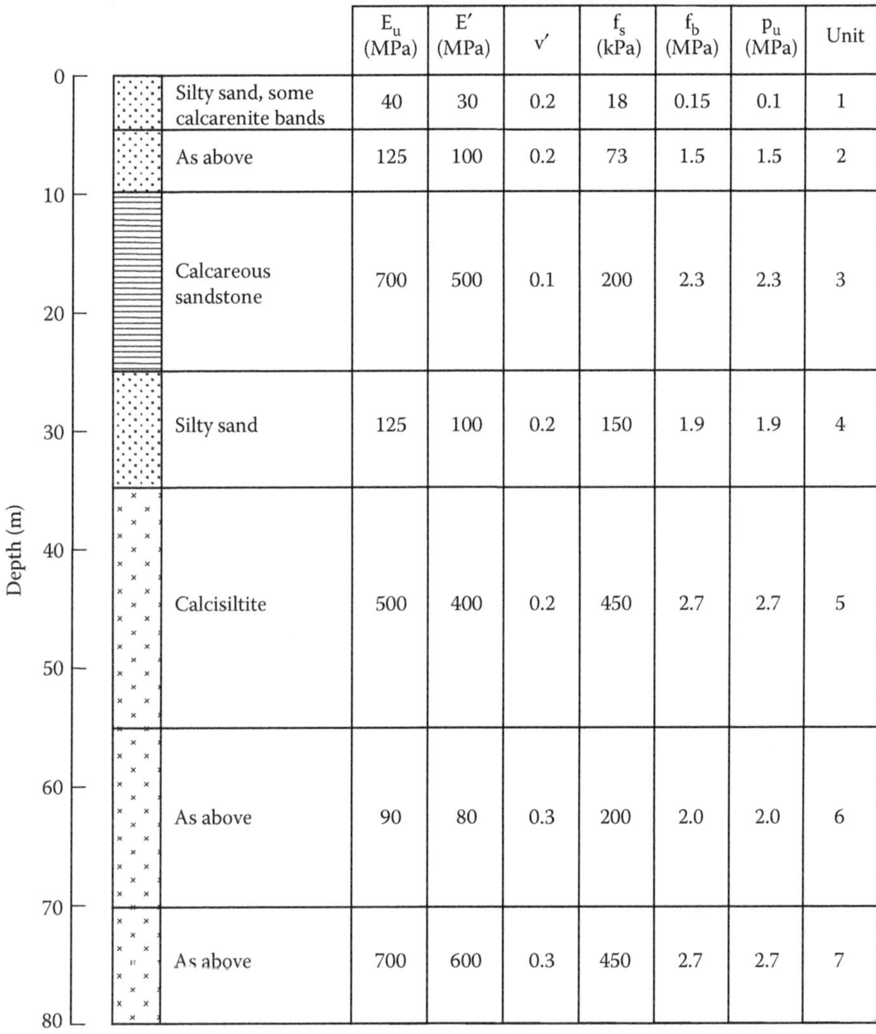

| | $E_u$ (MPa) | E′ (MPa) | v′ | $f_s$ (kPa) | $f_b$ (MPa) | $p_u$ (MPa) | Unit |
|---|---|---|---|---|---|---|---|
| Silty sand, some calcarenite bands | 40 | 30 | 0.2 | 18 | 0.15 | 0.1 | 1 |
| As above | 125 | 100 | 0.2 | 73 | 1.5 | 1.5 | 2 |
| Calcareous sandstone | 700 | 500 | 0.1 | 200 | 2.3 | 2.3 | 3 |
| Silty sand | 125 | 100 | 0.2 | 150 | 1.9 | 1.9 | 4 |
| Calcisiltite | 500 | 400 | 0.2 | 450 | 2.7 | 2.7 | 5 |
| As above | 90 | 80 | 0.3 | 200 | 2.0 | 2.0 | 6 |
| As above | 700 | 600 | 0.3 | 450 | 2.7 | 2.7 | 7 |



*Figure 15.4* Geotechnical model adopted for design. (Adapted from Poulos, H.G. and Davids, A.J. 2005. *Canadian Geotechnical Journal*, 42: 716–730.)

Comparisons for one of the compression tests that were performed are shown in Figure 15.5 for the load–deflection behaviour and in Figure 15.6 for the distribution of pile axial load with depth. The predictions for this pile and for the other pile tests that were performed were considered to be reasonable.

### 15.4.5 Settlement predictions for towers

Predictions were made for the settlement of the tower foundations. These predictions were made using various computer programs, and these were then compared with the measured vertical settlement of the towers during construction. A comparison of the predicted settlement and the measured settlement for the Hotel Tower is shown in Figure 15.7.

As can be seen from Figure 15.7, the predicted and measured settlements of the Hotel were very far from agreement, even though the predictions for the load tests on single piles

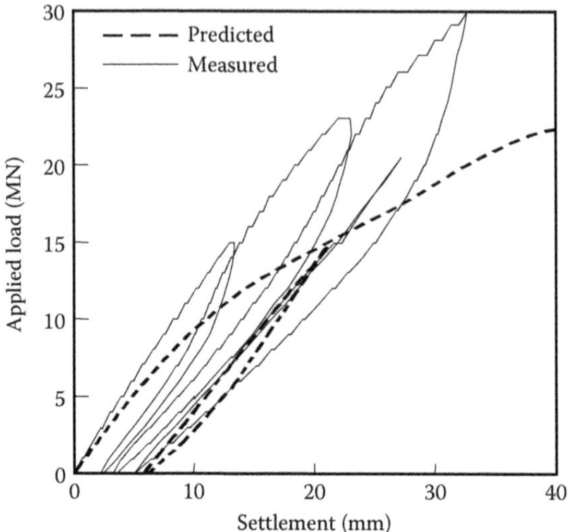


*Figure 15.5* Predicted and measured load–settlement behaviour for pile P3 (Hotel). (Adapted from Poulos, H.G. and Davids, A.J. 2005. *Canadian Geotechnical Journal*, 42: 716–730.)

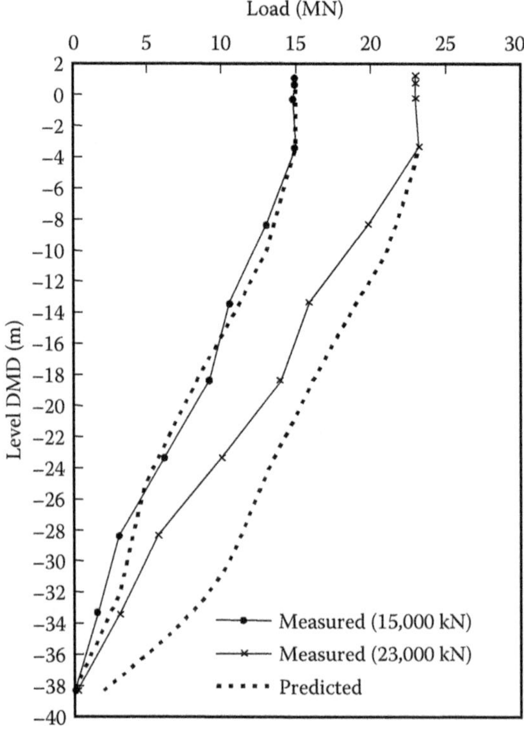


*Figure 15.6* Predicted and measured axial load distribution for pile P3 (Hotel). (Adapted from Poulos, H.G. and Davids, A.J. 2005. *Canadian Geotechnical Journal*, 42: 716–730.)

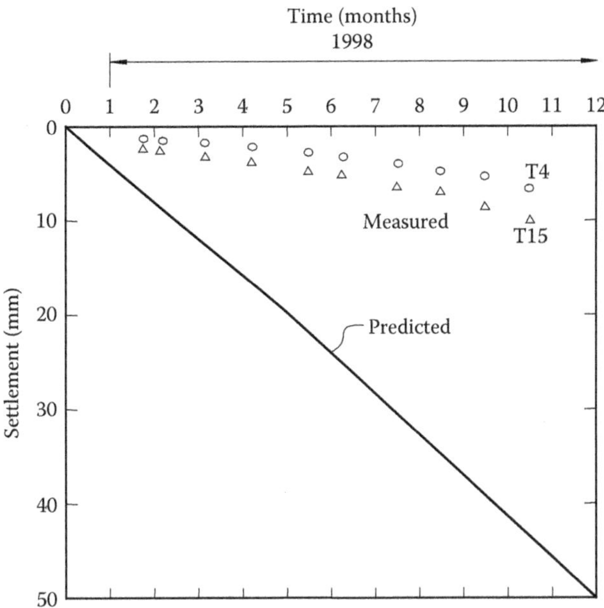


*Figure 15.7* Measured and predicted time–settlement behaviour for Hotel Tower. (Adapted from Poulos, H.G. and Davids, A.J. 2005. *Canadian Geotechnical Journal*, 42: 716–730.)

were similar to those measured. Two prime reasons for the larger prediction of settlement for the pile group were thought to be (i) the interaction effects of the piles were overestimated, and (ii) the stiffness adopted for the ground below RL-53 m, and the assumption that it remained constant with depth, was too conservative. The calculation of interaction among the piles can have a large influence on group settlements when there are a large number of piles, and so the interaction factors were re-assessed. Pile groups stress the ground to greater depths than single piles, and if the soil modulus at depth is different than assumed, this can lead to inaccuracies. Lower strain levels in the ground at depth mean that a higher modulus should be used, since the stiffness of the ground is strain dependent. These two reasons provided a credible explanation as to why the single pile predictions were reasonable, but the pile group predictions were not. When more careful account was taken of the effects on pile–pile interaction of the increasingly stiffer ground with depth below the pile toes, the computed final settlement of the Hotel Tower reduced from 138 to 43 mm, a more realistic assessment. Further details are provided by Poulos and Davids (2005).

## 15.4.6 Lessons learned

This example demonstrates the value of monitoring programs, especially when designing in new or unfamiliar ground conditions. The monitoring programme indicated that the modelling procedure for the piled raft had shortcomings, and that caution must be exercised in applying interaction factor methods to the prediction of the settlement of large pile groups. The experience gained from the Emirates project was used to improve the procedure for making settlement predictions for the Burj Khalifa, which was founded in similar materials (see Section 15.5).

## 15.5 CASE 3: THE BURJ KHALIFA, DUBAI

### 15.5.1 Introduction

The Burj Khalifa project in Dubai comprised the construction of a 160 storey high-rise tower, with a podium development around the base of the tower, including a 4–6 storey garage. The client for the project was Emaar, a leading developer based in Dubai. The Burj Khalifa Tower (originally denoted as the Burj Dubai prior to its completion and opening) is, as at 2016, the world's tallest building at 828 m. It is founded on a 3.7 m thick raft supported on bored piles, 1.5 m in diameter, extending approximately 45 m below the base of the raft. Figure 15.8 shows the completed tower. The ground surface is generally horizontal and the site levels are related to Dubai Municipality Datum (DMD).

The key challenges in this case were to undertake an economical foundation design for the world's tallest building, where the founding conditions were relatively weak rock and where significant wind loadings were to be resisted. A detailed description of the geotechnical aspects of this case is given by Poulos and Bunce (2008).

The architects and structural engineers for the project were Skidmore Owings and Merrill LLP (SOM) in Chicago. Hyder Consulting (UK) Ltd. (HCL) was appointed geotechnical consultant for the works and carried out the design of the foundation system, while an independent peer review was undertaken by Coffey Geosciences (Coffey). The process of foundation design and verification process is described below, together with the results of the pile load testing programs. The predicted settlements are then compared with those measured during construction.

The building was 'Y' shaped in plan, to reduce the wind forces on the tower and to keep the structure relatively simple and aid constructability. Baker et al. (2008) describe the

*Figure 15.8* The Burj Khalifa.

structural system as a 'buttressed core'. Each wing had its own high-performance concrete corridor walls and perimeter columns, and buttressed the others via a six-sided central core or hexagonal hub. As a consequence, the tower was very stiff laterally and torsionally. The structural aspects are described by Baker et al. (2008), while Smith (2008) provides an architectural perspective of the building. The structural design involved a 3D model consisting of the reinforced concrete walls, link beams, slabs, raft and piles, together with the steel structural steel system. Gravity, wind and seismic loadings were considered. According to Baker et al. (2008), under lateral wind loading, the building deflections were assessed to be well below commonly used criteria. Dynamic analyses indicated a period of 11.3 s for the first lateral mode of vibration, a period of 10.2 s for the second mode, with the fifth mode (torsional motion) having a period of 4.3 s.

The construction of the Burj Khalifa utilised advancements in construction techniques and material technology, using 80 and 60 MPa concrete with flyash, the higher strength being used for the lower portion of the structure. The walls were formed using an automatic self-climbing formwork system, and the circular nose columns were formed with steel forms, while the floor slabs were poured on to special formwork. The wall reinforcement was fabricated on the ground in 8 m sections to allow for rapid placement. The central core and slabs were cast first, in three sections: the wing walls and slabs then followed, and after them, the wing nose and slabs followed. Concrete was pumped by specially designed pumps, capable of pumping to heights of 600 m in a single stage. A special GPS system was developed to monitor the verticality of the structure during construction.

### 15.5.2 Geotechnical investigation and testing program

The geotechnical investigation was carried out in four phases as follows:

- *Phase 1*: 23 boreholes, in situ SPTs, 40 pressuremeter tests in three boreholes, installation of four standpipe piezometers, laboratory testing, specialist laboratory testing and contamination testing
- *Phase 2*: Three geophysical boreholes with cross-hole and tomography geophysical surveys carried out between three new boreholes and one existing borehole
- *Phase 3*: Six boreholes, in situ SPTs, 20 pressuremeter tests, installation of two standpipe piezometers and laboratory testing
- *Phase 4*: One borehole, in situ SPTs, cross-hole geophysical testing in three boreholes, down-hole geophysical testing in one borehole and laboratory testing

The drilling was carried out using cable percussion techniques with follow-on rotary drilling methods to depths between 30 and 140 m below ground level. The quality of core recovered in some of the earlier boreholes was somewhat poorer than that recovered in later boreholes, and therefore the defects noted in the earlier rock cores may not have been representative of the actual defects present in the rock mass. Phase 4 of the investigation was targeted to assess the difference in core quality and this indicated that the differences were probably related to the drilling fluid used and the overall quality of drilling.

Disturbed and undisturbed samples and split spoon samples were obtained from the boreholes. Undisturbed samples were obtained using double tube core barrels (with Coreliner) and wire line core barrels, producing core varying in diameter between 57 and 108.6 mm. SPTs were carried out at various depths in the boreholes and were generally carried out in the overburden soils, and in weak rock or soil bands encountered within the rock strata.

Pressuremeter testing, using an OYO Elastmeter, was carried out in five boreholes between depths of 4–60 m below ground level, and typically below the Tower footprint.

The geophysical survey comprised cross-hole seismic survey, cross-hole tomography and down-hole geophysical survey. The main purpose of the geophysical survey was to complement the borehole data and provide a check on the results obtained from borehole drilling, in situ testing and laboratory testing.

The cross-hole seismic survey was used to assess compression (P) and shear (S) wave velocities through the ground profile. Cross-hole tomography was used to develop a detailed distribution of P-wave velocity in the form of a vertical seismic profile of P wave with depth, and to highlight any variations in the nature of the strata between boreholes. Down-hole seismic testing was used to determine shear (S) wave velocities through the ground profile.

The geotechnical laboratory testing program consisted of two broad classes of test:

- Conventional tests, including moisture content, Atterberg limits, particle size distribution, specific gravity, UCS, point load index, direct shear tests and carbonate content tests.
- Sophisticated tests, including stress path triaxial, resonant column, cyclic undrained triaxial, cyclic simple shear and CNS direct shear tests. These tests were undertaken by a variety of commercial, research and university laboratories in the United Kingdom, Denmark and Australia.

### 15.5.3 Geotechnical conditions

The ground conditions comprised a horizontally stratified subsurface profile which was complex and highly variable, due to the nature of deposition and the prevalent hot arid climatic conditions. Medium dense to very loose granular silty sands (marine deposits) were underlain by successions of very weak to weak sandstone layers, interbedded with very weakly cemented sand, gypsiferous fine-grained sandstone/siltstone and weak to moderately weak conglomerate/calcisiltite.

Groundwater levels were generally high across the site, at about +0.0 m DMD (approximately 2.5 m below ground level). The ground conditions encountered in the investigation were consistent with the available geological information.

The ground profile and derived geotechnical design parameters assessed from the investigation data are summarised in Table 15.3. Values of Young's modulus derived by various means are plotted in Figure 15.9. Non-linear stress–strain responses were derived for each strata type using the results from the SPTs, the pressuremeter, the geophysics and the standard and specialist laboratory testing. An allowance for degradation of the mass stiffness of the materials was incorporated in the derivation of the non-linear stress–strain curves used in the numerical design analyses.

An assessment of the potential for degradation of the stiffness of the strata under cyclic loading was carried out through a review of the CNS and cyclic triaxial specialist test results, and also using the computer program SHAKE91 (Idriss and Sun, 1991) for potential degradation under earthquake loading. The results indicated that there was a modest potential for degradation of the mass stiffness of the materials, but limited potential for degradation at the pile–soil interface.

### 15.5.4 Foundation design

An assessment of the foundations for the structure was carried out and it was clear that piled foundations would be appropriate for both the Tower and Podium construction. An initial assessment of the pile capacity was carried out using the following design

*Table 15.3* Geotechnical model and design parameters for the Burj Khalifa site

| *Stratum* | *Sub-strata* | *Subsurface material* | *Level at top of stratum (m DMD)* | *Thickness (m)* | *UCS (MPa)* | *Undrained Modulus* $E_u$ (MPa)* | *Drained Modulus* $E'$ (MPa)* | *Ult. comp. shaft friction $f_s$ (kPa)* |
|---|---|---|---|---|---|---|---|---|
| 1 | 1a | Medium dense silty sand | +2.50 | 1.50 | – | 34.5 | 30 | – |
| | 1b | Loose to very loose silty sand | +1.00 | 2.20 | – | 11.5 | 10 | – |
| 2 | 2 | Very weak to moderately weak calcarenite | −1.20 | 6.10 | 2.0 | 500 | 400 | 350 |
| 3 | 3a | Medium dense to very dense sand/silt with frequent sandstone bands | −7.30 | 6.20 | – | 50 | 40 | 250 |
| | 3b | Very weak to weak calcareous sandstone | −13.50 | 7.50 | 1.0 | 250 | 200 | 250 |
| | 3c | Very weak to weak calcareous sandstone | −21.00 | 3.00 | 1.0 | 140 | 110 | 250 |
| 4 | 4 | Very weak to weak gypsiferous sandstone/calcareous sandstone | −24.00 | 4.50 | 2.0 | 140 | 110 | 250 |
| 5 | 5a | Very weak to moderately weak calcisiltite/conglomeritic calcisiltite | −28.50 | 21.50 | 1.3 | 310 | 250 | 285 |
| | 5b | Very weak to moderately weak calcisiltite/conglomeritic calcisiltite | −50.00 | 18.50 | 1.7 | 405 | 325 | 325 |
| | 6 | Very weak to weak calcareous/conglomerate strata | −68.50 | 22.50 | 2.5 | 560 | 450 | 400 |
| | 7 | Weak to moderately weak claystone/siltstone interbedded with gypsum layers | −91.00 | >46.79 | 1.7 | 405 | 325 | 325 |

Note: $E_u$ and $E'$ values relate to relatively large strain levels in the strata below the structure.

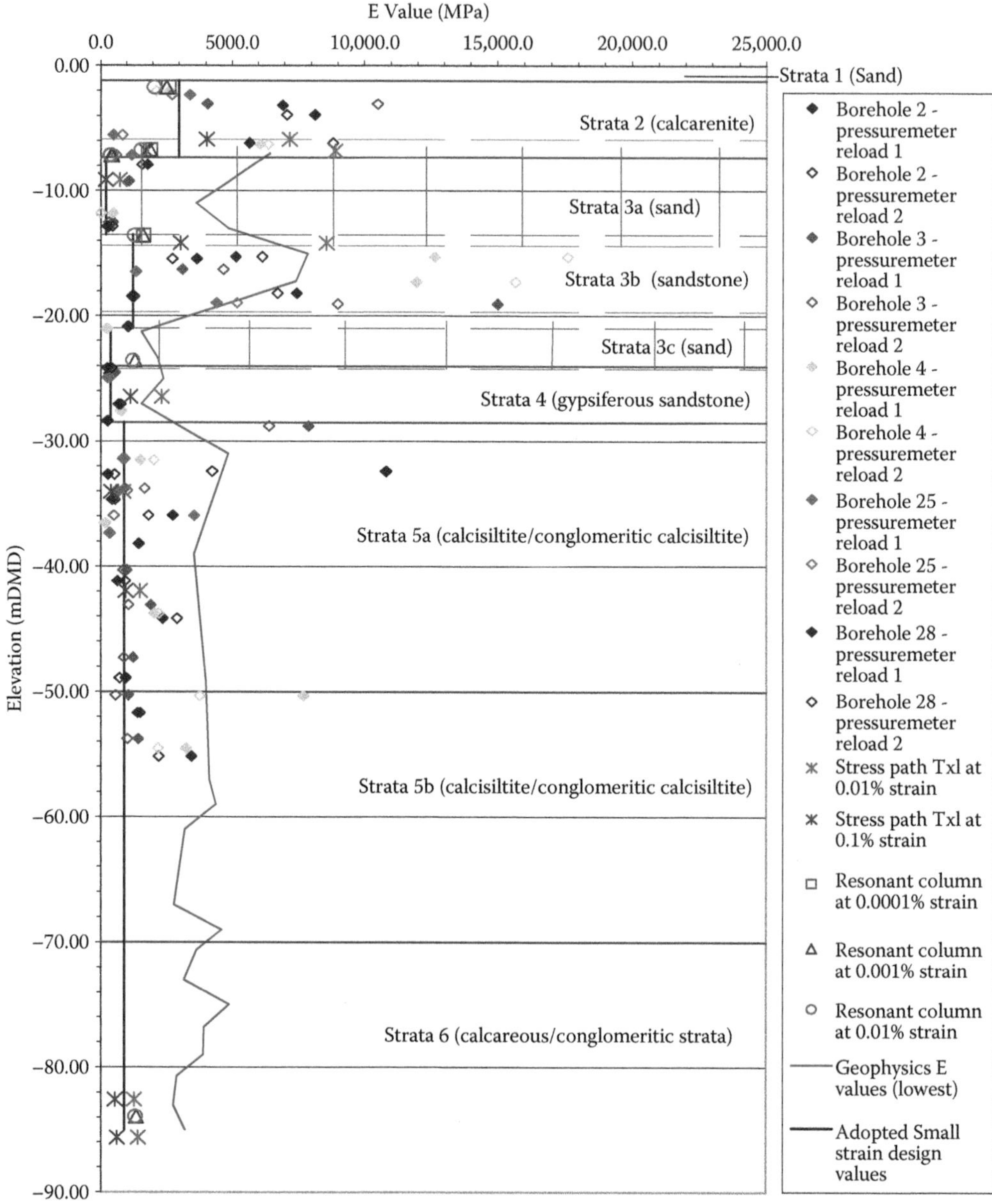


*Figure 15.9* Modulus values versus elevation. (Adapted from Poulos, H.G. and Bunce, G. 2008. Foundation design for the Burj Dubai – The world's tallest building. *Proceedings of the 6th International Conference on Case Histories in Geotechnical Engineering,* Paper 1.47, Arlington, VA, CD volume. Courtesy of the Curators of the University of Missouri.)

recommendations given by Horvath and Kenney (1979), as presented by Burland and Mitchell (1989):

$$\text{Ultimate unit shaft resistance } f_s = 0.25(q_u)^{0.5}\text{ MPa} \quad (15.1)$$

where $f_s$ is in MPa and $q_u$ is the uniaxial compressive strength in $MN/m^2$.

The adopted ultimate compressive unit shaft friction values for the various site rock strata are tabulated in Table 15.3. The ultimate unit pile skin friction of a pile loaded in tension was taken, conservatively, as half the ultimate unit shaft resistance of a pile loaded in compression. The initial computer analyses, using the commercially available program ABAQUS, indicated that the strains in the strata were within the initial small-strain region of the non-linear stress–strain curves developed for the materials. The secant elastic modulus values at small-strain levels were therefore adopted for the validation and sensitivity analyses carried out using the programs PIGLET and REPUTE. A non-linear analysis was carried out via the program VDISP, using the non-linear stress–strain curves developed for the materials.

Linear and non-linear analyses were carried out to obtain predictions for the load distribution in the piles and for the settlement of the raft and podium. The assessed pile capacities were provided to the structural designers, who then supplied details on the layout, number and diameter of the piles. Tower piles were 1.5 m diameter and 47.45 m long with the tower raft founded at –7.55 mDMD. The podium piles were 0.9 m diameter and 30 m long with the podium raft being founded at –4.85 mDMD. The thickness of the raft was 3.7 m. Loadings were provided by SOM and comprised eight load cases, including four load cases for wind and three for seismic conditions.

The estimated vertical dead load was about 4380 MN (including the net raft weight and uplift pressures), while the live load was approximately 300 MN.

The settlements from the finite element (FE) model and from VDISP were converted from those for a flexible pile cap to those for a rigid pile cap for comparison with the REPUTE and PIGLET models, using the following approximate equation:

$$\delta_{rigid} = \frac{1}{2}(\delta_{centre} + \delta_{edge})_{flexible} \quad (15.2)$$

The computed settlements are shown in Table 15.4, which shows that the settlements from the FEA model correlated acceptably well with the results obtained from REPUTE, PIGLET and VDISP.

The maximum and minimum pile loadings were obtained from the FE analysis for all loading combinations. The maximum loads were at the corners of the three 'wings' and were of the order of 35 MN, while the minimum loads were within the centre of the group and were of the order of 12–13 MN. Figure 15.10 shows contours of the computed maximum axial load. The impact of cyclic loading on the pile was an important consideration and in order to address this, the load variation above or below the dead load plus live load cases was determined. The maximum load variation was found to be less than 10 MN.

SOM carried out an analysis of the pile loads and a comparison of the results indicated that although the maximum pile loads were similar, the distribution was different. The SOM calculations indicated that the largest pile loads were in the central region of the Tower piled

*Table 15.4* Computed settlements

| | | *Settlement (mm)* | |
|---|---|---|---|
| *Analysis method* | *Load case* | *Rigid* | *Flexible* |
| FE-ABAQUS | Tower only (DL + LL) | 56 | 66 |
| REPUTE | Tower only (DL + LL) | 45 | – |
| PIGLET | Tower only (DL + LL) | 62 | – |
| VDISP | Tower only (DL + LL) | 46 | 72 |

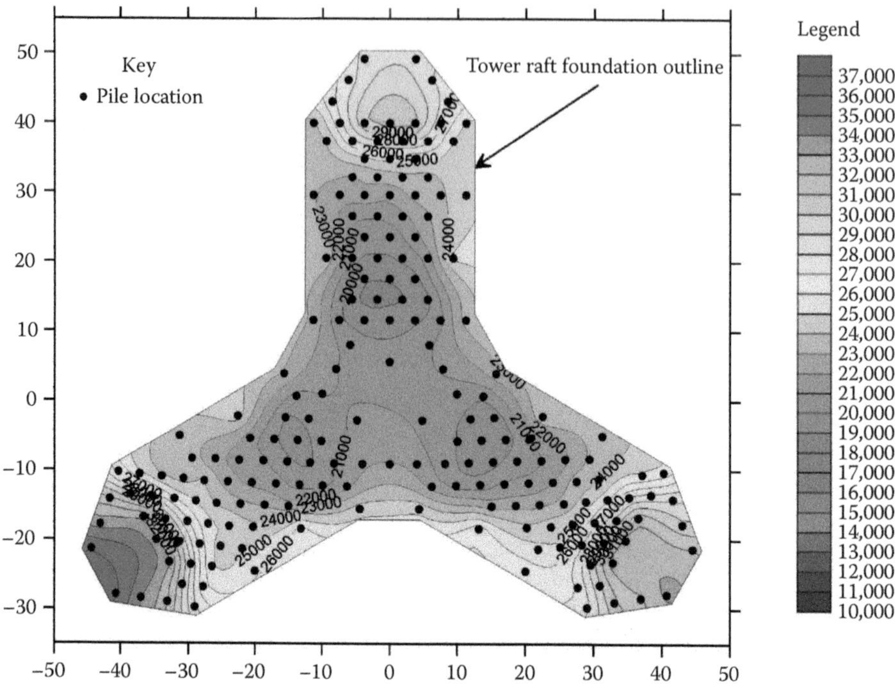


*Figure 15.10* Contours of maximum axial load (kN). (Adapted from Poulos, H.G. and Bunce, G. 2008. Foundation design for the Burj Dubai – The world's tallest building. *Proceedings of the 6th International Conference on Case Histories in Geotechnical Engineering*, Paper 1.47, Arlington, VA, CD volume. Courtesy of the Curators of the University of Missouri.)

raft and decreasing toward the edges. However, the FE analyses indicated the opposite where the largest pile loads were concentrated toward the edges of the pile group reducing toward the centre of the group. Similarly, the PIGLET and REPUTE standard pile group analyses carried out indicated that the largest pile loads were concentrated toward the edge of the pile cap. This almost certainly resulted from the implicit assumption in these analyses that the raft is rigid.

The difference between the pile load distributions from SOM and those from the geotechnical analyses could be attributed to a number of reasons:

- The FE, REPUTE and PIGLET models take account of the pile–soil–pile interaction whereas SOM modelled the soil as springs connected to the raft and piles using an S-frame analysis.
- The FE analysis modelled the soil/rock using non-linear responses, compared to the linear spring stiffnesses assumed in the SOM analysis.
- The specified/assumed superstructure stiffening effects on the foundation response were modelled more accurately in the SOM analysis.

The actual pile load distribution was expected to be somewhere between the two models, depending on the impact of the different modelling approaches.

### 15.5.5 Overall stability assessment

The minimum centre-to-centre spacing of the piles for the tower was 2.5 times the pile diameter. A check was therefore carried out to ensure that the Tower foundation was stable both vertically and laterally, assuming that the foundation acted as a block comprising the piles and soil/rock. A factor of safety of slightly less than 2 was assessed for vertical block movement, excluding base resistance of the block while a factor of safety of greater than 2 was determined for lateral block movement. A factor of safety of approximately 5 was obtained against overturning of the block.

### 15.5.6 Liquefaction assessment

An assessment of the potential for liquefaction during a seismic event at the Burj Dubai site was carried out using the Japanese Road Association Method and the method of Seed et al. (1984). Both approaches gave similar results and indicated that the marine deposits and sand to 3.5 m below ground level (from +2.5 m DMD to −1.0 m DMD) could potentially liquefy. However, the foundations of the Podium and Tower structures were below this level. Consideration was however required in the design and location of buried services and shallow foundations which were within the top 3.5 m of the ground. Occasional layers within the sandstone layer between −7.3 m DMD and −11.75 m DMD could also potentially liquefy. However, taking into account the imposed confining stresses at the foundation level of the Tower, this potential liquefaction was considered to have a negligible effect on the design of the Tower foundations. The assessed reduction factor to be applied to the soil strength parameters, in most cases, was found to be equal to 1.0, and hence liquefaction would have a minimal effect upon the design of the Podium foundations. However, consideration was given in the design to potential down-drag loads on pile foundations constructed through the liquefiable strata.

### 15.5.7 Independent verification analyses

The geotechnical model used in the verification analyses was assessed independently on the basis of the available information and experience gained from the nearby Emirates project described in Section 15.4, and is summarised in Table 15.5. In general, this model was rather more conservative than the original model employed by HCL for the design. In particular, the ultimate end bearing capacity was reduced together with the Young's modulus in several of the upper layers, and the presence was assumed of a stiffer layer, with a modulus of

*Table 15.5* Summary of geotechnical model for independent verification analyses

| *Stratum number* | *Description* | *RL range DMD* | *Undrained modulus $E_u$ (MPa)* | *Drained modulus $E'$ (MPa)* | *Ultimate skin friction (kPa)* | *Ultimate end bearing (MPa)* |
|---|---|---|---|---|---|---|
| 1a | Med. dense silty sand | +2.5 to +1.0 | 30 | 25 | – | – |
| 1b | Loose–v. loose silty sand | +1.0 to −1.2 | 12.5 | 10 | – | – |
| 2 | Weak–mod. weak calcarenite | −1.2 to −7.3 | 400 | 325 | 400 | 4.0 |
| 3 | V. weak calc. sandstone | −7.3 to −24 | 190 | 150 | 300 | 3.0 |
| 4 | V. weak-weak sandstone/ calc. sandstone | −24 to −28.5 | 220 | 175 | 360 | 3.6 |
| 5A | V. weak–weak–mod. weak calcisiltite/conglomerate | −28.5 to −50 | 250 | 200 | 250 | 2.5 |
| 5B | V. weak–weak–mod. weak calcisiltite/conglomerate | −50 to −70 | 275 | 225 | 275 | 2.75 |
| 6 | Calcareous siltstone | −70 and below | 500 | 400 | 375 | 3.75 |

1200 MPa below RL-70 m DMD, to allow for the fact that the strain levels in the ground decrease with increasing depth.

The following three-stage approach was employed for the independent verification process:

- The commercially available computer program FLAC was used to carry out an axi-symmetric analysis of the foundation system for the tower. The foundation plan was represented by a circle of equal area, and the piles were represented by a solid block containing piles and soil. The axial stiffness of the block was taken to be the same as that of the piles and the soil between them. The total dead plus live loading was assumed to be uniformly distributed. The soil layers were assumed to be Mohr Coulomb materials, with the modulus values as shown in Table 15.5, and values of cohesion taken as 0.5 times the estimated UCS. The main purpose of this analysis was to calibrate and check the second, and more detailed, analysis, using an in-house simplified computer program (PIGS) for pile group analysis.
- A group analysis was carried out for the tower alone, to check the settlement with that obtained by FLAC. In this analysis, the piles were modelled individually, and it was assumed that each pile was subjected to its nominal working load of 30 MN. The stiffness of each pile was computed via the program DEFPIG (Poulos, 1990), allowing for contact between the raft section above the pile and the underlying soil. The pile stiffness values were assumed to vary hyperbolically with increasing load level, using a hyperbolic factor ($R_f$) of 0.4.
- Finally, an analysis of the complete tower-podium foundation system was carried out, considering all 926 piles in the Tower and Podium system. Again, each of the piles was subjected to its nominal working load.

#### *15.5.7.1 FLAC and PIGS results for the tower alone*

Because of the difference in shape between the actual foundation and the equivalent circular foundation, only the maximum settlement was considered for comparison purposes. The following results were obtained for the central settlement:

- FLAC analysis, using an equivalent block to represent the piles: 72.9 mm
- Pile group (PIGS) analysis, modelling all 196 piles: 74.3 mm

Thus, despite the quite different approaches adopted, the above computed settlements were in good agreement with those of the designer. It should be noted that, as found with the Emirates project, the computed settlement was influenced by the assumptions made regarding the ground properties below the pile tips. For example, if in the PIGS analysis the modulus of the ground below RL-70 m DMD was taken as 400 MPa (rather than 1200 MPa), the computed settlement at the centre of the tower would increase to about 96 mm.

#### *15.5.7.2 PIGS results for tower and podium*

Figure 15.11 shows the contours of computed settlement for the entire tower and podium area. It can be seen that the maximum settlements are concentrated in the central area of the tower.

Figure 15.12 shows the settlement profile across a section through the centre of the tower. The notable feature of this figure is that the settlements reduce rapidly outside the tower area, and become in the order of 10–12 mm for much of the podium area.

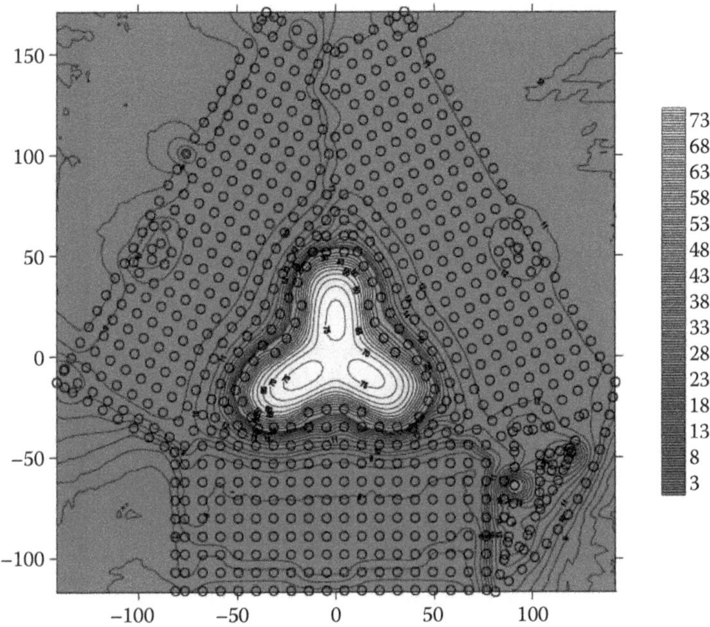


*Figure 15.11* Computed settlement contours for tower and podium. (Adapted from Poulos, H.G. and Bunce, G. 2008. Foundation design for the Burj Dubai – The world's tallest building. *Proceedings of the 6th International Conference on Case Histories in Geotechnical Engineering*, Paper 1.47, Arlington, VA, CD volume. Courtesy of the Curators of the University of Missouri.)

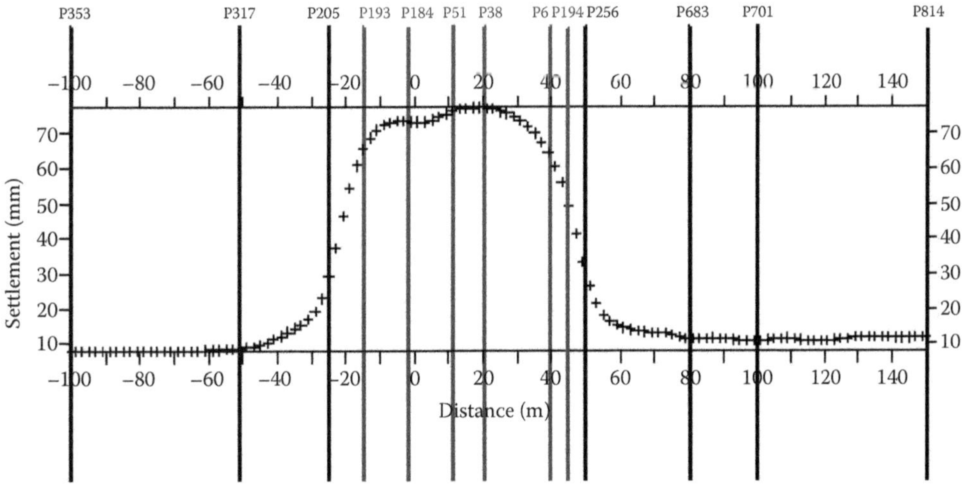


*Figure 15.12* Computed settlement across section through centre of tower. (Adapted from Poulos, H.G. and Bunce, G. 2008. Foundation design for the Burj Dubai – The world's tallest building. *Proceedings of the 6th International Conference on Case Histories in Geotechnical Engineering*, Paper 1.47, Arlington, VA, CD volume. Courtesy of the Curators of the University of Missouri.)

#### 15.5.7.3 Cyclic loading effects

The possible effects of cyclic loading were investigated via the following means:

- Cyclic triaxial laboratory tests
- Cyclic direct shear tests
- Cyclic CNS laboratory tests
- An independent theoretical analysis carried out by the independent verifier

The cyclic triaxial tests indicated that there was some potential for degradation of stiffness and accumulation of excess pore pressure, while the direct shear tests indicated a potential reduction in residual shear strength, although these were carried out using large strain levels which were not representative of the likely field conditions.

The CNS tests indicated that there was not a significant potential for cyclic degradation of skin friction, provided that the cyclic shear stress remained within the anticipated range.

The independent analysis of cyclic loading effects was undertaken using the approach described by Poulos (1988a), and implemented via a computer program SCARP (Static and Cyclic Axial Response of Piles). This analysis involved a number of simplifying assumptions, together with parameters that were not easily measured or estimated from available data. As a consequence, the analysis was indicative only. Since the analysis of the entire foundation system was not feasible with SCARP, only a typical pile (assumed to be a single isolated pile) with a diameter of 1.5 m and a length of 48 m was considered. The results were used to explore the relative effects of the cyclic loading, with respect to the case of static loading.

It was found that a loss of capacity would be experienced when the cyclic load exceeded about ±10 MN. The maximum loss of capacity (due to degradation of the skin friction) was in the order of 15%–20%. The capacity loss was relatively insensitive to the mean load level, except when the mean load exceeded about 30 MN. It was predicted that, at a mean load equal to the working load and under a cyclic load of about 25% of the working load, the relative increase in settlement for 10 cycles of load would be about 27%.

The indicative pile forces, calculated from the ABAQUS FE analysis of the structure, suggested that cyclic loading of the Burj Tower foundation would not exceed ±10 MN. Thus, it seemed reasonable to assume that the effects of cyclic loading would not significantly degrade the axial capacity of the piles, and that the effects of cyclic loading on both capacity and settlement were unlikely to be significant.

### 15.5.8 Pile load testing

Two programs of static load testing were undertaken for the Burj Khalifa project:

- Static load tests on seven trial piles prior to foundation construction.
- Static load tests on eight works piles, carried out during the foundation construction phase (i.e. on about 1% of the total number of piles constructed).

In addition, dynamic pile testing was carried out on 10 of the works piles for the tower and 31 piles for the podium, that is, on about 5% of the total works piles. Sonic integrity testing was also carried out on a number of the works piles. Attention here is focussed on the static load tests.

#### 15.5.8.1 Preliminary pile testing program

The details of the piles tested within this program are summarised in Table 15.6. The main purpose of the tests was to assess the general load–settlement behaviour of piles of the

*Table 15.6* Summary of pile load tests – Preliminary pile testing

| *Pile no.* | *Pile diameter (m)* | *Pile length (m)* | *Side grouted?* | *Test type* |
|---|---|---|---|---|
| TP1 | 1.5 | 45.15 | No | Compression |
| TP2 | 1.5 | 55.15 | No | Compression |
| TP3 | 1.5 | 35.15 | Yes | Compression |
| TP4 | 0.9 | 47.10 | No | Compression (cyclic) |
| TP5 | 0.9 | 47.05 | Yes | Compression |
| TP6 | 0.9 | 36.51 | No | Tension |
| TP7A | 0.9 | 37.51 | No | Lateral |

anticipated length below the tower, and to verify the design assumptions. Each of the test piles was different, allowing various factors to be investigated, as follows:

- The effects of increasing the pile shaft length
- The effects of shaft grouting
- The effects of reducing the shaft diameter
- The effects of uplift (tension) loading
- The effects of lateral loading
- The effect of cyclic loading

The piles were constructed using polymer drilling fluid, rather than the more conventional bentonite drilling fluid. The use of the polymer led to piles whose performance exceeded expectations. Strain gauges were installed along each of the piles, enabling detailed evaluation of the load transfer along the pile shaft, and the assessment of the distribution of mobilised skin friction with depth along the shaft. The reaction system provided for the axial load tests consisted of four or six adjacent reaction piles (depending on the pile tested), and these reaction piles had the potential to influence the results of the pile load tests via interaction with the test pile through the soil. The possible consequences of this interaction are discussed subsequently.

### 15.5.8.2 Ultimate axial load capacity

None of the six axial pile load tests appears to have reached its ultimate axial capacity, at least with respect to geotechnical resistance. The 1.5 m diameter piles (TP1, TP2 and TP3) were loaded to twice the working load, while the 0.9 m diameter test piles TP4 and TP6 were loaded to 3.5 times the working load, and TP5 was loaded to 4 times working load. With the exception of TP5, none of the other piles showed any strong indication of imminent geotechnical failure. Pile TP5 showed a rapid increase in settlement at the maximum load, but this was attributed to structural failure of the pile itself. From a design viewpoint, the significant finding was that, at the working load, the factor of safety against geotechnical failure appeared to be in excess of 3, thus giving a comfortable margin of safety against failure, especially as the raft would also provide additional resistance to supplement that of the piles.

### 15.5.8.3 Ultimate shaft friction

From the strain gauge readings along the test piles, the mobilised skin friction distribution along each pile was evaluated. Figure 15.13 summarises the ranges of skin friction deduced

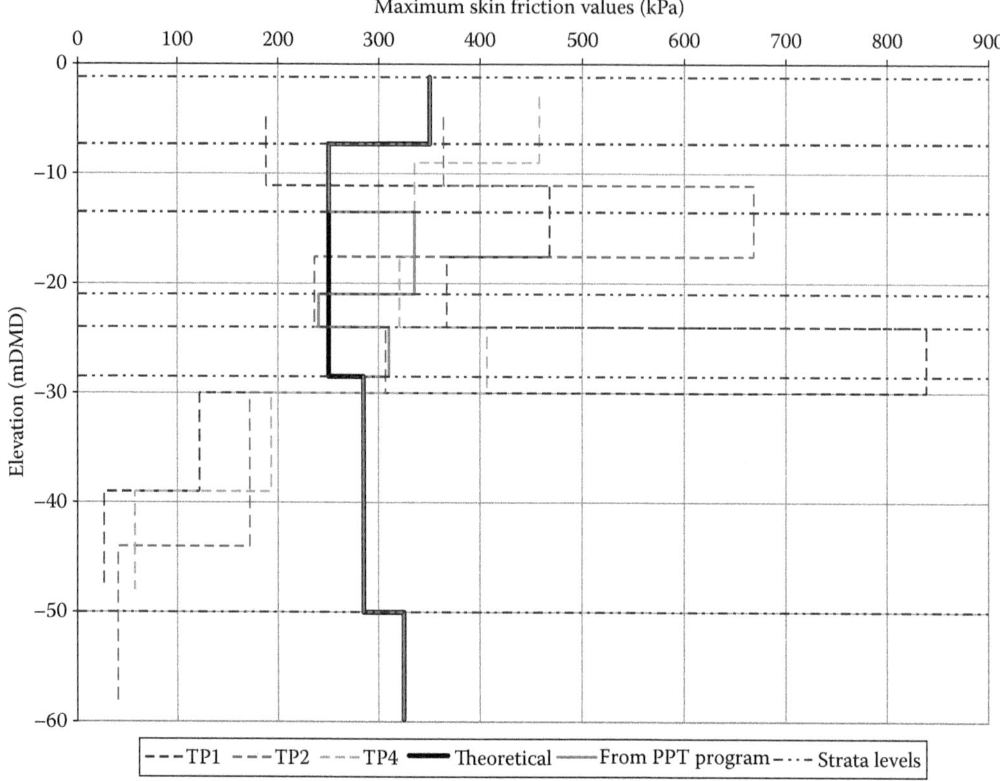


*Figure 15.13* Measured and design values of shaft friction. (Adapted from Poulos, H.G. and Bunce, G. 2008. Foundation design for the Burj Dubai – The world's tallest building. *Proceedings of the 6th International Conference on Case Histories in Geotechnical Engineering*, Paper 1.47, Arlington, VA, CD volume. Courtesy of the Curators of the University of Missouri.)

from the measurements, together with the original design assumptions and the modified design recommendations made after the preliminary test results were evaluated. The following observations were made:

1. The skin friction values down to about RL-30 m DMD appeared to be ultimate values, that is, the available skin friction had been fully mobilised.
2. The skin friction values below about RL-30 m DMD did not appear to have been fully mobilised, and thus were assessed to be below the ultimate values.
3. The original assumptions appeared to be comfortably conservative within the upper part of the ground profile.
4. Shaft grouting appeared to enhance the skin friction developed along the pile.

Because the skin friction in the lower part of the ground profile did not appear to have been fully mobilised, it was recommended that the original values (termed the 'theoretical ultimate unit skin friction') be used in the lower strata. It was also recommended that the 'theoretical' values in the top layers (Strata 2 and 3a) be used because the presence of the casing in the tests would probably have given skin friction values that may have been too low. For Strata 3b, 3c and 4, the minimum measured skin friction values were used for the final design.

#### 15.5.8.4 Ultimate end bearing capacity

None of the load tests was able to mobilise any significant end bearing resistance, because the skin friction appeared to be more than adequate to resist loads well in excess of the working load. Therefore, no conclusions could be reached about the accuracy of the estimated end bearing component of pile capacity. For the final design, the length of the piles was increased where the proposed pile toe levels were close to or within the gypsiferous sandstone layer (Stratum 4).

This was the case for the 0.9 m diameter podium piles. It was considered prudent to have the pile toes founded below this stratum, to allow for any potential long-term degradation of engineering properties of this layer (e.g. via solution of the gypsum) that could reduce the capacity of the piles.

#### 15.5.8.5 Load-settlement behaviour

Table 15.7 summarises the measured pile settlements at the working load and at the maximum test load, and the corresponding values of pile head stiffness (load/settlement). The following observations are made:

- The measured stiffness values were relatively large, and were considerably in excess of those anticipated
- As expected, the stiffness was greater for the larger diameter piles
- The stiffness of the shaft grouted piles (TP3 and TP5) was greater than that of the corresponding ungrouted piles

#### 15.5.8.6 Effect of reaction piles

On the basis of the experience gained in the nearby Emirates project, it had been expected that the pile head stiffness values for the Burj Dubai piles would be somewhat less than those for the Emirates Towers, in view of the apparently inferior quality of rock at the Burj Dubai site.

This expectation was certainly not realised, and it is possible that the improved performance of the piles in the present project may be attributable, at least in part, to the use of polymer drilling fluid, rather than bentonite, in the construction process. However, it was also possible that at least part of the reason for the apparently high stiffness values was related to the interaction effects of the reaction piles. As discussed in Chapter 13, when applying a compressive load to the test pile, the reaction piles experience a tension and a consequent uplift, which tends to reduce the settlement of the test pile. Thus, the apparent high stiffness of the pile may not reflect the true stiffness of the pile beneath the structure.

*Table 15.7* Summary of pile load test results – Axial loading

| *Pile number* | *Working load (MN)* | *Max. load (MN)* | *Settlement at W. load (mm)* | *Settlement at max. load (mm)* | *Stiffness at W. load (MN/m)* | *Stiffness at max. load (MN/m)* |
|---|---|---|---|---|---|---|
| TP1 | 30.13 | 60.26 | 7.9 | 21.3 | 3,819 | 2,834 |
| TP2 | 30.13 | 60.26 | 5.6 | 16.8 | 5,429 | 3,576 |
| TP3 | 30.13 | 60.26 | 5.8 | 20.2 | 5,213 | 2,977 |
| TP4 | 10.1 | 35.07 | 4.5 | 26.6 | 2,260 | 1,317 |
| TP5 | 10.1 | 40.16 | 3.6 | 27.5 | 2,775 | 1,463 |
| TP6 | −1.0 | −3.5 | −0.65 | −4.9 | 1,536 | 717 |

An analysis of the effects of the reaction piles on the settlement of pile TP1 revealed that the presence of the reaction piles could reduce the settlement at the working load of 30 MN by 30%. In other words, the real stiffness of the piles might be only about 70% of the values measured from the load test. This would then reduce the stiffness to a value which is more in line with the stiffness values experienced in the Emirates project, where the reaction was provided by a series of inclined anchors that would have had a very small degree of interaction with the test piles.

#### *15.5.8.7 Uplift versus compression loading*

On the basis of the tension test on pile TP6, the ultimate skin friction in tension was taken as 0.5 times that for compression. It is customary to allow for a reduction in skin friction for piles in granular soils or rocks subjected to uplift. As set out in Chapter 7, De Nicola and Randolph (1993) developed a theoretical relationship between the tensile and compressive skin friction values, and showed that this relationship depends on the Poisson's ratio of the pile, the relative stiffness of the pile to the soil, the interface friction characteristics and the pile length to diameter ratio. This theoretical relationship was applied to the Burj Khalifa case, and the calculated ratio of tension to compression skin friction was about 0.6, which was reasonably consistent with the initial assumption of 0.5 made in the design.

#### *15.5.8.8 Cyclic loading effects*

In all of the axial load tests, a relatively small number of cycles of loading was applied to the pile after the working load was reached. Table 15.8 summarises the test results inferred from the load–settlement data. The settlement after cycling was related to the settlement for the first cycle, both settlements being at the maximum load of the cycling process. It can be seen that there was an accumulation of settlements under the action of the cyclic loading, but that this accumulation was relatively modest, given the relatively high levels of mean and cyclic stress that were applied to the pile (in all cases, the maximum load reached was 1.5 times the working load).

These results were consistent with the assessments made during design that cyclic loading effects were unlikely to be significant for this building.

#### *15.5.8.9 Lateral loading*

One lateral load test was carried out, on pile TP7A, with the pile being loaded to twice the working load (50 t). At the working lateral load of 25 t, the lateral deflection was about 0.47 mm, giving a lateral stiffness of about 530 MN/m, a value which was consistent with

*Table 15.8* Summary of displacement accumulation for cyclic loading

| *Pile number* | *Mean load/$P_w$* | *Cyclic load/$P_w$* | *No. of cycles (N)* | $S_N/S_1$ |
|---|---|---|---|---|
| TP1 | 1.0 | ±0.5 | 6 | 1.12 |
| TP2 | 1.0 | ±0.5 | 6 | 1.25 |
| TP3 | 1.0 | ±0.5 | 6 | 1.25 |
| TP4 | 1.25 | ±0.25 | 9 | 1.25 |
| TP5 | 1.25 | ±0.25 | 6 | 1.3 |
| TP6 | 1.0 | ±0.5 | 6 | 1.1 |

Note: $P_w$ = working load; $S_N$ = settlement after N cycles; $S_1$ = settlement after one cycle.

the designer's predictions using the program ALP (Oasys, 2001). An analysis of lateral deflection was also carried out by the independent verifier using the program DEFPIG. In this latter analysis, the Young's modulus values for lateral loading were assumed to be 30% less than the values for axial loading, while the ultimate lateral pile–soil pressure was assumed to be similar to the end bearing capacity of the pile, with allowances being made for near-surface effects. These calculations indicated a lateral movement of about 0.7 mm at 25 t load, which was larger than the measured deflection, but of a similar order.

Thus, pile TP7A appeared to perform better than anticipated under the action of lateral loading, mirroring the better-than-expected performance of the test piles under axial load. However, there may again have been some effect of the reaction system used for the test, as the reaction block developed a surface shear which would tend to oppose the lateral deflection of the test pile.

#### *15.5.8.10 Works pile testing program*

A total of eight works pile tests were carried out, including two 1.5 m diameter piles and six 0.9 m diameter piles. All piles were tested in compression, and each pile was tested approximately 4 weeks after construction. The piles were tested to a maximum load of 1.5 times the working load.

The following observations were made from the test results:

- The pile head stiffness of the works piles was generally larger than for the trial piles.
- None of the works piles reached failure, and indeed, the load–settlement behaviour up to 1.5 times the working load was essentially linear, as evident from the relatively small difference between the stiffness values at the working load and at 1.5 times the working load. In contrast, the relative difference between the two stiffnesses was considerably greater for the preliminary trial piles.

At least three possible explanations could be offered for the greater stiffness and improved load–settlement performance of the trial piles:

1. The level of the bottom of the casing was higher for the works piles than for the trial piles (about 3.5–3.6 m higher), thus leading to a higher skin friction along the upper portion of the shaft
2. A longer period between the end of construction and testing of the works piles (about 4 weeks, versus about 3 weeks for the trial piles)
3. Natural variability of the strata

Cyclic loading was undertaken on two of the works piles, and it was observed that there was a relatively small amount of settlement accumulation due to the cyclic loading, and certainly less than that observed on TP1 or the other trial piles (see Table 15.8). The smaller amount of settlement accumulation could be attributed to the lower levels of mean and cyclic loading applied to the works piles (which were considered to be more representative of the design condition) and also to the greater capacity that the works piles seemed to possess. Thus, the results of these tests reinforced the previous indications that the cyclic degradation of capacity and stiffness at the pile–soil interface was negligible.

#### *15.5.8.11 Summary of pile testing outcomes*

Both the preliminary test piling program and the tests on the works piles provided very positive and encouraging information on the capacity and stiffness of the piles. The measured pile head stiffness values were well in excess of those predicted. The interaction effects between the test piles and the reaction piles may have contributed to the higher apparent pile head stiffnesses, but the piles nevertheless exceeded expectations. The capacity of the piles also appeared to be in excess of the predicted values, although none of the tests fully mobilised the available geotechnical resistance. The works piles performed even better than the preliminary trial piles, and demonstrated almost linear load–settlement behaviour up to the maximum test load of 1.5 times working load.

Shaft grouting appeared to have enhanced the load–settlement response of the piles, but it was assessed that shaft grouting would not need to be carried out for this project, given the very good performance of the ungrouted piles.

The inferences from the pile load test data were that the design estimates of capacity and settlement may have been conservative, although it was recognised that the overall settlement behaviour (and perhaps the overall load capacity) would be dependent not only on the individual pile characteristics, but also on the characteristics of the ground within the zone of influence of the structure.

### 15.5.9 Settlement performance during construction

The settlement of the Tower raft was monitored after completion of concreting. A summary of the settlements to February 2008 in Wing C is shown in Figure 15.14 which also shows the final predicted settlement profile from the original design. At that time, the majority of

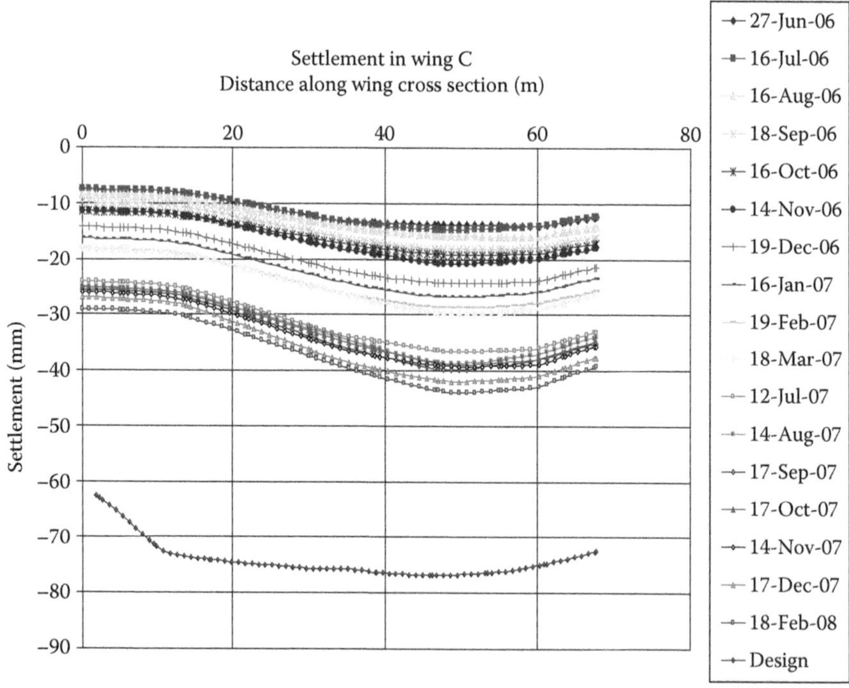


*Figure 15.14* Measured and predicted settlements for Wing C, Burj Khalifa. (Adapted from Poulos, H.G. and Bunce, G. 2008. Foundation design for the Burj Dubai – The world's tallest building. *Proceedings of the 6th International Conference on Case Histories in Geotechnical Engineering*, Paper 1.47, Arlington, VA, CD volume. Courtesy of the Curators of the University of Missouri.)

the dead loading would have been applied to the foundation, and the maximum settlement measured was about 43 mm. It will be seen that the measured settlements are less than those predicted during the design process However, there remained some dead and live load to be applied to the foundation system, and it should also be noted that the monitored figures do not include the impact of the raft, cladding and live loading which would be in excess of 20% of the overall mass. Extrapolating for the full dead plus live load, it was anticipated that the final settlement would be of the order of 55–60 mm, which was comfortably less than the predicted final settlement of about 70–75 mm.

Russo et al. (2013) have carried out a careful re-assessment of the settlement analyses, taking into account such factors as the structure stiffness, the interpretation of the preliminary pile tests, and the effects of the reaction piles in the load tests. They found that the total predicted maximum settlement would then be reduced to about 52 mm.

Figure 15.15 shows contours of measured settlement. The general distribution is similar to that predicted by the various analyses.

To put the foundation settlements into perspective, the computed shortening of the structure itself after 30 years was estimated to be about 300 mm (Baker et al., 2008), which is substantially greater than the foundation settlements.

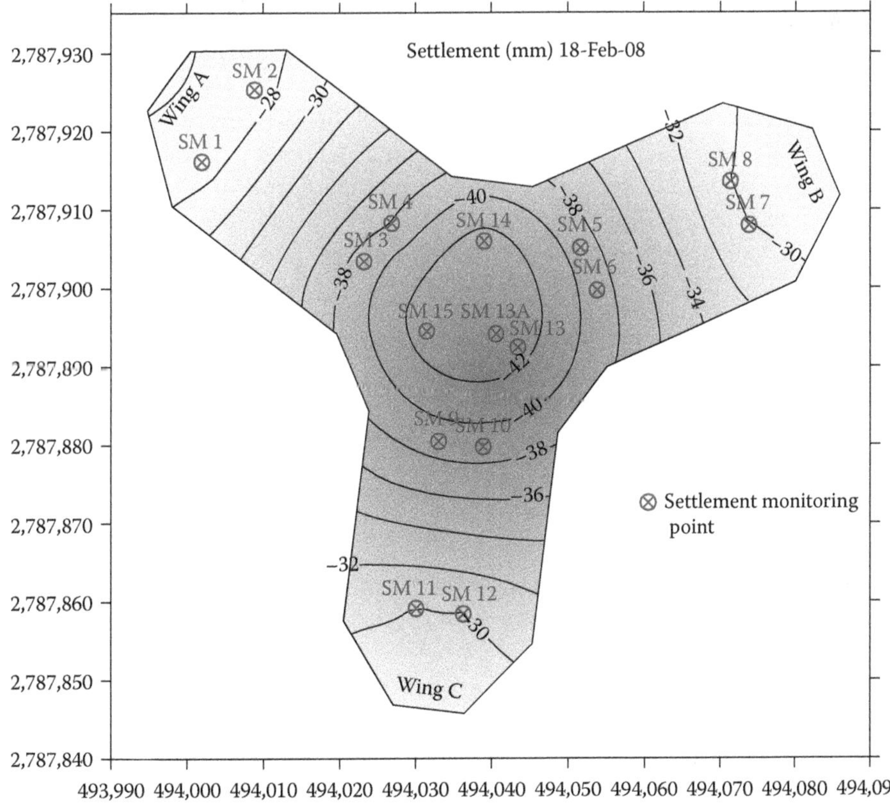


*Figure 15.15* Contours of measured settlement as of February 2008. (Adapted from Poulos, H.G. and Bunce, G. 2008. Foundation design for the Burj Dubai – The world's tallest building. *Proceedings of the 6th International Conference on Case Histories in Geotechnical Engineering*, Paper 1.47, Arlington, VA, CD volume. Courtesy of the Curators of the University of Missouri.)

### 15.5.10 Summary and lessons learned

For the Burj Khalifa, the maximum settlement predicted by ABAQUS for the tower and podium foundation compared reasonably well with the maximum settlement estimated by the PIGS analysis carried out during the independent verification process.

It was assessed that there was a potential for a reduction in axial load capacity and stiffness of the foundation strata under cyclic loading, but based on the pile load test data, laboratory tests and on theoretical analyses, it would appear that the cyclic degradation effects at the pile–soil interface were relatively small.

Both the preliminary test piling program and the tests on the works piles provided very positive and encouraging information on the capacity and stiffness of the piles.

The measured pile head stiffness values were well in excess of those predicted, and those expected on the basis of the experience with the nearby Emirates Towers. However, the interaction effects between the test piles and the reaction piles may have contributed to the higher apparent pile head stiffnesses. The capacity of the piles also appeared to be in excess of that predicted, and none of the tests appeared to have fully mobilised the available geotechnical resistance.

The works piles performed even better than the preliminary trial piles, and demonstrated almost linear load–settlement behaviour up to the maximum test load of 1.5 times working load.

The settlements measured during construction were consistent with, but comfortably smaller than, those predicted. Overall, the performance of the piled raft foundation system exceeded expectations.

As with previous high-rise projects, the Burj Khalifa involved close interaction between the structural and geotechnical designers in designing the piled raft foundation for a complex and significant high-rise structure. Such interaction has some major benefits in avoiding over-simplification of geotechnical matters by the structural engineer, and over-simplification of structural matters by the geotechnical engineer.

## 15.6 CASE 4: INCHEON 151 TOWER, SOUTH KOREA

### 15.6.1 Introduction

A 151 storey super high-rise building project was planned to be constructed on reclaimed land, located on soft marine clay in Songdo, Korea. This building is illustrated in Figure 15.16 and is described in detail by Badelow et al. (2009) and Abdelrazaq et al. (2011); thus, only a brief summary is presented here.

The challenges in this case related to a very tall building, sensitive to differential settlements, to be constructed on a site with complex geological conditions.

### 15.6.2 Ground conditions and geotechnical model

The Incheon area has extensive sand/mud flats and near-shore intertidal areas. The site lay entirely within an area of reclamation, comprising approximately 8 m of loose sand and sandy silt, constructed over approximately 20 m of soft to firm marine silty clay, referred to as the upper marine deposits (UMD). These deposits were underlain by approximately 2 m of medium dense to dense silty sand, referred to as the lower marine deposits (LMD), which was underlain by residual soil and a profile of weathered rock.

The lithological rock units present under the site were referred to locally as 'soft rock', and comprised granite, granodiorite, gneiss (interpreted as possible roof pendant

*Figure 15.16* Incheon 151 Tower (artist's impression).

metamorphic rocks) and aplite. The rock materials within about 50 m from the surface had been affected by weathering which reduced their strength to a very weak rock or a soil-like material. This depth increased where the bedrock was intersected by closely spaced joints, and sheared and crushed zones that were often related to the existence of the roof pendant sedimentary/metamorphic rocks. The geological structures at the site were complex and comprised geological boundaries, sheared and crushed seams, possibly related to faulting movements and jointing.

From the available borehole data for the site, inferred contours were developed for the surface of the 'soft rock' founding stratum within the tower foundation footprint. These are reproduced in Figure 15.17. It can be seen that there was a potential variation in level of the top of the soft rock (the pile founding stratum) of up to 40 m across the foundation.

For design, the footprint of the tower was divided into eight zones which were considered to be representative of the variation of ground conditions, and geotechnical models were developed for each zone. Appropriate geotechnical parameters were selected for the various strata based on the available field and laboratory test data, together with experience of similar soils on adjacent sites. One of the critical design issues for the tower foundation was the performance of the soft UMD under lateral and vertical loading, and hence careful consideration was given to the selection of parameters for this stratum. Typical parameters adopted for the initial foundation design are presented in Table 15.9.

### 15.6.3 Foundation layout

The foundation comprises a raft and piles supporting columns and core walls. The numbers and layout of piles, and the pile size, were obtained from a series of trial analyses through collaboration between the geotechnical engineer and the structural designer. The pile depth was determined by the geotechnical engineer, considering the performance and capacity of piles. The pile layout was selected from the various options considered, and comprised 172 2.5 m diameter bored piles, socketed into the soft rock layer and connected to a 5.5 m thick raft. The layout is presented in Figure 15.18.

*Figure 15.17* Inferred contours of top of soft rock – Incheon Tower.

## 15.6.4 Loadings

Typical loads acting on the tower were as follows:

Vertical dead plus live load: Pz(DL + LL) = 6622 MN
Horizontal wind loads: Px(WL) = 149 MN Py(WL) = 115 MN
Horizontal earthquake loads: Px(E) = 105 MN Py(E) 105 MN
Wind load moments: Mx(WL) = 12578 MN m My(WL) = 21173 MN m
Wind load torsional load: Mz(WL) = 1957 MN m

The vertical loads (DL + LL) and overturning moments (Mx, My) were represented as vertical load components at column and core locations. The load combinations, as provided by the structural designer, were adopted throughout the geotechnical analysis, and 24 wind load combinations were considered.

*Table 15.9* Summary of geotechnical parameters

| *Strata* | $E_v$ *(MPa)* | $E_h$ *(MPa)* | $f_s$ *(kPa)* | $f_b$ *(MPa)* |
|---|---|---|---|---|
| UMD | 7–15 | 5–11 | 29–48 | – |
| LMD | 30 | 21 | 50 | – |
| Weathered soil | 60 | 42 | 75 | – |
| Weathered rock | 200 | 140 | 500 | – |
| Soft rock (above EL-50 m) | 300 | 210 | 750 | 12 |
| Soft rock (below EL-50 m) | 1700 | 1190 | 750 | 12 |

Note: $E_v$ = vertical modulus; $f_s$ = ultimate shaft friction; $E_h$ = horizontal modulus; $f_b$ = ultimate end bearing.

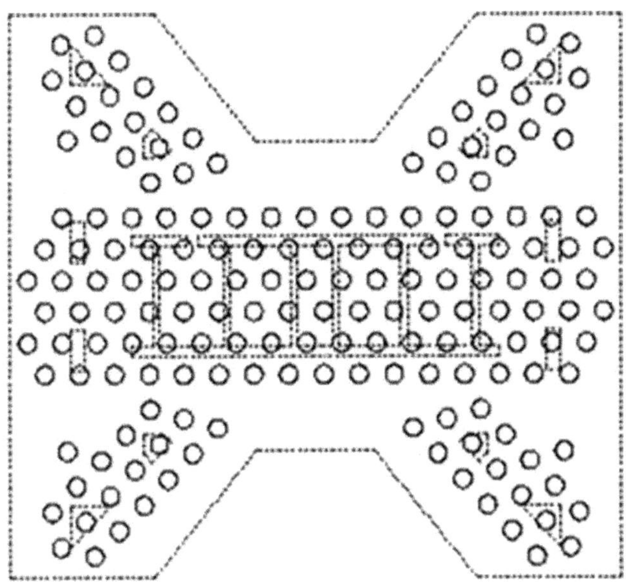

*Figure 15.18* Pile layout plan.

### 15.6.5 Assessment of pile capacities

The geotechnical capacities of piles were estimated from the shaft friction and end bearing capacities, and the required pile length was based on these geotechnical capacities. For a large pile group founding in weak rock, the overall settlement behaviour of the pile group could control the required pile lengths rather than the overall geotechnical capacity. In this case, the soft rock layer was considered to be a more appropriate founding stratum than the overlying weathered rock, in particular the soft rock below EL-50 m. This was because this stratum provided a more uniform stiffness and therefore was likely to result in a more consistent settlement behaviour of the foundation. The basic guidelines to establish the pile founding depth were

- Minimum socket length in soft rock = 2 diameters
- Minimum toe level = EL-50 m

The pile depths required to control settlement of the tower foundation were greater than those required to provide the necessary geotechnical capacity. The pile design parameters for the weathered/soft rock layer are shown in Table 15.10 and were estimated on the basis of the pile test results in the adjacent site and the ground investigation data such as pressuremeter tests and rock core strength tests.

*Table 15.10* Ultimate capacities for pile analysis

| *Material* | *Ultimate friction* $f_s$ *(kPa)* | *Ultimate end bearing* $f_b$ *(MPa)* |
|---|---|---|
| Weathered rock | 500 | 5 |
| Soft rock | 750 | 12 |

### 15.6.6 Assessment of vertical pile behaviour

The vertical pile head stiffness for each of the 172 foundation piles under serviceability loading conditions (DL + LL) was assessed using in-house computer programs CLAP and GARP. CLAP, a development of the DEFPIG program, was used to assess the geotechnical capacities, interaction factors and stiffness values for each pile type under serviceability loading for input into the group assessment. CLAP computed the distributions of axial and lateral deflections, rotations and axial and lateral loads and moments, at the top of a group of piles, subjected to a combination of vertical loads, lateral loads, moments, and torsion. GARP (Small and Poulos, 2007) was used to assess the group foundation behaviour of the Tower.

Individual pile vertical stiffness values were computed, and it was found that the outer piles were stiffer. The analysis was non-linear, and therefore the higher stiffness values for the outer piles degraded more rapidly under loading than the central piles. It was considered that foundation behaviour could be simulated more realistically by using the individual pile stiffness values, rather than an average value for all piles within the group. Lower and upper bound estimates of pile stiffness values were provided to the structural engineers to include in their analyses, in order to capture the upper and lower bound behaviour of the raft foundation and the potential impact on the tower superstructure.

### 15.6.7 Predicted settlements

The overall settlement of the foundation system was estimated during all three stages of design, using the available data at that stage, and relevant calculation techniques. Table 15.11 summarises the predicted maximum settlements, and indicates that the very simple Category 2 equivalent pier estimate during the first stage was of a similar order to that predicted from more refined Category 3 estimates carried out during the later stages of design.

### 15.6.8 Assessment of lateral pile behaviour

One of the critical design issues for the tower foundation was the performance of the pile group under lateral loading. Therefore, several numerical analysis programs were used in order to validate the predictions of lateral behaviour obtained. The numerical modelling packages used in the analyses were

- 3D FE computer program PLAXIS 3D Foundation
- Computer program DEFPIG (Poulos, 1990)
- The in-house computer program CLAP
- 3D FE structural analysis programs (MIDAS, ETABS, SAFE) that included the effect of soil–structure interaction

*Table 15.11* Summary of predicted settlements

| *Design stage* | *Method* | *Predicted settlement (mm)* | *Remarks* |
|---|---|---|---|
| 1 (preliminary) | Equivalent pier | 75 | Average settlement |
| 2 (detailed) | Program GARP | 67 | Maximum, taking account of all eight ground profiles |
| 3 (final) | Program PLAXIS3D | 56 | Maximum, adopting a single representative ground profile |

*Table 15.12* Summary of lateral stiffness of pile group and raft

| *Horizontal load (MN)* | *Pile group disp. (mm)* | *Lateral pile stiffness (MN/m)* | *Lateral raft stiffness (MN/m)* | *Total lateral stiffness (MN/m)* |
|---|---|---|---|---|
| 149 (x dirn.) | 17 | 8,760 | 198 | 8958 |
| 115 (y dirn.) | 14 | 8,210 | 225 | 8435 |

PLAXIS 3D provided an assessment of the overall lateral stiffness of the foundation. The programs DEFPIG and CLAP were used to assess the lateral stiffness provided by the pile group, assuming that the raft is not in contact with the underlying soil. A separate calculation was carried out to assess the lateral stiffness of the raft and basement. Table 15.12 presents the computed lateral stiffness for the piled raft foundation obtained from these analyses.

### 15.6.9 Assessment of pile group rotational stiffness

An assessment of the rotational spring stiffness values at selected pile locations within the foundation was undertaken using the in-house computer program CLAP. To assess the rotational spring constant at each pile location, the average dead load, horizontal load (x and y direction) and moment (about the x, y and z axes) were applied to each pile head. The passive resistance of the soil surrounding the raft, and the friction between the soil and the raft base, were not included in the analysis as it was assessed that the base friction of the raft footing and the passive resistance of the soil on the raft would be relatively small when compared to the lateral resistance of the piles. Table 15.13 presents a summary of the assessed rotational spring stiffness values obtained from the analysis for four piles considered to represent the range of values for different piles within the pile foundation.

The overall torsional stiffness of the piled raft was assessed using the computer program PLAXIS 3D Foundation. A schematic of the PLAXIS model analysed is given in Figure 15.19. The overall torsional stiffness of the piled raft estimated using PLAXIS was 10,750,000 MN m/radian, which was approximately equivalent to 16 mm displacement at the edge of the raft for the applied torsional moment of 1956 MN m applied at the centre of the raft.

### 15.6.10 Cyclic loading due to wind action

Wind loading for the tower structure was quite severe, and therefore in order to assess the effect of low frequency cyclic wind loading, an assessment was made on the basis of

*Table 15.13* Rotational spring constants including horizontal loads applied at the pile heads

| *Pile* | | *Pile head angular rotation (rad)* | *Pile head rotational spring stiffness (MN m/rad)* |
|---|---|---|---|
| 3 | Maximum | 0.094 | 2680 |
| | Minimum | 0.036 | 1380 |
| 27 | Maximum | 0.144 | 1750 |
| | Minimum | 0.056 | 903 |
| 70 | Maximum | 0.126 | 2000 |
| | Minimum | 0.049 | 1030 |
| 78 | Maximum | 0.187 | 1350 |
| | Minimum | 0.073 | 700 |

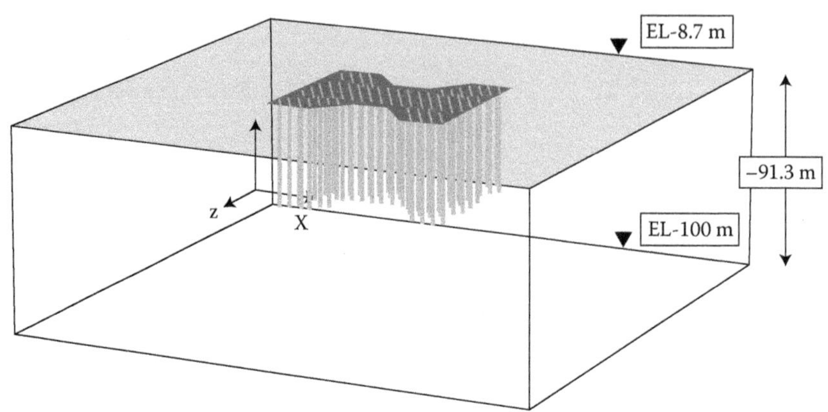


*Figure 15.19* Schematic of PLAXIS 3D model.

the criterion suggested in Equation 7.4 of Chapter 7. The factor η was selected to be 0.5, based on experience with similar projects. To assess the half-amplitude of cyclic axial wind induced load, the difference in pile load between the following load cases was computed.

- CASE A: 0.75(DL + LL)
- CASE B: 0.75(DL + LL + $WL_x$ + $WL_y$)

where DL is the dead load, LL is live load, $WL_x$ is vertical load resulting from x-component of wind, $WL_y$ is vertical load resulting from y-component of wind.

The difference in axial load between the two load cases was the half-amplitude of the cyclic load ($S_c$*). Table 15.14 summarises the results of the cyclic loading assessment and Figure 15.20 shows the assessed factor for each pile within the foundation system. The assessment indicates that, for all piles, the ratio of cyclic load to pile shaft capacity was less than 0.5, and so degradation of shaft capacity due to cyclic loading was unlikely to occur.

### 15.6.11 Pile load tests

A total of five pile load tests were undertaken, four on vertically loaded piles via the O-cell procedure, and one on a laterally loaded pile jacked against one of the vertically loaded test piles. For the vertical pile test, two levels of O-cells were installed in each pile, one at the pile tip and another located between the weathered rock layer and the soft rock layer. The cell movement and pile head movement were measured by LVWDTs in each of four locations, and the pile strains were recorded by the strain gauges attached to the vertical steel bars. The monitoring system is shown schematically in Figure 15.21.

The double cell test system was planned to obtain more accurate and detailed data for the main bearing layer, and so the typical test was performed in two stages as shown in

*Table 15.14* Summary of cyclic loading assessment

| *Quantity* | *Value* |
|---|---|
| Maximum half-amplitude cyclic axial wind load $S_c$* (MN) | 29.2 |
| Maximum ratio $\eta = S_c^*/R_{gs}^*$ | 0.43 |
| Cyclic loading criterion satisfied? | Yes |

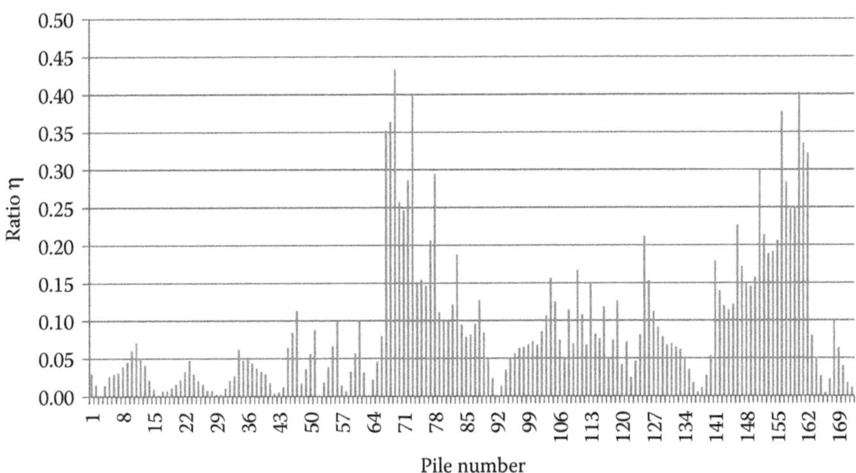


*Figure 15.20* Results of cyclic loading analysis.

Figure 15.22. The Stage 1 test was focused on the friction capacity of weathered rock and the movement of soft rock socket and pile shaft in the weathered rock layer, while Stage 2 focussed on the friction and end bearing capacities of the soft rock, with the upper O-cell open to separate the soft rock socket from the remaining upper pile section.

The vertical test piles were loaded up to a maximum one way load of 150 MN in about 30 incremental stages, in accordance with ASTM recommended procedures. The dynamic loading–unloading test was carried out at the design loading ranges by applying 20 load cycles to obtain the dynamic characteristics of the pile rock socket.

A borehole investigation was carried out at each test pile location to confirm the ground conditions and confirm the pile length and soft rock socket depth of 5–6 m before piling work commenced, and also to properly match the test results to the actual ground strata. A summary of the vertical pile test results is shown in Table 15.15, which is based on the pile test analysis performed by the Load Test Corporation.

Test Pile 3 (TP3) results are not shown herein due to construction defects identified in the pile via sonic logging tests (Poulos et al., 2013a); thus, these test results were ignored

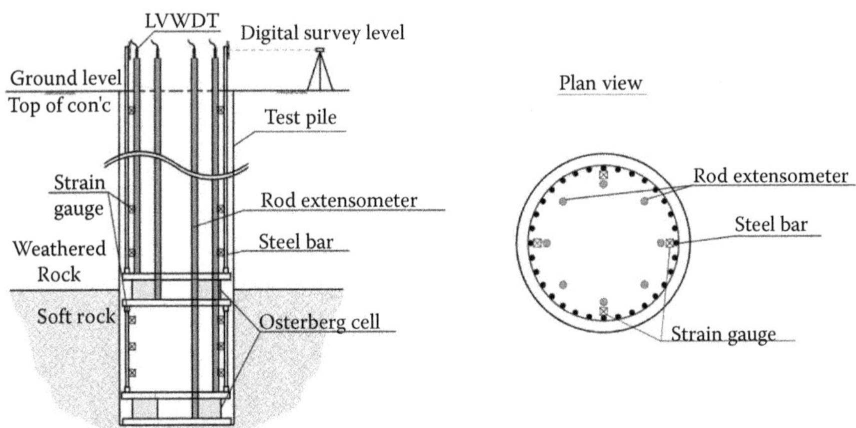


*Figure 15.21* Schematic of monitoring for vertical pile load test.

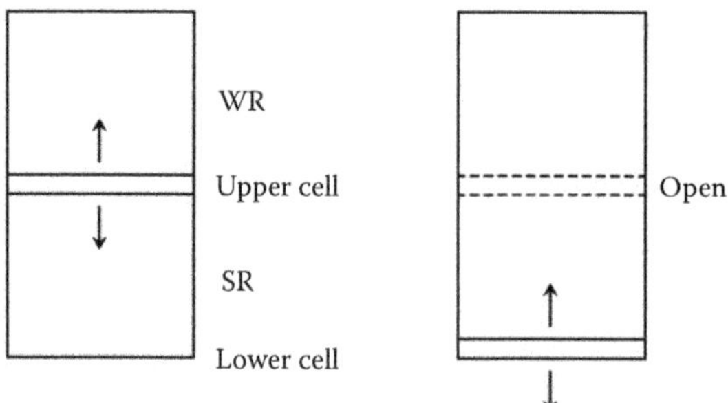


*Figure 15.22* Typical procedure of O-cell test.

in obtaining the average results. While the overall performance of the test piles exceeded expectations, TP3 highlighted the possibility that variability in rock elevation within a short distance could affect the overall quality of the pile and may require careful assessment, during construction, of the pile excavation and the quality of the rock at all levels. The pile testing program also demonstrated that the foundation system could still be optimised, given the higher than anticipated shaft and base resistances that were obtained in the other pile tests.

A lateral pile load test was also performed after excavation of about 8 m of the upper soil to simulate a similar ground condition and performance as designed for the tower foundation. Both the test pile (TP5) and the reaction pile (TP4) were monitored by inclinometers to obtain the lateral displacement along the pile depth, and strain gauges were installed to obtain the stress in the pile section, and eventually the bending moment distribution along the pile shaft. An LVWDT was used for each pile head displacement measurement. A schematic diagram of the monitoring system is shown in Figure 15.23.

The lateral test pile was subjected to a maximum lateral load of 2.7 MN. The dynamic load–unload test was carried out at 900, 1350 and 1800 kN by applying 20 cycles to obtain the lateral dynamic performance of the pile, especially within the marine clay layer. The load–pile head displacement relationship from the lateral pile test is shown in Figure 15.24. The result indicated that the lateral stiffness of the pile was greater than expected during the initial loading stage, presumably due to the repeated loading condition and also due to the over-consolidated ground conditions arising from excavation. The stiffer behaviour under cyclic loading is summarised in Table 15.16. This stiffer pile behaviour could potentially be considered in the final structural design of the tower foundation system, as well as for the predicted pile group movement.

*Table 15.15* Summary of vertical pile test results (allowable pile bearing capacities)

| | | | *Pile test* | | | |
|---|---|---|---|---|---|---|
| *Strata* | | *Design value* | *TP1* | *TP2* | *TP4* | *Aver.* |
| Soft rock | End bearing (MPa) | 4.0 | 6.3 | 9.0 | 9.2 | 8.1 |
| | Friction (kPa) | 350 | 743 | 897 | 663 | 767 |
| Weathered rock | Friction (kPa) | 250 | 357 | 527 | 178 | 354 |

Note: FOS = 3 is applied for end bearing from ultimate or test load; FOS = 2 for shaft friction from yield loading point.

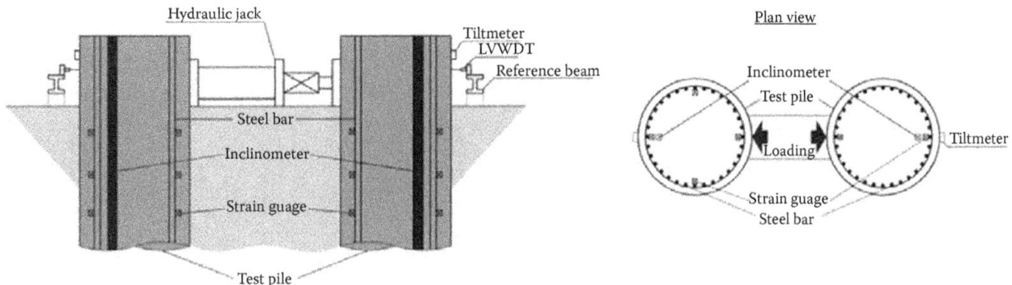


*Figure 15.23* Schematic of monitoring for lateral pile load test.

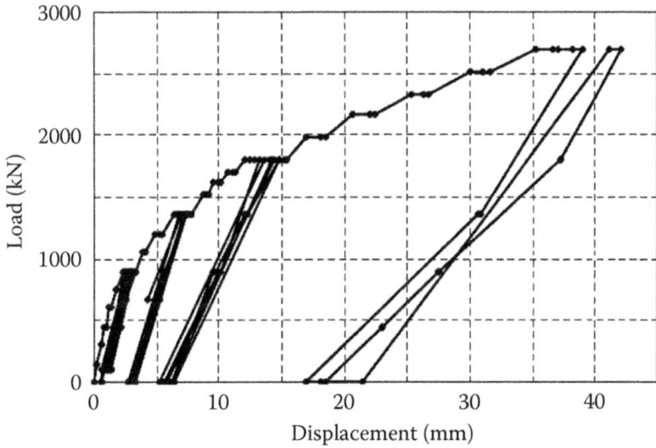


*Figure 15.24* Load versus displacement curve TP5.

### 15.6.12 Summary and lessons learned

This case involved the design and testing process of a piled raft foundation system for a super high-rise building to be located within reclaimed area in Songdo, Korea. The design process involved three principal phases, namely concept design, the main design phase, the final design/study phase, together with the vertical and lateral load testing programs.

The use of a suite of commercially available and in-house computer programs allowed the detailed analysis of the large group of piles to be undertaken, incorporating pile–soil–pile interaction effects, varying pile lengths and varying ground conditions in the foundation design. An independent FE analysis using readily available commercial programs was used to include the effect of soil–structure interaction and to include the impact of the foundation system on the overall behaviour of the tower.

*Table 15.16* Lateral stiffness of the test pile

| Design stiffness (MN/m) | *Measured secant stiffness of test pile (MN/m)* | | | |
|---|---|---|---|---|
| | *Static* | | *Dynamic* | |
| | *0–900 kN* | *900–1350 kN* | *0–900 kN* | *900–1350 kN* |
| 86–120 | 294 | 97 | 488 | 326 |

The final design process was extended in order to obtain the actual response of the ground and the piles due to various loadings. From the results of pile load tests carried out in the post-design period, the prediction of pile behaviour could be refined and the pile capacities could be updated, and could potentially lead to a more cost-effective foundation design.

An extensive high-quality vertical and lateral pile testing program was developed and performed for the project, and it was shown that the pile behaviour and capacities were higher than expected, so that it could be beneficial to revise some of the more conservative assumptions made in the design.

## 15.7 CASE 5: TOWER ON KARSTIC LIMESTONE, SAUDI ARABIA

### 15.7.1 Introduction

Karstic limestone is relatively widespread around the world, including many parts of the Middle East. The identification of cavities in karstic limestone often creates, at best, a sense of anxiety among foundation designers, who may then proceed to take extreme measures to overcome the perceived dangers and high risks associated with the proximity of cavities to a foundation system.

For a high-rise project in Jeddah, Saudi Arabia, involving a tower over 390 m high, potentially karstic conditions were identified in some parts of the site. Figure 15.25 shows an architectural rendering of the tower which is of a twisted form. A piled raft foundation system was developed for this tower, as it was considered that such a system would allow the raft to redistribute load to other piles in the group if cavities caused a reduction of capacity or stiffness in some piles within the group.

A brief description of the foundation design aspects of the project is presented below, and then a post-design investigation is described for the assessment of the consequences on foundation performance of cavities being present within the underlying limestone.

*Figure 15.25* Architectural rendering of tower in Jeddah, Saudi Arabia.

The key challenges in this project were to assess whether the adverse effects on foundation performance of cavities within the limestone would be within acceptable limits, or whether special treatment would be required to provide an adequate foundation system. A more complete description of this case is given by Poulos et al. (2013b).

## 15.7.2 Geological and geotechnical conditions

The city of Jeddah is located within the Makkah quadrangle in the southern part of the Hijaz geographic province in Saudi Arabia. Eastward of the flat, low-lying coastal plain are the Sarawat mountains that culminate in a major erosional escarpment that has resulted from uplift associated with Red Sea rifting. The underlying reefoidal limestone is considered to be a Quaternary deposit and is raised in some locations to about 3–5 m above mean sea level, and is underlain by silty sand and gravel.

The reefoidal limestone is the dominant deposit in the Jeddah area. All the available boreholes indicate the presence of coastal coralline limestone (coral reef deposits) which contain fresh shells and are typically cavernous in nature. Above these limestone deposits is a surficial soil layer which consists mainly of aeolian sands and gravels that were deposited in Holocene times.

A plan of the site showing borehole locations is presented in Figure 15.26. Originally, 12 boreholes were drilled to depths of between 40 and 75 m, and subsequently, two deeper boreholes were drilled to 100 m. The borehole data showed that the soil profile consists mainly of coralline limestone deposits that were highly fractured, and could contain cavities. SPTs carried out in the boreholes showed that the coralline limestone was dense to very

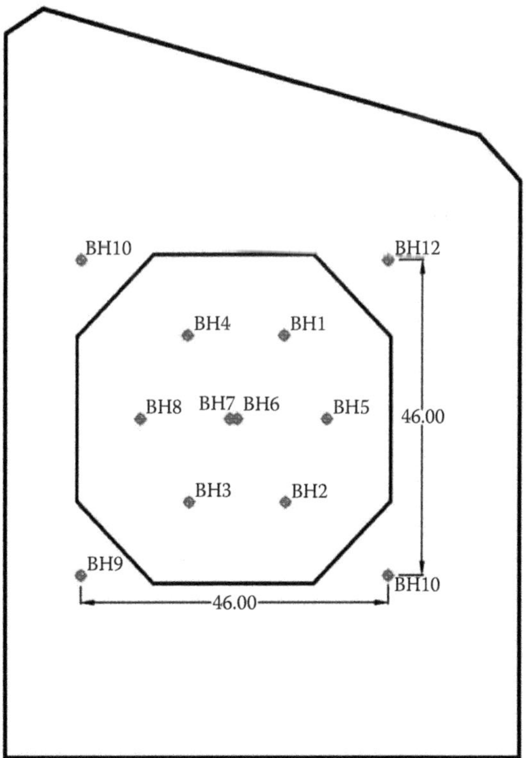


*Figure 15.26* Site plan and borehole locations.

dense. Figure 15.27 shows the stratigraphy derived from a typical borehole, BH05. Features of this particular borehole are the low RQD values of the recovered core samples, the low values of total core recovery (TCR), especially below a depth of about 25 m, the occasional presence of small cavities, and the presence of what appeared to be very loose sediments between about 55 and 63 m below ground surface. It was possible that the process of drilling may have affected the cores and made them appear to be weaker than they were in reality. The groundwater table ranged between 2.1 and 3.8 m below ground surface.

Cross-hole seismic testing was carried out at boreholes BH07 and BH08, and distributions with depth of P-wave velocity and shear wave velocity were obtained. These distributions indicated increasing velocities with depth up to about 20 m, with relatively little systematic increase at greater depths. There was no evidence of a hard layer within the depths investigated, and this conclusion was consistent with the borehole data.

### 15.7.3 Geotechnical model

The quantitative data from which engineering properties could be estimated was relatively limited, and included the following:

1. Unconfined compression test
2. Shear wave velocity data
3. Pressuremeter testing
4. SPT data in the weaker strata

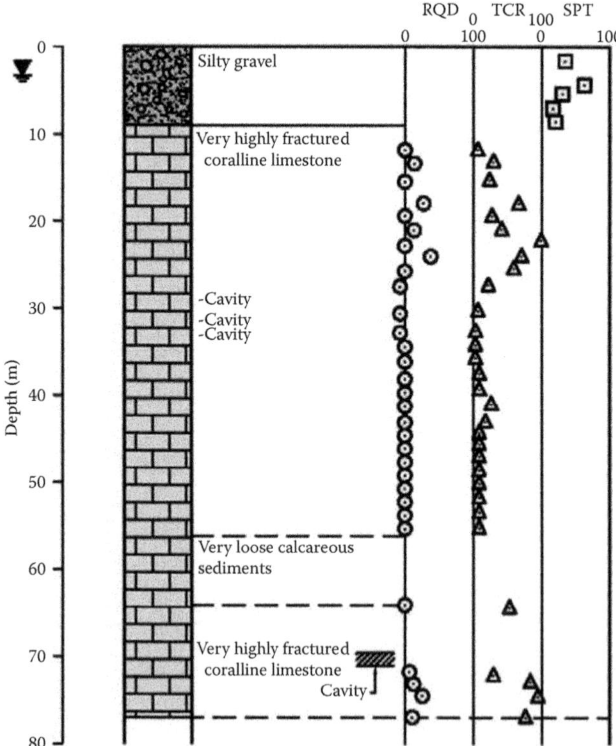


*Figure 15.27* Details of BH05. (Adapted from Poulos, H.G., Small, J.C. and Chow, H.S.W. 2013b. Foundation design for high-rise towers in karstic ground. In: J.L. Withiam, K-K. Phoon and M.H. Hussein (Eds.), *Foundation Engineering in the Face of Uncertainty*, ASCE GSP229, pp. 720–731. Courtesy of ASCE.)

Use was made of these data to assess the engineering properties which were required for the settlement analysis, primarily the Young's modulus of the ground deposits (long-term drained values), the ultimate distribution of pile shaft friction with depth and the ultimate pile end bearing capacity. The values adopted for the analyses are summarised in Table 15.17, and the procedures adopted to assess each of these parameters are described briefly below.

#### 15.7.3.1 Long-term Young's modulus

The assessment of this parameter was critical as it greatly influenced the predicted settlement. Three different methods of assessment were used:

1. Modulus values from the PMTs
2. Values correlated to UCS via the correlation $E_s' = 100$ UCS, where $E_s'$ is long-term Young's modulus
3. Values derived from the small-strain Young's modulus values obtained from shear wave velocity measurements, but scaled by a factor of 0.2 to allow for the effects of practical strain levels, as discussed in Chapter 6

Figure 15.28 compares the values obtained from each of these three approaches. On the basis of these data, the following assumptions were originally made:

1. From the surface to a depth of 20 m, an average long-term Young's modulus (for vertical loading), $E_s'$, is 150 MPa
2. From 20 to 50 m, $E_s' = 200\,\text{MPa}$
3. From 50 to 70 m, $E_s' = 400\,\text{MPa}$
4. Below 70 m, $E_s' = 1000\,\text{MPa}$, which reflects the greater stiffness expected because of the smaller levels of strain within the ground at greater depths

Subsequent to these initial assessments, a load test was undertaken using the O-cell technique. The pile head stiffness derived from this test was considerably larger than that implied by the initially selected values of Young's modulus. Accordingly, the initially selected values were multiplied by a factor of 3 for the final settlement prediction.

#### 15.7.3.2 Ultimate pile shaft friction and end bearing

Use was made of correlations between the ultimate shaft friction, $f_s$, and end bearing, $f_b$, with UCS. For the reefoidal coral deposits, the following conservative relationship was used for the assessment of ultimate shaft friction $f_s$:

$$f_s = 0.1(\text{UCS})^{0.5}\,\text{MPa} \tag{15.3}$$

where UCS is the unconfined compressive strength (MPa).

*Table 15.17* Soil properties used for tower analysis

| *Depth at bottom of geo-unit (m)* | *Description of geo-unit* | *$E_v$ (MPa)* | *$f_s$ (MPa)* | *$f_b$ (MPa)* |
|---|---|---|---|---|
| 20 | Coralline limestone (1) | 450 | 0.2 | 2 |
| 50 | Coralline limestone (2) | 600 | 0.2 | 9.8 |
| 70 | Coralline limestone (3) | 1200 | 0.35 | 9.8 |
| 100 | Coralline limestone (4) | 3000 | 0.4 | 9.8 |

Note: $E_v$ = modulus of soil for vertical pile response; $f_s$ = limiting pile shaft skin friction; $f_b$ = limiting pile base load.

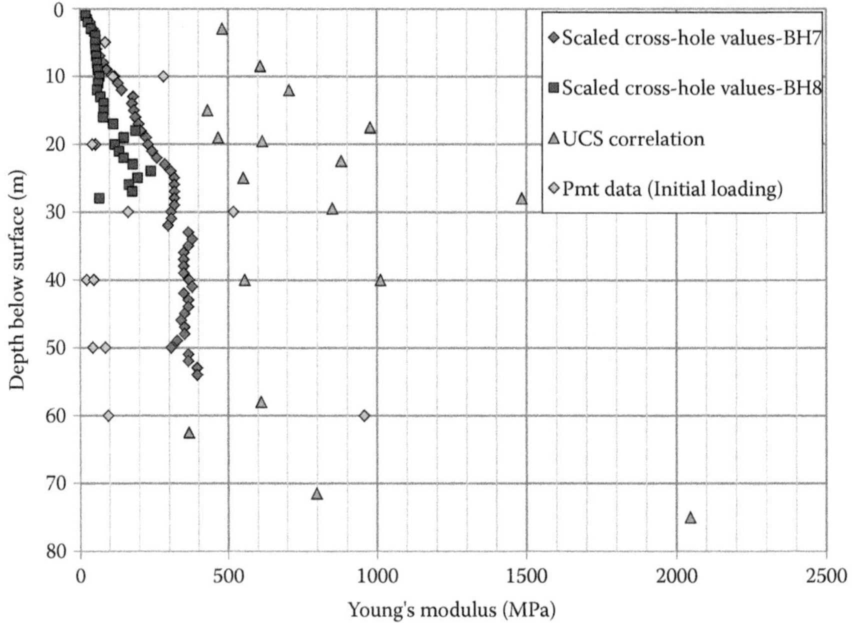


*Figure 15.28* Young's modulus values derived from various sources. (Adapted from Poulos, H.G., Small, J.C. and Chow, H.S.W. 2013b. Foundation design for high-rise towers in karstic ground.In: J.L. Withiam, K-K. Phoon and M.H. Hussein (Eds.), *Foundation Engineering in the Face of Uncertainty*, ASCE GSP229, pp. 720–731. Courtesy of ASCE.)

The average ultimate shaft friction for the upper 50 m was thus taken to be 0.2 MPa (200 kPa). The subsequent pile load test revealed that this was a conservative estimate of shaft friction, as values of about 500 kPa were mobilised along some portions of the test pile, with an average value of about 310 kPa.

The following correlation for end bearing capacity, suggested by Zhang and Einstein (1998), was employed:

$$f_b = 4.8(UCS)^{0.5} \text{ MPa} \tag{15.4}$$

On this basis, for an average UCS of 4 MPa, $f_b$ was 9.6 MPa. This value assumed that there were no cavities in the area of influence of the base of a pile.

### 15.7.4 Tower foundation details

Figure 15.29 shows the foundation layout for the tower. The basement of the building was to be located at shallow depth above the water table. The raft beneath the tower was taken to be 5.5 m thick and was to be supported on 145 bored piles 1.5 m in diameter. A pile length of 40 m was assessed to be required to support the stated working load of 22 MN per pile, based on a factor of safety of about 2.4. For the analyses described herein, only the central 5.5 m thick raft and 40 m long piles were analysed. The total vertical load for serviceability conditions was specified as 2859 MN.

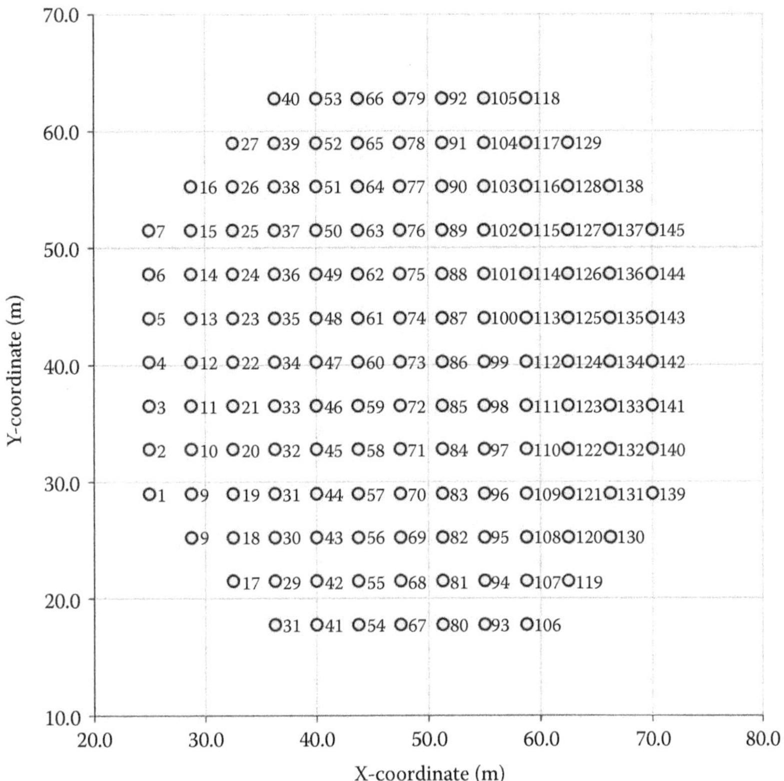


*Figure 15.29* Pile layout for tower.

### 15.7.5 Foundation analyses for design

At the design stage, analyses were undertaken using the computer program GARP (Small and Poulos, 2007). The complete foundation system was divided into 2095 elements with 6484 nodes, and no account was taken in this present analysis of the stiffness of the superstructure. From the GARP analysis, the maximum settlement was predicted to be approximately 50 mm.

### 15.7.6 Study of effects of cavities on foundation performance

The initial analyses assumed that no significant cavities existed below the pile toes. If cavities were to be found during construction, then it would be necessary to reassess the performance of the foundation system and make provision for grouting of the cavities if this was deemed to be necessary. Thus, subsequent to the foundation design, a further series of analyses was undertaken to investigate the possible effects of cavities on the settlements and also on the raft bending moments and pile loads. For these analyses, the commercially available program PLAXIS 3D was used.

Figure 15.30 shows the pile group and the raft as modelled by the 3D FE software Plaxis 3D. The raft was octagonal in shape and 5.5 m thick while the piles were 40 m long and 1.5 m diameter and were laid out on a rectangular grid at 3.75 m centre to centre spacings. In plan, the raft was 47.5 m wide and 47.5 m high (from flat to flat of the octagon).

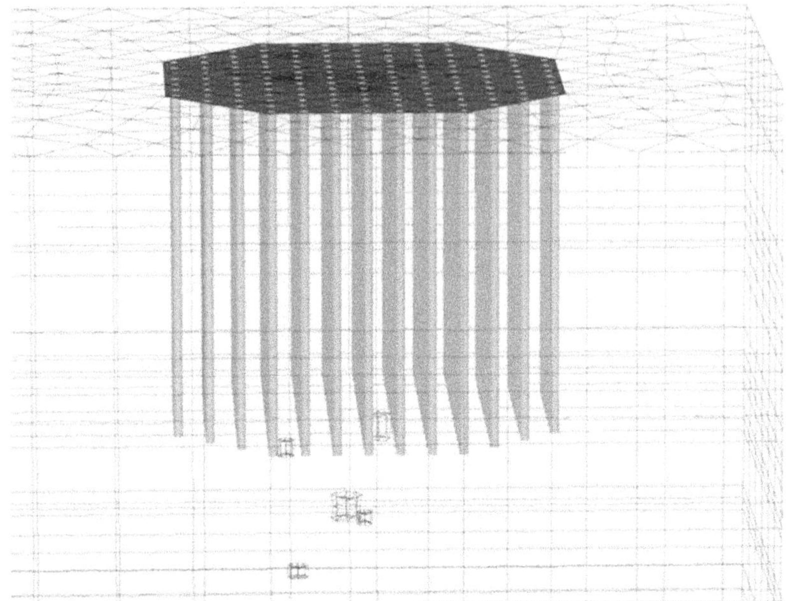

*Figure 15.30* PLAXIS 3D FE mesh for the piled raft. (Adapted from Poulos, H.G., Small, J.C. and Chow, H.S.W. 2013b. Foundation design for high-rise towers in karstic ground. In: J.L. Withiam, K-K. Phoon and M.H. Hussein (Eds.), *Foundation Engineering in the Face of Uncertainty*, ASCE GSP229, pp. 720–731. Courtesy of SEAGS.)

Firstly, the effect of a single cavity at different locations along the centre line of the raft at different depths was examined. The cavity was introduced into the FE mesh at the depths shown in Table 15.18, and was taken as being 3 m wide by 2 m deep.

It may be seen from the table that the vertical displacement of the raft does not change much when the cavity is within the pile group (i.e. at a depth of less than 40 m). However, when the cavity is below the toe of the piles at about 50–60 m depth, the deflection reaches its maximum value, which is about 8 mm greater than the no-cavity case.

#### 15.7.6.1 Random cavities beneath the piled raft

Generally the locations of cavities beneath the foundation are not known, and only cavities found in specific boreholes can be identified. It was therefore of interest to gauge the effect of

*Table 15.18* Deflection of central Point of raft for a single cavity at various depths along centreline

| *Depth of cavity (m)* | *Max. raft displacement (mm)* |
|---|---|
| 0 | 55.7 |
| 20 | 55.5 |
| 40 | 56.7 |
| 50 | 58.0 |
| 60 | 58.4 |
| 70 | 55.9 |
| 80 | 55.8 |
| 90 | 55.7 |
| 100 | 55.7 |

boreholes at random locations and of random sizes. To do this, a random number generator was used to select a random number between 0 and 1 and then this was used to obtain the location and size of the cavity. A different scaling was used for selecting a given location or size, for example, the X-coordinate of the centre of the cavity was scaled so that it had to lie within the confines of the raft, and the depth was scaled so that its centre lay within 70 m depth below the surface.

The number of randomly placed cavities was limited to five for each of the cases listed in Table 15.19. A new 3D mesh had to be generated for each case because the location and sizes of the cavities changed. One example of the location of the cavities is shown in Figure 15.31.

Results of the analyses are presented in Table 15.19, where it may be seen that the vertical deflection of the central point of the raft changed from 65 mm for Case 3 to 74 mm for Case 2, a range of 9 mm. Thus, multiple cavities resulted in an increase in maximum settlement of about 24 mm as compared with the no-cavity case. The piled raft system therefore appeared to be effective in smoothing out the effect of the cavities on the overall settlement of the foundation.

For Case 3 the vertical settlement contours of the raft are shown in Figure 15.32. It may be seen from the plot that the raft tilted due to the effect of the cavities, and that the maximum settlement was about 68 mm. This is because, in this case, the larger cavities were to the bottom left of the raft.

#### *15.7.6.2 Pile loads for random cavities*

The effect that the random set of cavities has on the loads in the piles may be seen from the plots of Figure 15.33a (pile 73 at centre of raft) and 15.33b (pile 142 at edge of raft).

*Table 15.19* Effects of randomly selected cavities

| | *Cavity location (centre)* | | *Depth below raft* | | | |
|---|---|---|---|---|---|---|
| *Case* | *X (m)* | *Y (m)* | *Top of cavity, Z1 (m)* | *Bottom of cavity, Z2 (m)* | *Diameter of cavity (m)* | *Raft displacement (mm)* |
| 1 | 1.875 | 0 | 40 | 43 | 3 | 72 |
| | −1.875 | −1.875 | 50 | 53 | 4 | |
| | 0 | 7.5 | 50 | 51.5 | 2 | |
| | −9.25 | 0 | 43 | 45 | 2 | |
| | −7.5 | −15 | 61.5 | 63 | 1.25 | |
| 2 | 11 | 13 | 34 | 35 | 2 | 74 |
| | 10 | 20 | 44 | 45 | 2 | |
| | −2 | 4 | 49 | 51 | 4 | |
| | −10 | −9 | 53 | 55 | 4 | |
| | 3 | 16 | 28 | 31 | 3 | |
| 3 | −13 | 10 | 48 | 51 | 4 | 68 |
| | −7 | 2 | 23 | 25 | 3 | |
| | 13 | −10 | 41 | 44 | 3 | |
| | 16 | 11 | 69 | 71 | 1 | |
| | 16 | −2 | 44 | 47 | 2 | |
| 4 | 2 | −7 | 59 | 62 | 2 | 65 |
| | 15 | 7 | 39 | 41 | 4 | |
| | −19 | −7 | 50 | 52 | 4 | |
| | −6 | −12 | 66 | 68 | 2 | |
| | 0 | 4 | 38 | 39 | 1 | |

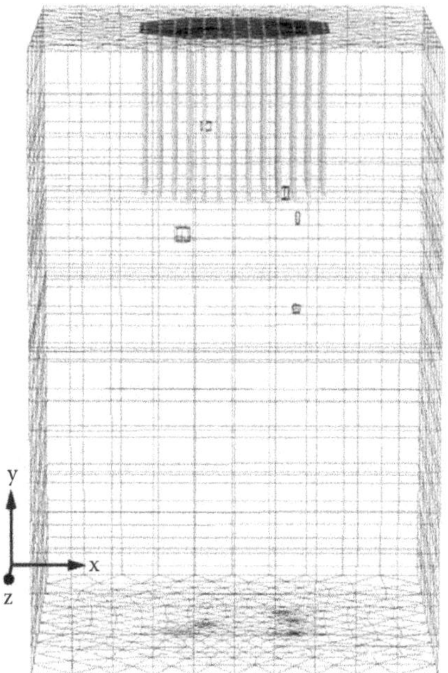


*Figure 15.31* Example of the location of randomly placed cavities in the FE mesh. (Adapted from Poulos, H.G., Small, J.C. and Chow, H.S.W. 2013b. Foundation design for high-rise towers in karstic ground. In: J.L. Withiam, K-K. Phoon and M.H. Hussein (Eds.), *Foundation Engineering in the Face of Uncertainty.* ASCE GSP229, pp. 720–731. Courtesy of SEAGS.)

The plots are presented for the case of no cavities in the foundation, and Case 2 (of Table 15.19) where there are five randomly placed cavities in the foundation. It may be seen from the figures that there was not a great deal of change in the axial load, with the load general decreasing in the centre pile and increasing in the edge pile for the locations of cavities in this example.

#### *15.7.6.3 Moments in raft for random cavities*

Computed moments in the raft are shown for the case of no cavities (Figure 15.34a) and for a set of random cavities (Case 2 of Table 15.19) in Figure 15.34b. The maximum and minimum moments are shown in Table 15.20.

The minimum moment (the largest absolute value) increased to 26,190 kN m/m from 23,120 kN m/m when cavities were present. This represented an increase of about 13% in the largest moment in the raft. Thus for design purposes, it was possible to make allowance for the effects of cavities by increasing the moment capacity of the raft by about 10%–15%.

### 15.7.7 Summary and lessons learned

From this post-design investigation of the piled raft foundation system for the tall tower in Jeddah, it was demonstrated that the consequences of cavities, while not insignificant, may not be as serious as might be feared, because of the inherent redundancy of the piled raft foundation system. The maximum settlement was increased by about 24 mm when multiple cavities were present, while the maximum bending moment was increased by about 13%.

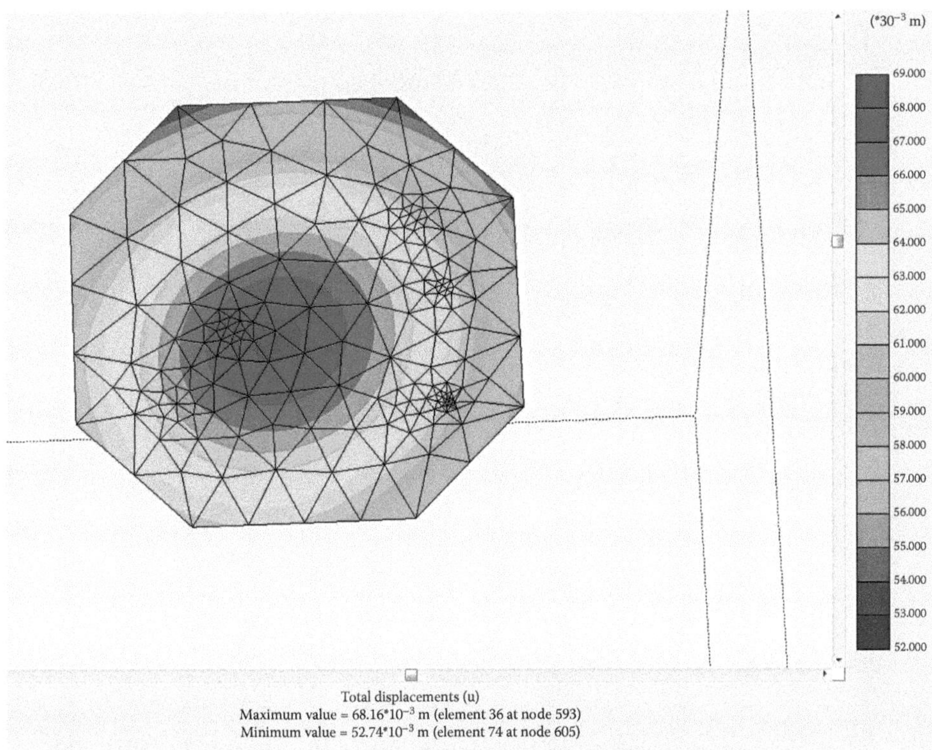


*Figure 15.32* Computed settlement contours (Case 3). Maximum settlement is 68 mm. (Adapted from Poulos, H.G., Small, J.C. and Chow, H.S.W. 2013b. Foundation design for high-rise towers in karstic ground. In: J.L. Withiam, K-K. Phoon and M.H. Hussein (Eds.), *Foundation Engineering in the Face of Uncertainty*, ASCE GSP229, pp. 720–731. Courtesy of SEAGS.)

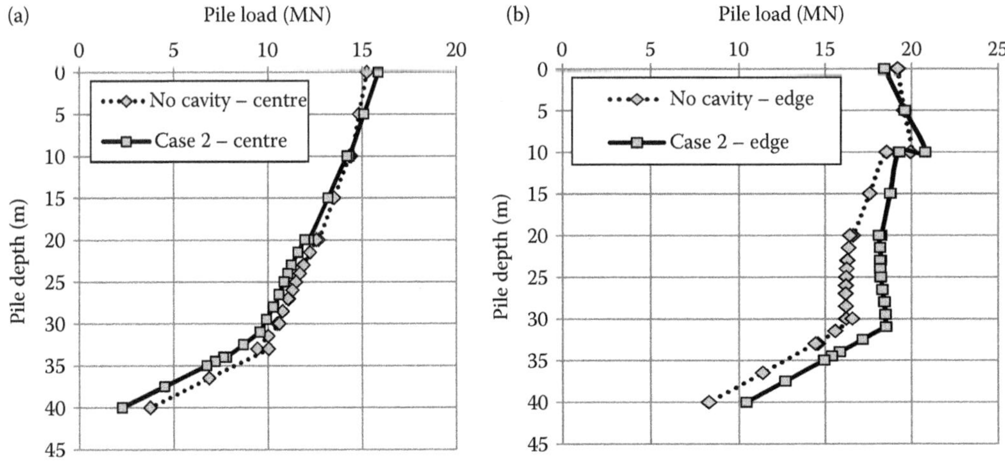


*Figure 15.33* Axial load with depth in edge and centre piles (with and without cavities): (a) centre pile and (b) edge pile. (Adapted from Poulos, H.G., Small, J.C. and Chow, H.S.W. 2013b. Foundation design for high-rise towers in karstic ground. In: J.L. Withiam, K-K. Phoon and M.H. Hussein (Eds.), *Foundation Engineering in the Face of Uncertainty*. ASCE GSP229, pp. 720–731. Courtesy of SEAGS.)

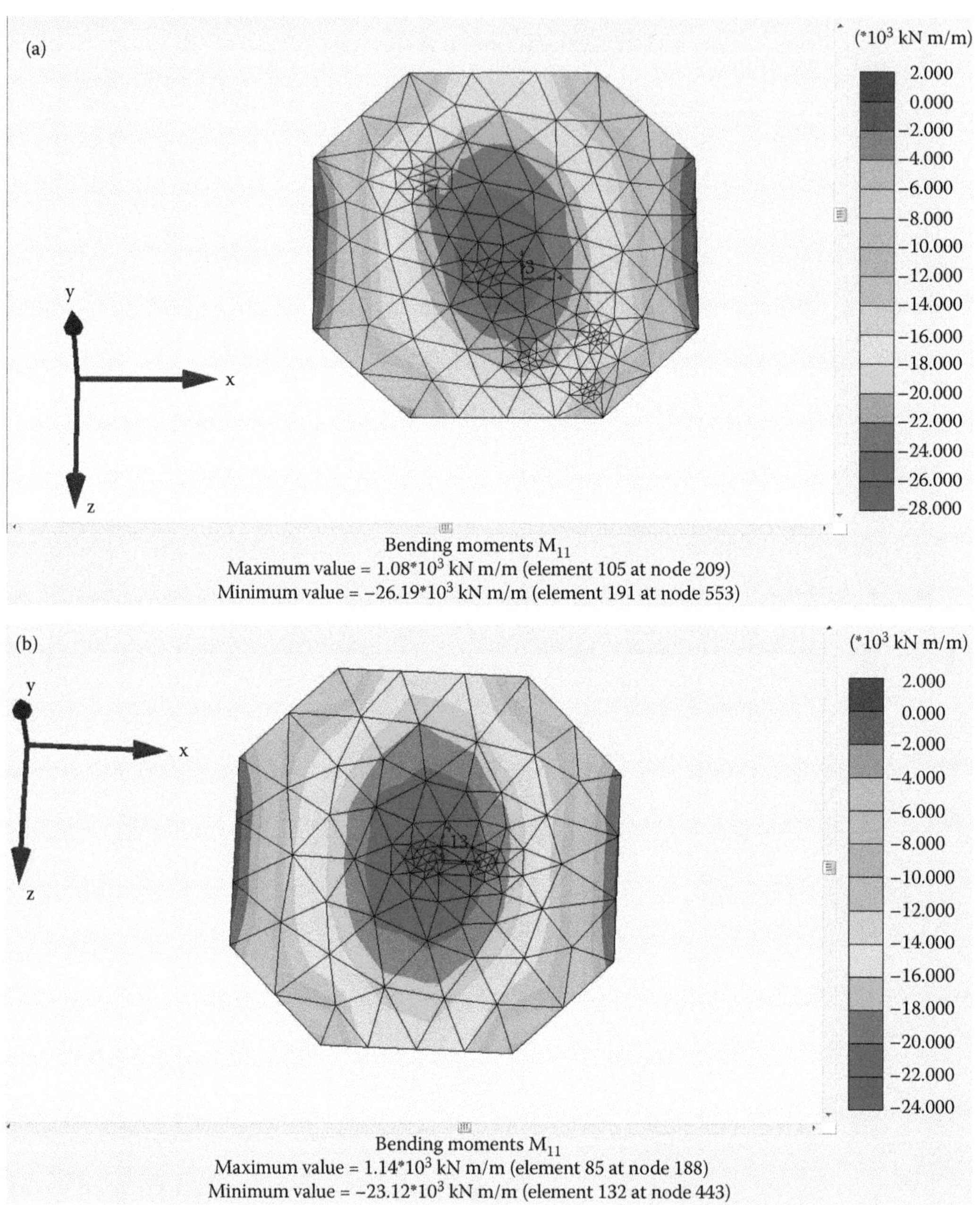


*Figure 15.34* (a) Moments in raft for no foundation cavities. (b) Moments in raft for Case 2 set of cavities. (Adapted from Poulos, H.G., Small, J.C. and Chow, H.S.W. 2013b. Foundation design for high-rise towers in karstic ground. In: J.L. Withiam, K-K. Phoon and M.H. Hussein (Eds.), *Foundation Engineering in the Face of Uncertainty*, ASCE GSP229, pp. 720–731. Courtesy of SEAGS.)

*Table 15.20* Maximum moment in raft

| *Problem* | *Maximum moment (kN m/m run)* | *Minimum moment (kN m/m run)* |
|---|---|---|
| No cavities | 1140 | −23,120 |
| Case 2 cavities | 1080 | −26,190 |

While the analyses undertaken were insufficient to enable a quantitative assessment of risk to be made, they did enable a good appreciation to be gained of the sensitivity of the computed foundation response to the presence of random cavities. Clearly, using redundant foundation systems may not only reduce the risks associated with building towers on karstic limestone, but may also provide a more economical foundation than using very long foundation piles in an attempt to carry foundation loads through the karstic zones.

# References


Abagnara, V., Poulos, H.G. and Small, J.C. 2012. Comparison of two piled raft analysis programs. *Proceedings of the 11th Australia-New Zealand Conference on Geomechanics*, Melbourne Australia, pp. 326–331.

Abdelrazaq, A. 2011. Validating the dynamics of the Burj Khalifa. *CTBUH Journal*, (2): 18–23.

Abdelrazaq, A. 2012. Validating the structural behavior and response of Burj Khalifa. *International Journal of High-Rise Buildings*, 1(1): 37–51.

Abdelrazaq, A., Baker, W.F., Chung, K.R., Wang, I. and Yom, K.S. 2004. Integration of design and construction of the tallest building in Korea, Tower Palace III. *CTBUH 2004 Seoul Conference*, Seoul, Korea.

Abdelrazaq, A., Poulos, H.G., Badelow, F. and Sung H-K. 2011. Foundation design and pile testing program for the 151 story Incheon Tower in a reclamation area, Incheon, Korea. *International Conference on Tall Buildings & Urban Habitat (CTBUH)*, Paper TS21-02, Seoul, Korea.

Abdoun, T., Dobry, R. and O'Rourke, T.D. 1997. Centrifuge and numerical modeling of soil-pile interaction during earthquake-induced liquefaction and lateral spreading. *Observation and Modelling in Numerical Analysis and Model Tests in Dynamic Soil-Structure Interaction Problems – Proceedings of Sessions Held in Conjunction with Geotechnical-Logan*, Logan, UT, pp. 76–90.

Abdrabbo, F. and Abouseeda, H. 2002. Effect of construction procedures on the performance of bored piles. *Deep Foundations*, 2: 1438–1454.

Abdrabbo, F. and El-Hansy, R. 1994. Pile load test is a proof test, but? *Proceedings of International Conference on Design and Construction of Deep Foundations*, Orlando, FL, USA, Vol. 2, pp. 493–512.

Ahmadi, M.M. and Kouhaki, B.M. 2016. New and simple equations for ultimate bearing capacity of a strip footing on two-layered clays: Numerical study. *International Journal of Geomechanics, ASCE*, August, Paper 06015014.

Alaghmandan, M., Bahrami, P. and Elnimeiri, M. 2014. The future trend of architectural form and structural system in high-rise buildings. *Architecture Research*, 4(3): 55–62.

Alaghmandan, M., Elnimeiri, M., Krawczyk, R.J. and von Buelow, P. 2016. Modifying tall building form to reduce along-wind effect. *CTBUH Journal*, (2): 34–39.

Ali, M.M. and Al-Kodmany, K. 2012. Tall buildings and urban habitat of the 21st century: A global perspective. *Buildings*, 2: 384–423.

Ali, M.M. and Moon, K.S. 2007. Structural developments in tall buildings: Current trends and future prospects. *Architectural Science Review*, 50.3: 205–223.

Ambraseys, N.N. and Sarma, K. 1967. The response of earth dams to strong earthquakes. *Geotechnique*, 17(3): 181–213.

Ameratunga, J., Sivakugan, N. and Das, B.M. 2016. *Correlations of Soil and Rock Properties in Geotechnical Engineering*. Springer, New Delhi.

Amini, A., Fellenius, B., Sabbagh, M., Naesgaard, E. and Buehler, M.M. 2008. Pile loading tests at Golden Ears Bridge. *Proceeding of the Canadian Geotechnical Conference*, Edmonton.

Amirsoleymani, T. 1991a. Elimination of excessive differential settlement by different methods. *Proceedings of the 9th Asian Regional Conference on Soil Mechanics and Foundation Engineering*, Bangkok, Vol. 2, pp. 351–354.

Amirsoleymani, T. 1991b. Removing soil layers under foundations to rotate buildings. *Proceedings of the 9th Asian Regional Conference on Soil Mechanics and Foundation Engineering*, Bangkok, Vol. 1, pp. 201–204.

Andrus, R.D. and Stokoe, K.H. 2000. Liquefaction resistance of soils from shear wave velocity. *Journal of Geotechnical and Geoenvironmental Engineering, ASCE*, 126(11): 1015–1025.

Ansal, A. and Pinto, P.S. 1999. Strong ground motions and site amplification. *Proceedings of the 2nd International Conference on Earthquake Geotechnical Engineering*, Lisbon, pp. 879–894.

API. 2000. *Recommended Practice for Planning, Designing and Constructing Fixed Offshore Platforms-Working Stress Design*. American Petroleum Institute, Washington, DC, RP2A-WSD.

Arya, S., O'Neill, M. and Pincus, G. 1979. *Design of Structures and Foundations for Vibrating Machines*. Gulf Publishing Company, Houston.

AS 2159. *1995. Piling – Design and Installation*. Standards Australia, Sydney, Australia.

AS 2159. *2009. Piling – Design and Installation*. Standards Australia, Sydney, Australia.

AS1170.0. 2002. *Structural Design Actions – Part 0: General Principles*. Standards Australia, Sydney, Australia.

AS1170.2. 2011. *Structural Design Actions – Wind Actions*. Standards Australia, Sydney, Australia.

AS1170.4. 2007. *Structural Design Actions – Earthquake Actions*. Standards Australia, Sydney, Australia.

ASCE/SEI. 2012. *Wind Tunnel Testing for Buildings and Other Structures*. American Society of Civil Engineers, Chicago, IL.

Ascher, K. 2011. *The Heights – Anatomy of a Skyscraper*. The Penguin Press, New York.

ASTM D1143 – 81 (re-approved 1994). *Standard Test Method for Piles under Static Axial Compressive Load*. ASTM.

ASTM D3966 –90 (re-approved 1995). *Standard Test Method for Piles under Lateral Loads*. ASTM.

ASTM D5882 – 07. *Standard Test Method for Low Strain Impact Integrity Testing of Deep Foundations*. ASTM.

ASTM D6760 – 02. *Standard Test Method for Integrity Testing of Concrete Deep Foundations by Ultrasonic Crosshole Testing*. ASTM.

Atkinson, J.H. 2007. *The Mechanics of Soils and Foundations*. Taylor & Francis, London.

Atkinson, J.H. and Potts, D.M. 1977. Stability of a shallow circular tunnel in cohesionless soil. *Geotechnique*, 27(2):203–15.

Attewell, P.B. and Woodman, J.P. 1982. Predicting the dynamics of ground settlement and its derivatives by tunnelling in soil. *Ground Engineering*, 15: 13–22.

Badelow, F., Kim, S., Poulos, H.G. and Abdelrazaq, A. 2009. Foundation design for a tall tower in a reclamation area. In: F.T.K. Au (Ed.), *Proceedings of the 7th International Conference on Tall Buildings*, Hong Kong. Research Publishing, pp. 815–823.

Baker, W.F., Korista, D.S. and Lawrence C. 2008. Engineering the world's tallest—Burj Dubai. *CTBUH 8th World Congress*, Dubai.

Balasubramaniam, A.W. and Brenner, R.P. 1981. Consolidation and settlement of soft clay, Chapter 7. In: E.W. Brand and R.P. Brenner (Eds.), *Soft Clay Engineering*. Elsevier, Amsterdam.

Banerjee, P.K. and Driscoll, R.M. 1976. Three-dimensional analysis of raked pile groups. *Proceedings of the Institution of Civil Engineers, Part 2*, 61: 653–671.

Bea, R.G. 1980. Dynamic response of piles in offshore platforms. In: *Proceedings of the Specialty Conference on Dynamic Response of Pile Foundations—Analytical Aspects. ASCE, Geotechnical Engineering Division*, pp. 80–109.

Beredugo, Y.O. and Novak, M. 1972. Coupled horizontal and rocking vibrations of embedded footings. *Canadian Geotechnical Journal*, 9: 411–497.

Berezantzev, V.G., Khristoforov, V. and Golubkov, V. 1961. Load bearing capacity and deformation of piled foundations. *Proceedings of the, 5th International Conference on Soil Mechanics and Foundation Engineering*, Paris, Vol. 2, pp. 11–15.

Bermingham, P., Ealy, C.D. and White, J.K. 1994. A comparison of Statnamic and static field tests at seven FHWA sites. In: *International Conference on Design and Construction of Deep Foundations*, FHWA, Orlando, Vol. 2, pp. 616–630.

Bernal, J.B. and Reese, L.C. 1984. *Drilled Shafts and Lateral Pressure of Fresh Concrete*. Geotechnical Engineering Report GR B4-4, Geotechnical Engineering Centre, Bureau of Engineering Research, University of Texas, Austin, 126p.

Bhattacharya, S. and Bolton, M. 2004. Errors in design leading to pile failures during seismic liquefaction. *Proceedings of the 5th International Conference on Case Histories in Geotechnical Engineering*, Paper No. 12A-12, New York.

Bjerrum, L. and Eide, O. 1956. Stability of strutted excavations in clay. *Geotechnique*, 6(1): 32–47.

Bjerrum, L., Johannesson, I.J. and Eide, O. 1969. Reduction of skin friction in piles to rock. *Proceedings of the 7th International Conference on Soil Mechanics and Foundation Engineering*, Mexico City, Vol. 2, pp. 27–34.

Bjerrum, L. and Landva, A. 1966. Direct simple shear tests on a Norwegian quick clay. *Geotechnique*, 16(1): 1–20.

Boggs, D. and Dragovich, J. 2006. The nature of wind loads and dynamic response. In: J.M. Bracci (Ed.), *Performance-Based Design of Concrete Buildings for Wind Loads*. American Concrete Institute, Montreal, Quebec, pp. 15–44.

Bogard, D., Montreal, Quebec, and Matlock, H. 1983. In: Procedures for analysis of laterally loaded pile groups in soft clay. In: S.G. Wright (Ed.), *Proceedings of the Conference on Geotechnical Practice in Offshore Engineering*, ASCE, Austin, TX, pp. 499–535.

Bolton, A. 1994. *Structural Dynamics in Practice*. McGraw-Hill, Maidenhead, UK.

Bommer, J.J. and Martinez-Pereira, M. 1999. The effective duration of earthquake strong motion. *Journal of Earthquake Engineering*, 3(2): 127–172.

Bommer, J.J. and Stafford, P.J. 2009. Eismic hazard and earthquake actions, Chapter 2. In: A.Y. Elghazouli (Ed.), *Seismic Design of Buildings to Eurocode 8*. Spon Press, Abingdon.

Booker, J.R. and Poulos, H.G. 1976. Analysis of creep settlement of pile foundations. *Journal of Geotechnical Engineering Division, ASCE*, 102(GT1): 1–14.

Boscardin, M.D. and Cording, E.J. 1989. Building response to excavation-induced settlement. *Journal of Geotechnical Engineering, ASCE*, 115: 1–21.

Bourne-Webb, P.J., Amatya, B., Soga, K., Amis, T. Davidson, C. and Payne, P. 2009. Energy pile test at Lambeth College, London: Geotechnical and thermodynamic aspects of pile response to heat cycles. *Geotechnique*, 59(3): 237–248.

Bowles, J.E. 1988. *Foundation Analysis and Design*. 4th Ed. McGraw-Hill, New York.

Bowman, E.T. and Soga, K. 2003. Creep, ageing and microstructural change in dense granular material. *Soils and Foundations, Japanese Geotechnical Society*, 43(4): 107–117.

Brandl, H. 1989. Underpinning. Special lecture D. *Proceedings of the 12th International Conference on Soil Mechanics and Foundation Engineering*, Rio de Janeiro, Vol. 4, pp. 2227–2258.

Brandl, H. 2006. Energy foundations and other thermo-active ground structures. *Geotechnique*, 56(2): 81–122.

Bransby, M.F. 1999. Selection of p-y curves for the design of single laterally loaded piles. *International Journal for Numerical and Analytical Methods in Geomechanics*, 23(15): 1909–1926.

Bray, J.D. and Sancio, R.B. 2006. Assessment of the liquefaction susceptibility of fine-grained soils. *Journal of Geotechnical and Geoenvironmental Engineering*, 132(9): 1165–1177.

Brettman, T. and Duncan, J.M. 1996. Computer application of CLM lateral load analysis to piles and drilled shafts. *Journal of Geotechnical Engineering, ASCE*, 122(6): 496–498.

Briaud, J-L. and Tucker, L.M. 1996. *Design and construction guidelines for downdrag on uncoated and bitumen coated piles*. NCHRP Report, Texas A&M University.

Broms, B.B. 1964a. Lateral resistance of piles in cohesionless soils. *Journal of SMFD, ASCE*, 90(SM3): 123–156.

Broms, B.B. 1964b. Lateral resistance of piles in cohesive soils. *Journal of SMFD, ASCE*, 90(SM2): 27–63.

Broms, B.B. 1966. Methods of calculating the ultimate bearing capacity of piles – A summary. *Sols Soils*, 18–19: 21–32.

Brown, D.A. 2004. Zen and the art of drilled shaft construction: The pursuit of quality. *Geotechnical Special Publication, ASCE*, 124: 19–33.

Brown, J.D. and Meyerhof, G.G. 1969. Experimental study of bearing capacity in layered clays. *Proceedings of the 7th International Conference on Soil Mechanics and Foundation Engineering*, Mexico, Vol. 2, pp. 45–51.

Bruce, D.A. 1994. Small-diameter cast-in-place elements for load-bearing and in situ earth reinforcement, Chapter 6. In: P.P. Xanthakos, L.W. Abramson and D.A. Bruce (Eds.), *Ground Control and Improvement*, John Wiley, New York.

BTS. 2011. *Monitoring Underground Construction – A Best Practice Guide.* British Tunnelling Society, ICE Publishing, London.

Budhu, M. 2011. *Soil Mechanics Fundamentals.* John Wiley, Chichester.

Budhu, M. and Al-Karni, A. 1993. Seismic bearing capacity of soils. *Geotechnique*, 43(1): 181–187.

Budhu, M. and Davies, T.G. 1987. Nonlinear analysis of laterally loaded piles in cohesionless soils. *Canadian Geotechnical Journal*, 24: 21–39.

Budhu, M. and Davies, T.G. 1988. Analysis of laterally loaded piles in soft clays. *Journal of Geotechnical Engineering, ASCE*, 114(1): 21–39.

Buisman, A.S. 1936. Results of long duration settlement tests. *Proceedings of the 1st ICSMFE*, Cambridge, MA, Vol. 1, pp. 103–106.

Bull, J.W. (Ed.). 2012. *ICE Manual of Structural Design: Buildings.* Institution of Civil Engineers, London.

Burd, H.J. and Frydman, S. 1997. Bearing capacity of plane-strain footings on layered soils. *Canadian Geotechnical Journal*, 34: 241–253.

Burland, J.B. 1973. Shaft friction on piles in clay-a simple fundamental approach. *Ground Engineering*, 6(3): 30–42.

Burland, J.B. 1995. Piles as settlement reducers. *18th Italian Congress on Soil Mechanics*, Keynote Address, Pavia, Italy.

Burland, J.B. 2004. The leaning tower of Pisa revisited. State of the Art and Practice, *4th International Conference on Case Histories in Geotechnical Engineering*, Paper No. 2, New York.

Burland, J.B. and Burbridge, M.C. 1985. Settlements of foundations on sand and gravel. *Proceedings of the ICE*, Vol. 1, pp. 1325–1381.

Burland, J.B. and Mitchell, J.M. 1989. Piling and deep foundations. *Proceedings of International Conference on Piling and Deep Foundations*, London.

Burland, J.B., Chapman, T., Skinner, H. and Brown, M. 2012. *ICE Manual of Geotechnical Engineering*, Vol. 2, Chapter 67. ICE Publishing, London.

Burland, J.B., Standing, J.R. and Jardine, F.M. 2002. Assessing the risk of building damage due to tunnelling – Lessons from the Jubilee Line extension. *Proceedings of the 2nd International Conference on Soil Structure Interaction in Urban Engineering*, Zurich, Vol. 1, pp. 11–38.

Burton, M., Buttgereit, V. and Cammelli, S. 2014. Future of wind engineering for tall and super-tall buildings, Chapter 11. In: A.R. Tamboli (Ed.), *Tall and Supertall Buildings, Planning and Design.* McGraw-Hill Education, New York, pp. 361–401.

Bustamante, M. and Gianeselli, L. 1982. Pile bearing capacity prediction by means of static penetrometer CPT. *Proceedings of the ESOPT II*, Amsterdam, Vol. 2, pp. 492–500.

Button, S.J. 1953. The bearing capacity of footings on a two-layer cohesive subsoil. *Proceedings, 3rd International Conference on Soil Mechanics and Foundation Engineering*, Zurich, Vol. 1, pp. 332–335.

CACA. 1991. *Concrete Design Handbook*, 2nd Ed. Cement and Concrete Association of Australia, Sydney, Australia.

Carter, J.P., Desai, C.S., Potts, D.M., Schweiger, H.M. and Sloan, S.W. 2000. Computing and computer modelling in geotechnical engineering. *GeoEng2000*, November, Melbourne, Australia. Technomic Publishing Co., Lancaster, PA, pp. 157–252.

Cavallaro, A., Maugeri, M. and Mazzarella, R. 2001. Tatic and dynamic properties of Leighton Buzzard sand from laboratory tests. *Proceedings of the 4th International Conference on Recent Advances in Geotechnical Earth Engineering and Soil Dynamics*, Paper No. 1.13, San Diego, CA.

Celebi, E., Goktepe, F. and Karahan, N. 2012. Non-linear finite element analysis for prediction of seismic response of buildings considering soil-structure interaction. *Natural Hazards and Earth System Science*, 12: 3495–3505.

Chen, W.F. and Davidson, H.L. 1973. Bearing capacity determination by limit analysis. *Journal of Soil Mechanics and Foundations Division, ASCE*, 99: 433–449.

Chen, L.T. and Poulos, H.G. 1996. Some aspects of pile response near an excavation. *Proceedings of 7th Australia and New Zealand Conference on Geomechanics*, Adelaide, pp. 604–609.

Chen, L.T. and Poulos, H.G. 1997. Piles subjected to lateral soil movements. *Journal of Geotechnical and Geoenvironmental Engineering, ASCE*, 123(9): 802–811.

Chen, L.T. and Poulos, H.G. 1999. Design charts for analysis of piles subjected to lateral soil movements. In: N. Vitharana and R. Colman (Eds.), *Proceedings of 8th Australia and New Zealand Conference on Geomechanics*, Hobart, Australia, pp. 367–373.

Chen, L.T. and Poulos, H.G. 2001. Approximation of lateral soil movements for analyzing lateral pile response. *Proceedings of the 20th Annual Seminar of Geotechnical Division, The Hong Kong Institution of Engineers*, Hong Kong, pp. 14–23.

Chen, L.T., Poulos, H.G. and Loganathan, N. 1999. Pile responses caused by tunnelling. *Journal of Geotechnical and Geoenvironmental Engineering, ASCE*, 125(3): 207–215.

Chew, Y.L.M. 2012. *Construction Technology for Tall Buildings*, 4th Ed. World Scientific, Singapore.

Chin, F.K. 1970. Estimation of the ultimate load of pile from tests not carried to failure. *Proceedings of the 2nd SE Asian Conference on Soil Mechanics and Foundation Engineering*, Singapore, pp. 81–92.

Chin, J.T. and Poulos H.G. 1992. Cyclic axial pile loading analyses: A comparative study. *Computers and Geotechnics*, 13(3): 137–158.

Cho, N-J. and Kulhawy, F.H. 1995. The undrained behaviour of drilled shaft foundations subjected to static inclined loading. *Journal of the Korean Geotechnical Society*, 11(3): 91–111.

Choi, H.S. 2009. Super tall building design approach. Presentation on behalf of Thornton Thomasetti.

Choubane, B., Wu, C-L. and Tia, M. 1996. Coarse aggregate effects on elastic moduli of concrete. *Transportation Research Record: Journal of the Transportation Research Board*, 1547: 29–34.

Chow, F.C., Jardine, R.J., Brucy, F. and Nauroy, J.F. 1997. Time related increases the shaft capacities of driven piles in sand. *Geotechnique* 47(2): 353–361.

Chow, F.C., Jardine, R.J., Brucy, F. and Nauroy, J.F. 1998. Effects of time on capacity of pipe piles in dense marine sand. *Journal of Geotechnical and Geoenvironmental Engineering, ASCE*, 124(3): 254–264.

Chow, H., Small, J.C. and Poulos, H.G. 2010. Pile cap effects on lateral response of pile groups and piled rafts. *Proceedings of Geo 2010*, Calgary, Canada.

Chow, H.S.W. and Poulos, H.G. 2015. The significance of raft flexibility in pile group and piled raft design. *Proceedings of the 15th ANZ Conference Geomechanics*, Wellington.

Chow, Y.K., Chin, J.T. and Lee, S.L. 1990. Negative skin friction on pile groups. *International Journal for Numerical and Analytical Methods in Geomechanics*, 14(1): 75–91.

Chu, Y.K. 1994. A failure case study of island method excavation in soft clay. *Proceedings of the International Conference on Design and Construction of Deep Foundations*, Orlando, FL, Vol. 3, pp. 1216–1230.

Chu, J., Ivanov, V., He, J., Naemi, M., Li, B. and Stabnikov, V. 2011. Development of microbial geotechnology in Singapore. *Proceedings of Geo-Frontiers 2011 Conference, ASCE Conference Proceedings*, pp. 397–416, doi: 10.1061/41165.

Chua, T.S., Tan, B.C. and Yu, T.L. 1996. Deep excavation work near MRT structures. *Proceedings of the 12th Southeast Asian Geotechnical Conference*, Kuala Lumpur, pp. 453–457.

Ciliz, S., Ozkan, M.Y. and Cetin, K.O. 2007. Effect of basin edge effects on the dynamic response of soil deposits. *Proceedings of the 4th International Conference on Earthquake Geotechnical Engineering*, Paper No. 1309, Thessaloniki, Greece.

CIRIA. 2000. *Groundwater Control – Design and Practice.* CIRIA 515, Spon, London.

CIRIA. 2003. *C580, Embedded Retaining Walls – Guidelines for Economic Design*, RP29. CIRIA London.

Clausen, C.J.F., Aas, P.M. and Karlsrud, K. 2005. Bearing capacity of driven piles in sand, the NGI approach. *Proceedings of the ISFOG*, Perth, pp. 677–681.

Clayton, C.R.I. 2001. Managing geotechnical risk: Time for change? *Proceedings of the ICE, Geotechnical Engineering*, Vol. 149(1), pp. 3–11.

Clayton, C.R.I., Matthews, M.C. and Simons, N.E. 2005. *Site Investigation*, 2nd Ed. Thomas Telford, London.

Clayton, C.R.I., Milititsky, J. and Woods, R.I. 1993. *Earth Pressure and Earth-Retaining Structures*, 2nd Ed. Blackie Academic & Professional, London.

Clough, G.W. and O'Rourke, T.D. 1990. Construction induced movements of in-situ walls. *Design and Performance of Earth Retaining Structures*, ASCE GSP No. 25, pp. 439–470.

Clough, G.W. and Schmidt, B. 1981. Excavations and tunnelling, Chapter 8. In: E.W. Brand and R.P. Brenner (Eds.), *Soft Clay Engineering*. Elsevier, Amsterdam, pp. 569–634.

Clough, G.W., Smith, E.M. and Sweeney, B.P. 1989. Movement control of excavation support systems by iterative design. *Proceedings of the ASCE, Foundation Engineering: Current Principles and Practices*, Evanston, IL, Vol. 2, pp. 869–884.

Coe, J.T. and Kermani, B. 2016. Comparison of borehole ultrasound and borehole radar in evaluating the length of two unknown bridge foundations. *Journal of the Deep Foundations Institute*, 10(1): 8–24.

Coffey. 2007. *CLAP User's Manual*. Coffey Geotechnics, Sydney.

Comodromos, E.M. and Bareka, S.V. 2005. Evaluation of negative skin friction effects in pile foundations using 3D nonlinear analysis. *Computers and Geotechnics*, 32: 210–221.

Comodromos, E.M. and Bareka, S.V. 2008. Effects of pile-to-pile interaction on piled raft foundations. In: A.J. Brennan and J.A. Knappett (Eds.), *Foundations: Proceedings of the Second BGA International Conference on Foundations*, ICOF2008. IHS BRE Press, Dundee, Scotland.

Comodromos, E.M. and Papadopoulou, M.C. 2012. Response evaluation of horizontally loaded fixed-head pile groups in clayey soils. *Géotechnique*, 62(4): 329–339.

Comodromos, E.M. and Papadopoulou, M.C. 2013. Effects from diaphragm wall installation to surrounding soil and adjacent buildings. *Computers and Geotechnics*, 53: 106–121.

Comodromos, E.M. and Pitilakis, K.D. 2009. Response evaluation of horizontally loaded fixed-head pile groups using 3-D nonlinear analysis. *International Journal for Numerical and Analytical Methods in Geomechanics*, 29(6): 597–625.

Comodromos, E.M., Papadopoulou, M.C. and Laloui, L. 2016. Contribution to the design methodologies of piled raft foundations under combined loadings. *Canadian Geotechnical Journal*, 53: 1–19.

Costanzo, D. and Lancellotta, R. 1998. A note on pile interaction factors. *Soils and Foundations*, 38(4): 251–253.

Cox, H.V. 1966. *A Review of the Geotechnical Characteristics of the Recent Marine Clay in Southeast Asia*. Asian Institute of Technology, Research Report No. 6.

Cox, W., Reese, L.C. and Grubbs, B.R. 1974. Field testing of laterally loaded piles in sand. *Proceedings of the 6th Offshore Technology Conference*, Dallas, TX, pp. 459–472.

Craighead, G. 2009. *High-Rise Security and Fire Life Safety*, 3rd Ed., Butterworth-Heinemann, Burlington, MA.

Cubrinovski, M. 2006. Pile response to lateral spreading of liquefied soils. *NZGS 2006 Symposium: Earthquakes and Urban Development*, Nelson, NZ, pp. 127–142.

Cubrinovski, M. 2008. Key parameters in pseudo-static analysis of piles in liquefying sand. *Australian Geomechanics*, 43(1): 61–66.

Cubrinovski, M. and Ishihara, K. 2004. Simplified method for analysis of piles undergoing lateral spreading in liquefied soils. *Soils and Foundations*, 44(5): 119–133.

Cubrinovski, M., Ishihara, K. and Poulos, H.G. 2009. Pseudo-static analysis of piles subjected to lateral spreading. *Bulletin of the New Zealand Society for Earthquake Engineering*, 42(1): 28–38.

Cubrinovski, M., Kokusho, T. and Ishihara. K. 2006. Interpretation from large-scale shake table tests on piles subjected to spreading of liquefied soils. In: M. Cubrinovski and M.J. Pender (Eds.), *Proceedings of the New Zealand Workshop on Geotechnical Earthquake Engineering*, University of Canterbury, NZ, CD volume.

CUR. 1996. *Building on Soft Soils*. CUR Centre for Civil Engineering, CRC Press, The Netherlands.

D'Appolonia, D.J., Poulos, H.G. and Ladd, C.C. 1971. Initial settlement of structures on clay. *JSMFD, ASCE*, 97(SM10): 1359–1397.

Davies, A., Galsworthy, J.K., Gibbons, M., Irwin, P.A., Morava, B., Sifton, V., Tang, V. and Wu, H. 2014. Wind engineering for tall and supertall buildings, Chapter 10. In: A.R. Tamboli (Ed.), *Tall and Supertall Buildings, Planning and Design*. McGraw-Hill Education, New York, pp. 289–360.

Davies, T.G. and Budhu, M. 1986. Non-linear analysis of laterally loaded piles in stiff clay. *Geotechnique*, 36(4): 527–538.

Davis, E.H. and Booker, J.R. 1971. The bearing capacity of strip footings from the standpoint of plasticity theory. *Proceedings of the First Australia-New Zealand Conference on Geomechanics*, Melbourne, Australia, Vol. 1, pp. 276–282.

Davis, E.H. and Booker, J.R. 1973. The effect of increasing strength with depth on the bearing capacity of clays. *Géotechnique*, 23: 551–563.

Davis, E.H. and Poulos, H.G. 1972. Rate of settlement under three dimensional conditions. *Géotechnique*, 22(1): 95–114.

Davis, E.H. and Poulos, H.G. 1963. Triaxial testing and three-dimensional settlement analysis. *Proceedings of the 4th Australia-New Zealand Conference on Soil Mechanics and Foundation Engineering*, Adelaide, p. 233.

Davis, E.H. and Poulos, H.G. 1968. The use of elastic theory for settlement prediction under three-dimensional conditions. *Geotechnique*, 18:67–91.

Davisson, M.T. 1970. Lateral load capacity of pile groups. *Highway Research Record* 333, 104–112.

De Nicola, A. and Randolph, M.F. 1993. Tensile and compressive shaft capacity of piles in sand. *Journal of Geotechnical Engineering, ASCE*, 119(12): 1952–1973.

De Sanctis, L. and Mandolini, A. 2006. Bearing capacity of piled rafts on soft clay soils. *Journal of Geotechnical and Geoenvironmental Engineering, ASCE*, 132(12): 1600–1610.

De Seixas, A. De Seixas, J.R. and De Seixas, J.J. 2006. Control and monitoring of buildings foundation – Applications in very high buildings structure. *Shaping the Change, XXIII FIG Congress*, Paper T78, Munich, Germany, 8–13 October, 21pp.

Decourt, L. 1982. Prediction of the bearing capacity of piles based exclusively on N values of the SPT. *Proceedings of the ESOPT II*, Amsterdam, Vol. 1, pp. 29–34.

Decourt, L. 1995. Prediction of load-settlement relationships for foundations on the basis of the SPT-T. *Ciclo de Conferencias International "Leonardo Zeevaert", UNAM*, Mexico, pp. 85–104.

DeJong, J.T., Mortensen, B.M., Martinez, B.C. and Nelson, D.C. 2010. Bio-remediated soil improvement. *Ecological Engineering*, 36: 197–210.

DeJong, J.T., Mortensen, B., Soga, K. and Kavazanjian, E. 2011. Harnessing in-situ biogeochemical systems for "natural" ground improvement. *Geo-Strata, ASCE GeoInstitute*, July/August 2011, 36–39.

Desai, C.S. 1974. Numerical design-analysis for piles in sands. *Journal of Geotechnical Engineering Division, ASCE*, 100(GT6): 613–635.

Dezi, F., Carbonari, S. and Leoni, G. 2010. Kinematic bending moments in pile foundations. *Soil Dynamics and Earthquake Engineering*, 30: 119–132.

Dezi, F. and Poulos, H. 2016. Kinematic bending moments in square pile groups. *International Journal of Geomechanics*, doi: 10.1061/(ASCE)GM.1943-5622.0000747, 04016066.

DFI. 2015. *Deep Foundations Magazine*, What is sustainability?, November/December Issue: 111.

Di Laora, R., Mandolini, A. and Mylonakis, G. 2012. Insight on kinematic bending of flexible piles in layered soil. *Soil Dynamics and Earthquake Engineering*, 43: 309–322.

Ding, J.M. and Wu, H.L. 2014. Current situation and research of structural design for super high-rise buildings in China. *IES Journal Part A, Civil Engineering*, 7(2): 114–120.

Dobry, R., Gazetas, G. 1988. Imple methods for dynamic stiffness and damping of floating pile groups. *Geotechnique*, 38(4): 557–574.

Dobry, R. and Ladd, R.S. 1980. Discussion to "Soil liquefaction and cyclic mobility evaluation for level ground during earthquakes", by H.B. Seed. *Journal of Geotechnical Engineering Division, ASCE*, 106(GT6): 720–724.

Dobry, R. and O'Rourke, M.J. 1983. Discussion on 'Seismic response of end-bearing piles' by Flores–Berrones, R. and Whitman, R.V. *Journal of Geotechnical Engineering Division*, 109(5): 778–781.

Dotson, D. and Tarquino, F. 2003. A creative solution to problems with foundation construction in karst. *Proceedings of the 9th Multidisciplinary Conference on Sinkholes and the Engineering and Environment Impacts of Karst*, Huntsville, AL, September.

Douglas, J. 2002. *A comprehensive worldwide summary of strong-motion attenuation relationships for peak ground acceleration and spectral ordinates (1969 to 2000).* Errata and Additions to ESEE Report No. 01-1. Department of Civil and Environmental Engineering Department Report, Imperial College, London, UK.

Dowrick, D.J. 2009. *Earthquake Resistant Design and Risk Reduction*, 2nd Ed., John Wiley, Chichester.

Duncan, J.M. 1979. Site characterization for analysis. In: C.H. Dowding (Ed.), *Site Characterization and Exploration*. ASCE, New York, pp. 70–84.

Duncan, J.M., Evans, L.T. and Ooi, P.S.K. 1994. Lateral load analysis of single piles and drilled shafts. *Journal of Geotechnical Engineering Division, ASCE*, 120(6): 1018–1033.

Duncan, J.M., Wright, S.G. and Wong, K.S. 1990. Slope stability during rapid drawdown. *Seed Memorial Symposium Proceedings*, BiTech Publishers, Vancouver, BC, Canada, Vol. 2, pp. 235–272.

Dunnicliff, J. 1993. *Geotechnical Instrumentation for Monitoring Field Performance.* John Wiley & Sons, New York.

Dyvik, R., Berre, T., Lacasse, T. and Raadim, B. 1987. Comparison of truly undrained and constant volume direct simple shear tests. *Geotechnique*, 37(1): 3–10.

El Naggar, M.H. 2001. Dynamics of foundations, Chapter 12. In: R.K. Rowe (Ed.), *Geotechnical and Geoenvironmental Engineering Handbook*. Kluwer Academic Publishers, Boston.

El Naggar, M.H. 2014. DYNA6.1 software. Western Geotechnical Research Centre, University of Western Ontario, Canada.

El Naggar, M.H. and Novak, M. 1995. Nonlinear lateral interaction in pile dynamics. *Soil Dynamics and Earthquake Engineering*, 14(2): 141–157.

El Naggar, H. and El Naggar, M.H. 2007. Simplified approximate approach to group effect in pile dynamics. *Proceedings of the 4th International Conference on Earthquake Geotechnical Engineering*, Paper No. 1325, Thessaloniki, ISSMGE.

El Sharnouby, B. and Novak, M. 1985. Static and low-frequency response of pile groups. *Canadian Geotechnical Journal*, 22: 79–94.

Elnaggar, H.A. and Krizek, R.J. 1970. Statistical approximation for consolidation settlement. *Highway Research Record*, 323, 87–96.

Elnimeiri, M. and Almusharaf, A. 2010. Structure and architectural form of tall buildings. *Proceedings of the International Conference on Sustainable Building Asia*, SB10, Seoul, pp. 54–61.

Engin, H.K., Septanika, E.G. and Brinkgreve, R.E.J. 2008. Estimation of pile group behaviour using embedded elements. *Proceedings of the 12th International Conference on IACMAG*, Goa, pp. 3231–3238.

England, M. and Fleming, W.G.K. 1994. Review of foundation testing methods and procedures. *Proceedings of the Institution of the Civil Engineers: Geotechnical Engineering*, Vol. 107(3): pp. 135–142.

Ensoft. 2014. *DYNAPILE*. Ensoft, Austin, TX.

Ensoft. 2015. *GROUP8 & LPILE*. Ensoft, Austin, TX.

Fahey, M. and Carter, J.P. 1993. A finite element study of the pressuremeter using a nonlinear elastic plastic model. *Canadian Geotechnical Journal*, 30(2): 348–362.

Fellenius, B.H. 1991. Pile foundations, Chapter 13. In: H-YFang (Ed.), *Foundation Engineering Handbook*, 2nd Ed., Van Nostrand Reinhold, New York, pp. 511–536.

Fellenius, B.H. 1989. Unified design of piles and pile groups. *Transportation Research Board, Washington, TRB Record*, 1169, 75–82.

Fellenius, B.H. 1998. Recent advances in the design of piles for axial loads, dragloads, downdrag, and settlement. *Proceedings of the ASCE Seminar*, New York, April, 19p.

Fellenius, B.H. 2004. Unified design of piled foundations with emphasis on settlement analysis. *ASCE Special Publication GSP*, 125, 253–275.

Fellenius, B.H. 2016. *Basics of Foundation Design*, Electronic Ed. www.Fellenius.net, 451p.

Finno, R.P. 2014. Performance monitoring of geotechnical structures. *Geostrata, Geo-Institute ASCE*, March–April, 14–16.

Finno, R.J., Blackburn, J.T. and Robocki, T.F. 2007. Three-dimensional effects for supported excavations in clay. *Journal of Geotechnical and Geoenvironmental Engineering, ASCE*, 133(1): 30–36.

Finno, R.J., Laurence, S.A., Allauh, N.F. and Harahap, I.S. 1991. Analysis of performance of pile groups adjacent to deep excavation. *Journal of Geotechnical Engineering, ASCE*, 117(6): 934–955.

Fleming, W.G.K. 1992. A new method for single pile settlement prediction and analysis. *Geotechnique*, 42(3): 411–425.

Fleming, W.G.K. and Sliwinski, Z.J. 1977. *The Use and Influence of Bentonite in Bored Pile Construction.* CIRIA Report No. PG3. Construction Industry Research & Information Association, London, 93pp.

Fleming, W.G.K., Weltman, A.J., Randolph, M.F. and Elson, W.K. 1985. *Piling Engineering.* Surrey University Press, Halsted Press.

Fleming, W.G.K., Weltman, A.J. and Randolph, M.F. 1992. *Piling Engineering*, 2nd Ed. Halsted Press. New York.

Fleming, W.G.K., Weltman, A.J., Randolph, M.F. and Elson, W.K. 2009. *Piling Engineering*, 3rd Ed. Spon Press, London.

Focht, J.A. 1967. Discussion on paper by Coyle and Reese. *Journal of the Soil Mechanics and Foundations Division, ASCE*, 93(SM1): 133–138.

Focht, J.A., Khan, F.R. and Gemeinhardt, J.P. 1978. Performance of one shell plaza deep mat foundation. *Journal of Geotechnical Engineering Division, ASCE*, 104(GT5): 593–608.

Focht, J.A. and Koch, K.J. 1973. Rational analysis of the lateral performance of offshore pile groups. *Proceedings of the 5th OTC*, Paper 1896, Houston, Vol. 2, pp. 701–708.

Fookes, P.G. 1997. Geology for engineers: The geological model, prediction and performance. *Quarterly Journal of Engineering Geology and Hydrogeology*, 30(4):293–424.

Foti, S., Lai, C.G., Rix, G.J. and Strobbia, C. 2014. *Surface Wave Methods for Near-Surface Site Characterization.* CRC Press, Boca Raton, FL.

FPS. 2006. *Handbook on Pile Load Testing.* Federation of Piling Specialists, UK.

Frank, R. and Magnan, J-P. 1995. Cone penetration testing in France: National report. In: *Proceedings of the CPT 95*, Swedish Geotechnical Society, Linkoping, Vol. 3, pp. 147–156.

Franke, E., El-Mossallamy, Y. and Wittmann, P. 2000. Calculation methods for raft foundations in Germany, Chapter 12. In: J.A. Hemsley (Ed.), *Design Applications of Raft Foundations*, Thomas Telford, London.

Franke, E., Lutz, B. and El-Mossallamy, Y. 1994. Measurements and numerical modelling of high-rise building foundations on Frankfurt clay. *Vertical and Horizontal Deformation of Foundations and Embankments, ASCE Geotechnical Special Publication No. 40*, 2: 1325–1336.

Fraser, R.A. and Wardle, L.J. 1976. Numerical analysis of rectangular rafts on layered foundations. *Géotechnique*, 26(4): 613–630.

Fujioka, T. and Yamada, K. 1994. The development of a new pile load testing system. In: *Proceedings of the International Conference on Design and Construction of Deep Foundations*, FHWA, Orlando, Vol. 2, pp. 670–684.

Gaba, A.R., Simpson, B., Powrie, W. and Beadman, D.R. 2003. *Embedded Retaining Walls – Guidance for Economic Design*, Report No. C580. CIRIA, London, UK.

Gallagher, P.M. 2000. Passive site remediation for mitigation of liquefaction risk. PhD thesis, Virginia Polytechnic Institute and State University, Blacksburg, VA.

Gallagher, P.M., Conlee, C.T. and Rollins, K.M. 2007. Full-scale field testing of colloidal silica grouting for mitigation of liquefaction risk. *Journal of Geotechnical and Geoenvironmental Engineering, ASCE*, 133(2): 186–196.

Gallagher, P.M. and Lin, Y. 2009. Colloidal silica transport through liquefiable porous media. *Journal of Geotechnical and Geoenvironmental Engineering, ASCE*, 135(11): 1702–1712.

Garlanger, J.E. 1972. The consolidation of soils exhibiting creep under constant effective stress. *Geotechnique*, 22(1): 71–78.

Gaull, B.A., Michael-Leiba, M.O. and Rynn, J.M.W. 1990. Probabilistic earthquake risk maps of Australia. *Australian Journal of Earth Sciences*, 37: 169–187.

Gazetas, G. 1991. Foundation vibrations, Chapter 15. In: H.S. Fang (Ed.), *Foundation Engineering Handbook*, 2nd Ed. Chapman & Hall, New York, pp. 563–593.

Gazetas, G., Fan, K., Tazoh, T. and Shimizu, K. 1993. Seismic response of the pile foundation of Ohba-Ohashi bridge. *Proceedings of the 3rd International Conference on Case Histories in Geotechnical Engineering*, St. Louis, pp. 1803–1809.

GEO. 1996. *Pile Design and Construction*. Publication 1/96, Geotechnical Engineering Office, Civil Engineering Department, Hong Kong.

Geocentrix. 2013. *REPUTE v2*. Geocentrix, Banstead, Surrey, UK.

Geosolve. 2013. WALLAP. http://www.geosolve.co.uk/wallap1.htm.

Gibson, R.E. and Lo, K.Y. 1961. A theory of consolidation for soils exhibiting secondary consolidation. *Acta Polytechnica Scandinavica*, 296.

Giroud, J-P. 1973. *Tables pour le calcul des foundations*, 3 volumes. Dunod, Paris (in French).

Glisic, B., Inaudi, D., Lau, J.M. and Fong, C.C. 2013. Ten-year monitoring of high rise building columns using long-gauge fiber optic sensors. *Smart Material and Structures*, 22(0055030): 1–15.

Goble, G.G. 1994. Pile driving – An international state-of-the-art. In: *Proceedings of the International Conference on Design and Construction of Deep Foundations*, FHWA, Orlando, Vol. 1, pp. 1–26.

Goble, G.G. and Rausche, F. 1970. Pile load test by impact driving. *Highway Research Record*, 333: 123–129.

Goh, A.T., The, C.I. and Wong, K.S. 1996. The response of vertical piles to ground movements from adjacent braced excavations. *Proceedings of the 12th Southeast Asian Geotechnical Conference*, Kuala Lumpur, Vol. 1, pp. 403–409.

Goldberg, D.T., Jaworski, W.E. and Gordon, M.D. 1976. *Lateral Support Systems and Underpinning*, Vol. 1. Report FHA-RD-75-128, Federal Highway Administration, Washington, DC.

Golder, H.Q. and Osler, J.C. 1968. Settlement of a furnace foundation, Sorel, Quebec. *Canadian Geotechnical Journal*, 5(1): 46–56.

Goldsworthy, J.S. 2006. Quantifying the risk of geotechnical site investigations. PhD thesis, University of Adelaide, Australia.

Griffiths, D.V. 1982. Computation of bearing capacity factors using finite elements. *Geotechnique*, 32(3): 195–202.

Guo, W.D. 1997. Analytical and numerical analyses for piles. PhD thesis, University of Western Australia.

Guo, W.D. 2000. Visco-elastic load transfer models for axially loaded piles. *International Journal for Numerical and Analytical Methods in Geomechanics*, 24(2): 135–163.

Guo, W.D. and Randolph, M.F. 1997. Vertically loaded piles in non-homogeneous soils. *International Journal for Numerical and Analytical Methods in Geomechanics*, 21: 507–532.

Guo, W.D. and Randolph, M.F. 1999. An efficient approach for the settlement of pile groups. *Geotechnique*, 49(2); 161–179.

Gutenberg, B. and Richter, C.F. 1954. *Seismicity of the Earth and Associated Phenomena*. Princeton University Press, New Jersey.

Haberfield, C.M. 2013. Tall Tower Foundations – From concept to construction. In: K.K. Phoon et al. (Eds.), *Advances in Foundation Engineering*. Research Publishing Services, Singapore, pp. 33–65.

Hamada, J., Aso, N., Atsunari Hanai, A. and Yamashita, K. 2014. *Seismic observation of piled raft foundation subjected to unsymmetrical Earth pressure*. Takenaka Technical Research Report No. 70.

Hardin, B.O. and Drnevich, V.P. 1972. Shear modulus and damping in soils: Design equations and curves. *Journal of Soil Mechanics and Foundations Division, ASCE*, 98(SM7): 6678–6692.

Hasancebi, N. and Ulusay, R. 2007. Empirical correlations between shear wave velocity and penetration resistance for ground shaking assessments, *Bull Eng Geol Environ*, 66: 203–213.

Hayward Baker. 2003. Project summary on warehouse remediation. Promotional literature.

Hemsley, J.A. 2000. *Design Applications of Raft Foundations*. Thomas Telford, London.

Hirayama, H. 1994. Secant Young's modulus from N-value or Cu considering strain levels. In: S. Shibuya and T. Mitachi (Eds.), *Pre-failure Deformation of Geomaterials*. A.A. Balkema, Rotterdam, pp. 247–252.

Hobbs, N.B. 1957. Unusual necking of cast-in-place concrete piles. *Proceedings of the 4th International Conference on Soil Mechanics and Foundation Engineering*, London, Vol. 3, pp. 40–42.

Hoek, E. and Diederichs, M.S. 2006. Empirical estimation of rock mass modulus. *International Journal of Rock Mechanics and Mining Sciences, Elsevier*, 43(2006): 203–215.

Holmes, J.D. 2015. *Wind Loading of Structures*, 3rd Ed., CRC Press, Boca Raton, FL.

Holtz, R.D. and Kovacs, W.D. 2010. *An Introduction to Geotechnical Engineering*. Prentice-Hall, Englewood Cliffs, NJ.

Horikoshi, K., Randolph, M.F. 1998. A contribution to optimum design of piled rafts. *Geotechnique*, 48(3):301–17.

Horvath, R. and Kenney, T.C. 1979. Shaft resistance of rock-socketed drilled piers. *Presented at ASCE Annual Convention*, Atlanta, GA, preprint No. 3698.

Houlsby, G.T. and Wroth, C.P. 1983. Calculation of stresses on shallow penetrometers and footings. *Proceedings of the IUTAM Symposium*, Delft, pp. 107–112.

Hull, T.S. 1987. The behaviour of laterally loaded piles. PhD thesis, University of Sydney.

Hunt, R.E. 2005. *Geotechnical Investigation Handbook*, 2nd Ed. Taylor & Francis, Boca Raton.

Hwang, J.H., Li, J.C.C. and Liang, N. 2003. On methods for interpreting bearing capacity from a pile load test. *Journal of Geotechnical Engineering, SEAGS*, 34(1): 27–39.

ICE. 1991. *Inadequate Site Investigation*. Thomas Telford, London.

Idriss, I.M. and Boulanger, R.W. 2006. Semi-empirical procedures for evaluating liquefaction potential during earthquakes. *Soil Dynamics and Earthquake Engineering*, 26(2–4): 115–130.

Idriss, I.M. and Sun, J.I. 1991. SHAKE91: A computer program for conducting equivalent linear seismic response analyses of horizontally layered soil deposits. Center for Geotechnical Modeling, University of California, Davis.

Inaudi, D. and Glisic, B. 2007. Overview of fiber optic sensing techniques for geotechnical instrumentation and monitoring. *Geotechnical News*, September, 4–8.

Ishibashi, I. and Zhang, X. 1993. Unified dynamic moduli and damping ratios of sand and clay. *Soils and Foundations*, 33(1): 182–191.

Ishihara, K. 1997. Geotechnical aspects of the 1995 Kobe earthquake. Terzaghi Oration, *Proceedings of the 14th International Conference on Soil Mechanics and Foundation Engineering*, Hamburg, Vol. 4, pp. 2047–2073.

Ishihara, K. and Cubrinovski, M. 1998. Performance of large-diameter piles subjected to lateral spreading of liquefied soils. *Proceedings of the 13th Southeast Asian Geotechnical Conference, Keynote Lecture*, Taipei, Vol. 2, pp. 13–26.

Ishihara, K. and Cubrinovski, M. 2004. Case studies of pile foundations undergoing lateral spreading in liquefied deposits. *Proceedings of the 5th International Conference on Case Histories in Geotechnical Engineering*, Paper SOAP 5. New York. CD volume.

Ishihara, K. and Yoshimine, M. 1992. Evaluation of settlements in sand deposits following liquefaction during earthquakes. *Soils and Foundations*, 32(1): 173–188.

ISSMGE. 2013. In: R. Katzenbach and D. Choudhury (Eds.), *ISSMGE Combined Pile-Raft Foundation Guidelines*. ISSMGE, Technishe Universitat, Darmstadt.

Itasca. 2015. *FLAC3D Manual Version 5.01*. Itasca Consulting Group, Minneapolis, MN.

Iwasaki, T., Arakawa, T. and Tokida, K-I. 1984. Simplified procedure for assessing soil liquefaction during earthquakes. *Soil Dynamics and Earthquake Engineering*, 3(1): 49–58.

Jamiolkowski, M. 2001. The leaning tower of Pisa – End of an odyssey, Terzaghi Oration. *15th International Conference on Soil Mechanics and Geotechnical Engineering*, Istanbul, Vol. 4, pp. 2979–2996.

Janbu, N. 1963. Soil compressibility as determined by oedometer and triaxial tests. *Proceedings of the European Conference on Soil Mechanics and Foundation Engineering*, Wiesbaden, Vol. 1, pp. 19–25.

Jardine, R.J. and Chow, F.C. 1996. *New Design Methods for Offshore Piles*. MTD Pub. 96/103, Marine Technology Directorate, London, UK.

Jardine, R.J., Chow, F.C., Overy, R.F. and Standing, J.R. 2005. *ICP Design Methods for Driven Piles in Sands and Clays*. Thomas Telford, London.

Jardine, R.J., Lehane, B.M., Smith, P.R. and Gildea, P.A. 1995. Vertical loading experiments on rigid pad foundations at Bothkennar. *Geotechnique*, 45(4): 573–597.

Jardine, R.J., Standing, J.R. and Chow, F.C. 2006. Some observations of the effects of time on the capacity of piles driven in sand. *Geotechnique*, 56(4): 227–244.

JGS. 1998. *Remedial Measures Against Liquefaction*. Japanese Geotechnical Society (JGS), Balkema, Rotterdam.

Johnson, K.R. 2016. Analyzing thermal integrity profiling data for drilled shaft evaluation. *Journal of the Deep Foundations Institute*, 10(1): 25–40.

Juang, C., Schuster, M., Ou, C. and Phoon, K. 2011. Fully probabilistic framework for evaluating excavation-induced damage potential of adjacent buildings. *Journal of Geotechnical and Geoenvironmental Engineering*, 130–139, doi: 10.1061/(ASCE)GT.1943-5606.0000413.

Juang, C.H., Jiang, T. and Andrus, R.D. 2002. Assessing probability based method for liquefaction evaluation. *Journal of Geotechnical and Geoenvironmental Engineering, ASCE*, 128(7): 580–589.

Katzenbach, R., Leppia, S. and Choudhury, D. 2016. *Foundation Systems for High-Rise Structures*. CRC Press, Boca Raton, FL, LLC.

Katzenbach, R., Arslan, U. and Moorman, C. 2000. Piled raft foundation projects in Germany, Chapter 13. In: J.A. Hemsley (Ed.), *Design Applications of Raft Foundations*. Thomas Telford, London, pp. 323–391.

Katzenbach, R., Arslan, U., Moormann, C. and Reul, O. 1998. Piled raft foundation – Interaction between piles and raft. *Darmstadt Geotechnics*, Darmstadt University of Technology, 4: 279–296.

Kavazanjian, E. and De Jong, J. 2016. Biogeotechnical mitigation of earthquake-induced soil liquefaction. *Geostrata*, November/December: 62–67.

Kay, J.N. 1976. Safety factor evaluation for single piles in sand. *Journal of Geotechnical Engineering Division, ASCE*, 102(GT10): 1093–1108.

Kayabali, K. 1996. Soil liquefaction evaluation using shear wave velocity. *Engineering Geology*, 44: 121–127.

Kaynia, A.M. 1982. *Dynamic stiffness and seismic response of pile groups. MIT Research Report R82-03*, MIT, Cambridge, MA.

Kaynia, A.M. and Kausel, E. 1982. Dynamic behaviour of pile groups. *Proceedings of the 2nd International Conference on Numerical Methods in Offshore Piling*, University of Texas, Austin, pp. 509–532.

Kayvani, K. 2014. Design of high-rise buildings: Past, present and future. In: S.T. Smith (Ed.), *Proceedings of the 23rd Australasian Conference on Mechanics of Structures and Materials (ACMSM23)*, Southern Cross University, pp. 14–20.

Khan, F.R. 1969. Recent structural systems in steel for high-rise buildings. *Proceedings of the British Constructional Steelwork Association Conference on Steel in Architecture*. British Constructional Steelwork Association, London, pp. 24–26.

Kimura, M., Adachi, T., Kamei, H. and Zhang, F. 1995. 3-D finite element analysis of the ultimate behaviour of laterally loaded cast-in-place concrete piles. *Proceedings of the 15th International Conference on Numerical Models in Geomechanics*, Davos, Switzerland, pp. 589–594.

Kinemetrics. 2013. *Building Instrumentation Requirements for Seismic Monitoring*. Application Note #73, Kinemetrics Inc., Pasadena, CA.

Kitiyodom, P., Matsumoto, T. and Kanefusa, N. 2004. Influence of reaction piles on the behaviour of a test pile in static load testing. *Canadian Geotechnical Journal*, 41(3): 408–420.

Klar, A., Baker, R. and Frydman, S. 2004. Seismic soil-pile interaction in liquefiable soil. *Soil Dynamics and Earthquake Engineering*, 24(8): 551–564.

Klimis, N., Anastasiadis, A., Gazetas, G. and Apostolou, M. 2004. Liquefaction risk assessment and design of pile foundations for highway bridge. *Proceedings of the 13th World Conference on Earthquake Engineering*, Paper No. 2973, Vancouver, Canada.

Knappett, J.A. and Craig, R.F. 2012. *Craig's Soil Mechanics*, 8th Ed. Taylor & Francis, London.

Knott, D.L., Newman, F.B., Rojas-Gonzalez, L.F. and Gray, R.E. 1993. Foundation engineering practice for bridges in karstic areas in Pennsylvania. In: B.F. Beck (Ed.), *Applied Karst Geology.* Balkema, Rotterdam, pp. 225–230.

Kokusho, T. and Matsumoto, M. 1997. Nonlinear site response during the Hyogoken-Nambu earthquake recorded by vertical arrays in view of seismic zonation methodology. *Seismic Behaviour of Ground and Geotechnical Structures, Proceedings of Special Technical Session on Earthquake Geotechnical Engineering during 14th International Conference on Soil Mechanics and Foundation Engineering*, pp. 61–69.

Kolk, H.J., Baaijens, A.E. and Senders, M. 2005. Design criteria for pipe piles in silica sand. *Proceedings of the ISFOG*, Perth, Australia, pp. 711–716.

Komiya, K. Soga, K., Akagi, H, Hagiwara, T. and Bolton, M.D. 1999. Finite element modelling of excavation and advancement processes of a shield tunnelling machine. *Soils and Foundations*, 39(3): 37–52.

Kraft, L.M., Ray, R.P. and Kagawa, T. 1981. Theoretical t-z curves. *Journal of Geotechnical Engineering, ASCE*, 107(GT11): 1543–1561.

Krishnan, R., Gazetas, G. and Velez, A. 1983. Static and dynamic lateral deflections of piles in non-homogeneous stratum. *Geotechnique*, 23(3): 307–325.

Kulhawy, F.K. 1984. Limiting tip and side resistance: Fact or fallacy? In: J.R. Meyer (Ed.), *Analysis and Design of Piled Foundations.* ASCE, New York, NY, pp. 80–98.

Kulhawy, F.H. and Chen, Y-J. 1995. A thirty year perspective on Broms' lateral loading models, as applied to drilled shafts. *Broms Symposium*, Nanyang University, Singapore, pp. 225–242.

Kulhawy, F.H. and Mayne, P.W. 1990. *Manual on Estimating Soil Properties for Foundation Design.* Report EL-6800, Electric Power Research Institute, Palo Alto.

Kulhawy, F.K. and Phoon, K. K. 1993. Drilled shaft side resistance in clay soil to rock. *Design and Performances of Deep Foundations, ASCE, Special Publication*, 38: 172–183.

Kumar, J. and Mohan Rao, V.B.K. 2002. Seismic bearing capacity factors for spread foundations. *Geotechnique*, 52(2): 79–88.

Kung, G.T.C., Juang, C.H., Hsaio, E.C.L. and Hashash, Y.M.A. 2007. Simplified model for wall deflection and ground-surface settlement caused by braced excavation in clays. *Journal of Geotechnical and Geoenvironmental Engineering, ASCE*, 133(6): 731–747.

Kurian, N.P. 2006. *Shell Foundations.* Universities Press, Hyderabad.

Kuwabara, F. 1989. An elastic analysis for piled raft foundations in a homogeneous soil. *Soils and Foundations*, 28(1): 82–92.

Kuwabara, F. and Poulos, H.G. 1989. Downdrag forces in a group of piles. *Journal of Geotechnical Engineering Division, ASCE*, 115(6): 806–818.

Ladd, C.C. and Foott, R. 1974. New design procedure for stability of soft clays. *Journal of Geotechnical Engineering Division, ASCE*, 100(GT7): 763–786.

Lam, C. and Jefferis, S.A. 2014. The use of polymer solutions for deep excavations: Lessons from far east experience. *HKIE Transactions*, 21(4): 263–271.

Lam, C., Jefferis, S.A. and Martin, C.M. 2014. The effects of polymer and bentonite support fluids on concrete-sand interface strength. *Geotechnique*, 64(1): 28–39.

Lam, N.T.K., Sinadinovski, C., Koo, R.C.K. and Wilson, J.L. 2003. Peak ground velocity modelling for Australian earthquakes. *International Journal of Seismology and Earthquake Engineering*, 5(2): 11–22.

Lam, T.S.K. and Johnston, I.W. 1982. A constant normal stiffness direct shear machine. *Proceedings of the 7th SE Asian Conference on Soil Engineering*, Hong Kong, pp. 805–820.

Lambe, T.W. 1964. Methods of estimating settlement *Journal of SMFD, ASCE*, 90(SM5): 47–71.

Lambe, T.W. and Whitman, R.V. 1979. *Soil Mechanics*, SI Ed. John Wiley, New York.

Latotszke, J., Konong, D. and Jessberger, H.L. 1997. Effects of reaction piles in axial pile tests. *Proceedings of the 14th International Conference on SMFE*, Hamburg, Vol. 2, pp. 1097–1101.

Lee, R.G., Turner, A.J. and Whitworth, L.J. 1994. Deformations caused by tunneling beneath a piled structure. *Proceedings of the XIII International Conference on Soil Mechanics and Foundation Engineering*, New Delhi, University Press, London, pp. 873–878.

Lee, S.L., Chow, Y.K., Karunaratne, G.P. and Wong, K.Y. 1988. Rational wave equation model for pile-driving analysis. *Journal of Geotechnical Engineering, ASCE*, 114(3): 306–325.

Lees, A. 2016. *Geotechnical Finite Element Analysis*. ICE Publishing, London.

Lehane, B.M., Gaudin, C. and Schneider, J.A. 2005. Shaft capacity of model piles buried in dense sand. *Geotechnique*, 55(10): 709–720.

Leung, C.F., Chow, Y.K. and Shen, R.F. 2000. Behaviour of pile subject to excavation-induced soil movement. *Journal of Geotenchnial and Geoenvironmental Engineering*, 126: 947–954.

Levy, J.F. 1970. Sonic pulse method of testing cast-in-situ concrete piles. *Ground Engineering*, 3(3): 17–19.

Liao and Roesset, J.M. 1997. Identification of defects in piles through dynamic testing. *International Journal for Numerical and Analytical Methods in Geomechanics*, 21(4): 277–291.

Li, H-N., Li, D-S. and Song, O-B. 2004. Applications of fiber optic sensors to health monitoring in civil engineering. *Engineering Structures*, 26: 1647–1657.

Likitlersuang, S., Surarak, C., Suwansawat, S., Wanatowski, D., Oh, E. and Balasubramaniam, A. 2014. Simplified finite element modelling for tunnelling-induced settlements. *Geotechnical Research, ICE*, 1(4): 133–152.

Linkimer, L. 2008. Relationship between peak ground acceleration and modified Mercalli Intensity in Costa Rica. *Revista Geologica de America Central*, 38: 81–94.

Liu, J. 2004. Grouting and dewatering in balancing settlement of a building. In: S. Prakash (Ed.), *Proceedings of the 5th International Conference on Case Histories*, Paper No. 1.15, CD volume, University of Missouri, Missouri, Rolla.

Liyanapathirana, S. and Poulos, H.G. 2002. Numerical simulation of soil liquefaction due to earthquake loading. *Soil Dynamics and Earthquake Engineering*, 22: 511–523.

Liyanapathirana, S. and Poulos, H.G. 2005. Pseudostatic approach for seismic analysis of piles in liquefying soil. *Journal of Geotechnical and Geoenvironmental Engineering, ASCE*, 131(12): 1480–1487.

Loganathan, N. and Flanagan, R.F. 2001. Prediction of tunnelling-induced ground movements – Assessment and evaluation. *Tunnelling Conference*, TUCS, Singapore.

Loganathan, N. and Poulos, H.G. 1998. Analytical prediction for tunneling-induced ground movements in clays. *J. Georch. Engrg ASCE*, 124(9): 846–856.

Loganathan, N. and Poulos, H.G. 1999. Tunnelling-induced ground deformations and their effects on adjacent piles. *Proceedings of the 10th Australian Tunnelling Conference*, Melbourne, pp. 241–250.

Loganathan, N. and Poulos, H.G. 2002. Centrifuge modelling: Tunnelling-induced ground movements and pile behaviour. *Proceedings of the 28th ITA Conference*, Sydney, Australia, CD volume.

Loganathan, N., Poulos, H.G. and Stewart, D.P. 2000. Centrifuge model testing of tunnelling-induced ground and pile deformation. *Geotechnique*, 50(3): 283–294.

Loganathan, N., Poulos, H.G. and Xu, K.J. 2001. Ground and pile-group responses to tunnelling. *Soils and Foundations*, 41(1): 57–67.

Long, M. 2001. Database for retaining wall and ground movements due to deep excavations. *Journal of Geotechnical and Geoenvironmental Engineering, ASCE*, 127(3): 203–224.

Look, B. 2007. *Handbook of Geotechnical Investigations and Design Tables*. Taylor & Francis, London.

Love, J.P., Burd, H.J., Milligan, G.W.E. and Houlsby, G.T. 1987. Analytical and model studies of reinforcement of a layer of granular fill on a soft clay subgrade. *Canadian Geotechnical Journal*, 24: 611–622.

Luan, L., Liu, L. and Li, Y. 2015. Numerical simulation for the soil-pile-structure interaction under seismic loading. *Mathematical Problems in Engineering*, 2015: Article ID 959581.

Lumbantoruan, P.M.H. 2005. Probabilistic post-liquefaction shear strength analyses of cohesionless soil deposits: Application to the Koecali (1999) and Duzce (1999) earthquakes. MSc thesis, Virginia Polytechnic Institute and State University.

Luo, Z., Chen, Y. and Liu, Y. 2000. Monitoring the dynamic characteristics of tall buildings by GPS technique, *Geospatial Information Science*. Taylor & Francis, 3(4): 61–66.

Madabhushi, G., Knappett, J. and Haigh, S. 2010. *Design of Pile Foundations in Liquefiable Soils.* Imperial College Press, London.

Maheshwari, B.K., Truman, K.Z., El Naggar, M.H. and Gould, P.L. 2004. Three dimensional nonlinear analysis for seismic soil-pile-structure interaction. *Soil Dynamics and Earthquake Engineering*, 24(4): 343–356.

Maiorano, R.M.S., de Sanctis, L., Aversa, S. and Mandolini, A. 2009. Kinematic response analysis of piled foundations under seismic excitation. *Canadian Geotechnical Journal*, 46: 571–584.

Mair, R.J. 2008. Tunnelling and geotechnics: New horizons. 46th Rankine Lecture. *Geotechnique*, 58(9): 695–736.

Mair, R.J., Gunn, M.J. and O'Reilly, M.P. 1983. Ground movements around shallow tunnels in soft clays. *Proceedings of the 10th International Conference Soil Mechanics and Foundation Engineeing*, Helsinki, pp. 323–328.

Mair, R.J. and Taylor, R.N. 1997. Theme lecture: Bored tunnelling in the urban environment. *Proceedings of the 14th International Conference of. Soil Mechanics and Foundation Engineering*, Hamburg. Balkema, Rotterdam, Vol. 4, pp. 2353–2385.

Mair, R.J., Taylor, R.N. and Bracegirdle, A. 1993. Subsurface settlement profiles above tunnels in clays. *Geotechnique*, 43, 315–320.

Mair, R.J., Taylor, R.N. and Burland, J.B. 1996. Prediction of ground movements and assessment of risk of building damage due to bored tunnelling. In: R.J. Mair and R.N. Taylor (Eds.), *Proceedings of the International Symposium on Geotechnical Aspects of Underground Construction in Soft Ground*, London. Balkema, Rotterdam, pp. 713–718.

Makarchian, M. and Poulos, H.G. 1996. Simplified method for design of underpinning piles. *Journal of Geotechnical Engineering, ASCE*, 122(9): 745–751.

Mana, A.I. and Clough, G.W. 1981. Prediction of movements for braced cuts in clay. *Jnl. Geot. Eng. Divn., ASCE*, 107(GT6): 759–777.

Mandolini, A. 2003. Design of piled rafts foundation: Practice and development. In: W.F. van Impe (Ed.), *Deep Foundations on Bored and Auger Piles.* Millpress, Rotterdam, pp. 59–80.

Mandolini, A. 2012. Detailing new advancements in deep foundations. *Presentation to Underground Infrastructure and Deep Foundations Qatar*, Doha, IQPC/DFI.

Mandolini, A., Russo, G. and Viggiani, C. 2005. Pile foundations: Experimental investigations, analysis and design. *Proceedings of the 16th International Conference on Soil Mechanics and Geotechnical Engineering*, Osaka, Vol. 1, pp. 177–213.

Mandolini, A. and Viggiani, C. 1997. Settlement of piled foundations. *Géotechnique*, 47(4): 791–816.

Marchetti, S. 1982. Detection of liquefiable sand layers by means of quasi static penetration tests. *Proceedings of the 2nd European Symposium on Penetration Testing*, Vol. 2, Balkema, Rotterdam, Netherlands, pp. 689–695.

Marcuson, W.F., Hynes, M.E. and Franklin, A.G. 1990. Evaluation and use of residual strength in seismic safety analysis of embankments. *Earthquake Spectra*, 6(3): 529–572.

Marfella, G., Richardson, S. and Vaz-Serra, P. 2016. The logic of rapid extrusion produces the "jumping" Phoenix. *CTBUH Journal*, (2): 26–32.

Marr, W.A. 2000. Why monitor geotechnical performance? *49th Geotechnical Conference in Minnesota*, Minneapolis.

Martinelli, M., Burghignoli, A. and Callisto, L. 2016. Dynamic response of a pile embedded in a layered soil. *Soil Dynamics and Earthquake Engineering*, 87: 16–28.

Matlock, H. 1970. Correlations for design of laterally loaded piles in soft clay. *Proceedings of the 2nd OTC*, Houston, Vol. 1, pp. 577–594.

Matlock, H. and Foo, S.H.C. 1980. Axial analysis of piles sing a hysteretic and degrading soil model. In: *Numerical Methods in Offshore Piling*, ICE, London, pp. 127–134.

Matlock, H., Foo, S.H.C. and Bryant, L.M. 1978. Simulation of lateral pile behavior under earthquake loading. *Earthquake Engineering and Soil Dynamics, ASCE Special Conference*, Pasadena, CA, Vol. 2, pp. 600–619.

Matlock, H., Ingram, W.B., Kelley, A.E. and Bogard, D. 1980. Field tests on the lateral load behavior of pile groups in soft clay. *Proceedings of the 12th OTC*, Paper 3871, Houston, pp. 163–174.

Matsuda, H. and Ohara, S. 1991. Geotechnical aspects of earthquake-induced settlement of a clay layer. *Marine Geotechnology*, 9: 179–206.

Matsui, T., Ohara, H. and Ito, T. 1980. Cyclic stress-strain history and shear characteristics of clay. *Journal of Geotechnical Engineering, ASCE*, 106(GT10): 1101–1120.

Matsumoto, T. and Tsuzuki, M. 1994. Statnamic tests on steel pipe piles driven in a soft rock. In: *Proceedings of the International Conference on Design and Construction of Deep Foundations*, FHWA, Orlando, Vol. 2, pp. 586–600.

Maugeri, M. and Carrubba, P. 1997. Microzonation of ground motion during the 1980 Irpinia earthquake at Calabritto Italy. In: P. Seco e Pinto (Ed.), *Proceedings of the Conference on Seismic behaviour of Ground and Structures*. A.A. Balkema, Rotterdam, pp. 81–96.

Mayne, P.W. 1995. Profiling yield stresses in clays by in-situ tests. In: *Transportation Research Record 1479*. National Academy Press, Washington, DC, pp. 43–50.

Mayne, P.W. 2001. Stress-strain-strength-flow parameters from enhanced in-situ tests. In: P.P. Rahardjo and T. Lunne (Eds.), *Proceedings of the International Conference on In-situ Measurement of Soil Properties and Case Histories*, Bali, Indonesia, pp. 27–48.

Mayne, P.W. and Holtz, R.D. 1988. Profiling stress history from piezocone soundings. *Soils and Foundations*, 28(1): 16–28.

Mayne, P.W. and Poulos, H.G. 1999. Approximate displacement influence factors for elastic shallow foundations. *Journal of Geotechnical and Geoenvironmental Engineering, ASCE*, 125(6): 453–460.

Mayne, P.W. and Rix, G.J. 1993. Gmax-qc relationships for clays. *Geotechnical Testing Journal, ASTM*, 16(1): 54–60.

Mayne, P.W. and Schneider, J.A. 2001. Evaluating axial drilled shaft response by seismic cone. *Foundation and Ground Improvement*, GSP No. 113, ASCE, Reston, VA, pp. 655–669.

Mayne, P.W., Coop, M.R., Springman, S.M. and Huang, A-B. 2009. Geomaterial behaviour and testing. *Proceedings of the 17th International Conference on Soil Mechanics and Geotechnical Engineering*, Theme Lecture, Alexandria, Vol. 4, pp. 2777–2872.

McCabe, B.A. and Shiel, B.B. 2014. Pile group settlement estimation: suitability of nonlinear interaction factors. *International Journal of Geomechanics*, doi: 10.1061/(ASCE)GM.1943-5622.0000395, 04014056.

McClelland, B., Focht, J.A. and Emrich, W.J. 1969. Problems in design and installation of offshore piles. *Journal of SMFD, ASCE*, 95(SM6): 1419–1514.

McKenzie, N.P. and Pender, M.J. 1996. Representative shear modulus for shallow foundation seismic soil-structure interaction. *Proceedings of the 11th World Conference on Earthquake Engineering*, Acapulco, Mexico, Paper 931.

Meigh, A.C. and Wolski, W. 1979. Design parameters for weak rocks. *Proceedings of the 7th European Conference on SMFE*, Brighton, Vol. 5, pp. 57–77.

Melis, M. and Rodriguez Ortiz, J.M. 2001. Consideration of the stiffness of buildings in the estimation of subsidence damage by EPB tunnelling in the Madrid Subway. In: F. Jardine (Ed.), *Proceedings of the International Conference on Response of Buildings to Excavation-Induced Ground Movements*. CIRIA SP201, London, pp. 387–394.

MELT. 1993. *Regles techniques de conception et de calcul des fondations des ouvrages de genie civil*. Cahier des Clauses Techniques Generales applicables aux Marches Publics de Travaux, Fascicule 62 – Titre V, Ministere de l'Equipement du Logement et des Transports, Paris.

Mendis, P., Ngo, T., Haritos, N., Hira, A., Samali, B. and Cheung, J. 2007. Wind loading on tall buildings. *EJSE Special Issue, Loading on Structures*, 3: 41–54.

Merifield, R.S., Lyamin, A.V. and Sloan, S.W. 2006. Limit analysis solutions for the bearing capacity of rock masses using the generalised Hoek-Brown criterion. *International Journal of Rock Mechanics and Mining Sciences*, 43: 920–937.

Mesri, G. and Godlewski, P.M. 1977. Time- and stress-compressibility interrelationshi. *Journal of Geotechnical Engineering, ASCE*, 103: 417–430.

Mesri, G., Lo, D.O.K. and Feng, T.W. 1994. Settlement of embankments on soft clays. *Proceedings of Settlement '94, ASCE Special Pubublication No. 40*, 1: 8–56.

Meyerhof, G.G. 1956. Penetration tests and bearing capacity of cohesionless soils. *Journal of Soil Mechanics and Foundations Division, ASCE*, 82(SM1): 1–19.

Meyerhof, G.G. 1959. Compaction of sands and bearing capacity of piles. *Journal of Soil Mechanics and Foundations Division, ASCE*, 85(SM6): 1–29.

Meyerhof, G.G. 1974. Ultimate bearing capacity of footings on sand layer overlaying clay. *Canadian Geotechnical Journal*, 11: 223–229.

Meyerhof, G.G. 1976. Bearing capacity and settlement of pile foundations. *Journal of Geotechnical Engineering Division, ASCE*, 102(GT3): 195–228.

Meyerhof, G.G. 1995. Behaviour of pile foundations under special loading conditions. 1994 R.M. Hardy Keynote Address, *Canadian Geotechnical Journal*, 32(2): 204–222.

Meyerhof, G.G. and Adams, J.I. 1968. The ultimate uplift capacity of foundations. *Canadian Geotechnical Journal*, 5(4): 525–544.

Meyerhof, G.G. and Hanna, A.M. 1978. Ultimate bearing capacity of foundations on layered soils under inclined load. *Canadian Geotechnical Journal*, 15: 565–572.

Michalowski, R.L. and Lei, Shi 1995. Bearing capacity of footings over two-layer foundation soils. *Journal of Geotechnical Engineering, ASCE*, 121: 421–427.

Middendrop, P., Birmingham, P. and Kuiper, B. 1992. Statnamic load testing of foundation piles. *Proceedings of the 4th International Conference on the Application. of Stress Wave Theory to Piles*, Balkema, The Hague, pp.581–588.

Middendorp, P. and van Weele, P.J. 1986. Application of characteristic stress wave method to offshore practice. *Proceedings of the 3rd International Conference on Numerical Methods in Offshore Piling*, Nantes, France.

Mikami, T. and Nishizawa, T. 2015. Health monitoring of high-rise building with fiber optic sensor (SOFO). *International Journal of High Rise Buildings*, 4(1): 27–37.

Mikola, R.G., Candia, G. and Sitar, N. 2014. Seismic earth pressures on retaining structures and basement walls. *10th US National Conference on Earthquake Engineering*, Anchorage, Alaska.

Miller, G.A. and Lutenegger, A.J. 1997. Predicting pile skin friction in overconsolidated clay. *Proceedings of the 14th International Conference on Soil Mechanics and Foundation Engineering*, Hamburg, Vol. 2, pp. 853–856.

Mitchell, J.K. 2008. Aging of sand – A continuing enigma? *6th International Conference on Case Histories in Geotechnical Engineering*, Paper SOAP 11. Missouri University of Science & Technology, Arlington, VA.

Mitwally, H. and Novak, M. 1987. Response of offshore towers with pile interaction. *Journal of Engineering Mechanics, ASCE*, 113(7): 1065–1083.

Moh, Z.C. 1994. Current deep foundation practice in Taiwan and Southeast Asia. In: *Proceedings, International Conference on Design and Construction of Deep Foundations*, FHWA, Orlando, Vol. 1, pp. 236–259.

Mohamad, H. 2008. Distributed fibre optic strain sensing of geotechnical structures. PhD thesis, University of Cambridge.

Moldovan, A.R. and Popa, A. 2012. Finite element modelling for tunnelling excavation. *Acta Technica Napocensis: Civil Engineering and Architecture*, 55(1): 98–113.

Moller, S.C. and Vermeer, P.A. 2008. On numerical simulation of tunnel installation. *Tunnelling and Underground Space Technology*, 23(4): 461–475.

Monaco, P. and Marchetti, S. 2007. Evaluating liquefaction potential by seismic dilatometer (SDMT) accounting for aging/stress history. In *Paper presented to 4th International Conference on Earthquake Geotechnical Engineering*, Thessaloniki, Greece.

Monaco, P., Marchetti, S., Totani, G. and Calabrese, M. 2005. Sand liquefiability assessment by flat dilatometer test (DMT). *Proceedings of the XVI International Conference on Soil Mechanics and. Geotechnical Engineering*, Osaka, Vol. 4, pp. 2693–2697.

Moon, K.S. 2008. Material saving design strategies for tall building structures. *CTBUH 8th World Congress*, Dubai (available on CTBUH website).

Moorman, C. and Katzenbach, R. 2002. Three-dimensional effects of deep foundations with rectangular shape. *Proceedings of the 2nd International Conference on Soil Structure Interaction in Urban Engineering*, Zurich, Vol. 1, pp. 134–142.

Morrison, J. 2000. Raft foundations for two Middle East tower blocks, Chapter 6. In: J.A. Hemsley (Ed.), *Design Applications of Raft Foundations*. Thomas Telford, London, pp. 155–172.

Morton, J.D. and King, K.H. 1979. Effect of Tunnelling on the bearing capacity and settlement of piled foundations. In: M.J. Jones (Ed.), *Proceedings of the Tunnelling '79.*, IMM, London, pp. 57–58.

Mroueh, H. and Shahrour, I. 2002. Three-dimensional finite element analysis of the interaction between tunnelling and pile foundation. *International Journal for Numerical and Analytical Methods in Geomechanics*, 26: 217–230.

Murchison, J.M. and O'Neill, M.W. 1984. Evaluation of p-y relationships in cohesionless soil. In: J.R. Meyer (Ed.), *Analysis and Design of Pile Foundations.* ASCE, New York, pp. 174–191.

Mylonakis, G. 2001. Simplified model for seismic pile bending at soil layer interfaces. *Soils and Foundations*, 41(4): 47–58.

Mylonakis, G, Gazetas, G. 1998. Dynamic vertical vibration and additional distress of grouped piles in layered soil. *Soils and Foundations*, 38(1): 1–14.

Mylonakis, G., Kloukinas, P. and Papatonopoulos, C. 2007. An alternative to the Mononobe-Okabe equation for seismic earth pressures. *Soil Dynamics and Earthquake Engineering*, 27(10): 957–969.

Nauroy, J.F., Brucy, F., Le Tirant, P. and Kervadec, J-P. 1986. Design and installation of piles in calcareous formations. *Proceedings of 3rd Internation Conference on Numerical Methods in Offshore Piling,* Nantes, Inst. Francais Du Petrole, pp. 461–480.

NAVFAC 1982. DM-7.03. Soil dynamics and special design aspects. In: *Handbook*, USA Dept. Defense, Washington.

Negro, A., Karlsrud, K., Srithar, S., Ervin, M. and Vorster, E. 2009. Prediction, monitoring and evaluation of performance of geotechnical structures. *Proceedings of the 17th International Conference on Soil Mechanics and Geotechnical Engineering*, Alexandria, Egypt, Vol. 4, pp. 2930–3005.

Nelson, J.D. and Miller, D.J. 1992. *Expansive Soils: Problems and Practice in Foundation and Pavement Engineering.* Wiley Interscience, New York.

Ng, C.W.W. and Lei, G.H. 2003. Performance of long rectangular barrettes in granitic saprolites. *Journal of Geotechnical and Geoenvironmental Engineering*, 129(8): 685–696.

Ng, C.W.W., Simons, N. and Menzies, B. 2004. *A Short Course in Soil-Structure Engineering of Deep Foundations, Excavations and Tunnels.* Thomas Telford, London.

Nikolaou, S., Mylonakis, G., Gazetas, G. and Tazoh, T. 2001. Kinematic pile bending during earthquakes: Analysis and filed measurements. *Geotechnique*, 51(4): 425–440.

Nordlund, R.L. 1963. Bearing capacity of piles in cohesionless soils. *Journal of the Soil Mechanics and Foundation Division, ASCE*, 89(SM3): 1–35.

Novak, M. 1974. Dynamic stiffness and damping of piles. *Canadian Geotechnical Journal*, 11(4): 574–598.

Novak, M. 1977. Vertical vibration of floating piles. *Journal of the Engineering Mechanics Division, ASCE*, 103(EM1): 153–168.

Novak, M. 1987. State of the art in analysis and design of machine foundations. In: *Soil Structure Interaction.* Elsevier/CML Publ, New York, pp. 171–192.

Novak, M. and Beredugo, Y. O. 1972. Vertical vibration of embedded footings. *Journal of the Soil Mechanics and Foundation Division, ASCE*, 98(SM12): 1291–1310.

Novak, M. and El Sharnouby, B. 1983. Stiffness constants of single piles. *Journal of Geotechnical Engineering, ASCE* 109(7): 961–974.

Novak, M. and Janes, M. 1989. Dynamic and static response of pile groups. *Proceedings of the 12th International Conference on SMFE*, Rio de Janeiro, Vol. 2, pp. 1175–1178.

Novak, M. and Mitwally, H. 1990. Random response of offshore towers with pile-soil-pile interaction. *Journal of Offshore Mechanics and Arctic Engineering*, 112: 35–41.

Novak, M. and Sheta, M. 1982. Dynamic response of piles and pile groups. *Proceedings of the 2nd International Conference on Numerical Methods in Offshore Piling*, University of Texas, Austin, pp. 489–507.

O'Neill, M.W. 2001. Side resistance in piles and drilled shafts. 34th Terzaghi Lecture, *Journal of Geotechnical and Geoenvironmental Engineering*, 127(1): 1–16.

O'Neill, M.W., Ghazzaly, O.I. and Ha, H.B. 1977. Analysis of three-dimensional pile groups with nonlinear soil response and pile-soil-pile interaction. *Proceedings of the 9th OTC*, Paper 2838, Houston, pp. 245–256.

O'Neill, M.W. and Hassan, K.M. 1994. Drilled shafts: Effects of construction on performance and design criteria. In: *Proceedings of the International Conference on Design and Construction of Deep Foundations*, FHWA, Orlando, Vol. 1, pp. 137–187

O'Neill, M.W., Hawkins, R.A. and Mahar, L.J. 1982. Load transfer mechanisms in piles and pile groups. *Journal of Geotechnical Engineering, ASCE*, 108(GT12): 1605–1623.

O'Reilly, M.P. and New, B.M. 1982. Settlements above tunnels in the U.K. – their magnitude and prediction. *Tunnelling*, 82: 173–181.

Oasys. 2001. *ALP Version 19.2*. Oasys Ltd., London.

Ochoa, M. and O'Neill, M.W. 1989. Lateral pile interaction factors in submerged sand. *Journal of Geotechnical Engineering, ASCE*, 115(3): 359–378.

Ohara, S. and Matsuda, H. 1988. Study on the settlement of saturated clay layer induced by cyclic shear. *Soils and Foundations*, 28(3): 103–113.

Ohsaki, Y. 1969. The effects of local soil conditions upon earthquake damage. *Proceedings of the Special Session No. 2, 7th International Conference on Soil Mechanics and Foundation Engineering*, Mexico City, pp. 3–32.

Ohsaki, Y. 1976. *Foundation structures for tall buildings. IABSE Congress Report*, ETH, Zurich.

Ohta, Y. and Goto, N. 1978. Empirical shear wave velocity equations in terms of characteristic soil indexes. *Earthquake Engineering and Structural Dynamics*, 6: 167–187.

Okabe, T. 1973. *Report of measured negative friction on double pipe piles and pile group*. Report C-0261, Japan National Railway (in Japanese).

Okamoto, S. 1984. *Introduction to Earthquake Engineering*, 2nd Ed. University of Tokyo Press, Japan.

Okur, V. and Ansal, A. 2001. Dynamic characteristics of clays under irregular cyclic loadings. In: A.M. Ansal (Ed.), *TC4 satellite conference on "Lessons Learned from Recent Strong Earthquakes". 15th Intenational Conference on Soil Mechanics and Geotechnical Engineering*, Istanbul, pp. 267–270.

Olgun, C., Abdelaziz, S. and Martin, J. 2012. Long-term performance and sustainable operation of energy piles. In: W.K.O. Chong, J. Gong, J. Chang and M.K. Siddiqui (Eds.), *ICSDEC 2012*, ASCE, Fort Worth, TX, pp. 534–542.

Oliveira, D.A.F. and Wong, P.K. 2014. Use of embedded pile elements in 3D modelling of piled-raft foundations. *Australian Geomechanics Journal*, 46(3): 9–19.

Ooi, P.S.K. and Duncan, J.M. 1994. Lateral load analysis of groups of piles and drilled shafts. *Journal of Geotechnical Engineering, ASCE*, 120(6): 1034–1050.

Ooi, L.H. and Carter, J.P. 1987. A constant normal stiffness direct shear device for static and cyclic loading. *Geotechnical Testing Journal, ASTM*, 10: 3–12.

Osterberg, J.O. 1979. Failures in exploration programs. In: C.H. Dowding (Ed.), *Site Characterization and Exploration*. ASCE, New York, pp. 3–9.

Osterberg, J. 1989. New device for load testing driven and drilled shafts separates friction and end bearing. *Proceedings of the International Conference on Piling and Deep Foundations*, London, pp. 421–427.

Ottaviani, M. 1975. Three-dimensional finite element analysis of vertically loaded pile groups. *Geotechnique*, 25(2): 159–174.

Papadopoulou, M.C. and Comodromos, E.M. 2010. On the response prediction of horizontally loaded pile groups in sands. *Computers and Geotechnics*, 37(7–8): 930–941.

Parker, D. and Wood, A. 2013. *The Tall Buildings Reference Book*. Routledge, New York.

Patrizi, P. and Burland, J.B. 2001. Developments in the design of driven piles in clay in terms of effective stress. *Rivista Italiana di Geotecnica*, 3: 35–49.

Payan, M., Senetakis, K., Khoshghalb, A. and Khalili, N. 2016. Influence of particle shape on small-strain damping ratio of dry sands. *Geotechnique*, 66(7): 610–616.

PDI. 2016. *SQUID Device. Promotional Literature*. Pile Dynamics Inc.

Peaker, K.R. 1984. Lakeview Tower: Case history of foundation failure. *Proceedings of the International, Conference on Case Histories in Geotechnical Engineering*, University of Missouri-Rolla, Rolla, MO, pp. 7–13.

Pearlman, S.L., Walker, M.P. and Boscardin, M.D. 2004. Deep underground basements for major urban building construction. *GeoSupport 2004*, ASCE, Orlando, FL, pp. 545–560.

Peck, R.B. 1969. Deep excavations and tunneling in soft ground. *Proceedings of the 7th International Conference on Soil Mechanics and Foundation Engineering*, State-of-the-Art-Report, Mexico City, State-of-the-Art-Volume, pp. 225–290.

Pedler, I.V. and Whiteley, R.J. 2000. Applying innovative geophysical methods to assist identification and repair of road collapses. *Proceedings of the GeoEng 2000*, Paper G0739, Melbourne.

Pells, P.J.N. and Turner, R.M. 1980. End bearing on rock with particular reference to sandstone. In: *Proceedings of the International Conference on Structural Foundations on Rock*. A.A. Balkema, Rotterdam, Vol. 1, pp. 181–190.

Perau, E.W. 1997. Bearing capacity of shallow foundations. *Soils and Foundations*, 37: 77–83.

Poh, T.Y., Goh, A.T.C. and Wong, I.H. 2001. Ground movements associated with wall construction: Case histories. *Journal of Geotechnical and Geoenvironmental Engineering, ASCE*, 127(12): 1061–1069.

Popa, A., Lacatus, F., Rebeleanu, V. and Taria, O. 2001. Underpinning of buildings by means of jet grouted piles. *Proceedings of the 15th International Conference on Soil Mechanics and Geotechnical Engineering*, Istanbul, Vol. 3, pp. 1835–1838.

Potts, D.M. and Zdravkovic, L. 2001. *Finite Element Analysis in Geotechnical Engineering*, 2 volumes). Thomas Telford, London.

Poulos, H.G. 1967. Stresses and displacements in an elastic layer underlain by a rough rigid base. *Geotechnique*, 17: 378–410.

Poulos, H.G. 1971a. The behaviour of laterally-loaded piles: I. single piles. *Journal of Soil Mechanics and Foundations Division, ASCE*, 97(SM5): 711–731.

Poulos, H.G. 1971b. The behaviour of laterally-loaded piles: II. Pile groups. *Journal of Soil Mechanics and Foundations Division, ASCE*, 97(SM5): 733–751.

Poulos, H.G. 1975. Lateral load-deflection prediction for pile groups. *Journal of Geotechnical Engineering Division, ASCE*, 101(GT1): 19–34.

Poulos, H.G. 1979a. Settlement of single piles in non-homogeneous soils. *Journal of Geotechnical Engineering Division, ASCE*, 105(GT5): 627–641.

Poulos, H.G. 1979b. An approach for the analysis of offshore pile groups. In: *Numerical Methods in Offshore Piling*, ICE, London, pp. 119–126.

Poulos, H.G. 1982a. Developments in the analysis of static and cyclic lateral response of piles. *4th International Conference on Numerical Methods in Geomechanics*, Edmonton, Vol. 3, pp. 1117–1135.

Poulos, H.G. 1982b. Single pile response to cyclic lateral load. *Journal of Geotechnical Engineering Division, ASCE*, 108(GT3): 355–375.

Poulos, H.G. 1985. Ultimate lateral pile capacity in two-layer soil. *Geotechnical Engineering, SEAGS*, 16(1): 25–38.

Poulos, H.G. 1987a. Analysis of residual stress effects in piles. *Journal of Geotechnical Engineering, ASCE*, 113(3): 216–229.

Poulos, H.G. 1987b. Piles and piling, Chapter 52. In: F.G. Bell (Ed.), *Ground Engineer's Reference Book*. Butterworths, London.

Poulos, H.G. 1988a. The mechanics of calcareous sediments. John Jaeger Memorial Lecture. In: *Australian Geomechanics*, Special Ed. Aust. Geomechanics Society, Sydney, Australia, pp. 8–41.

Poulos, H.G. 1988b. Cyclic stability diagram for axially loaded piles. *Journal of Geotechnical Engineering, ASCE*, 114(8): 877–895.

Poulos, H.G. 1989. Pile behaviour - Theory and application. 29th Rankine Lecture. *Géotechnique*, 39(3): 65–415.

Poulos, H.G. 1990. *DEFPIG User's Manual*. Centre for Geotechnical Research, University of Sydney, Sydney.

Poulos, H.G. 1991a. Analysis of piled strip foundations. In: G. Beer et al. (Eds.), *Computer Methods and Advances in Geomechanics*. Balkema, Cairns, Australia, Vol. 1, pp. 182–191.

Poulos, H.G. 1991b. *ERLS User's Manual*. Coffey Partners International, Sydney.

Poulos, H.G. 1993a. Settlement of bored pile groups. In: W.F. van Impe (Ed.), *Proceedings of BAP II*, Ghent. Balkema, Rotterdam, pp. 103–117.

Poulos, H.G. 1993b. Effect of earthquakes on settlements and axial pile response in clays. *Australian Civil Engineering Transactions, IEAust*, CE35(1): 43–48.

Poulos, H.G. 1994a. An approximate numerical analysis of pile-raft interaction. *International Journal for Numerical and Analytical Methods in Geomechanics*, 18: 73–92.

Poulos, H.G. 1994b. Settlement prediction for driven piles and pile groups. *Special Technical Publication 40, ASCE*, 2: 1629–1649.

Poulos, H.G. 1994c. Effect of pile driving on adjacent piles in clay. *Canadian Geotechnical Journal*, 31(6): 856–867.

Poulos, H.G. 1994d. Analysis and design of piles through embankments. *Proceedings of the International Conference on Design and Construction of Deep Foundations*, Orlando, Vol. 3, pp. 1403–1421.

Poulos, H.G. 1995. Design of reinforcing piles to increase slope stability. *Canadian Geotechnical Journal*, 32(5): 808–818.

Poulos, H.G. 1997a. Piles subjected to negative friction: A procedure for design. *Geotechnical Engineering, SEAGS*, 28(1): 23–44.

Poulos, H.G. 1997b. Behaviour of pile groups with defective piles. In: *Proceedings of the 14th International Conference on Soil Mechanics and Foundation Engineering*, Hamburg. Balkema, Rotterdam, Vol. 2, pp. 871–876.

Poulos, H.G. 1997c. Failure of a building supported on piles. In: T.W. Hulme and Y.S. Lau (Eds.), *Proceedings of the International Conference on Foundation Failures*, Singapore. Institution of Engineers, Singapore, pp. 53–66.

Poulos, H.G. 1998. The effects of ground movements on pile foundations. *Darmstadt Technical University*, 4(2): 237–254.

Poulos, H.G. 1999. The design of piles with particular reference to the Australian piling code. *Australian Geomechanics*, 34(4): 25–39.

Poulos, H.G. 2000a. Design of slope stabilizing piles. In: N. Yagi, T. Yamagami and J-C. Jiang (Eds.), *Slope Stability Engineering*. Balkema, Rotterdam, Vol. 1, pp. 83–100.

Poulos, H.G. 2000b. Pile testing – From the designer's viewpoint. In: O. Kusakabe, F. Kuwabara and T. Matsumoto (Eds.), *Proceedings of the 2nd International Statnamic Seminar*, Tokyo. A.A. Balkema, Rotterdam, pp. 3–21.

Poulos, H.G. 2001a. Pile foundations, Chapter 10. In: R.K. Rowe (Ed.), *Geotechnical and Geoenvironmental Engineering Handbook*. Kluwer Academic Press, Boston, pp. 261–304.

Poulos, H.G. 2001b. Piled raft foundations – Design and applications. *Geotechnique*, 51(2): 95–113.

Poulos, H.G. 2002. Prediction of settlement of building due to tunnelling operations. *International Symposium on Underground Construction*, Toulouse.

Poulos, H.G. 2003. Analysis of soil extraction for correcting uneven settlement of pile foundations. In: C.F. Leung et al. (Eds.), *Proceedings of the 12th Asian Regional Conference on SMGE*, Singapore, Vol. 1. World Scientific, New Jersey, pp. 653–656.

Poulos, H.G. 2005a. Pile behavior – Consequences of geological and construction imperfections. 40th Terzaghi Lecture. *Journal of Geotechnical and Geoenvironmental Engineering, ASCE*, 131(5): 538–563.

Poulos, H.G. 2005b. The influence of construction side effects on existing pile foundations. *Geotechnical Engineering, SEAGS*, 36(1): 51–67.

Poulos, H.G. 2005c. Piled raft and compensated piled raft foundations for soft soil sites. In: C. Vipulanandan and F.C. Townsend (Eds.), *Advances in Designing and Testing Deep Foundations*, Austin, TX, Geotechnical Special Publication, GSP. 129. ASCE, pp. 214–234.

Poulos, H.G. 2006a. Pile group settlement estimation – Research to practice. In: R.L. Parsons et al. (Eds.), *Foundation Analysis and Design, Innovative Methods*, Keynote Paper. ASCE, Shanghai, China. GSP No. 153, pp. 1–22.

Poulos, H.G. 2006b. Ground Movements – A hidden source of loading on deep foundations. 2006 John Mitchell Lecture, In: *Proceedings of the 10th International Conference on Piling and Deep Foundations*. DFI, Amsterdam, pp. 2–19.

Poulos, H.G. 2006c. Use of stiffness inserts in pile groups and piled rafts. *Geotechnical Engineering, ICE*, 159(GE3): 153–160.

Poulos, H.G. 2008a. A practical design approach for piles with negative friction. *Proceedings of the Institution of Civil Engineers, Geotechnical Engineering*, 161(GE1): 19–27.

Poulos, H.G. 2008b. Simulation of the performance and remediation of imperfect pile groups. In: W.F. van Impe and P.O. van Impe (Eds.), *Deep Foundations on Bored and Auger Piles*, BAP V. Taylor & Francis, London, pp. 143–154.

Poulos, H.G. 2009. Tall buildings and deep foundations – Middle East challenges. *Terzaghi Oration, Proceedings of the 17th International Conference on Soil Mechanics and Geotechnial Engineering*, Alexandria, Egypt, 5–9 October, Vol.4, pp. 3173–3205.

Poulos, H.G. 2013a. Tall building foundation design – The 151 storey Incheon Tower. In: S. Prakash (Ed.), *Proceedings of the 7th International Conference on Case Histories in Geotechnical Engineering*, Wheeling IL, State-of-the-Art and Practice Paper No. SOAP 8, CD volume, ISBN: 1-887009-17-5.

Poulos, H.G. and Bunce, G. 2008. Foundation design for the Burj Dubai – The world's tallest building. *Proceedings of the 6th International Conference on Case Histories in Geotechnical Engineering*, Paper 1.47, Arlington, VA, CD volume.

Poulos, H.G. and Chen, L. 1996. Pile response due to unsupported excavation-induced lateral soil movements. *Canadian Geotechnical Journal*, 33: 670–677.

Poulos, H.G. and Chen L. 1997. Pile response due to excavation-induced lateral soil movement. *Journal of Geotechnical and Geoenvironmental Engineering, ASCE*, 123(2): 94–99.

Poulos, H.G. and Davids, A.J. 2005. Foundation design for the Emirates Twin Towers, Dubai. *Canadian Geotechnical Journal*, 42: 716–730.

Poulos, H.G. and Davis, E.H. 1974. *Elastic Solutions for Soil and Rock Mechanics*. John Wiley, New York.

Poulos, H.G. and Davis, E.H. 1980. *Pile Foundation Analysis and Design*. John Wiley, New York.

Poulos, H.G. and Hull, T.S. 1989. The role of analytical geomechanics in foundation engineering. In: F.H. Kulhawy (Ed.), *Foundation Engineering – Principles & Practices*. ASCE, Evanston, IL Vol. 2, pp. 1578–1606.

Poulos, H.G. and Randolph, M.F. 1983. Pile group analysis: A study of two methods. *Journal of Geotechnical Engineering, ASCE*, 109(3): 355–372.

Poulos, H.G., Badelow, F. and Powell, G.E. 2003. A theoretical study of constructive application of excavation for foundation correction. In: F.M. Jardine (Ed.), *Proceedings of the International Conference on Response of Buildings to Excavation-Induced Ground Movements, Special Publication*, London, CIRIA, Vol. 199, pp. 469–484.

Poulos, H.G., Badelow, F., Tosen, R., Abdelrazaq, A. and Kim, S.H. 2013a. Identification of test pile defects in a super-tall building foundation. *Proceedings of the 18th International Conference on Soil Mechanics and Foundation Engineering*, Paris, Vol. 4, pp. 2823–2826.

Poulos, H.G., Carter, J.C. and Small, J.C. 2001. Foundations and retaining structures – Research and practice. *Theme Lecture, Proceedings of the 15th International Conference on Soil Mechanics and Geotechnical Engineering*, Istanbul, Balkema, Vol. 4, pp. 2527–2606.

Poulos, H.G., Small, J.C. and Chow, H. 2011. Piled raft foundations for tall buildings. *Geotechnical Engineering, SEAGS*, 42(2): 78–84.

Poulos, H.G., Small, J.C. and Chow, H.S.W. 2013b. Foundation design for high-rise towers in karstic ground. In: J.L. Withiam, K-K. Phoon and M.H. Hussein (Eds.), *Foundation Engineering in the Face of Uncertainty*. ASCE GSP229, pp. 720–731.

Poulos, H.G., Small, J.C., Ta, L.D., Sinha, J. and Chen, L. 1997. Comparison of some methods for analysis of piled rafts. *Proceedings of the 14th International Conference on Soil Mechanics and Foundation Engineering*, Hamburg. Balkema, Rotterdam, Vol. 2, pp. 1119–1124.

Powers, J.P., Corwin, A.B., Schmall, P.C. and Kaeck, W.E. 2007. *Construction Dewatering and Groundwater Control: New Methods and Applications*, 3rd Ed. John Wiley & Sons, New York.

Prakash, S. 1962. Behavior of pile groups subjected to lateral load. *PhD dissertation*, Department of Civil Engineering, University of Illinois.

Prakash, S. and Kumar, S. 1996. Nonlinear lateral pile deflection prediction in sands. *Journal of Geotechnical Engineering, ASCE*, 122(2): 130–138.

Prakoso, W.A. and Kulhawy, F.H. 2001. A contribution to piled raft foundation design. *Journal of Geotechnical and Geoenvironmental Engineering, ASCE*, 127(1): 17–24.

Prandtl, L. 1921. Uber die Eindringungsfestigkeit plastische Baustoffe und die Festigkeit von Schneiden. *Zeitschrift für Angewandte Mathematik und Mechanik*, 1: 15–20.

Pressley, J.S. and Poulos, H.G. 1986. Finite elment analysis of mechanisms of pile group behaviour. *International Journal for Numerical and Analytical Methods in Geomechanics*, 10: 213–221.

Puller, M. 2003. *Deep Excavations: A Practical Manual.* Thomas Telford, London.

Randolph, M.F. 1981. The response of flexible piles to lateral loading. *Geotechnique*, 31(2): 247–259.

Randolph, M.F. 1983. Design considerations for offshore piles. *ASCE Special Conference on Geotechnical Practice in Offshore Engineering*, Austin, pp. 422–439.

Randolph, M.F. 1992. Settlement of pile groups. *Lecture No. 4, Modern Methods of Pile Design, University of Sydney*, School of Civil and Mining Engineering.

Randolph, M.F. 1994. Design methods for pile groups and piled rafts. *Proceedings of the 13th ICSMFE*, State of the Art Report, New Delhi, Vol. 5, pp. 61–82.

Randolph, M.F. 1996. PIGLET. Analysis and design of pile groups. In: *Users' Manual*, University of Western Australia, Perth, Australia.

Randolph, M.F. 2003. Science and empiricism in pile foundation design. 43rd Rankine Lecture, *Geotechnique*, 53(10): 847–875.

Randolph, M.F. and Clancy, P. 1993. Efficient design of piled rafts. In: W.F. van Impe (Ed.), *Deep Foundations on Bored and Auger Piles*. Balkema, Rotterdam, pp. 119–130.

Randolph, M.F. and Simons, H. 1986. An improved soil model for one-dimensional pile driving analysis. *Proceedings of the 3rd International Conference on Numerical Methods in Offshore Piling*, Nantes, France, pp. 3–17.

Randolph, M.F. and Wroth, C.P. 1978. Analysis of deformation of vertically loaded piles. *Journal of Geotechnical Engineering Division, ASCE*, 104(GT12): 1465–1488.

Randolph, M.F. and Wroth, C.P. 1979. An analysis of the vertical deformation of pile groups. *Geotechnique*, 29(4): 423–439.

Rasouli, R., Hayashi, K. and Zen, K. 2016. Controlled permeation grouting method for mitigation of liquefaction. *Journal of Geotechnical and Geoenvironmental Engineering, ASCE*, Paper 04016052, 142(111).

Rausche, F., Goble, G.G. and Likins, G. 1985. Dynamic determination of pile capacity. *Journal of Geotechnical Engineering, ASCE*, 111(GT3): 367–383.

Reddy, A.S. and Srinivasan, R.J. 1967. Bearing capacity of footings on layered clays. *Journal of Soil Mechanics and Foundations Division, ASCE*, 93: 83–99.

Reese, L.C. 1978. Design and construction of drilled shafts. *Journal of Geotechnical Engineering Division, ASCE*, 104(GT1): 95–116.

Reese, L.C. and O'Neill, M.W. 1988. *Drilled Shafts: Construction Procedures and Design Methods.* Pub. No. FHWA-H1-88-042, US Department of Transportation.

Reese, L.C., Cox, W.R. and Koop, F.D. 1974. Analysis of laterally loaded piles in sand. *Proceedings of the 6th OTC*, Paper 2080, Houston, pp. 473–483.

Reese, L.C., Cox, W.R. and Koop, F.D. 1975. Field testing and analysis of laterally loaded piles in stiff clay. *Proceedings of the 7th OTC*, Paper 2312, Houston, pp. 671–690.

Reese, L.C. and Welch, R.C. 1975. Lateral loading of deep foundations in stiff clay. *Journal of Geotechnical Engineering Division, ASCE*, 101(GT7): 633–649.

Reissner, H. 1924. Zum erddruckproblem. In: C.B. Biezeno and J.M. Burgers (Eds.), *Proceedings of the 1st International Congress on Applied Mechanics*, Delft. Technische Boekhandel en Drukkerii J Waltman Jr., Delft, The Netherlands, pp. 295–311.

Reul, O. and Randolph, M.F. 2003. Pile rafts in overconsolidated clay: Comparison of in-situ measurements and numerical analyses. *Geotechnique*, 53(3): 301–315.

Richards, T.D. 2005. Diaphragm walls. *Presented to Central PA Geotechnical Conference*, Hershey, PA. Nicholson Construction Company.

Richart, F.E., Hall, J.R. and Woods, R.D. 1970. *Vibrations of Soils and Foundations*. Prentice-Hall, Englewood Cliffs, NJ.

Robertson, P.K. 2004. Evaluating soil liquefaction and post-earthquake deformations using the CPT. *International Conference on Site Characterization 2*, Porto, Portugal.

Rollins, K.M., Clayton, R.J. and Mikesell, R.C. 1997. Ultimate side friction of drilled shafts in gravels. *Proceedings of the 14th International Conference on Soil Mechanics and Foundation Engineering*, Hamburg, Vol. 2, pp. 1021–1024.

Rowe, R.K. and Davis, E.H. 1982. The behaviour of anchor plates in clay. *Geotechnique*, 32(1): 9–23.

Rowe, R.K. and Lee, K.M. 1992. Subsidence owing to tunnelling. II: Evaluation of a prediction technique. *Canadian Geotechnical Journal*, 29: 941–954.

Russo, G. 1998. Numerical analysis of piled rafts. *International Journal for Numerical and Analytical Methods in Geomechanics*, 22(6): 477–493.

Russo, G., Abagnara, V., Poulos, H.G. and Small, J.C. 2013. Re-Assessment of foundation settlements for the Burj Khalifa, Dubai. *Acta Geotecnica*, 8(1): 3–15.

Russo, G. and Viggiani, C. 2008. Piles under horizontal load: An overview. In: M.J. Brown et al. (Eds.), *Proceedings of the 2nd BGA Conference on Foundations, ICOF 2008*. HIS BRE Press, Dundee, Scotland, pp. 61–79.

Rynn, J.M.W. 1986. The seismicity of Queensland and northeastern New South Wales 1866 through 1985. *Proceedings of the 8th Australian Geological Convention; Geological Society of Australia; Earth Resources in Time and Space*, Abstracts – Geological Society of Australia, Adelaide, Vol. 15, pp. 170–171.

Sales, M., Small, J.C. and Poulos, H.G. 2010. Compensated piled rafts in clayey soils: Behaviour, measurements, and predictions. *Canadian Geotechnical Journal*, 47: 327–345.

Saravanan, V.K. 2011. Cost effective and sustainable practices for piling construction in the UAE. MSc thesis, Heriot Watt University.

Sato, K., Higuchi, S. and Matsuda, T. 2004. A study of the effect of countermeasures for pile foundation under lateral flow caused by ground liquefaction. *Proceedings of the 13th World Conference on Earthquake Engineering*, Paper No. 452, Vancouver.

Sawada, S. 2004. A simplified equation to approximate natural period of layered ground on the elastic bedrock for seismic design of structures. *Proceedings of the 13th World Conference on Earthquakes*, Paper No. 1100, Vancouver.

Schmertmann, J.H. 1970. Static cone to compute static settlement over sand. *Journal of Soil Mechanics and Foundations Division, ASCE*, 96(3): 1011–1043.

Schnaid, F. 2009. *In situ Testing in Geomechanics*. Taylor & Francis, Abingdon, UK.

Schneider, J., Xu, X. and Lehane, B. 2008. Database assessment of CPT-based design methods for axial capacity of driven piles in siliceous sands. *Journal of Geotechnical and Geoenvironmental Engineering, ASCE*, 134(9): 1227–1244.

Schneider, H.R. 1997. Definition and determination of characteristic soil properties. *In: Proceedings of the 14th International Conference on Soil Mechanics and Foundation Engineering*, Hamburg. Balkema, Rotterdam, Vol. 4, pp. 2271–2274.

Schofield, A.N. and Wroth, C.P. 1968. *Critical State Soil Mechanics*. McGraw-Hill, London.

Scholl, R.E. 1984. Brace dampers: An alternative structural system for improving the earthquake performance of buildings. In: *Proceedings of the 8th World Conference on Earthquake Engineering*, Prentice-Hall, San Francisco, Vol. 5, pp. 1015–1022.

Schultze, E. and Sherif, G. 1973. Prediction of settlements from evaluated settlement observations for sand. *Proceedings of the 8th ICSMFE*, Moscow, Vol. 1.3, pp. 225–230.

Schwamb, T., Soga, K., Elshafie, M.Z.E.B. and Mair, R.J. 2016. Considerations for monitoring of deep circular excavations. *Geotechnical Engineering, ICE*, Paper 1500063, 160(GE6): 477–493.

Schweiger, H.F. 1998. Results from two geotechnical benchmark problems. In: A. Cividini (Ed.), *Proceedings of the 4th European Conference on Numerical Methods in Geotechnical Engineering*, Stuttgart, Germany, pp. 645–654.

Seed, H.B. and Booker, J.R. 1977. Stabilization of potentially liquefiable sand deposits. *Journal of Geotechnical Engineering, ASCE*, 103(GT7): 757–768.

Seed, H.B. and Idriss, I. M. 1969. Influence of soil conditions on ground motions during earthquakes. *Journal of Soil Mechanics and Foundations Division, ASCE*, 95: 99–137.

Seed, H.B. and Idriss, I.M. 1982. *Ground Motions and Liquefaction during Earthquakes*. Earthquake Engineering Research Institute, ISBN 0-943198-24-0.

Seed, H.B., Idriss, I.M. and Kiefer, F.W. 1969. Characteristics of rock motions during earthquakes. *Journal of Soil Mechanics and Foundations Division, ASCE*, 95(SM5): 1199–1218.

Seed, H.B. and Harder, L.F. 1990. SPT-based analysis of cyclic pore pressure generation and undrained residual strength. In: J.M. Duncan (Ed.), *Proceedings of the H. Bolton Seed Memorial Symposium*, University of California, Berkeley, Vol. 2, pp. 351–376.

Seed, H.B. and Whitman, R.V. 1970. Design of earth retaining structures for dynamic loads. *ASCE Specialty Conference, Lateral Stresses in the Ground and Design of Earth Retaining Structures*, Cornell University, Ithaca, NY. 103-147.

Seed, R.B. et al. 2003. Recent advances in soil liquefaction engineering: A unified and consistent framework. *Keynote Presentation, 26th Annual ASCE Los Angeles Geotechnical Spring Seminar*, Long Beach, CA.

Selvadurai, A.P.S. 1979. *Elastic Analysis of Soil-Foundation Interaction*. Elsevier Publishing Co., New York.

Senetakis, K., Anastasiadis, A. and Pitilakis, K. 2013. Normalized shear modulus reduction and damping ratio curves of quartz sand and rhyolitic crushed rock. *Soils and Foundations*, 53(6): 879–893.

Shen, M., Chen, Q., Zhang, J., Gong, W. and Juang, C.H. 2016. Predicting liquefaction probability based on shear wave velocity: An update. *Bulletin of Engineering Geology and the Environment*, 75: 1199–1214.

Sica, S., Mylonakis, G. and Simonelli, A.L. 2011. Transient kinematic pile bending in two-layer soil. *Soil Dynamics and Earthquake Engineering*, 31: 891–905.

Sim, R. and Weismantle, P. 2014. Kingdom tower, Chapter 8. In: A.R. Tamboli (Ed.), *Tall and Supertall Buildings*. McGraw-Hill Education, New York, pp. 189–216.

Simons, N.E., Menzies, B. and Matthews, M.C. 2002. *A Short Course in Geotechnical Site Investigation*. Thomas Telford, London.

Sinha, J. 1996. Piled raft foundations subjected to swelling and shrinking soils. PhD thesis, University of Sydney, Australia.

Skempton, A.W. 1953. Discussion: piles and piled foundations, settlement of pile foundations. *Proceedings of the 3rd International Conference on Soil Mechanics and Foundation Engineering*, Zurich, Vol. 3, p. 172.

Sloan, S.W. 2012. Geotechnical stability analysis. 51st Rankine Lecture, *Geotechnique*, 63(7): 531–571.

Small, J.C. 2016. *Geomechanics in Soil, Rock and Environmental Engineering*. CRC Press, Boca Raton, FL.

Small, J.C. and Poulos, H.G. 2007. A method of analysis of piled rafts. In: J. Ameratunga, B. Taylor and M. Patten (Eds.), *Proceedings of the 10th ANZ Conference on Geomechanics*, Brisbane. Australian Geomechanics Society, Vol. 2, pp. 602–607.

Smith, A. 2008. Burj Dubai: Designing the world's tallest. *CTBUH 8th World Congress*, Dubai.

Sommer, H. 1993. Development of locked stresses and negative shaft resistance at the piled raft foundation Messeturm Frankfurt/Main. In: W.F. van Impe (Ed.), *Deep Foundations on Bored and Auger Piles BAP II*. Balkema, Rotterdam, pp. 347–349.

Sommer, H., Wittmann, P. and Ripper, P. 1985. Piled raft foundation of a tall building in Frankfurt clay. *Proceedings of the 11 ICSMFE*, San Francisco, Vol. 4, pp. 2253–2257.

Stain, R.T. and Williams, H.T. 1991. Interpretation of sonic coring results – A research project. In: *Proceedings of the 4th International Conference on Piling and Deep Foundations*, Stresa, Balkema, Rotterdam, Vol. 1, pp. 633–640.

Stark, T.D. and Mesri, G. 1992. Undrained shear strength of sands for stability analysis. *Journal of Geotechnical Engineering, ASCE*, 118(11): 1727–1747.

Stas, C.V. and Kulhawy, F.H. 1984. *Critical evaluation of design methods for foundations under axial uplift and compression loading*. Report for EPRI, No. EL-3771, Cornell University.

Submaneewong, C. and Teparaksa, W. 2009. Performance of T-shape barrette pile against lateral force in Bangkok subsoils. *Geotechnical Engineering, SEAGS*, December, 247–256.

Sullivan, W.R., Reese, L.C. and Fenske, C.W. 1980. Unified method for analysis of laterally loaded piles in clay. In: *Numerical Methods in Offshore Piling*, ICE, London, pp. 135–146.

Ta, L.D. and Small, J.C. 1996. Analysis of piled raft systems in layered soils. *International Journal for Numerical and Analytical Methods in Geomechanics*, 20: 57–72.

Tabesh, A. and Poulos, H.G. 2001. Pseudostatic approach for seismic analysis of single piles. *Journal of Geotechnical and Geoenvironmental Engineering, ASCE*, 127(9): 757–765.

Tabesh, A. and Poulos, H.G. 2007. Design charts for seismic analysis of single pile in clay. *Geotechnical Engineering, ICE*, 160(GE2): 85–96.

Taiebat, H. and Carter, J.P. 2000. Numerical studies of the bearing capacity of shallow footings on cohesive soil subjected to combined loading. *Géotechnique*, 50(4): 409–18.

Tamboli, A.R. 2014. *Tall and Supertall Buildings – Planning and Design*. McGraw-Hill Education, New York.

Tchepak, S. 1998. Pile testing. *Mini-symposium on Recent Developments in Piling Practice in Sydney*, Australian Geomechanics Society, Sydney, Chapter.

Teh, C.I. and Wong, K.S. 1995. Analysis of downdrag on pile groups. *Geotechnique*, 45(2): 191–207.

Teo, P.L. and Wong, K.S. 2012. Application of the Hardening Soil model in deep excavation analysis. *IES Journal, Part A: Civil and Structural Engineering*, 5(3): 152–165.

Teparaksa, W., Thasnanipan, N. and Anwar, M.A. 1999. Base grouting of wet process bored piles in Bangkok subsoil. *Proceedings of the 11th Asian Regional Conference on Soil Mechanics and Geotechnical Engineering*, Seoul, Vol. 1, pp. 269–272.

Terzaghi, K. 1943. *Theoretical Soil Mechanics*. John Wiley, New York.

Terzaghi, K. and Peck, R.B. 1967. *Soil Mechanics in Engineering Practice*. John Wiley, New York.

Thasnanipan, N., Maung, A.W. and Baskaran, G. 2000. Diaphragm wall and barrette construction for Thiam Ruam MIT station box, MRT Chaloem Ratchamongkhon line, Bangkok. *Proceedings of the GeoEng2000, International Conference on Geotechnical and Geological Engineering*, Melbourne, Australia, 19–24 November.

Thorburn, S. and Thorburn, J.Q. 1977. *Review of Problems Associated with the Construction of Cast in Place Piles*. CIRIA Report PG2, Storeys Gate, London.

TMR. 2015. *Geotechnical Design Standard – Minimum Requirements*. Department of Transport and Main Roads, Queensland, Australia.

Tokimatsu, K. and Seed, H.B. 1987. Evaluation of settlements in sands due to earthquake shaking. *Journal of Geotechnical Engineering, ASCE*, 113(8): 861–878.

Tokimatsu, K., Suzuki, H. and Sato, M. 2005. Effects of inertial and kinematic interaction on seismic behaviour of pile with embedded foundation. *Soil Dynamics and Earthquake Engineering*, 25: 753–762.

Tomlinson, M.J. 2001. *Foundation Design and Construction*, 7th Ed. Prentice-Hall, New Jersey.

Tomlinson, M.J. 2004. *Pile Design and Construction*, 4th Ed. Spon, London.

Towhata, I. 2008. *Geotechnical Earthquake Engineering*. Springer, Berlin.

Tradigo, F., Pisano, F. and di Prisco, C. 2016. On the use of embedded pile elements for the numerical analysis of disconnected piled rafts. *Computers and Geotechnics*, 72: 89–99.

Tsai, K.W., Chao, C.S. and Chou, K.T. 1991. Tilted high-rise building corrected by grouting. *Proceedings of the 11th European Conference on Soil Mechanics and Foundations Engineering*, Florence, Vol. 2, pp. 623–624.

Tschuchnigg, F. and Schweiger, H.F. 2013. Comparison of deep foundation systems using 3D finite element analysis employing different modeling techniques. *Geotechnical Engineering, SEAGS*, 44(3): 40–46.

Tsukamoto, Y. and ishihara, K. 2010. Analysis on settlement of soil deposits following liquefaction during earthquakes. *Soils and Foundations*, 50(2): 399–411.

Turner, M.J. 1997. *Integrity Testing in Piling Practice*. Report 144, CIRIA, London.

Twine, D. and Roscoe, H.C. 1999. *Temporary Propping of Deep Excavations-Guidance on Design.* CIRIA, C517.

UFC. 2004. Dewatering and groundwater control. US Department of Defence, document UFC 3-220-05.

Van der Stoel, A.E.C. 2002. Grouting for pile foundation improvement: Full scale test results. *Proceedings of the 2nd International Conference on Soil Structure Interaction in Urban Engineering*, Zurich, Vol. 1, pp. 219–224.

Van Impe, W.F. 1991. Deformation of deep foundations. *Proceedings of the Tenth European Conference on Soil Mechanics and Foundation Engineering*, Florence, Vol. 3, pp. 1031–1062.

van Meurs, G., van der Zon, W., Lambert, J., van Ree, D., Whiffin, V. and Molendijk, W. 2006. *The Challenge to Adapt Soil Properties.* Delft University of Technology.

Varun 2006. A simplified model for lateral response of caisson foundations. *MSc Civil Engineering thesis*, Georgia Institute of Technology.

Varun, Assimaki, D. and Gazetas, G. 2009. A simplified model for lateral response of large diameter caisson foundations linear elastic formulation. *Soil Dynamics and Earthquake Enginerring*, 29: 268–291.

Vermeer, P.A., Möller, S.C. and Ruse, N. 2003. On the application of numerical analysis in tunnelling. *Post Proceedings 12th Asian Regional Conference on Soil Mechanics and Geotechnical Engineering (12 ARC)*, Singapore, 4–8 August, Vol. 2, pp. 1539–1549.

Vesic, A.S. 1969. Experiments with instrumented pile groups in sand. *ASTM*, STP 444: 177–222.

Vessia, G. and Venisti, N. 2011. Liquefaction damage potential for seismic hazard evaluation in urbanized areas. *Soil Dynamics and Earthquake Engineering*, 31: 1094–1105.

Vidalie, J.F. 1977. *Relations entre les proprieties physicochemiques des sols compressibles.* Rapport de Rechereche No. 65, LPC, Paris.

Viggiani, C. 1998. Pile groups and piled rafts behaviour. In: *Deep Foundations on Bored and Auger Piles, BAP III*, W.F. van Impe and W. Haegman (Eds.), Balkema, Rotterdam, 77–90.

Viggiani, C., Mandolini, A. and Russo, G. 2012. *Piles and Pile Foundations.* Spon Press, London.

Villaverde, R. 2009. *Fundamental Concepts of Earthquake Engineering.* CRC Press, Boca Raton.

Vucetic, M. and Dobry, R. 1991. Effect of soil plasticity on cyclic response. *Journal of Geotechnical Engineering, ASCE*, 117(1): 89–107.

Wang, W.D., Xu, Z.H. and Li, Q. 2015. Design and case histories of large deep excavations in complex urban environment in Shanghai. *ISSMGE Bulletin*, 9(6): 18–39.

Wesley, L.D. 2010. *Fundamentals of Soil Mechanics for Sedimentary and Residual soils.* John Wiley, Hoboken, NJ.

Whiteley, R.J. 1983. Recent developments in the application of geophysics to geotechnical investigation. In: M.C. Ervin (Ed.), *In-situ Testing for Geotechnical Investigation.* Balkema, Rotterdam, pp. 87–110.

Whiteley, R.J. 2000. Seismic imaging of ground conditions from buried conduits & boreholes. In: *Proceedings of Paris 2000: Petrophysics Meets Geophysics*, Paper B-15, Paris, 6–8 November, 2000, EAGE.

Whiteley, R.J. and Pedler, I.V. 1994. Geophysical assessment of epsom road. In: *Proceedings of the 2nd National Conference on Trenchless Technology, Sydney*, Australian Society for Trenchless Technology, 8pp.

Whitman, R.V. and Richart, F.E., Jr. 1967. Design procedures for dynamically loaded foundations. *Journal of Soil Mechanics and Foundations Division, ASCE*, 93(6): 169–193.

Whittle, K. 2016. A case study on designing superslim in Melbourne. *Shenzhen-Guangzhou-Hong Kong Conference CTBUH*, Hong Kong, pp. 1208–1215.

Wightman, W.E., Jalinoos, F., Sirles, P. and Hanna, K. 2003. *Application of Geophysical Methods to Highway Related Problems.* Federal Highway Administration, Central Federal Lands Highway Division.

Wilson, D.W., Boulanger, R.W. and Kutter, B.L. 1999. Lateral resistance of piles in liquefying sand. *Geotechnical Special Publication*, 88: 165–179.

Wolf, J.P. 1994. *Foundation Vibration Analysis Using Simple Physical Models.* Prentice-Hall, Englewood Cliffs, NJ.

Wong, K.S. and Teh, C.I. 1996. Special load test for piles subjected to negative skin friction. *Proceedings of the 12th SE Asian Geotechnical Conference*, Kuala Lumpur, pp. 399–402.

Wong, S.C. and Poulos, H.G. 2006. Approximate pile-to-pile interaction factors between two dissimilar piles. *Computers and Geotechnics*, 32: 613–618.

Wu, G. and Liam Finn, W.D. 1997. Dynamic nonlinear analysis of pile foundations using finite element method in the time domain. *Canadian Geotechnical Journal*, 34: 44–52.

Xanthakos, P.A. 1994. *Slurry Walls as Structural Systems*. McGraw-Hill, New York.

Xu, K.J. and Poulos, H.G. 2000. General elastic analysis of piles and pile groups. *Internation Journal for Numerical and Analytical Methods in Geomechanics*, 24: 1109–1138.

Xu, K.J. and Poulos, H.G. 2001. 3-D elastic analysis of vertical piles subjected to passive loadings. *Computers and Geotechnics*, 28: 349–375.

Xu, X.T., Schneider, J.A. and Lehane, B.M. 2008. Cone penetration test (CPT) methods for end-bearing assessment of open- and closed-ended driven piles in siliceous sand. *Canadian Geotechnical Journal*, 45(1): 1130–1141.

Yamashita, K., Hashiba, T. and Ito, H. 2012. Settlement and load sharing behaviour of a piled raft subjected to strong seismic motion. *Proceedings of the 2012 ANZ Conference on Geomechanics*, Melbourne, pp. 1514–1519.

Yamashita, K., Kakurai, M. and Yamada, T. 1994. Investigation of a piled raft foundation on stiff clay. *Proceedings of the 13th International Conference on Soil Mechanics and Foundation Engineering*, New Delhi, Vol. 2, pp. 543–546.

Yamashita, K., Yamada, T. and Kakurai, M. 1998. Method for analysis piled raft foundations. *Proceedings of the Deep Foundations on Bored and Auger Piles*, Ghent, Belgium, pp. 457–464.

Yasuda, S. and Yamaguchi, I. 1985. Dynamic soil properties of undisturbed samples. *20th Annual Convention of JSSMFE*, Nagoya, pp. 539–542.

Yi, L. 2004. Finite element study on static pile load testing. *MEng thesis*, National University of Singapore.

Yong, K.Y., Lee, F.H. and Liu, X.K. 1996. Three dimensional finite element analysis of deep excavations in marine clay. *Proceedings of the 12th Southeast Asian Geotechnical Conference*, Kuala Lumpur, Vol. 1, pp. 435–440.

Youd, L., Hansen, C.M. and Bartlett, S.F. 2002. Revised multilinear regression equations for prediction of lateral spread displacement. *Journal of Geotechnical and Geoenvironmental Engineering, ASCE*, 128(12): 1007–1017.

Zeevaert, L. 1957. Compensated friction pile foundation to reduce settlement of buildings on the highly compressible volcanic clay of Mexico City. *Proceedings of the 4th International Conference on SMFE*, London, Vol. 2, pp. 81–86.

Zhang, B.Q. and Small, J.C. 1994. Finite layer analysis of soil-raft-structure interaction. *Proceedings of the XIII International Conference on Soil Mechanics and Foundation Engineering*, New Delhi, India, 5–10 January, Vol. 2, pp. 587–590.

Zhang, H.H. and Small, J.C. 2000. Analysis of capped pile groups subjected to horizontal and vertical loads. *Computers and Geotechnics*, 26(1): 1–21.

Zhang, L. and Einstein, H.H. 1998. End bearing capacity of drilled shafts in rock. *Journal of Geotechnical and Geoenvironmental Engineering, ASCE*, 124(7): 574–584.

Zhang, L. and Ng, A.M.Y. 2006. Limiting tolerable settlement and angular distortion for building foundations. In: *Geotechnical Special Publication No. 170, Probabilistic Applications in Geotechnical Engineering*, ASCE (on CD Rom).

Zheng, C., Kouretzis, G.P., Ding, X., Liu, H. and Poulos, H.G. 2016. Three-dimensional effects in low strain integrity testing of piles: Analytical solution. *Canadian Geotechnical Journal*, 53(2): 187–195.

Zhou, F., Wang, X-d., Lin, C. and Chen, J. 2016. Application of deformation adjusters in piled raft foundations. *Geotechnical Engineering, ICE*, Paper 1500187, 160(GE6): 527–540.

Zhou, J., Bao, L. and Qian, P. 2014. Study on structural efficiency of supertall buildings. *International Journal of High-Rise Buildings*, 3(3): 185–190.

Zou, Y. 1996. Ein neues verfahren zum aufrichten geneigter gebaude. *Bautechnik*, 73(7): 437–442.

# Index